T0234864

Lecture Notes in Computer Science 14233

The series Lecture Notes in Computer Science (LNCS), including its subseries Lecture Notes in Artificial Intelligence (LNAI) and Lecture Notes in Bioinformatics (LNBI), has established itself as a medium for the publication of new developments in computer science and information technology research, teaching, and education.

LNCS enjoys close cooperation with the computer science R & D community, the series counts many renowned academics among its volume editors and paper authors, and collaborates with prestigious societies. Its mission is to serve this international community by providing an invaluable service, mainly focused on the publication of conference and workshop proceedings and postproceedings. LNCS commenced publication in 1973.

Gian Luca Foresti · Andrea Fusiello ·
Edwin Hancock
Editors

Image Analysis and Processing – ICIAP 2023

22nd International Conference, ICIAP 2023
Udine, Italy, September 11–15, 2023
Proceedings, Part I

Editors
Gian Luca Foresti
University of Udine
Udine, Italy

Andrea Fusiello
University of Udine
Udine, Italy

Edwin Hancock
University of York
York, UK

ISSN 0302-9743 ISSN 1611-3349 (electronic)
Lecture Notes in Computer Science
ISBN 978-3-031-43147-0 ISBN 978-3-031-43148-7 (eBook)
https://doi.org/10.1007/978-3-031-43148-7

This Springer imprint is published by the registered company Springer Nature Switzerland AG
The registered company address is: Gewerbestrasse 11, 6330 Cham, Switzerland

Paper in this product is recyclable.

Preface

The International Conference on Image Analysis and Processing (ICIAP) is a biennial scientific meeting promoted by the Italian Association for Computer Vision, Pattern Recognition and Machine Learning (CVPL - formerly GIRPR), the Italian IAPR Member Society. The 22nd International Conference on Image Analysis and Processing (ICIAP 2023) was held in Udine, Italy, from 11 to 15 September 2023, in the prestigious venue of Palazzo di Toppo – Garzolini – Wasserman. It was co-organised by the Department of Informatics, Mathematics and Physics (DMIF) and the Polytechnic Department of Engineering and Architecture (DPIA) of the University of Udine, and sponsored by ST Microelectronics.

The conference traditionally covers topics related to theoretical and experimental areas of Computer Vision, Image Processing, Pattern Recognition and Machine Learning, with emphasis on theoretical aspects and applications. Keeping with this trend, ICIAP 2023 focused on the following areas: Pattern Recognition, Machine Learning and Deep Learning, 3D Computer Vision and Geometry, Image Analysis: Detection and Recognition, Video Analysis & Understanding, Biomedical and Assistive Technology, Digital Forensics and Biometrics, Multimedia, Cultural Heritage, Robot Vision and Automotive, Shape Representation, Recognition and Analysis, Augmented and Virtual Reality, Geospatial Analysis, and Computer Vision for UAVs.

The ICIAP 2023 main conference received 144 paper submissions from all over the world. The selection process, guided by the three Programme Chairs, resulted in the final selection of 92 high-quality manuscripts, with an overall acceptance rate of 64%.

To ensure the quality of papers ICIAP 2023 implemented a two-round review process. Each submission was managed by two Area Chairs and reviewed by at least three reviewers. Papers were selected through a double-blind peer review process, considering originality, significance, clarity, soundness, relevance and technical content.

The main conference programme included 24 oral presentations, 68 posters and three invited talks by leading experts in computer vision and pattern recognition: Danijel Skočaj (University of Ljubljana), Andrew Fitzgibbon (Graphcore), and Tomas Pajdla (CTU in Prague).

ICIAP 2023 also included 4 tutorials and hosted 15 workshops and 2 competitions, on topics of great relevance with respect to the state of the art. An industrial poster session was organised to bring together papers written by scientists working in industry and with a strong focus on application.

Several awards were presented during the ICIAP 2023 conference. The Eduardo Caianiello award was attributed to the best paper authored or co-authored by at least one young researcher. A Best Paper Award dedicated to Prof. Alfredo Petrosino was also assigned after a careful selection made by an ad hoc appointed committee.

The success of ICIAP 2023 is due to the contribution of many people. Special thanks go to all the reviewers and Area Chairs for their hard work in selecting the papers. Our thanks also go to the organising committee for their tireless efforts, advice and support.

We hope that you will find the papers in this volume interesting and informative, and that they will inspire you to further research in the field of image analysis and processing.

September 2023

Gian Luca Foresti
Andrea Fusiello
Edwin Hancock

Organization

General Chairs

Gian Luca Foresti	University of Udine, Italy
Andrea Fusiello	University of Udine, Italy
Edwin Hancock	University of York, UK

Program Chairs

Michael Bronstein	University of Oxford, UK
Barbara Caputo	Politecnico Torino, Italy
Giuseppe Serra	University of Udine, Italy

Steering Committee

Virginio Cantoni	University of Pavia, Italy
Luigi Pietro Cordella	University of Napoli Federico II, Italy
Rita Cucchiara	University of Modena-Reggio Emilia, Italy
Alberto Del Bimbo	University of Firenze, Italy
Marco Ferretti	University of Pavia, Italy
Gian Luca Foresti	University of Udine, Italy
Fabio Roli	University of Cagliari, Italy
Gabriella Sanniti di Baja	ICAR-CNR, Italy

Workshop Chairs

Federica Arrigoni	Politecnico Milano, Italy
Lauro Snidaro	University of Udine, Italy

Tutorial Chairs

Christian Micheloni	University of Udine, Italy
Francesca Odone	University of Genova, Italy

Publications Chairs

Claudio Piciarelli	University of Udine, Italy
Niki Martinel	University of Udine, Italy

Publicity/Social Chairs

Matteo Dunnhofer	University of Udine, Italy
Beatrice Portelli	University of Udine, Italy

Industrial Liaison Chair

Pasqualina Fragneto	STMicroelectronics, Italy

Local Organization Chairs

Eleonora Maset	University of Udine, Italy
Andrea Toma	University of Udine, Italy
Emanuela Colombi	University of Udine, Italy
Alex Falcon	University of Udine, Italy
Andrea Brunello	University of Udine, Italy

Area Chairs

Pattern Recognition

Raffaella Lanzarotti	University of Milano, Italy
Nicola Strisciuglio	University of Twente, The Netherlands

Machine Learning and Deep Learning

Tatiana Tommasi	Politecnico Torino, Italy
Timothy M. Hospedales	University of Edinburgh, UK

3D Computer Vision and Geometry

Luca Magri	Politecnico Milano, Italy
James Pritts	CTU Prague, Czech Republic

Image Analysis: Detection and Recognition

Giacomo Boracchi	Politecnico Milano, Italy
Mårten Sjöström	Mid Sweden University, Sweden

Video Analysis and Understanding

Elisa Ricci	University of Trento, Italy

Shape Representation, Recognition and Analysis

Efstratios Gavves	University of Amsterdam, The Netherlands

Biomedical and Assistive Technology

Marco Leo	CNR, Italy
Zhigang Zhu	City College of New York, USA

Digital Forensics and Biometrics

Alessandro Ortis	University of Catania, Italy
Christian Riess	Friedrich-Alexander University, Germany

Multimedia

Francesco Isgrò	University of Napoli Federico II, Italy
Oliver Schreer	Fraunhofer HHI, Germany

Cultural Heritage

Lorenzo Baraldi	University of Modena-Reggio Emilia, Italy
Christopher Kermorvant	Teklia, France

Robot Vision and Automotive

Alberto Pretto	University of Padova, Italy
Henrik Andreasson	Örebro University, Sweden
Emanuele Rodolà	Sapienza University of Rome, Italy
Zorah Laehner	University of Siegen, Germany

Augmented and Virtual Reality

Andrea Torsello	University of Venezia Ca' Foscari, Italy
Richard Wilson	University of York, UK

Geospatial Analysis

Enrico Magli	Politecnico Torino, Italy
Mozhdeh Shahbazi	University of Calgary, Canada

Computer Vision for UAVs

Danilo Avola	University of Roma Sapienza, Italy
Parameshachari B. D.	Nitte Meenakshi Institute of Technology, India

Brave New Ideas

Marco Cristani	University of Verona, Italy
Hichem Sahbi	Sorbonne University, France

Endorsing Institutions

International Association for Pattern Recognition (IAPR)
Italian Association for Computer Vision, Pattern Recognition and Machine Learning (CVPL)

Contents – Part I

Contents – Part II

Image Retrieval in Semiconductor Manufacturing

Giuseppe Gianmarco Gatta[1,2](✉), Diego Carrera[1], Beatrice Rossi[1], Pasqualina Fragneto[1], and Giacomo Boracchi[2]

[1] STMicroelectronics, Geneva, Switzerland
{diego.carrera,beatrice.rossi,pasqualina.fragneto}@st.com
[2] Politecnico di Milano, Milan, Italy
giuseppegianmarco.gatta@st.com, giacomo.boracchi@polimi.it

Abstract. Content-Based Image Retrieval has a lot of applications in the industry, where large collections of data from manufacturing need to be automatically queried e.g. for quality inspection purposes. In this work we design an image retrieval solution over IMAGO, a dataset of Transmission Electron Microscopy (TEM) images of nano-sized silicon structures collected in the production site of STMicroelectronics, in Agrate Brianza, Italy. Image retrieval in imago is challenging because: *i)* only a limited portion of images are provided with labels, namely type of semiconductor structure, *ii)* most images refer to unseen classes that are not represented in the training set, and *iii)* images of the same class can be acquired at different magnification levels of the electronic microscope. Our main contribution is the design of a deep-learning based image retrieval system that leverages a training procedure that alternates between siamese loss, assessed on annotated samples, and reconstruction loss, assessed on unlabelled samples. Our solution exploits the whole information in the IMAGO dataset, and our experiments confirm we can successfully retrieve images of unseen classes that exhibit the same structure of the query ones. Our solution is currently deployed in STMicroelectronics production sites.

Keywords: Content-Based Image Retrieval · Siamese Networks · Autoencoders

1 Introduction

Pushed by a steadily increasing demand for chips and memories, semiconductor industries have been struggling to produce smaller electronic components to improve performance, power efficiency and device reliability, and at the same time, reduce production costs, time and wastes. Transmission Electron Microscopy (TEM) plays a key role in the design and production stage of semiconductors, as this can yield highly magnified images (up to 2 million times). TEM images are customarily used to perform the physical and compositional characterization of specimens and to analyze faults and nano-sized structures. In the production site of STMicroelectronics in Agrate Brianza, hundreds of TEM images are acquired everyday and collected in a large dataset called IMAGO, containing around two million of entries acquired during different production stages and modalities, thus exhibiting variable quality and diverse magnifications

G. L. Foresti et al. (Eds.): ICIAP 2023, LNCS 14233, pp. 1–13, 2023.
https://doi.org/10.1007/978-3-031-43148-7_1

and resolutions. Only a fraction of IMAGO images are associated to a known class, namely type of *semiconductor structure*, respectively *locos*, *sti*, *spacers*, *tccv* and *cell* (Fig. 1(a)–(d); cell images are not reported to preserve confidentiality). The vast majority of IMAGO contains *unknown* classes that were never annotated, see Fig. 1(e).

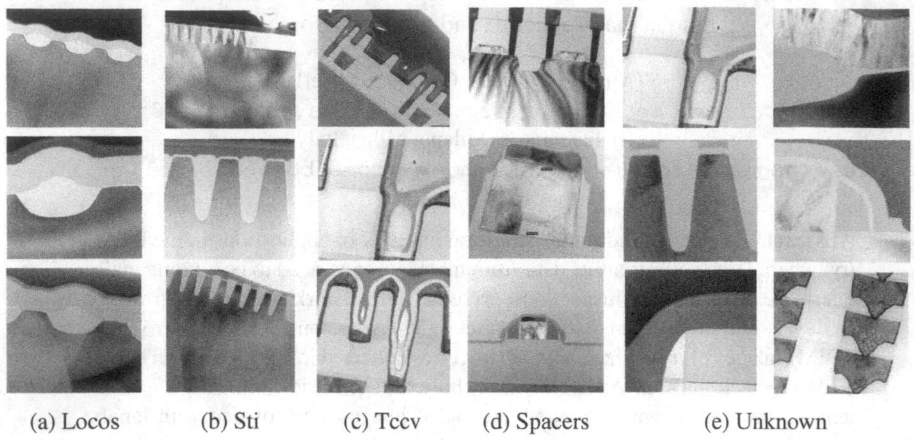

(a) Locos (b) Sti (c) Tccv (d) Spacers (e) Unknown

Fig. 1. Examples of images from IMAGO. In columns (a)–(d), images belonging to known classes, each acquired at a different magnification (spanning from 0.1 µm to 10 nm). In column (e), images belonging to unknown classes. Note the large variability in content, resolution, illumination and noise level even in images belonging to the same class.

Image retrieval from IMAGO raises a series of unique challenges since *i*) only a limited portion of IMAGO images are provided with labels and *ii*) images of the same class can be acquired at different magnification levels, as illustrated in Fig. 1. This setup prevents both methods assuming that images of the same class preserve a few distinctive visual features (e.g. up to a change of perspective, noise, as in [2]), and solution purely based on siamese neural networks that requires labels for training [3,6,11]. A valid strategy is to resort to unsupervised learning, thus training an autoencoder to extract an embedding of the images [9,13,16,17]. However, image retrieval based on autoencoders might fail when images belong to the same class but are visually dissimilar, as these might have embeddings that are very far apart, resulting in poor variability in the retrieved images.

In this work we present a solution able to retrieve IMAGO images depicting the same structure of a given query, disregarding whether this belongs to a class annotated in the training set or not. Our solution is based on a deep neural network trained by alternating the minimization of two losses: a (supervised) triplet loss typical of siamese neural networks and an (unsupervised) reconstruction loss typical of autoencoders. In this way, we exploit the whole information in the IMAGO dataset obtaining a single neural network trained, potentially, over the entire dataset. After training, images are retrieved by simply performing similarity search in the embedding space. Our system, thanks to the proposed training strategy, is able to effectively retrieve images even of unknown classes, i.e., representing structures that have never been annotated, without overfitting the small training set of labelled images. Our experiments prove that our solution leads to outstanding retrieval results, outperforming three baselines built on the same backbone of our solution in terms of precision and mean average precision. We also show, through a leave-one-class-out experiment, that we can achieve stable results when querying unknown structures, and that our method provides a good variability in the scale of the retrieved images, exploiting not only the visual similarity between images but also the semantic information provided by the labels. The proposed solution is currently running in the production site of STMicroelectronics in Agrate Brianza (Italy) helping engineers to set up new technologies, diagnose and source manufacturing problems, and train inexperienced personnel.

2 Related Works

We consider *Content-Based Image Retrieval* methods, which enable searching through a large database of images from a query image. The mainstream approach to content-based image retrieval consists in extracting feature vectors from the query image and then assessing its similarity with the feature vectors extracted from the dataset where to retrieve images from. The advent of Deep Learning introduced a paradigm shift in image retrieval, and early methods based on hand-crafted features like color, textures, shapes [7] have been replaced by data-driven features [5]. A critical aspect in this regard is the loss function used to train the models, which boils down to defining a similarity measure in the feature space. *Metric learning* [8] and both *supervised* and *unsupervised* methods have been proposed to learn embedding in the such that similar input samples are close together in the latent space, while dissimilar ones are far apart. Supervised methods leverage a set of labelled inputs to learn such embedding, while unsupervised ones aim at learning an embedding that reconstructs the input. Since IMAGO contains partially annotated data, we cannot employ a completely supervised or unsupervised approach, but rather look for a balance between the two. In this sense, our problem can be formulated as a *semi-supervised* image retrieval problem.

Among the solution closest to ours we mention siamese networks trained with triplet loss [3], which have been widely used in image retrieval where training data are provided with labels. Examples of this approach are [3], where a siamese network is trained to retrieve medical images, [6], which is applied to public image datasets as MNIST, and [11], where the siamese network is applied to a public dataset of photographs (Flickr15k). Unfortunately, training a siamese network might be unfeasible when a large amount of images belongs to unknown classes, as in the IMAGO dataset. Unsupervised learning can address this problems and autoencoders are particularly suited in the IMAGO settings. Examples of autoencoders used in image retrieval are [9, 13, 17], which were all trained over public datasets like Cifar-10 and ImageNet. All these datasets do not present the peculiarities of IMAGO, including the large variation in terms of scale and, most importantly, these unsupervised solutions ignore any annotation about similarity or class membership that our experiments demonstrate is very relevant for IMAGO.

3 Problem Formulation

Let $\mathcal{D}$ denote the IMAGO dataset, which contains images acquired at different scales. Each image $\mathbf{x} \in \mathcal{D}$ is associated with a class, corresponding to a specific structure $y \in \mathcal{Y} = \{y_1, \ldots, y_C\}$, where C denotes the total number of structures in $\mathcal{D}$ (C is possibly very large and unknown). We address the problem of retrieving from $\mathcal{D}$, TEM images belonging to the same class of a given query image $\mathbf{q} \in \mathcal{D}$, namely selecting K images $\{\mathbf{x}_1, \ldots, \mathbf{x}_K\}$ from $\mathcal{D}$ such that each $\mathbf{x}_i$ belongs to the same class as $\mathbf{q}$.

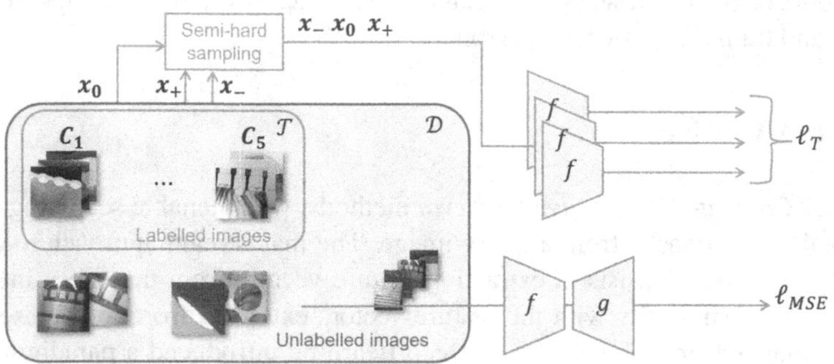

Fig. 2. The proposed training procedure: we train a siamese network by minimizing the triplet loss ℓ_T in (1), where triplets are sampled by semi-hard mining among annotated sampes in $\mathcal{T}$. Then, we alternate the autoencoder training by concatenating to f a decoder g. All the samples in $\mathcal{D}$ are employed to optimize the reconstruction loss ℓ_{MSE} in (3).

We assume that only a small subset $\mathcal{T}$ of the dataset $\mathcal{D}$ containing annotated samples is provided for training. Moreover, classes represented in $\mathcal{T}$ cover a small subset of $\mathcal{Y}$, which we will refer to as the known classes $\mathcal{S} \subset \mathcal{Y}$. Our goal is to retrieve images either when they belong to the known classes in $\mathcal{S}$ or when they refer to an unknown class $\mathcal{U} = \mathcal{Y} \backslash \mathcal{S}$.

4 Proposed Solution

We solve the retrieval of TEM images over the IMAGO dataset by learning an embedding function $f\colon \mathcal{D} \rightarrow \mathbb{R}^M$ that maps an image $\mathbf{x} \in \mathcal{D}$ into a feature vector $f(\mathbf{x}) \in \mathbb{R}^M$. We learn the embedding f through a neural network so that images from the same class correspond to nearby feature vectors, while images from different classes are mapped in feature vectors that are far apart. Then, given a query image $\mathbf{q}$, we retrieve images having the closest feature vectors to $f(\mathbf{q})$. Peculiarity of our solution is the training procedure described in Sect. 4.1, while the image retrieval itself is rather standard and illustrated in Sect. 4.2. Implementation details are in Sect. 4.3.

4.1 Learning the Embedding Function

Figure 2 illustrates the proposed training procedure for the network f, where we employ both labelled and unlabelled images in $\mathcal{D}$, including samples from unknown classes in $\mathcal{U}$. At first, we build a siamese network by replicating f into three branches. This siamese network is fed with triplets of annotated training samples drawn from $\mathcal{T}$, selected using semi-hard mining [12] and minimizing the triplet loss ℓ_T that we will illustrate in (1). A network trained exclusively on supervised samples however would behave unpredictably on samples from unknown classes in $\mathcal{U}$. Therefore, we adopt a reconstruction loss ℓ_{MSE} in (3) to train our network f on all the samples in $\mathcal{D}$. To this purpose, we build an autoencoder, where f acts as an encoder and we adopt an auxiliary network g as a decoder. The training procedure consists in alternating, for a maximum number of iterations, the optimization of these two losses as follows:

- train the siamese network by minimizing the triplet loss ℓ_T over $\mathcal{T}$ for an epoch;
- train the autoencoder (thus both f and g) by minimizing the reconstruction loss ℓ_{MSE} over $\mathcal{D}$ for an epoch.

After training the decoder g is discarded. In what follows we detail how these two steps are performed.

Training the Siamese Network. We minimize the triplet loss [10, 17] computed over of triplets of samples drawn from the annotated training set $\mathcal{T}$. Given a randomly sampled anchor $\mathbf{x}_0 \in \mathcal{T}$, we extract a positive and a negative sample, $\mathbf{x}_+$ and $\mathbf{x}_-$, respectively such that $\mathbf{x}_0$ and $\mathbf{x}_+$ belong to the same class, while $\mathbf{x}_-$ belongs to a different one. The triplet loss over $(\mathbf{x}_0, \mathbf{x}_+, \mathbf{x}_-)$ is defined as

$$\ell_T(\mathbf{x}_0, \mathbf{x}_+, \mathbf{x}_-) = \max(\|f(\mathbf{x}_0) - f(\mathbf{x}_+)\|_2^2 - \|f(\mathbf{x}_0) - f(\mathbf{x}_-)\|_2^2 + R, 0), \tag{1}$$

where the parameter $R > 0$ plays the role of a margin which measures how "far" we want samples that do not belong to the same class to be in our feature space. By optimizing f to minimize the triplet loss, we ensure that $f(\mathbf{x}_0)$ and $f(\mathbf{x}_+)$ are near-by, while $f(\mathbf{x}_0)$ and $f(\mathbf{x}_-)$ are pushed apart.

Different strategies can be employed to sample triplets $(\mathbf{x}_0, \mathbf{x}_+, \mathbf{x}_-)$ from the training set and in particular we adopt semi-hard mining [12]. In semi-hard mining, once the anchor $\mathbf{x}_0$ and the positive sample $\mathbf{x}_+$ have been drawn, the negative sample $\mathbf{x}_-$ is selected so that the distance in the feature space between $\mathbf{x}_0$ and $\mathbf{x}_-$ is greater than the distance between $\mathbf{x}_0$ and $\mathbf{x}_+$, but not large enough to nullify the loss function (1). In practice, $\mathbf{x}_-$ is selected such that:

$$\|f(\mathbf{x}_0) - f(\mathbf{x}_+)\|_2^2 < \|f(\mathbf{x}_0) - f(\mathbf{x}_-)\|_2^2 < \|f(\mathbf{x}_0) - f(\mathbf{x}_+)\|_2^2 + R. \tag{2}$$

If no of such $\mathbf{x}_-$ is available in $\mathcal{T}$, we select $\mathbf{x}_-$ as the negative sample closest to $\mathbf{x}_0$, as explained in [12]. The other popular alternative to construct triples, hard-mining [14], selects negatives $\mathbf{x}_-$ that are instead closer to $\mathbf{x}_0$ rather than $\mathbf{x}_+$.

Training the Autoencoder. Autoencoders consist of two subnetworks, an *encoder* and a *decoder*. The encoder learns to represent each input sample in a latent space, while the decoder learns to reconstruct the input from its latent representation. Here, we adopt f as the encoder, thus the latent representation of an input image $\mathbf{x}$ is the feature vector $f(\mathbf{x})$ used for retrieval. We train a second neural network g as a decoder to minimizing the Mean Squared Error (MSE) reconstruction loss:

$$\ell_{MSE}(\mathbf{x}, g(f(\mathbf{x}))) = \|\mathbf{x} - g(f(\mathbf{x}))\|_2^2. \tag{3}$$

Since to minimize (3) we not require labels, we can exploit the entire dataset $\mathcal{D}$ to train our model, and in particular all the unlabelled samples in $\mathcal{D} \backslash \mathcal{T}$ that represent the largest part of IMAGO and that include images belonging to unknown classes in $\mathcal{U}$.

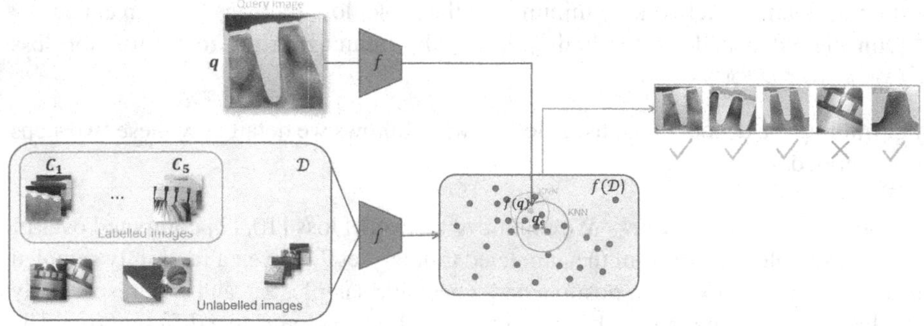

Fig. 3. Retrieval phase of our solution. After learning f, we compute and store the feature vectors for the entire dataset $\mathcal{D}$. At retrieval time, we compute the embedding of the query image, namely $f(\mathbf{q})$, and then we perform query expansion. The first K samples closest to the expanded query are returned.

4.2 Image Retrieval

We preliminary compute and store all the feature vectors for the dataset $\mathcal{D}$, namely $f(\mathcal{D}) = \{f(\mathbf{x}),\ \mathbf{x} \in \mathcal{D}\}$, thus enabling efficient searches in the dataset. Our procedure to retrieve K images is illustrated in Fig. 3 and detailed in Algorithm 1. Given a query image $\mathbf{q}$, we compute the embedding $f(\mathbf{q})$ (line 1), and instead of directly retrieving the K-nearest neighbor of $f(\mathbf{q})$, we perform *query expansion*, whose benefit are shown in [1]. In practice, we select the $\widetilde{K}$ closest feature vectors to $f(\mathbf{q})$ in $f(\mathcal{D})$. Typically $\widetilde{K}$ is very small [1] and we set $\widetilde{K} = 5$. Then, the retrieved $\widetilde{K}$ feature vectors $f(\mathbf{x}_i) \in \mathbb{R}^M$, $i = 1, \dots, \widetilde{K}$ (line 2) are used to compute the *expanded* feature vector $\widetilde{f(\mathbf{q})}$ as (line 3) as follows

$$\widetilde{f(\mathbf{q})} = \frac{1}{\widetilde{K} + 1} \left(f(\mathbf{q}) + \sum_{i=1}^{\widetilde{K}} f(\mathbf{x}_i) \right). \tag{4}$$

In practice, the expanded feature vector $\widetilde{f(\mathbf{q})}$ is obtained by averaging $f(\mathbf{q})$ with the $\widetilde{K}$ closest feature vectors to $f(\mathbf{q})$. The expanded feature vector $\widetilde{f(\mathbf{q})}$ is considered a better representative of the features that characterize the class of $\mathbf{q}$. As shown in Fig. 3, we finally select the K closest feature vectors to $\widetilde{f(\mathbf{q})}$ in $f(\mathcal{D})$ via K-NN, and consider the corresponding images in $\mathcal{D}$ as the retrieved images from query $\mathbf{q}$ (line 4).

4.3 Implementation Details

We adopt a pretrained VGG16 [15] as backbone architecture, where we replace the top fully connected layers with a global averaging pooling layer resulting in an output feature vectors of $M = 512$ components. Moreover, we add a normalization layer on top of the global averaging pooling, to set the output feature vector to have zero mean and standard deviation equals to one. This normalization improves the similarity search based on the Euclidean distance between feature vectors.

Algorithm 1. Image Retrieval Process

Require: Embedding of the dataset $f(\mathcal{D})$, query image $\mathbf{q}$, embedding $f : \mathcal{D} \longrightarrow \mathbb{R}^M$, integer $K \in \mathbb{N}$

Ensure: $\mathcal{R} \subseteq \mathcal{D}$ containing the first K retrieved images from query $\mathbf{q}$

1: Compute feature vector of the query $f(\mathbf{q}) \in \mathbb{R}^M$

2: Select from $f(\mathcal{D})$ the $\widetilde{K}$ nearest neighbors $f(\mathbf{x}_i) \in \mathbb{R}^M$ to $f(\mathbf{q})$, for $i = 1, \ldots, \widetilde{K}$

3: Compute the average $\widetilde{f(\mathbf{q})}$ of $f(\mathbf{q})$ with the $\widetilde{K}$ selected points

4: Select from $f(\mathcal{D})$ the K nearest neighbors to $\widetilde{f(\mathbf{q})}$ (and store the corresponding images in $\mathcal{R}$)

Before training the VGG16 backbone as described in Sect. 4.1, we preliminarly fine tune the network on $\mathcal{T}$ to perform classification on the IMAGO dataset on the known classes in $\mathcal{S}$. Then, we discard the last dense layer and use the resulting network f to train the siamese network and the autoencoder. The architecture of the decoder g is inspired to the symmetric VGG16 architecture, and corresponds to the f mirrored where we replace the pooling with upsampling layers. A customary procedure to enable g to reconstruct the input is to attach g after the last convolutional layer of f. Thus, the decoder g is not taking as input the feature vector extracted from f, but we nevertheless refer to the autoencoder as $g(f(\cdot))$ for the notation sake.

Finally, in all our training steps we adopt a data augmentation procedure that includes random shift, horizontal and vertical flips and change of scales. Moreover, we use the ADAM optimizer with default hyperparameters and a batch size of 32. As a preprocessing, all the images of $\mathcal{D}$ are resized to the resolution of 224×224 and pixel values are normalized by zero-centering each channel with respect to the ImageNet data set, as described in [15].

5 Experiments

To prove the effectiveness of our solution, we compare it against three image retrieval benchmarks in three different settings. In Sect. 5.4 we show that our solution achieves the best retrieval performance over the five annotated classes in $\mathcal{S}$. Then, in Sect. 5.5 we implement a leave-one-class-out procedure to assess the retrieval performance when querying unknown classes, and we show performances are in line with the previous experiment. Finally, in Sect. 5.6 we investigate the ability of methods to retrieve images of the same class at different scales, where we show that considering also siamese loss is beneficial for this task.

5.1 Dataset

For practical reasons, we test our solutions on a subset $\mathcal{D}$ of IMAGO which contains 35000 images, both labelled and unlabelled. We exclude from $\mathcal{D}$ a test set $\mathcal{Q}$ containing 368 labelled query images uniformly distributed among the five classes in $\mathcal{S}$. As before, we denote as $\mathcal{T}$ the labelled portion of $\mathcal{D}$ (containing about 2500 images).

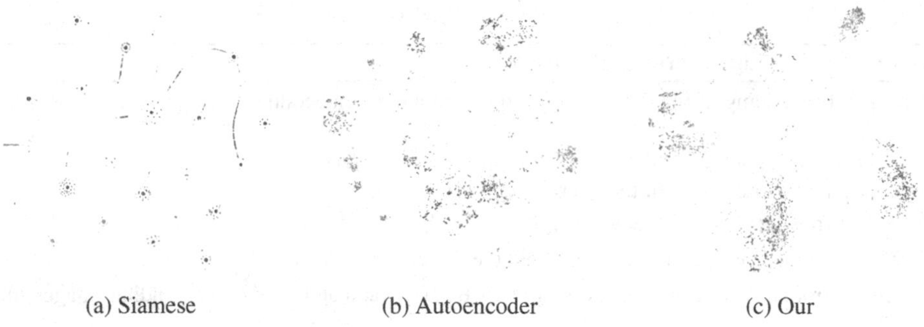

(a) Siamese (b) Autoencoder (c) Our

Fig. 4. Visual representation of the embedding of the labelled dataset $\mathcal{T}$ after performing a t-SNE on feature vectors obtained respectively by siamese, autoencoder and our solution and reducing their size from M to 2. Embedded images belonging to known classes are represented in the following colors: orange (locos), black (sti), purple (spacers), blue (tccv), and green (cell). (Color figure online)

5.2 Alternative Methods

We compare our solution against the following baselines:

- **Fine-tuned VGG16**: VGG16 backbone fine tuned on $\mathcal{T}$.
- **Siamese**: siamese network trained to minimize a triplet loss over semi-hard triplets sampled from $\mathcal{T}$.
- **Autoencoder**: autoencoder based on the VGG16 architecture, trained on $\mathcal{D}$ minimizing the reconstruction loss.

We did not consider more sophisticated backbones or Transformer-like architectures, as they are not compatible with the expected target platform. We remark that in all the solutions the retrieval is performed as explained in Sect. 4.2 and under equal conditions, employing the same queries $\mathcal{Q}$ and the same dataset to retrieve samples. Due to the supervised nature of VGG16 and siamese models, in these two cases the embedding f is trained employing only the labelled portion $\mathcal{T}$ of $\mathcal{D}$.

5.3 Figures of Merit

To assess the results of retrieval methods we consider *precision* and *mean average precision*, which are computed as follows. Given a query image $\mathbf{q}$ and the corresponding K retrieved images, the *precision* is the percentage of correctly retrieved samples, averaged over the query set. The *mean average precision at* K (MAP_K) [4] takes into account how many correctly retrieved images are among the first returned results rather than just K. In particular, we define $P(k, \mathbf{q})$ as the fraction of correctly retrieved samples among the first k images selected by the retrieval method, for $k = 1, \ldots, K$. Then, MAP_K is computed as:

$$MAP_K = \frac{\sum_{i=1}^{n}(AvgP(K, \mathbf{q}_i))}{n} \tag{5}$$

Table 1. Precision and Mean Average Precision at $K = 100$ performing the retrieval over $\mathcal{T}$.

	Precision	MAP_K
autoencoder	0.937	0.924
fine-tuned VGG16	0.930	0.918
our	**0.986**	**0.983**
siamese	0.962	0.950

where n is the size of the query set, and $AvgP(K, \mathbf{q}_i)$ denotes the average precision at K for the query $\mathbf{q}_i$ computed averaging precision over all the possible $k = 1, \ldots, K$:

$$AvgP(K, \mathbf{q}_i) = \frac{\sum_{k=1}^{K}(P(k, \mathbf{q}_i) \cdot \mathbb{1}\{y_k = y_i\})}{K}, \tag{6}$$

being $\mathbb{1}$ the indicator function.

5.4 Image Retrieval Performance on Known Classes

In this experiment, we assess the effectiveness of the considered solutions to retrieve query images belonging to known classes. For each query $\mathbf{q} \in \mathcal{Q}$, we retrieve $K = 100$ images from the set of labelled images $\mathcal{T}$ and we report in Table 1 the precision and mean average precision at K of all the methods. These results demonstrate that our solution retrieves the largest number of correct samples and also returns them in a better rank, where correct results come first. Keeping a high precision among the first retrieved samples is very important in an industrial scenario, where engineers might possibly look only at the most relevant results. The second best-performing solution is the siamese network which, like ours, leverages annotations. We speculate that the performance gap between our solution and the siamese network is due to the advantage of using unsupervised images during training.

To get a visual intuition of different embeddings, we perform a t-SNE on feature vectors corresponding to the dataset $\mathcal{D}$, reducing their dimension from M to 2. Figure 4 reports the distribution of the embedded samples obtained respectively by siamese, autoencoder and our solution, showing that our solution groups together vectors corresponding to images of the same class better than the other two methods, enabling better retrieval.

We also assess our solution performing the retrieval over the whole $\mathcal{D}$ instead of $\mathcal{T}$, thus including images from unknown classes. We consider the same query set $\mathcal{Q}$, and we ask experts to validate the retrieval results. Since performance assessment required visual inspection, we were not able to consider alternative methods, and we restricted to only $K = 30$ results. Our method achieves a precision of 0.93, confirming that our solution is very effective even in the more realistic scenario where the retrieval is performed over a larger set including images from unknown classes.

5.5 Retrieval of Images from Unknown Classes

In this experiment, we evaluate the effectiveness of retrieving query images belonging to unknown classes not represented in $\mathcal{T}$. To this purpose, we perform leave-one-out cross validation, excluding every time a specific class from supervised training. We then test the trained models on query images belonging to the excluded class only. It is worth however remarking that methods using unlabelled samples might have access to instances of the excluded class that appears unlabelled in $\mathcal{D}$. We again compare all the methods in terms of precision and mean average precision at $K = 100$, and we report results in Tables 2 and 3 respectively. Each column corresponds to an excluded class, and the figures of merit are averaged over query images belonging to the excluded class

Table 2. Precision of the considered methods trained using leave-one-class-out cross validation. The results are averaged only over the queries belonging to the excluded class.

	no cell	no locos	no spacers	no sti	no tccv	mean
autoencoder	**0.859**	0.290	**0.518**	**0.744**	0.452	0.573
fine-tuned VGG16	0.807	**0.540**	0.218	0.534	0.314	0.482
our	0.800	0.456	0.489	0.671	**0.845**	**0.652**
siamese	0.587	0.435	0.246	0.406	0.369	0.409

Table 3. Mean average precision at $K = 100$ for the cross-validation leave-one-out scenario on the considered models. The results are averaged only over the queries belonging to the excluded class.

	no cell	no locos	no spacers	no sti	no tccv	mean
autoencoder	**0.805**	0.101	**0.362**	**0.631**	0.270	0.434
fine-tuned VGG16	0.749	**0.444**	0.110	0.447	0.237	0.398
our	0.735	0.255	0.344	0.556	**0.773**	**0.533**
siamese	0.446	0.319	0.160	0.238	0.259	0.284

only. The column *mean* reports the corresponding figure of merit averaged over all the excluded classes. Remember that the siamese and the fine-tuned VGG16 networks have never seen a sample from the excluded class during training since their training completely rely on labelled samples. Not surprisingly, the best solutions are those leveraging unsupervised loss, namely the autoencoder and our solution, which however cannot take relevant advantage from the siamese loss during training.

5.6 Retrieval of Images at Different Scales

In our last experiment, we test models in retrieving images of the same class when there are substantial changes in appearance due to the magnification levels. It is in fact very important, according to STMicroelectronics engineers, for the system to return images selected at different scales. To this purpose, we measure the variability of the scales among the $K = 100$ retrieved images. In Table 4 we report the standard deviation of scale values computed from the correctly retrieved images, which is expected to be large when the model can retrieve images at different scales. As expected, the method that retrieves images with a wider range of scales is the siamese model, since it has been trained only on labelled samples. We speculate that training f as a siamese architecture pushes the network towards identifying specific semantic features of the images, which might appear at different scales. In contrast, the reconstruction loss can only rely on the visual similarity of the data. This perhaps prevents the autoencoder from retrieving images at different scales. Overall, our method performs comparably to the other solutions, and it is able to retrieve images with a significant scale variety.

Table 4. Standard deviation of the scale of the correctly retrieved images (the higher, the better). The results are averaged only over the queries belonging to the excluded class.

	no cell	no locos	no spacer	no sti	no tccv	mean
autoencoder	2.073	3.167	3.358	3.099	3.466	3.033
fine-tuned VGG16	1.804	1.863	2.639	2.700	1.907	2.182
our	1.817	2.667	4.001	4.210	3.193	3.178
siamese	2.456	2.940	3.058	4.483	3.198	3.227

6 Conclusions

In this paper, we address image retrieval in semiconductor manufacturing. The proposed solution consists in a new training procedure for deep neural networks, which alternates between the minimization of a triplet loss on annotated samples and the reconstruction loss on all the samples disregarding whether they are annotated or not. Our experiments demonstrate that our solution outperforms the considered alternatives on the IMAGO dataset. Even though the training procedure was designed to cope with the peculiarities of the IMAGO dataset, which is only partially labelled and where annotations cover only a fraction of the total number of classes, our solution can be in principle applied in other retrieval problems. Ongoing works focus on designing training procedure to further strengthen the network invariance to the severe scale changes characterizing these images. Moreover, we are going to address the experimental limitations of our work. In particular, we will test the method on a broader portion of the annotated dataset IMAGO, and assess its performances on the unlabeled portion of IMAGO by collecting feedback from domain experts.

References

1. Ahmed, A., Malebary, S.J.: Query expansion based on top-ranked images for content-based medical image retrieval. IEEE Access **8**, 194541–194550 (2020). https://doi.org/10.1109/ACCESS.2020.3033504
2. Balmachnova, E., Florack, L., ter Haar Romeny, B.: Feature vector similarity based on local structure. In: Sgallari, F., Murli, A., Paragios, N. (eds.) SSVM 2007. LNCS, vol. 4485, pp. 386–393. Springer, Heidelberg (2007). https://doi.org/10.1007/978-3-540-72823-8_33
3. Chung, Y., Weng, W.: Learning deep representations of medical images using Siamese CNNs with application to content-based image retrieval. CoRR abs/1711.08490 (2017). http://arxiv.org/abs/1711.08490
4. Cooper, W.S.: Expected search length: a single measure of retrieval effectiveness based on the weak ordering action of retrieval systems. Am. Doc. **19**(1), 30–41 (1968)
5. Dubey, S.R.: A decade survey of content based image retrieval using deep learning. IEEE Trans. Circ. Syst. Video Technol. **32**(5), 2687–2704 (2021). https://doi.org/10.1109/TCSVT.2021.3080920
6. Hoffer, E., Ailon, N.: Deep metric learning using triplet network. In: Feragen, A., Pelillo, M., Loog, M. (eds.) SIMBAD 2015. LNCS, vol. 9370, pp. 84–92. Springer, Cham (2015). https://doi.org/10.1007/978-3-319-24261-3_7
7. Kato, T.: Database architecture for content-based image retrieval. In: Jamberdino, A.A., Niblack, C.W. (eds.) Image Storage and Retrieval Systems, vol. 1662, pp. 112–123. International Society for Optics and Photonics, SPIE (1992). https://doi.org/10.1117/12.58497
8. Kaya, M., Bilge, H.Ş: Deep metric learning: a survey. Symmetry **11**(9), 1066 (2019)
9. Krizhevsky, A., Hinton, G.E.: Using very deep autoencoders for content-based image retrieval. In: ESANN, vol. 1, p. 2. Citeseer (2011)
10. Pandey, A., Mishra, A., Verma, V.K., Mittal, A., Murthy, H.: Stacked adversarial network for zero-shot sketch based image retrieval. In: Proceedings of the IEEE/CVF Winter Conference on Applications of Computer Vision, pp. 2540–2549 (2020)
11. Qi, Y., Song, Y.Z., Zhang, H., Liu, J.: Sketch-based image retrieval via Siamese convolutional neural network. In: 2016 IEEE International Conference on Image Processing (ICIP), pp. 2460–2464. IEEE (2016)
12. Schroff, F., Kalenichenko, D., Philbin, J.: FaceNet: a unified embedding for face recognition and clustering. In: Proceedings of the IEEE Conference on Computer Vision and Pattern Recognition, pp. 815–823 (2015)
13. Shen, Y., et al.: Auto-encoding twin-bottleneck hashing. In: Proceedings of the IEEE/CVF Conference on Computer Vision and Pattern Recognition, pp. 2818–2827 (2020)
14. Sheng, H., et al.: Mining hard samples globally and efficiently for person reidentification. IEEE Internet Things J. **7**(10), 9611–9622 (2020)
15. Simonyan, K., Zisserman, A.: Very deep convolutional networks for large-scale image recognition (2015)

16. Siradjuddin, I.A., Wardana, W.A., Sophan, M.K.: Feature extraction using self-supervised convolutional autoencoder for content based image retrieval. In: 2019 3rd International Conference on Informatics and Computational Sciences (ICICoS), pp. 1–5 (2019). https://doi.org/10.1109/ICICoS48119.2019.8982468
17. Wang, Y., Ou, X., Liang, J., Sun, Z.: Deep semantic reconstruction hashing for similarity retrieval. IEEE Trans. Circuits Syst. Video Technol. **31**(1), 387–400 (2020)

Continual Source-Free Unsupervised Domain Adaptation

Waqar Ahmed[1(✉)], Pietro Morerio[1], and Vittorio Murino[1,2]

[1] Pattern Analysis and Computer Vision, Istituto Italiano di Tecnologia, Genova, Italy
{waqar.ahmed,pietro.morerio,vittorio.murino}@iit.it
[2] Dipartimento di Informatica, University of Verona, Verona, Italy

Abstract. Source-free Unsupervised Domain Adaptation (SUDA) approaches inherently exhibit catastrophic forgetting. Typically, models trained on a labeled source domain and adapted to unlabeled target data improve performance on the target while dropping performance on the source, which is not available during adaptation. In this study, our goal is to cope with the challenging problem of SUDA in a continual learning setting, *i.e.,* adapting to the target(s) with varying distributional shifts while maintaining performance on the source. The proposed framework consists of two main stages: i) a SUDA model yielding cleaner target labels—favoring good performance on target, and ii) a novel method for synthesizing class-conditioned source-style images by leveraging only the source model and pseudo-labeled target data as a prior. An extensive pool of experiments on major benchmarks, *e.g.,* PACS, Visda-C, and DomainNet demonstrates that the proposed Continual SUDA (C-SUDA) framework enables preserving satisfactory performance on the source domain *without* exploiting the source data at all.

Keywords: Continual learning · Image synthesis · Source-free · Unsupervised domain adaptation

1 Introduction

Convolutional Neural Networks (CNNs) trained on a labeled *source* domain often fail to generalize well on a related but different *target* domain due to the well-known *domain shift* [18,25]. Since annotating data from a new domain is expensive and sometimes even impossible, Unsupervised Domain Adaptation (UDA) methods have been developed to address the drop in performance by exploiting *unlabelled* target data.

Conventional UDA methods address the adaptation task by *e.g.,* feature alignment [6], matching moments [20], or adversarial learning [24]. However, these methods typically require joint access to both *labeled source* and *unlabeled target* data during adaptation, making them unsuitable for most real-world scenarios, where source data is inaccessible (*e.g.,* due to data privacy or proprietary reasons). Nevertheless, the performance on the source often degrades after

G. L. Foresti et al. (Eds.): ICIAP 2023, LNCS 14233, pp. 14–25, 2023.
https://doi.org/10.1007/978-3-031-43148-7_2

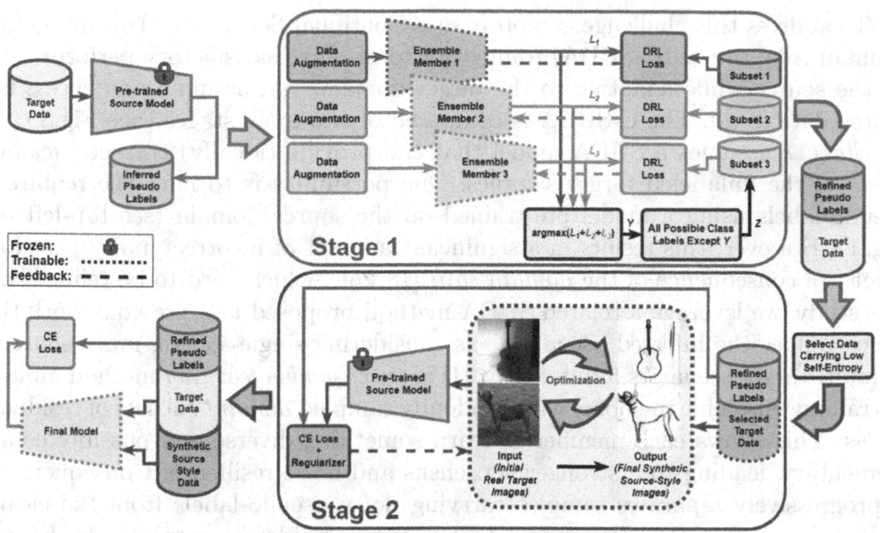

Fig. 1. Overview of the proposed method. We assume a pre-trained source model to infer pseudo-labels of the target set (top-left). *Stage 1* refines incorrect pseudo-labels to achieve Source-free Unsupervised Domain Adaptation (SUDA). *Stage 2* synthesizes source-style images to avoid catastrophic forgetting of the source, thus achieving Continual SUDA (C-SUDA). Finally, a single model is trained using real target and synthetic source images, each one associated with a refined pseudo-label (bottom-left).

adaptation to the target, even when using the source data during the training/adaptation process.

Recently proposed methods address UDA problem under a more realistic source-free assumption, *i.e.*, by using the pre-trained source model and unlabeled target domain data only [13,15,18]. However, due to absolutely no exposure to source data distribution by any means, these methods naturally undergo catastrophic forgetting, *i.e.*, the model adapted to the target experiences a substantial drop in performance if tested back on the source domain data.

This is actually a severe drawback in a practical realistic scenario. For example, a self-driving car company releases a recognition model trained on some proprietary data, *e.g.*, data collected in unknown urban regions, considered as a source. Thereafter, one will have to perform Source-free Unsupervised Domain Adaptation (SUDA) to obtain acceptable performance on a different target domain *e.g.*, rural or mountain regions. Yet, due to the unavoidable catastrophic forgetting issue, the adapted model will likely fail when deployed back in urban areas. The availability of multiple models trained on different domains would not work since it is clearly neither feasible nor scalable due to the typical limited hardware resources. Thus, it is desirable to preserve the model's performance on the source domain too for practical use cases.

To address this challenge, we propose a Continual Source-free Unsupervised Domain Adaptation (C-SUDA) framework to preserve satisfactory performance on the source while adapting to the target domain, yet, assuming no access to source data at all. The C-SUDA is composed of two main stages (See Fig. 1).

Stage 1 assumes a SUDA model that can provide (ideally) correct *pseudo-labels* of the unlabeled target samples. One possibility is to infer the required pseudo-labels using a model pre-trained on the source domain (see top-left of Fig. 1). However, this results in a significant amount of incorrect/noisy pseudo-labels—a consequence of the *domain shift* [18,25]—which need to be refined. In this study, we leverage a related SUDA method proposed in a previous work [1] which refines the inferred pseudo-labels considering single-source, multi-source, or multi-target scenarios indifferently. [1] is an *ensemble learning* method aimed at training ensemble members with randomly sampled *disjoint* subsets of residual labels. This allows each member to learn something diverse and possibly complementary, leading to a stronger consensus and noise resilience. Consequently, it progressively separates samples carrying noisy pseudo-labels from the clean ones, by assigning them lower and higher class confidence, respectively. Eventually, the low-confidence samples iteratively undergo a pseudo-label refinement process via reassignment. Note however that any of-the-shelf SUDA methods can be used here which provides clean pseudo-labels.

Stage 2 consists of an image synthesis process aimed at promoting continual learning, by generating synthetic source images that help in preserving satisfactory performance on the source domain. In particular, we leverage the fact that CNNs are capable of automatically discovering the rich underlying patterns hidden in the data (*e.g.*, the running average statistics stored in the Batch-Norm layers). The idea here is to employ feature distribution regularizers exploiting such rich information to generate class-conditioned (based on refined pseudo-labels obtained in Stage 1) source-style images. Only images with high confidence (low soft prediction entropy) are used in this stage. Specifically, we use target images as prior and optimize their statistics and style such that the transformed versions of the generated images resemble the source domain distribution and achieve sharp classification predictions when presented to the source model. We mix synthetic-source and real-target images for training the final (single) model that ensures good performance on both *real* source and target domains.

The proposed C-SUDA framework is fully-adaptive and can cope with single-source, multi-source, and multi-target UDA problems indifferently. With extensive experiments on various benchmarks carrying different amounts of inferred pseudo-labels noise, we show that our framework helps alleviating catastrophic forgetting when the model is tested back on the source domain(s). To summarise, the contributions of our work can be stated as follows:

- We propose a new, versatile Continual Source-free UDA (C-SUDA) framework, composed of two simple, yet effective stages, which attains state-of-the-art performance in both target and source domains. Notably, our method can equally face single-source, multi-source, and multi-target UDA scenarios indifferently.

- The second stage is a new image synthesis method that leverages pseudo-labeled target images and a pre-trained source model to generate high-fidelity class-conditional source-style images. Such synthetic images help preserving good performance on source domain(s) without using the real source samples.
- We validate our method on three well-known benchmarks, demonstrating the proposed C-SUDA's effectiveness and generalizability in diverse scenarios.

The remainder of the paper is organized as follows. Section 2 discusses related works. Section 3 introduces the proposed method. Section 4 presents the experimental setup and obtained results. Finally, conclusions are drawn in Sect. 5.

2 Related Work

Our work lies at the intersection of *Source-Free UDA* and *Continual Learning* (CL), whose related literature is discussed below. Also, we briefly review existing literature related to *Image Synthesis*, as synthesized source images constitute a core stage of the proposed pipeline.

Source-Free UDA. Recent years have seen growing interest in addressing the UDA problem in realistic source-free settings. A common approach is to leverage a pre-trained source model for either transferring the fixed source classifier to the target data employing information maximization and pseudo-labeling [15], or updating target model progressively by generating target-style samples through conditional generative adversarial networks [18], also by combining clustering-based regularization [13]. Similarly, to improve performance in domain-adaptive person re-identification tasks, [7] proposes a pseudo-label cleaning process with online refined soft pseudo-labels. Our proposed approach lies in this category of works, but we differ from previous methods by not requiring: i) a customized network, ii) to generate target-style data using *e.g.*, GAN-based models that require careful hyper-parameter tuning to reach stability, and iii) different techniques to tackle single-source, multi-source, and multi-target UDA.

Continual Learning. CL refers to learning by using many diverse data distributions (tasks) sequentially, while avoiding catastrophic forgetting [2,23]. Some recent works tackled the continual UDA problem such as, *e.g.*, [4], which proposes domain adversarial learning with sample replay. More recently, [26] employs a meta-learning strategy and domain randomization using heavy image manipulations. Despite the fact that these strategies are effective, they all require the access to source data. Similarly, [31] performs continual adaptation, but needs two auxiliary binary embedding layers to be specifically trained on source. These are stored and used during adaptation in order to overcome catastrophic forgetting, and this questions the source-free nature claimed by the method. On the contrary, we only assume the availability of standard CNN models pre-trained on the source, but do not require retraining of any auxiliary layers on source nor the access to partial source data to tackle the catastrophic forgetting.

Image Synthesis. In the deep learning era, this area refers to the generation of synthetic images, possibly indistinguishable from real ones, and generative adversarial networks (GAN) are among the most popular class of approaches adopted to date. GAN-inversion is an interesting research direction, in which an anchor image is used to guide a GAN to generate realistic images by inverting a pre-trained model [19]. Other works focus on network inversion that enables noise-to-image transformation by back-propagating gradients to the learnable input images like in, *e.g.*, [17], which introduces "dreaming" new visual features onto images, while [22] takes this approach a step further to generate more realistic images. In the context of Domain Generalization, [27] optimizes images in the pixels space to produce augmentations, extending the working domain of the classifier to unseen data distributions. Recently proposed Deep-Inversion [32] method assumes a uniform prior and backpropagates desired label to produce synthetic images using a regularizer based on feature matching for data-free knowledge distillation. We took inspiration from such work, but we realized that such synthetic images were not helpful in preserving source performance. Differently, we cast the problem as a conditional image-to-image translation task. The notable impact is that, with the refined pseudo-labels obtained in Stage 1, target samples can be converted to synthetic source-style images that are not only visually plausible, but also suitable for training a classifier.

3 The C-SUDA Method

Our method comprises two main stages (see Fig. 1). After inferring pseudo-labels of the target samples—using the source model, Stage 1 is devoted to *refining* the so called shift-noise [18] (affecting target samples) resulting in cleaner pseudo-labels. The strategy proposed in [1], detailed in Sect. 3.1, is sufficient to obtain state-of-the-art performance on the target set in the source-free setting, yet nothing prevents catastrophic forgetting. For this reason, the target set with refined pseudo-labels is exploited in Stage 2 as a prior for synthesizing source-style images which can "anchor" the model to its original performance on the source set, as detailed in Sect. 3.2.

Preliminaries. The goal of UDA is to adapt a model pre-trained on a labelled source domain $\mathcal{D}_s = \{(\boldsymbol{x}_s^i, y_s^i)\}_{i=1}^{N_s}$ on a different, yet related, unlabeled target domain $\mathcal{D}_t = \{\boldsymbol{x}_t^j\}_{j=1}^{N_t}$, which share the same label set, *i.e.*, $\mathcal{Y}_s = \mathcal{Y}_t$. We assume $\mathcal{D}_s$ is *never* available in a realistic source-free scenario, while of course a pre-trained source model $f_s(\cdot)$ is at our disposal. This can be used to infer pseudo-labels $\mathcal{P} = \{\tilde{y}^j\}_{j=1}^{N_t}$ for the target domain:

$$\tilde{y}^j = \operatorname{argmax} f_s(\boldsymbol{x}_t^j), \ j = 1 \ldots N_t \tag{1}$$

Clearly, $\mathcal{P}$ would be noisy, i.e., a significant amount of pseudo-labels is wrong due to domain shift. The following section discusses how to progressively filter out such noise and subsequently obtains a cleaner set $\mathcal{P}'$, which in turn translates into better accuracy on the target set.

3.1 Stage 1: Pseudo-label Refinement

Training a target model with cross-entropy given $\mathcal{P}$ as supervisory signal (i.e., using $\mathcal{P}$ as ground-truth) eventually results in overfitting noisy samples. For instance, in case of wrong pseudo-labels $\tilde{y}^j \neq y_t^j$ (y_t^j are *unknown* target labels), the model would undeniably try to maximize the probability of a sample belonging to the wrong class.

To mitigate such a problem, an Ensemble of classifiers can be used (see Fig. 1), which denotes the class of techniques of concurrently training multiple networks and, in its simplest form, averaging their output. More in detail, it adopts an idea of employing stochastically sampled *Disjoint Residual Labels (DRL)*: equally distributed disjoint subsets of complementary labels (spanning entire class-set except the given pseudo-label $\tilde{y}^j$) are used to back-propagate different feedback to each ensemble member. This helps ensemble members to learn different concepts and to achieve a strong consensus. For instance, in case of wrong pseudo-label $\tilde{y}^j$, the correct label y_t^j is wrongly provided as one of the complementary labels to only one member, while other members always learn from clean feedback. Thus, the loss is defined as:

$$\mathcal{L}_{DRL}(\mathcal{D}_t) = -\mathbb{E}_{x_t \sim \mathcal{D}_t} \frac{1}{N_{DRL}} \sum_{c=1}^{C} \mathbb{1}_{[c \in DRL]} log(1 - p^c \tag{2}$$

which is used to train each member *independently*. The predictions of the members are then late-fused via:

$$\boldsymbol{p}_e = \boldsymbol{\sigma}(\frac{1}{N_e \cdot N_a} \sum_{k=1}^{N_e} \sum_{l=1}^{N_a} f^{k,a}(\boldsymbol{x})) \tag{3}$$

where f^k is one of the N_e members and we use a moving average of N_a previous outputs ($N_a = 10$ for all the experiments).

With the growing number of epochs, noisy samples remain towards low confidence regime and clean samples obtain high confidence progressively. Consequently, pseudo-label refinement is achieved progressively during training in an adaptive manner, where the total noise is progressively reduced).

3.2 Stage 2: Image Synthesis for Continual Adaptation

As a result of Stage 1, a refined pseudo-label is associated to each target sample. Note that, at this point, some noise still remains in pseudo-labels, *i.e.*, some of them still do not correspond to the correct target label. Since we need to generate class-conditioned source-style images, by leveraging the source model together with $(\boldsymbol{x}_t^j, \tilde{y}_t^j)$ as a prior, we feed Stage 2 with only a subset of the target set which most likely guarantees less noise (see Fig. 1). More in detail, we compute prediction uncertainty of the target samples quantified by self-entropy [10] as:

$$H(\boldsymbol{x}_t) = -\sum \boldsymbol{p}(\boldsymbol{x}_t) log(\boldsymbol{p}(\boldsymbol{x}_t)), \tag{4}$$

where smaller entropy indicates more confident prediction (more details in supp. mat.). Based on this, we sort the target samples in ascending order and select only the first N_h samples from each class. These selected target samples are then optimized by minimizing the cross-entropy loss for the original source model and two feature distribution regularization terms. Note that here the source model (the only available information we have about the source domain) is kept frozen while we optimize pseudo-labeled target samples $(\boldsymbol{x}_t^j, \tilde{\boldsymbol{y}}_t^j)$ in the pixel space:

$$\boldsymbol{x} \leftarrow \boldsymbol{x} - \eta \nabla_{\boldsymbol{x}} \mathcal{L}(f_s(\boldsymbol{x}), \tilde{y}), \quad \boldsymbol{x}_0 = \boldsymbol{x}_t^j \tag{5}$$

$$\begin{aligned} \mathcal{L}(f_s(\boldsymbol{x}), \tilde{y}) = & \ell_{CE}((f_s(\boldsymbol{x})), \tilde{y}) + \\ & \lambda_{TV} \mathcal{R}_{TV}(\boldsymbol{x}) + \lambda_{BN} \mathcal{R}_{BN}(\boldsymbol{x}), \end{aligned} \tag{6}$$

$$\mathcal{R}_{TV}(\boldsymbol{x}) = \sum_{u,v} ((\boldsymbol{x}_{u,v+1} - \boldsymbol{x}_{uv})^2 + (\boldsymbol{x}_{u+1,v} - \boldsymbol{x}_{uv})^2)^{\frac{1}{2}},$$

$$\mathcal{R}_{BN} = \sum_{l,j} \parallel \mu_l(\boldsymbol{x}^j) - \mu_l \parallel + \parallel \sigma_l^2(\boldsymbol{x}^j) - \sigma_l^2) \parallel_2,$$

Here $\ell_{CE}(\cdot)$ is the cross-entropy loss, f_s is the frozen source model, λ_{TV} and λ_{BN} are scalar weights.

$\mathcal{R}_{TV}$ is a regularizer that penalizes the *Total Variation norm* approximated as finite pixel difference (u, v are pixel indexes) [16]: it provides more stable convergence and encourages $\boldsymbol{x}$ to consist of piece-wise uniform patches.

$\mathcal{R}_{BN}$ enforces batch-wise (j is the batch index) feature statistics similarities at all layers l, exploiting the BatchNorm running average parameters (μ_l, σ_l) stored in the source models, which implicitly capture the channel-wise means and variances of the original source images [32].

In principle, images can be generated by optimizing random noise [32]. But we verified that starting from the actual target images guarantees higher fidelity, realism, and most of all diversity in generated synthetic images (see Fig. 2). We also found that just a handful of synthetic source-style samples generated by our proposed method effectively help in preserving source performance.

Final Model. Once target data have been assigned pseudo-labels and source-style synthetic images are generated, we can proceed with the final model training (see Fig. 1, bottom left). We initialize the weights of the final trainable model with the source model. Subsequently, the weights of fully-connected (FC) layers are frozen while feature-extractor (FE) remains trainable. We train the final model with standard cross-entropy loss using both real target and synthetic source-style images together with corresponding refined pseudo-labels. Freezing FC layers not only help FE to learn the representation of synthetic source images most likely to be identical to the real source domain but also forces the target domain to get aligned with the source.

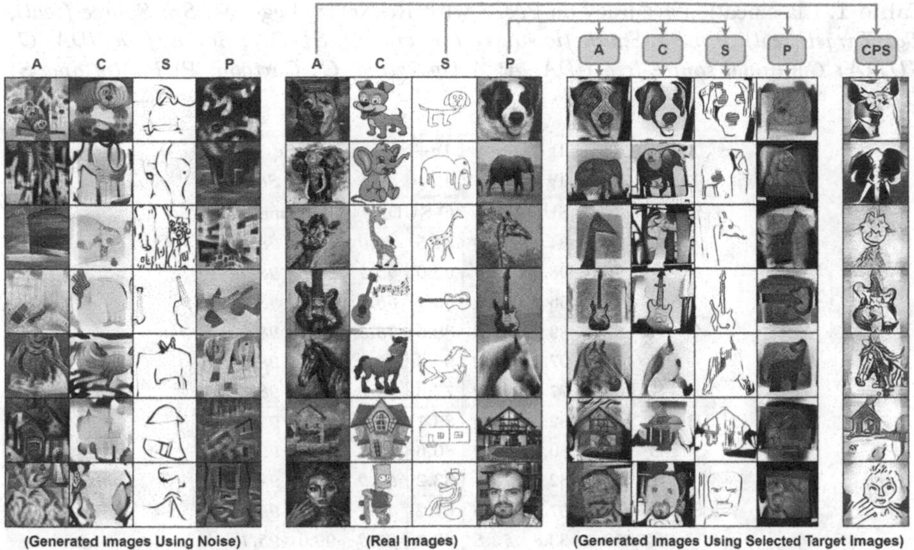

Fig. 2. Image Synthesis using PACS dataset. Source style images (*right*) are optimized from target images with refined pseudo-labels (*center*). We also provide an example of images synthesized from random noise (*left*). CPS exemplifies a multi-source case. Legend: ***A***: *Art-painting*, ***C***: *Cartoon*, ***P***: Photo, and ***S***: Sketch.

4 Experiments

For the image classification task, we evaluate performance of our method via extensive experiments on major UDA benchmarks including, PACS [12], VisDA-C [21], and DomainNet [20].

For stage 1, we evaluate the refined pseudo-labels from [1], which is based on the ensemble network comprising 3 members. We examine the effectiveness of stage 2 i.e., the usefulness of synthesized images in preserving good performance on the real source domain. Note that in stage 2, we have access to the pre-trained source model and the target images with associated refined pseudo-labels (output of stage 1) only. For single-source and multi-source continual source-free UDA, we synthesis a relevant source for each target. For multi-target case, one synthesized source is enough for all related targets.

In all experiments related to image synthesis, we synthesize 32 images per class—all together as one batch. The batch is initialized with real target samples carrying lowest self-entropy according to Eq. 4. We use *Adam* with a learning rate of $1e-1$ (with cosine annealing schedule) for images optimization. We set $\lambda_{TV} = 1e-4$, $\lambda_{BN} = 1e-2$, and batch receives $10K$ updates. Sample images are provided in Fig. 2.

4.1 Results

The reported results in Tables 1 and 2 present average accuracy of 3 runs. In Table 1, we report results for all the possible pairs. Also, we report upper-bound

Table 1. Classification accuracy on PACS with ResNet18. Legends: ***Sc:*** *Source (real),* ***Tg:*** *Target (real), SynSc: Synthetic source (generated),* ***SUDA:*** *Source-free UDA,* ***C-SUDA:*** *Continual source-free UDA,* ***A***: *Art-painting,* ***C***: *Cartoon,* ***P***: *Photo, and* ***S***: *Sketch.*

Train		Sc	Tg		Tg+*SynSc*		Tg+Sc	
Test		Tg	Tg	*Sc*	Tg	*Sc*	Tg	*Sc*
Sc	Tg		SUDA		C-SUDA		Baseline	
A	**C**	58.1	84.3	*63.0*	84.5	*81.0*	98.6	*98.1*
	P	96.0	98.4	*62.5*	98.0	*75.9*	99.5	*98.7*
	S	43.9	56.2	*17.9*	55.7	*65.5*	96.4	*98.6*
C	**A**	67.3	89.0	*61.9*	88.5	*78.5*	98.0	*98.2*
	P	85.6	97.2	*21.2*	96.4	*70.4*	98.9	*98.9*
	S	60.6	77.6	*46.9*	77.3	*71.2*	96.6	*98.8*
P	**A**	60.9	82.6	*87.9*	83.1	*92.0*	98.1	*99.1*
	C	24.8	80.5	*68.2*	80.6	*90.8*	99.1	*99.5*
	S	26.5	32.3	*13.4*	33.2	*90.9*	96.2	*98.2*
S	**A**	18.1	67.6	*42.6*	67.2	*74.7*	97.8	*96.7*
	C	32.6	83.8	*52.8*	83.9	*72.3*	99.0	*95.1*
	P	24.3	77.1	*17.9*	77.0	*72.6*	99.6	*96.7*
Avg.		49.9	77.2	*46.3*	77.1	*78.0*	98.1	*98.0*

Table 2. Classification accuracy on PACS with ResNet18. * results are taken from [8]. Legend: ***Sc→Tg***: *Inferred pseudo-labels,* ***SUDA:*** *Source-free UDA,* ***C-SUDA:*** *Continual source-free UDA, ⋆: Accuracy on real-source,* ***A***: *Art-Painting,* ***C***: *Cartoon,* ***P***: *Photo, and* ***S***: *Sketch.*

Multi-Target C-SUDA								Multi-Source C-SUDA					
Sc		***P***			***A***		Avg.	Sc	***C,P,S***	***A,P,S***	***A,C,S***	***A,C,P***	Avg.
Tg	***A***	***C***	***S***	***P***	***C***	***S***		Tg	***A***	***C***	***P***	***S***	
ADDA*	24.3	20.1	22.4	32.5	17.6	18.9	22.6	SIB [9]	88.9	89.0	98.3	82.2	89.6
DSN*	28.4	21.1	25.6	29.5	25.8	24.6	25.8	OML [11]	87.4	86.1	97.1	78.2	87.2
ITA*	31.4	23.0	28.2	35.7	27.0	28.9	29.0	RABN [29]	86.8	86.5	98.0	71.5	85.7
KD [3]	24.6	32.2	**33.8**	35.6	46.6	**57.5**	46.6	JiGen [5]	84.8	81.0	97.9	79.0	85.7
Sc→Tg		37.7			57.9		47.8	Sc→Tg	78.4	77.9	95.3	64.5	79.0
SUDA	**80.1**	**76.1**	25.9	**96.0**	**82.8**	49.8	**68.4**	SUDA	**90.8**	**89.5**	**98.8**	**85.2**	**91.1**
C-SUDA	**79.9**	**77.1**	25.1	**95.6**	**83.2**	47.6	**68.1**	C-SUDA	**89.5**	88.4	97.6	**84.6**	**90.0**
SUDA⋆		*56.5*			*47.8*		*52.2*	*SUDA*⋆	*69.1*	*62.2*	*34.1*	*35.5*	*50.2*
C-SUDA⋆		***91.2***			***74.1***		***82.7***	*C-SUDA*⋆	***81.9***	***78.5***	***68.9***	***66.1***	***73.8***

performance (Baseline), however, we skip this information in the rest of the tables. In Tables 2 and 3, the *Sc→Tg* row reports the amount of correct target pseudo-labels acquired using the frozen source model. Along with the performance of SUDA and C-SUDA on target, we also report the performance *degradation* on source due to source-free UDA (SUDA⋆) as well as the effectiveness of our method (C-SUDA⋆) in preserving the performance on the source.

In Table 2 *(left)*, we compare our method with the existing approaches addressing multi-target UDA on PACS. As can be noticed, with comparable

Table 3. Classification accuracy on DomainNet with ResNet101. Legend: ***C***: *Clipart*, ***I***: *Infograph*, ***P***: *Painting*, ***Q***:*Quickdraw*, ***R***: *Real*, and ***S***: *Sketch*.

Multi-Source C-SUDA							
Tg	***C***	***I***	***P***	***Q***	***R***	***S***	Avg.
MM [20]	58.6	26.0	52.3	6.3	62.7	49.5	42.6
OML [11]	62.8	21.3	50.5	15.4	64.5	50.4	44.1
CMSS [30]	64.2	28.0	53.6	16.0	63.4	53.8	46.5
DRT+ST [14]	71.0	**31.6**	**61.0**	12.3	**71.4**	60.7	51.3
Sc→Tg	68.5	23.6	53.5	17.6	65.9	55.2	47.4
SUDA	70.8	27.2	58.1	**24.1**	69.5	60.1	**51.6**
C-SUDA	**71.4**	26.5	57.1	**24.2**	67.9	59.0	51.0
*SUDA**	*29.6*	*30.7*	*35.3*	*7.9*	*35.1*	*36.7*	*29.2*
*C-SUDA**	***64.5***	***67.0***	***63.3***	***55.8***	***59.9***	***65.2***	***62.6***

Table 4. Classification accuracy on Visda-C with ResNet101.

Methods	*plane*	*bcycl*	*bus*	*car*	*horse*	*knife*	*mcycl*	*person*	*plant*	*skate*	*train*	*truck*	Avg.
Inferred	64.2	6.3	75.2	21.7	55.9	95.7	22.8	1.4	79.8	0.7	82.8	19.8	46.3
DADA [24]	92.9	74.2	82.5	65.0	90.9	**93.8**	87.2	74.2	89.9	71.5	86.5	48.7	79.8
SHOT [15]	94.3	**88.5**	80.1	57.3	93.1	94.9	80.7	80.3	91.5	89.1	86.3	58.2	82.9
A^2Net [28]	94.0	87.8	85.6	66.8	93.7	95.1	85.8	81.2	91.6	88.2	86.5	56.0	84.3
Sc→Tg	64.2	6.3	75.2	21.7	55.9	95.7	22.8	1.4	79.8	0.7	82.8	19.8	46.3
SUDA	**94.8**	68.1	**89.5**	**88.1**	86.5	90.4	**87.4**	**89.0**	53.2	81.5	**96.9**	**93.0**	**84.8**
C-SUDA	**94.9**	67.3	**89.2**	**87.8**	86.1	90.0	86.6	**88.7**	53.1	80.9	**96.5**	**94.6**	**84.6**
*SUDA**	*45.2*	*18.5*	*55.9*	*52.7*	*54.8*	*44.3*	*12.5*	*41.4*	*24.6*	*35.1*	*40.2*	*51.2*	*39.7*
*C-SUDA**	***47.6***	***21.4***	***58.2***	***54.3***	***61.1***	***49.5***	***27.9***	***41.9***	***44.8***	***36.2***	***43.1***	***55.4***	***45.1***

performance in 2 cases, our method achieves superior average accuracy. For multi-source UDA, we compare recent works in Table 2 *(right)*. Also in this framework, our method consistently outperforms existing methods, with only in one case getting lower, yet comparable, accuracy.

In Table 3, with comparable performance in one case, C-SUDA consistently outperforms existing methods despite the large number of classes and discrepancy across domains. Also in Table 4, the proposed method achieves state-of-the-art average accuracy on such a challenging benchmark.

5 Conclusions

This work proposes Continual Source-Free Unsupervised Domain Adaptation as a realistic adaptation scenario where source samples are not available, but the performance is to be preserved on the source domain. Our method is composed of two stages, a Source-free UDA technique based on pseudo-label refinement, and a procedure for synthesizing source-style images to avoid catastrophic forgetting. The proposed pipeline effectively solves the task by only assuming a pre-trained source model. We empirically demonstrate that our proposed method achieves state-of-the-art performance on major UDA benchmarks.

References

1. Ahmed, W., Morerio, P., Murino, V.: Cleaning noisy labels by negative ensemble learning for source-free unsupervised domain adaptation. In: IEEE/CVF Winter Conference on Applications of Computer Vision, pp. 1616–1625 (2022)
2. Bang, J., Kim, H., Yoo, Y., Ha, J.W., Choi, J.: Rainbow memory: continual learning with a memory of diverse samples. In: IEEE/CVF Conference on Computer Vision and Pattern Recognition, pp. 8218–8227 (2021)
3. Belal, A., Kiran, M., Dolz, J., Blais-Morin, L.A., Granger, E., et al.: Knowledge distillation methods for efficient unsupervised adaptation across multiple domains. Image Vis. Comput. **108**, 104096 (2021)
4. Bobu, A., Tzeng, E., Hoffman, J., Darrell, T.: Adapting to continuously shifting domains. In: ICLR (2018)
5. Carlucci, F.M., D'Innocente, A., Bucci, S., Caputo, B., Tommasi, T.: Domain generalization by solving jigsaw puzzles. In: IEEE Conference on Computer Vision and Pattern Recognition, pp. 2229–2238 (2019)
6. Chen, C., et al.: Progressive feature alignment for unsupervised domain adaptation. In: IEEE Conference on Computer Vision and Pattern Recognition, pp. 627–636 (2019)
7. Ge, Y., Chen, D., Li, H.: Mutual mean-teaching: pseudo label refinery for unsupervised domain adaptation on person re-identification. In: International Conference on Learning Representations (2020)
8. Gholami, B., Sahu, P., Rudovic, O., Bousmalis, K., Pavlovic, V.: Unsupervised multi-target domain adaptation: an information theoretic approach. IEEE Trans. Image Process. **29**, 3993–4002 (2020)
9. Hu, S.X., et al.: Empirical Bayes transductive meta-learning with synthetic gradients. In: International Conference on Learning Representations (2020)
10. Kim, Y., Cho, D., Han, K., Panda, P., Hong, S.: Domain adaptation without source data. IEEE Trans. Artif. Intell. **2**(6), 508–518 (2021). https://doi.org/10.1109/TAI.2021.3110179
11. Li, D., Hospedales, T.: Online meta-learning for multi-source and semi-supervised domain adaptation. In: Vedaldi, A., Bischof, H., Brox, T., Frahm, J.-M. (eds.) ECCV 2020. LNCS, vol. 12361, pp. 382–403. Springer, Cham (2020). https://doi.org/10.1007/978-3-030-58517-4_23
12. Li, D., Yang, Y., Song, Y.Z., Hospedales, T.M.: Deeper, broader and artier domain generalization. In: IEEE International Conference on Computer Vision, pp. 5542–5550 (2017)
13. Li, R., Jiao, Q., Cao, W., Wong, H.S., Wu, S.: Model adaptation: unsupervised domain adaptation without source data. In: IEEE/CVF Conference on Computer Vision and Pattern Recognition, pp. 9641–9650 (2020)
14. Li, Y., Yuan, L., Chen, Y., Wang, P., Vasconcelos, N.: Dynamic transfer for multi-source domain adaptation. In: IEEE/CVF Conference on Computer Vision and Pattern Recognition, pp. 10998–11007 (2021)
15. Liang, J., Hu, D., Feng, J.: Do we really need to access the source data? Source hypothesis transfer for unsupervised domain adaptation. In: III, H.D., Singh, A. (eds.) 37th International Conference on Machine Learning. Proceedings of Machine Learning Research, vol. 119, pp. 6028–6039. PMLR, 13–18 July 2020

16. Mahendran, A., Vedaldi, A.: Understanding deep image representations by inverting them. In: IEEE Conference on Computer Vision and Pattern Recognition (CVPR), June 2015
17. Mordvintsev, A., Olah, C., Tyka, M.: Inceptionism: going deeper into neural networks (2015)
18. Morerio, P., Volpi, R., Ragonesi, R., Murino, V.: Generative pseudo-label refinement for unsupervised domain adaptation. In: The IEEE Winter Conference on Applications of Computer Vision, pp. 3130–3139 (2020)
19. Pan, X., Zhan, X., Dai, B., Lin, D., Loy, C.C., Luo, P.: Exploiting deep generative prior for versatile image restoration and manipulation. IEEE Trans. Pattern Anal. Mach. Intell. **44**(11), 7474–7489 (2021)
20. Peng, X., Bai, Q., Xia, X., Huang, Z., Saenko, K., Wang, B.: Moment matching for multi-source domain adaptation. In: IEEE International Conference on Computer Vision, pp. 1406–1415 (2019)
21. Peng, X., Usman, B., Kaushik, N., Wang, D., Hoffman, J., Saenko, K.: VisDA: a synthetic-to-real benchmark for visual domain adaptation. In: IEEE Conference on Computer Vision and Pattern Recognition Workshops, pp. 2021–2026 (2018)
22. Santurkar, S., Ilyas, A., Tsipras, D., Engstrom, L., Tran, B., Madry, A.: Image synthesis with a single (robust) classifier. In: Wallach, H., Larochelle, H., Beygelzimer, A., d'Alché-Buc, F., Fox, E., Garnett, R. (eds.) Advances in Neural Information Processing Systems, vol. 32. Curran Associates, Inc. (2019)
23. Shi, Y., Yuan, L., Chen, Y., Feng, J.: Continual learning via bit-level information preserving. In: IEEE/CVF Conference on Computer Vision and Pattern Recognition, pp. 16674–16683 (2021)
24. Tang, H., Jia, K.: Discriminative adversarial domain adaptation. In: AAAI, pp. 5940–5947 (2020)
25. Torralba, A., Efros, A.A.: Unbiased look at dataset bias. In: IEEE/CVF Conference on Computer Vision and Pattern Recognition, pp. 1521–1528 (2011)
26. Volpi, R., Larlus, D., Rogez, G.: Continual adaptation of visual representations via domain randomization and meta-learning. In: IEEE/CVF Conference on Computer Vision and Pattern Recognition (CVPR), pp. 4443–4453 (2021)
27. Volpi, R., Namkoong, H., Sener, O., Duchi, J., Murino, V., Savarese, S.: Generalizing to unseen domains via adversarial data augmentation. In: 32nd International Conference on Neural Information Processing Systems, pp. 5339–5349 (2018)
28. Xia, H., Zhao, H., Ding, Z.: Adaptive adversarial network for source-free domain adaptation. In: IEEE/CVF International Conference on Computer Vision (ICCV), pp. 9010–9019, October 2021
29. Xu, J., Xiao, L., López, A.M.: Self-supervised domain adaptation for computer vision tasks. IEEE Access **7**, 156694–156706 (2019)
30. Yang, L., Balaji, Y., Lim, S.-N., Shrivastava, A.: Curriculum manager for source selection in multi-source domain adaptation. In: Vedaldi, A., Bischof, H., Brox, T., Frahm, J.-M. (eds.) ECCV 2020. LNCS, vol. 12359, pp. 608–624. Springer, Cham (2020). https://doi.org/10.1007/978-3-030-58568-6_36
31. Yang, S., Wang, Y., van de Weijer, J., Herranz, L., Jui, S.: Generalized source-free domain adaptation. In: IEEE/CVF International Conference on Computer Vision, pp. 8978–8987 (2021)
32. Yin, H., et al.: Dreaming to distill: data-free knowledge transfer via DeepInversion. In: IEEE/CVF Conference on Computer Vision and Pattern Recognition, pp. 8715–8724 (2020)

Self-Similarity Block for Deep Image Denoising

Edoardo Peretti[1(✉)], Diego Stucchi[1], Diego Carrera[2], and Giacomo Boracchi[1]

[1] DEIB, Politecnico di Milano, Milan, Italy
{edoardo.peretti,diego.stucchi,giacomo.boracchi}@polimi.it
[2] STMicroelectronics, Agrate Brianza, Italy
diego.carrera@st.com

Abstract. Non-Local Self-Similarity (NLSS) is a widely exploited prior in image denoising algorithms. The first deep Convolutional Neural Networks (CNNs) for image denoising ignored NLSS and were made of a sequence of convolutional layers trained to suppress noise. The first denoising CNNs leveraging NLSS prior were performing non-learnable operations outside the network. Then, pre-defined similarity measures were introduced and finally learnable, but scalar, similarity scores were adopted inside the network. We propose the Self-Similarity Block (SSB), a novel differentiable building block for CNN denoisers to promote the NLSS prior. The SSB is trained in an end-to-end manner within convolutional layers and learns a multivariate similarity score to improve image denoising by combining similar vectors in an activation map. We test SSB on additive white Gaussian noise suppression, and we show it is particularly beneficial when the noise level is high. Remarkably, SSB is mostly effective in image regions presenting repeated patterns, which most benefit from the NLSS prior.

1 Introduction

Digital sensors are affected by photon counting, thermal, and quantization noise, which needs to be suppressed to provide visually-pleasant images. As a matter of fact, denoising is a fundamental step in almost every image processing pipeline [25], since noise can affect subsequent processing. During the last few decades, image denoising algorithms heavily relied on statistical modeling of images and carefully crafted signal processing techniques to promote priors such as sparsity and Non-Local Self-Similarity (NLSS) [1,3]. The NLSS of natural images suggests that most small patches contain patterns that are repeated across the same image, possibly at non-adjacent locations. Recently, deep-learning-based methods adopted Convolutional Neural Networks (CNNs) to solve the denoising problem by directly learning a sophisticated denoising function from a dataset of noise-free images corrupted by artificially added noise. However, while several classic methods exploit the NLSS prior, most CNN denoisers act locally as DnCNN [26], even when they have a large receptive field.

G. L. Foresti et al. (Eds.): ICIAP 2023, LNCS 14233, pp. 26–38, 2023.
https://doi.org/10.1007/978-3-031-43148-7_3

Recent research has focused on designing new building blocks and training procedures enforcing traditional priors in deep denoisers. Traditional non-local operations used to promote NLSS, e.g., block matching, are not differentiable and must be performed outside the CNN [11], yielding sub-optimal performance. The few differentiable non-local blocks that can be employed inside a CNN rely on scalar hand-crafted similarity measures [17] or are based on the attention mechanism [22], which is defined over the entire feature, while NLSS is typically exploited relatively to a search neighborhood.

We propose the Self-Similarity Block (SSB), a novel differentiable block promoting NLSS in deep denoisers. In practice, the SSB exploits the similarity of activation vectors from different spatial locations in the same activation map to perform denoising. To this purpose, the SSB uses an *entirely learnable multivariate similarity score* in a search neighborhood, computed as a series of 1×1 convolutions mixing the information from different locations. Our SSB can be easily included after any layer of a CNN, as it does not modify the dimension of the input activation map. In particular, the SSB can be trained with the whole network or fine-tuned in a pre-trained network to boost its denoising performance. In this regard, we design Self-Similarity Networks (SSN), which consist of a sequence of convolutional blocks as in DnCNN [26], interleaved by SSBs.

Our experiments confirm that inserting an SSB in denoising CNNs improves their performance. Moreover, SSN achieves comparable performance with respect to state-of-the-art methods based on comparable convolutional architectures, including deep denoisers exploiting the NLSS prior, while outperforms them on high noise levels. The source code is available at https://github.com/edpere/SSN.

2 Related Work

Many image priors are exploited in denoising [4,16,18,19,30], being NLSS one of the most popular. Usually, similarity scores are computed over non-adjacent patches, and then used to average pixel intensities (NLM [1]), in collaborative filtering in transform domain (BM3D [3]), and in low-rank approximation [6].

Here, we focus on deep CNN denoisers trained on clean-noisy image pairs. These comprise RED [14], which is a convolutional encoder-decoder architecture, and DnCNN [26], which is a stack of convolutions, batch normalizations [8] and ReLU activations. FFDNet [27] extended DnCNN by introducing a noise level map as input and a reversible downscaling operator. The first attempts to leverage the NLSS prior in CNN denoisers consisted in enlarging the receptive fields. However, stacking more layers [14] or inserting downscaling operators [27] does not significantly increase the *effective* receptive field [13] of CNN denoisers, which explains why very deep architectures do not yield a significantly better denoising [20]. Therefore, several methods [10,11,24], inspired by BM3D [3], introduced block matching for grouping similar patches as a pre-processing step before CNN filtering. For example, NN3D [2] combines a local CNN denoiser with a traditional non-local filter in an iterative manner. In all this methods the non-local operations are predefined and not learned.

Interestingly, recent works investigate differentiable layers promoting the NLSS within activation maps, to enable end-to-end training alongside the network. N^{3}Net [17] proposes a differentiable relaxation of KNN, and then concatenates similar features in activation maps. One of the most relevant attempt to leverage NLSS within activation maps are the nonlocal networks [22], which implement the attention mechanism in images, drawing a parallel with NLM. However, this is a global operation which was not originally proposed for image restoration but for visual recognition tasks. This non local-block was later adapted to image denoising both in a recurrent [12] and in a sophisticated feed-forward [28] architecture comprising multiple branches. This approach can be generalized by graph neural networks [21].

All these non-local blocks operate on activation maps using a scalar similarity measure, like the Euclidean one. Remarkably, our proposed SSB learns a *multivariate similarity score* directly on training data, capturing different aspects of the similarity embedded in the activation vectors, and improving convolutional denoisers, as shown in our experiments.

3 Problem Formulation

We consider grayscale image denoising where the noisy image z is defined over a finite grid $X \subset \mathbb{Z}^2$ and described as follows:

$$z(x) = y(x) + \eta(x) \qquad \forall x \in X, \tag{1}$$

where y is the clean image, $\eta(\cdot) \sim \mathcal{N}(0, \sigma^2)$ is the *additive white Gaussian noise* (AWGN), and σ is the noise standard deviation. An image denoiser $\mathcal{D}_\theta$, depending on the set of parameters $\theta \in \Theta$, is a map that provides an estimate $\widehat{y} = \mathcal{D}_\theta(z)$ of the noise-free image y. We consider $\mathcal{D}_\theta$ to be a CNN trained in a supervised manner, from a training set $\{(y_j, z_j)\}_j$ of clean-noisy image pairs. Remarkably, AWGN denoisers can handle different noise models by variance stabilizing transforms [5], and the extension to color images is trivial when operating with CNNs.

4 Proposed Method

In this section, we present the Self-Similarity Block (SSB), a differentiable layer for CNN denoisers designed to exploit the NLSS prior. Traditional algorithms exploit the NLSS prior to estimate each reference patch by combining information from similar patches within a search neighborhood. As in [22], our intuition is to leverage NLSS among *activation vectors* of the same activation map returned by a convolutional layer. More specifically, we consider an activation vector $\mathbf{v} \in \mathbb{R}^d$ obtained by stacking all the values along the d channels from an activation map at a given spatial location. Each activation vector can be interpreted as a learned embedding of the corresponding region in the input image.

The SSB estimates the similarity among pairs of activation vectors in a search region of the same activation map through a trainable multivariate similarity score. This similarity is used to guide the filtering of the activation vectors and to compute the weights for the final average. To improve denoising performance, we adopt SSB within a residual mapping. This is custom in CNN denoisers [26], and in practice forces SSB to learn the noise realization to be removed from each input activation map. Therefore, our SSB can be easily inserted in any pre-trained CNN denoiser since *i)* it takes any 3D tensor as input and outputs one of the same size and *ii)* it can be initialized as the identity map to be fine-tuned.

In Sect. 4.1, we illustrate the *differentiable multivariate similarity score* adopted by the SSB, and in Sect. 4.2, we describe the implementation of the proposed end-to-end learnable layer. Finally, in Sect. 4.3, we present the Self-Similarity Network (SSN), a customizable CNN combining DnCNN and SSBs.

4.1 Non-Local Filtering Guided by Vector Similarities

Let $\mathbf{v} \in \mathbb{R}^d$ be a reference activation vector computed from a noisy image z at a specific convolutional layer. We perform denoising by a *learnable residual mapping* Φ promoting the NLSS as follows

$$\widehat{\mathbf{v}} = \mathbf{v} - \Phi\left(\mathbf{v} \mid \{\mathbf{v}^k\}_{k=1}^M\right), \tag{2}$$

where $\{\mathbf{v}^k\}_k \subset \mathbb{R}^d$ are M vectors in a search neighborhood of $\mathbf{v}$ from the same activation map. Since we adopt residual learning, the goal of Φ is to estimate the noise realization affecting $\mathbf{v}$, then subtracted to obtain a noise-free estimate $\widehat{\mathbf{v}}$. We define

$$\Phi\left(\mathbf{v} \mid \{\mathbf{v}^k\}_{k=1}^M\right) = A \sum_{k=1}^{M} \left(\alpha_k \mathbf{r}^k\right), \tag{3}$$

where $\{\mathbf{r}^k\}_k \subset \mathbb{R}^d$ are neighbor contributions, each depending on a single activation vector belonging to the search neighborhood, $A \in \mathbb{R}^{d\times d}$ is a learnable weight matrix, and the weights $\{\alpha_k\}_k \subset \mathbb{R}$ define the contribution of each $\mathbf{r}^k$, based on the similarity score between $\mathbf{v}$ and $\mathbf{v}^k$.

We define the similarity score $\mathbf{s}^k \in \mathbb{R}^n$ between two activation vectors $\mathbf{v}$ and $\mathbf{v}^k$ as a multivariate weighted combination of the two, namely:

$$\mathbf{s}^k = \mathrm{ReLU}\left(R\mathbf{v} + N\mathbf{v}^k\right), \tag{4}$$

where $R, N \in \mathbb{R}^{n\times d}$ are learnable matrices defining two linear embeddings, that are independent of the reference $\mathbf{v}$ or the neighbor $\mathbf{v}^k$. Since these operations are linear, we introduce non-linearity with a ReLU function. After training, we expect each component of $\mathbf{s}^k$ to capture some form of similarity between $\mathbf{v}$ and $\mathbf{v}^k$, as a deep learning alternative of the patch distance of NLM. Then, we use $\mathbf{s}^k$ to guide the computation of a *neighbor contribution* $\mathbf{r}^k$, which is defined by extracting relevant information from a neighbor $\mathbf{v}^k$ and the similarity $\mathbf{s}^k$. More precisely, we compute $\mathbf{r}^k \in \mathbb{R}^d$ as

$$\mathbf{r}^k = \mathrm{ReLU}\left(P\mathbf{v}^k + Q\mathbf{s}^k\right), \tag{5}$$

where $P \in \mathbb{R}^{d\times d}$ and $Q \in \mathbb{R}^{d\times n}$ are learnable linear embedding matrices, and we introduce non-linearity by the ReLU function. To conclude, each $\mathbf{r}^k$ contributes to the final residual in (3) by a weight α_k that depends on the similarity $\mathbf{s}^k$ between $\mathbf{v}$ and $\mathbf{v}^k$. More precisely, we define the weights $\{\alpha_k\}_k \subset \mathbb{R}$ in (3) as the inner product of the similarity scores $\{\mathbf{s}^k\}_k$ against a shared learnable vector $\mathbf{h} \in \mathbb{R}^n$, normalizing the results with the Softmax:

$$\alpha_k = \text{Softmax}\left(\mathbf{h}\cdot\mathbf{s}^1, \mathbf{h}\cdot\mathbf{s}^2, \ldots, \mathbf{h}\cdot\mathbf{s}^M\right)_k, \tag{6}$$

such that weights sum to 1. Finally, the residual computed as in (3) is subtracted from the input vector $\mathbf{v}$ as in (2).

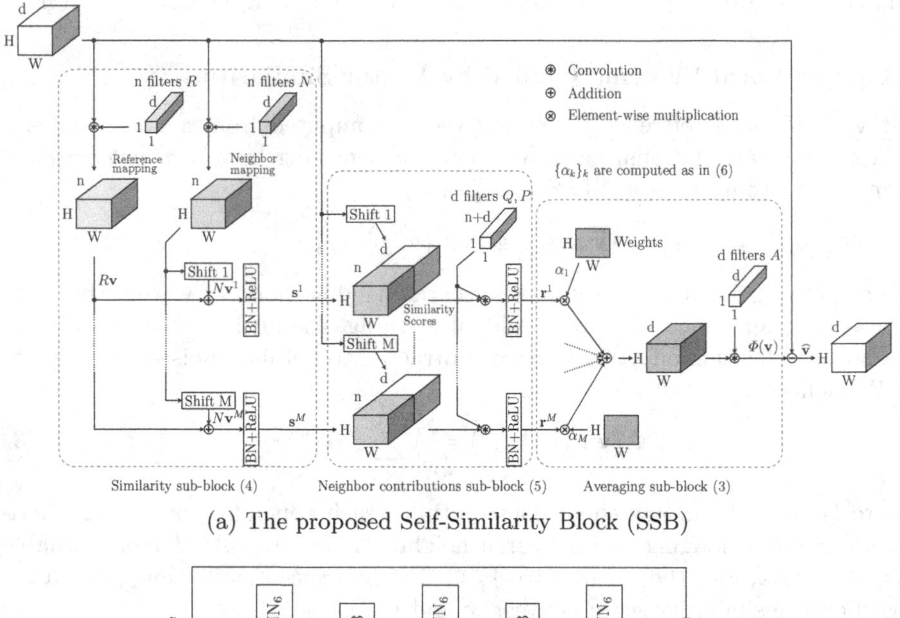

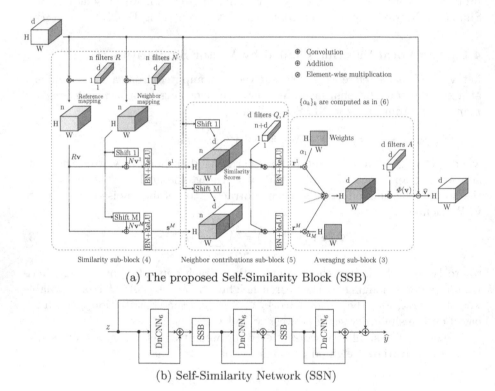

(a) The proposed Self-Similarity Block (SSB)

(b) Self-Similarity Network (SSN)

Fig. 1. (a) The Self-Similarity Block (SSB) exploits the NLSS prior for denoising by residual computation. It comprises three sub-blocks: *i)* similarity sub-block, that computes the scores $\{\mathbf{s}^k\}_k$, *ii)* neighbor contributions sub-block, which computes $\{\mathbf{r}^k\}_k$, and *iii)* averaging sub-block, aggregating all the neighbor contributions. SSB is employed with a skip connection to perform residual learning.(b) A Self-Similarity Network (SSN) consisting of 6-layer DnCNNs interleaved by SSBs.

4.2 Self-Similarity Block

Here, we illustrate the architecture of the SSB, which implements the denoising procedure described in Sect. 4.1. In particular, similarly to the search neighborhood adopted by NLM [1], for each reference vector $\mathbf{v}$, we define the neighbors $\{\mathbf{v}^k\}_{k=1}^M$ as the M vectors in a $(2\,\Omega+1)\times(2\,\Omega+1)$ grid centered around $\mathbf{v}$ separated by a stride w, where Ω is the half-neighborhood size. Moreover, we implement the matrix-by-vector multiplications of the linear embeddings in (3), (4) and (5) as layers of 1×1 convolutions against the rows of the corresponding matrices A, R, N, P, Q. As shown in Fig. 1a, SSB is made of three sub-blocks: a similarity sub-block implementing (4), a neighbor contributions sub-block implementing (5) and an averaging sub-block implementing (3).

The *similarity sub-block* (Fig. 1a, left) computes the similarity scores $\{\mathbf{s}^k\}_k$ (4) for all the reference/neighbor vector pairs. Noting that all the activation vectors $\mathbf{v}^k$ are, in turn, references and neighbors for other references, we decompose the similarity computation in two branches. One is responsible for the computation of $R\mathbf{v}$ for all $\mathbf{v}$ in the activation map, while the other computes $N\mathbf{v}^k$. We shift the input data to line up each neighbor $\mathbf{v}^k$ with the reference. Therefore, we consider M shift branches and compute all the similarity scores $\{\mathbf{s}^k\}_k$ at once by summing two properly shifted outputs of the branches. After the sum, we apply the Batch Normalization (BN) and the ReLU activation, which inserts non-linearity in an otherwise linear computation.

The *neighbor contributions sub-block* (Fig. 1a, center) computes the neighbor contributions $\{\mathbf{r}^k\}_k$ for every reference/neighbor vector pairs. The sum in (5) is computed by concatenating the similarity scores and the shifted neighbors and then performing d 1×1 convolutions. Again, we apply the BN and the ReLU to the results of this operation.

The *averaging sub-block* (Fig. 1a, right) combines the neighbor contributions in a weighted sum. We compute the weights $\{\alpha_k\}_k$ as in (6) and apply them to the contributions through a component-wise multiplication. The results are then summed up and convolved against the d rows of A, as in (3) to yield the final residuals $\Phi(\mathbf{v})$ for every $\mathbf{v}$ in the activation map. Finally, the residuals are subtracted from the input of the block via a skip connection.

We remark that inserting an SSB in a CNN does not significantly increase the number of trainable parameters, which are the matrices $R, N, Q \in \mathbb{R}^{n\times d}$, $P \in \mathbb{R}^{d\times d}$ and $A \in \mathbb{R}^{n\times n}$ and the vector $\mathbf{h} \in \mathbb{R}^n$, for a total of $n^2+d^2+3nd+n$ parameters. However, the number of operations increases significantly because of the M shifts performed to compute the similarity scores.

4.3 Network Architecture

We design the Self-Similarity Network (SSN) as a customizable denoising CNN architecture based on DnCNN [26] and our SSB. The architecture of SSN is reported in Fig. 1b and consists of instances of DnCNN interleaved by SSBs. Remarkably, the SSB takes a tensor of arbitrary size as input and returns an output of the same size. For this reason, it can be seamlessly inserted between layers of any CNN to leverage the NLSS. Moreover, if initialized as the identity transformation by setting $A = 0$, the SSB can be fine-tuned in pre-trained denoising networks to improve the denoising performance.

We denote by DnCNN_D a DnCNN of depth D, which consists of 1 Conv+ReLU followed by $(D-2)$ Conv+BN+ReLU layers and 1 Conv layer. Then, we denote by $m\text{SSN}_D$ a denoising CNN consisting of m instances of DnCNN_D interleaved by $m-1$ instances of SSB. Figure 1b represents the 3SSN_6 model that we adopt in our experiments, which consists of $3 \times \text{DnCNN}_6 + 2 \times \text{SSB}$. We point out that large SSN architectures are difficult to train, and a careful hyperparameter selection is needed to obtain satisfactory results. Moreover, the running time for an inference with 3SSN_6 is about 7.5 times that of the baseline DnCNN_{18}.

4.4 Differences with Non-Local Neural Networks

The proposed solution shares the same rationale underpinning the Non-Local Block (NLB) in [22], since they are both inspired by NLM [1] principles. This section therefore discusses the main differences between the two, while the experimental comparison is in Sect. 5.3. NLB is defined as

$$\widehat{\mathbf{v}}_i = \frac{1}{C} \sum_j f(\mathbf{v}_i, \mathbf{v}_j) g(\mathbf{v}_j), \tag{7}$$

where f computes a scalar similarity between two activation vectors, and g is a linear embedding. In practice, NLB estimates each latent vector $\widehat{\mathbf{v}}_i$ by a weighted average of the embedded features $g(\mathbf{v}_j)$ from the whole input, where the weights are the scalar similarities $f(\mathbf{v}_i, \mathbf{v}_j)$ representing the attention. The primary difference with SSB is that this computes a multivariate similarity score $\mathbf{s}^k$ (represented in pink in Fig. 1a, center). Moreover, the use of scores is different as $\mathbf{s}^k$ is mixed with the corresponding latent feature vector $\mathbf{v}^k$ and then averaged with learned weights α_k in the averaging sub-block (Fig. 1a, right). Our experiments in Sect. 5.2 shows that a multivariate similarity scores improves the denoising performance. Another difference is that SSB operates on a search neighbor, considering all the activation vectors belonging to a spatial neighborhood using a stride. Instead, the NLB considers all the activation vectors in the activation maps. This is a viable approach only for small maps (used in high-level tasks addressed in [22]), but it is computationally intractable in dense regression tasks like image restoration. In our experiments, we test a modified version of NLB that compares references only with activation vectors within a fixed-size search neighborhood, as in [12]. This version, which we denote as NLBr, considers all the activation vectors in the neighborhood without a stride. We additionally include a strided alternative, denoted NLBr+. Our experiments demonstrate that stride, which we also employ in SSB, is beneficial and that SSB outperforms both the improved variants of NLB at high noise levels.

5 Experiments

We analyze the effectiveness of the SSB in grayscale image denoising. First, we present the datasets, figures of merit, and competing methods employed in our experiments (Sect. 5.1). Then, we investigate how the configuration of SSN influences the denoising performance (Sect. 5.2) and compare a selected SSN against state-of-the-art CNN denoisers (Sect. 5.3).

5.1 Experimental Settings

Our experiments tackle the denoising problem with $\sigma \in \{25, 50, 70\}$ by training a distinct model for each noise level. We train all models on 80×80 patches randomly cropped from BSD400 [15], corrupted with Gaussian noise with fixed variance σ^2. As test sets, we adopt Set12, BSD68 [15], and Urban100 [7]. Training and testing images are strictly disjoint. We assess the denoising performance by the Peak Signal-to-Noise Ratio (PSNR) and the Structural Similarity (SSIM) index [23]. We consider the following methods in our experiments.

SSN: The architecture of an SSN consists of instances of DnCNN interleaved by SSBs. Figure 1b reports the architecture of the SSN that we use in the main experiment. We set the number of filters n in the similarity sub-block (Fig. 1a) as the number d of channels of the input. We train each SSN for 40 epochs using the Adam optimizer [9] with a mini-batch size of 16 and a weight decay factor of 10^{-6} on all convolutional weights except for those of SSB. We use a learning rate of 10^{-4} for the first 20 epochs, and then we divide it by 10 every 10 epochs. For a fair comparison, we adopt the ℓ_2 loss, which most deep denoisers use even though it is not the best choice for image restoration models [29].

Table 1. Denoising performance of 2SSN_9 for different values of half-neighborhood size Ω and stride w. These parameters determine the number of activation vectors M used by the SSB.

(Ω, w)	(2,2)	(4,2)	(6,2)	(5,5)	(10,5)	(15,5)
PSNR	30.15	30.19	30.25	30.18	30.30	30.38
SSIM	0.887	0.888	0.888	0.888	0.891	0.894
M	9	25	49	9	25	49

Table 2. Denoising performance of 2SSN_9 for different dimensions n of the similarity score. Adopting a multivariate similarity boosts the denoising performance.

n	1	2	4	8	16	32	64	128
PSNR	30.131	30.134	30.197	30.285	30.358	30.368	30.375	30.374
SSIM	0.886	0.886	0.889	0.891	0.893	0.893	0.894	0.893

Competing Methods: We compare SSN against a traditional denoising algorithm (BM3D [3]) and the following deep denoisers: DnCNN [26], NN3D [2] and N^{3}Net [17]. Since SSB aims to improve the denoising performance of CNN denoisers, we do not include recurrent networks or complex, branching, and computationally expensive convolutional or graph architectures. Furthermore, we train surrogates of SSN, replacing the proposed SSB with NLB [22]. We consider the setup and the results from [17] for all the methods. All the learning-based methods share the same training set as SSN and adopt comparable training procedures.

5.2 Assessing Different SSN Architectures

Many parameters affect the effectiveness of SSN, like the number of neighbors M, their distance to the reference, the dimension n of the similarity scores, and the number of SSBs employed by the network. Here, we compare the performance achieved by different configurations of SSN over Urban100 with $\sigma = 25$.

The first experiment considers the 2SSN$_9$, which contains a single SSB, with different values of half-neighborhood size Ω and stride w. Table 1 reports the number of neighbors M, and the PSNR and SSIM achieved by the SSN for various configurations of Ω and w. The results show that increasing M leads to better denoising performance because we have more chances to find similar vectors. Moreover, for a fixed M, considering a larger search neighborhood (i.e., a larger stride) is beneficial. In the following, we set $\Omega = 15$ and $w = 5$, which corresponds to a 31×31 search region containing $M = 49$ activation vectors.

The second experiment assesses the importance of adopting a multivariate similarity score. To this end, we train several instances of 2SSN$_9$ varying the parameter n, which defines the number of components of $\mathbf{s}_k$. With $n = 1$, the similarity of SSB is scalar and resembles the attention of NLB [22]. Table 2 shows a steady performance improvement when using multivariate similarity metrics, up to $n = 16$. Further increase of n yields a marginal performance boost.

Finally, we evaluate the denoising performance of SSN when increasing the number of SSB instances. Precisely, we consider SSNs with 1, 2, or 3 SSBs compared with two baselines DnCNN$_{18}$ and DnCNN$_{20}$ having the same number of convolutional layers. Table 3 shows that a single SSB already improves

Table 3. Denoising performance of SSN for different configurations.

Model	Description	PSNR	SSIM	# Params
DnCNN$_{18}$	$1 \times$ DnCNN$_{18}$	29.9917	0.885	594 113
DnCNN$_{20}$	$1 \times$ DnCNN$_{20}$	30.0051	0.884	668 225
2SSN$_9$	$2 \times$ DnCNN$_9$ + $1 \times$ SSB	30.3750	0.894	627 202
3SSN$_6$	$3 \times$ DnCNN$_6$ + $2 \times$ SSB	30.4238	0.895	660 291
4SSN$_5$	$4 \times$ DnCNN$_5$ + $3 \times$ SSB	30.4120	0.893	767 492

the denoising performance with respect to the DnCNN architecture, and a second block grants further improvement. However, when adding a third SSB, the network training becomes particularly difficult and more unstable, leading to a small decrease of the denoising performance. In the following, we adopt the $3SSN_6$ architecture, depicted in Fig. 1b.

5.3 Comparison Against State-of-the-Art

In this section, we compare the denoising performance of $3SSN_6$ (Fig. 1b) against the competing methods in Sect. 5.1. To fairly compare our method with NLB, we train similar architectures, where we replace our SSB with a NLBr or NLBr+ (defined in Sect. 4.4). The strided version NLBr+ adopts a selection of neighbors similar to SSB. Both NLBr and NLBr+ consider the same number of activation vectors of our $3SSB_6$ (i.e., $M = 49$).

Table 4 reports the PSNR achieved by the considered methods on the test sets corrupted by noise of standard deviation $\sigma \in \{25, 50, 75\}$. The results for the competing methods are from [17]. SSN outperforms the competitors on all datasets and noise levels. In particular, the improvement is more significant on Urban100, whose images present many repeated patterns and structures, suggesting that our SSB effectively promotes NLSS. Our SSN significantly outperforms the NLBr implemented as in [22]. In contrast, SSN achieves performance equivalent to NLBr+ for low σ, while we observe a significant improvement for high noise levels. This suggests that relying on a multivariate similarity becomes beneficial when the degradation is strong. More generally, exploiting the NLSS prior is beneficial to image denoising. Indeed SSN, N^3Net, and NLB consistently outperform DnCNN, which in turn achieves a higher average PSNR than BM3D, demonstrating the effectiveness of deep learning methods. To gain further insights into the effect of NLSS, Fig. 2 reports two denoised versions of a detail from Barbara (Set12), presenting a repeated fabric pattern. This example shows that SSN successfully recovers the fabric pattern also in low-contrast regions

Table 4. Denoising performance (PSNR) achieved by the considered methods on three datasets. The best results are in bold.

	σ	BM3D	DnCNN	NN3D	N^3Net	NLBr	NLBr+	SSN (ours)
Set12	25	29.96	30.44	30.45	30.55	30.51	30.61	**30.64**
	50	26.70	27.19	27.24	27.43	27.43	27.41	**27.46**
	70	25.21	25.56	25.61	25.90	25.71	25.82	**25.92**
BSD68	25	28.56	29.23	29.19	29.30	29.30	29.36	**29.37**
	50	25.63	26.23	26.19	26.39	26.31	26.40	**26.42**
	70	24.46	24.85	24.89	25.14	25.04	25.08	**25.16**
Urban100	25	29.71	29.97	30.09	30.19	30.19	30.40	**30.42**
	50	25.95	26.28	26.47	26.82	26.56	26.83	**26.92**
	70	24.27	24.36	24.53	25.15	24.73	25.02	**25.24**

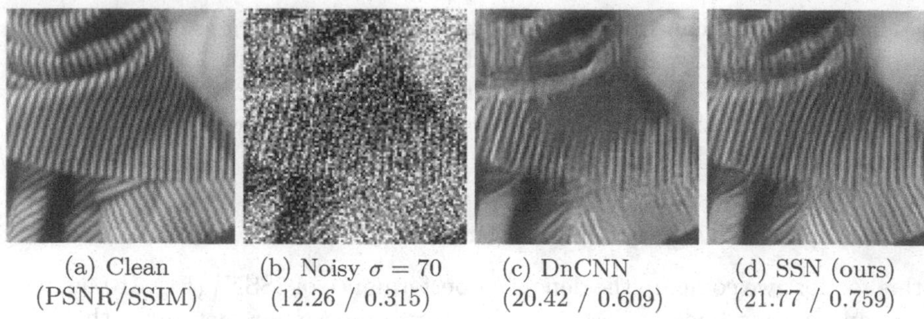

(a) Clean (PSNR/SSIM) (b) Noisy $\sigma = 70$ (12.26 / 0.315) (c) DnCNN (20.42 / 0.609) (d) SSN (ours) (21.77 / 0.759)

Fig. 2. Detail of Barbara denoised by DnCNN and SSN. Our method recovers repeated patterns also in low-contrast regions.

where DnCNN fails, demonstrating the effectiveness of our SSB at exploiting the NLSS of this texture.

6 Conclusions

We present SSB, a novel building block promoting the NLSS in CNN denoisers. Inspired by NLM, our block aggregates neighbor contributions from a search neighborhood, using weights that depend on the similarities with neighbors. The patch embedding, the multivariate similarity score, and the mixing function are learned during training. We show that SSB improves the performance of a baseline on AWGN suppression, particularly when σ is high.

Future work comprises training models for blind denoising, extending SSB to RGB images, which typically require adjustments in the network hyperparameters, as in [27]. Moreover, we will investigate the application of multi-head attention to image restoration, leveraging the connection between self-similarity and attention. Finally, we will address the computational efficiency problem by exploring alternative implementations to reduce the running time of SSB.

Acknowledgments. We gratefully acknowledge the support of NVIDIA Corporation with the four RTX A6000 GPUs granted through the Applied Research Accelerator Program to Politecnico di Milano.

References

1. Buades, A., Coll, B., Morel, J.: A non-local algorithm for image denoising. In: Proceedings of CVPR, vol. 2, pp. 60–65. IEEE (2005)
2. Cruz, C., Foi, A., Katkovnik, V., Egiazarian, K.: Nonlocality-reinforced convolutional neural networks for image denoising. IEEE Signal Process. Lett. **25**(8), 1216–1220 (2018)
3. Dabov, K., Foi, A., Katkovnik, V., Egiazarian, K.: Image denoising by sparse 3-D transform-domain collaborative filtering. IEEE TIP **16**(8), 2080–2095 (2007)

4. Elad, M., Aharon, M.: Image denoising via sparse and redundant representations over learned dictionaries. IEEE TIP **15**(12), 3736–3745 (2006)
5. Foi, A.: Clipped noisy images: Heteroskedastic modeling and practical denoising. Signal Process. **89**(12), 2609–2629 (2009)
6. Gu, S., Zhang, L., Zuo, W., Feng, X.: Weighted nuclear norm minimization with application to image denoising. In: Proceedings of CVPR, pp. 2862–2869 (2014)
7. Huang, J.B., Singh, A., Ahuja, N.: Single image super-resolution from transformed self-exemplars. In: Proceedings of CVPR, pp. 5197–5206. IEEE (2015)
8. Ioffe, S., Szegedy, C.: Batch normalization: accelerating deep network training by reducing internal covariate shift. In: Proceedings of ICML, pp. 448–456. PMLR (2015)
9. Kingma, D.P., Ba, J.: Adam: a method for stochastic optimization. arXiv preprint arXiv:1412.6980 (2014)
10. Lefkimmiatis, S.: Non-local color image denoising with convolutional neural networks. In: Proceedings of CVPR, pp. 3587–3596. IEEE (2017)
11. Lefkimmiatis, S.: Universal denoising networks: a novel CNN architecture for image denoising. In: Proceedings of the CVPR, pp. 3204–3213. IEEE (2018)
12. Liu, D., Wen, B., Fan, Y., Loy, C.C., Huang, T.S.: Non-local recurrent network for image restoration. In: Advances in NeurIPS, vol. 31 (2018)
13. Luo, W., Li, Y., Urtasun, R., Zemel, R.: Understanding the effective receptive field in deep convolutional neural networks. In: Advances in NeurIPS, vol. 29 (2016)
14. Mao, X., Shen, C., Yang, Y.: Image restoration using very deep convolutional encoder-decoder networks with symmetric skip connections. In: Advances in NeurIPS, vol. 29 (2016)
15. Martin, D., Fowlkes, C., Tal, D., Malik, J.: A database of human segmented natural images and its application to evaluating segmentation algorithms and measuring ecological statistics. In: Proceedings of ICCV, vol. 2, pp. 416–423. IEEE (2001)
16. Perona, P., Malik, J.: Scale-space and edge detection using anisotropic diffusion. IEEE TPAMI **12**(7), 629–639 (1990)
17. Plötz, T., Roth, S.: Neural nearest neighbors networks. In: Advances in NeurIPS, vol. 31 (2018)
18. Portilla, J., Strela, V., Wainwright, M.J., Simoncelli, E.P.: Image denoising using scale mixtures of Gaussians in the wavelet domain. IEEE TIP **12**(11), 1338–1351 (2003)
19. Roth, S., Black, M.J.: Fields of experts: a framework for learning image priors. In: Proceedings of CVPR, vol. 2, pp. 860–867. IEEE (2005)
20. Tai, Y., Yang, J., Liu, X., Xu, C.: MemNet: a persistent memory network for image restoration. In: Proceedings of ICCV, pp. 4539–4547. IEEE (2017)
21. Valsesia, D., Fracastoro, G., Magli, E.: Deep graph-convolutional image denoising. IEEE TIP **29**, 8226–8237 (2020)
22. Wang, X., Girshick, R., Gupta, A., He, K.: Non-local neural networks. In: Proceedings of CVPR, pp. 7794–7803. IEEE (2018)
23. Wang, Z., Bovik, A.C., Sheikh, H.R., Simoncelli, E.P.: Image quality assessment: from error visibility to structural similarity. IEEE TIP **13**(4), 600–612 (2004)
24. Yang, D., Sun, J.: BM3D-Net: a convolutional neural network for transform-domain collaborative filtering. IEEE Signal Process. Lett. **25**(1), 55–59 (2017)
25. Zhang, K., Li, Y., Zuo, W., Zhang, L., Van Gool, L., Timofte, R.: Plug-and-play image restoration with deep denoiser prior. IEEE TPAMI **44**(10), 6360–6376 (2021)
26. Zhang, K., Zuo, W., Chen, Y., Meng, D., Zhang, L.: Beyond a Gaussian denoiser: residual learning of deep CNN for image denoising. IEEE TIP **26**(7), 3142–3155 (2017)

27. Zhang, K., Zuo, W., Zhang, L.: FFDNet: toward a fast and flexible solution for CNN-based image denoising. IEEE TIP **27**(9), 4608–4622 (2018)
28. Zhang, Y., Li, K., Zhong, B., Fu, Y.: Residual non-local attention networks for image restoration. In: International Conference on Learning Representations (2019)
29. Zhao, H., Gallo, O., Frosio, I., Kautz, J.: Loss functions for image restoration with neural networks. IEEE TCI **3**(1), 47–57 (2016)
30. Zoran, D., Weiss, Y.: From learning models of natural image patches to whole image restoration. In: Proceedings of ICCV, pp. 479–486. IEEE (2011)

A Request for Clarity over the End of Sequence Token in the Self-Critical Sequence Training

Jia Cheng Hu, Roberto Cavicchioli(✉), and Alessandro Capotondi

University of Modena and Reggio Emilia, Modena, Italy
{jiachenghu,roberto.cavicchioli,alessandro.capotondi}@unimore.it

Abstract. The Image Captioning research field is currently compromised by the lack of transparency and awareness over the End-of-Sequence token (<Eos>) in the Self-Critical Sequence Training. If the <Eos> token is omitted, a model can boost its performance up to +4.1 CIDEr-D using trivial sentence fragments. While this phenomenon poses an obstacle to a fair evaluation and comparison of established works, people involved in new projects are given the arduous choice between lower scores and unsatisfactory descriptions due to the competitive nature of the research. This work proposes to solve the problem by spreading awareness of the issue itself. In particular, we invite future works to share a simple and informative signature with the help of a library called SacreEOS. Code available at: https://github.com/jchenghu/sacreeos.

1 Introduction

The standard training strategy of a modern Neural Image Captioning system includes a policy gradient method, called Self-Critical Sequence Training [27] (shortened as SCST) which is designed to maximize the evaluation score given to the outputs. In this work, we discuss the problems caused by the lack of transparency from the research community over the inclusion or omission of the End-of-Sequence token during the optimization. An easy-to-overlook implementation detail that can significantly increase the performance of any model despite yielding worse descriptions.

The lack of awareness of the impact of the End-of-Sequence (`<Eos>`) omission and the lack of explicit information on the SCST implementation during the reporting of results pose an obstacle to scientific progress as they make it challenging to compare established works and evaluate new ones. Our paper attempts to spread awareness about the issue and proposes a solution to increase transparency in future works. This paper is structured as follows: in Sect. 2, we discuss the problem of the End-of-Sequence omission and why it is a problem for the research community; in Sect. 3, we provide a qualitative and quantitative

This work was supported by the EU project 5GMETA, GA No 957360.

G. L. Foresti et al. (Eds.): ICIAP 2023, LNCS 14233, pp. 39–50, 2023.
https://doi.org/10.1007/978-3-031-43148-7_4

analysis of the issue and we sample some of the recent works in Image Captioning to demonstrate its pervasiveness and provide some practical examples of its impact; In Sect. 4, we propose a possible solution with the help of a Python library called SacreEOS; in Sect. 5, we mention some of the literature approaches, and, finally, we draw our conclusions in Sect. 6.

2 Problem Description

2.1 CIDEr Optimization

CIDEr [30] is an n-gram-based metric that evaluates the caption semantic content according to its similarities to the ground truths. Compared to the other metrics [2,3,16,24], it exploits the entire corpus of reference descriptions in the attempt of backing the evaluation with the consensus of the majority of people. In particular, each n-gram w_k in sequence Z is weighted according to the *tf-idf* term $g^n_k(Z)$ defined as:

$$\frac{h^n_k(Z)}{\sum\limits_{w_l \in \Omega} h^n_l(Z)} \cdot log(\frac{|I|}{\sum\limits_{I_i \in I} \min(1, \sum\limits_{q} h^n_k(V^i_q))}) \tag{1}$$

where Ω is the set possible n-grams in the corpus, I is the set of corpus images and $h^n_k(Z)$, $h^n_k(V^i_j)$ represent the number of occurrences of n-gram w_k in the sequence Z and in the j-th ground truth of image $I_i \in I$. The CIDEr and its alternative (CIDEr-D), compute the similarity between the candidate and reference description as the number of matching n-grams, weighted according to Eq. 1. We refer to [30] for additional details of the formula since they are unnecessary for the sake of the discussion.

The standard training practice of the Image Captioning model consists of a pre-training phase using the Cross-Entropy loss followed by a CIDer-D optimization by means of a policy gradient method called Self-Critical Sequence Training [27]. The latter minimizes the negative expected reward:

$$L_R(\theta) = -\mathbf{E}_{y_{1:T} \sim p_\theta}[r(y_{1:T})] \tag{2}$$

where r is the CIDEr function, and its gradient is approximated as follows:

$$\nabla_\theta L_R(\theta) \approx -(r(y^s_{1:T}) - r(y^b_{1:T}))\nabla_\theta log\ p_\theta(y^s_{1:T}) \tag{3}$$

where $y^s_{1:T}$ are the sampled captions and $y^b_{1:T}$ are the base predictions.

2.2 The End-of-Sequence Token in SCST

Two properties are desirable in an image description: completeness and correctness. While the first goal is pursued by the reward maximization, the SCST algorithm provides no explicit control over the latter, which is instead implicitly encouraged by the sequentiality of the decoding process. A token predicted at a specific time step also determines the most likely n-grams in the following ones. Since all n-grams are extracted from linguistically correct references, the final description will be correct, at least locally. Unfortunately, the CIDEr score does not consider a sentence's global correctness, and this aspect can be easily exploited by the SCST if not carefully implemented. In particular, the algorithm is allowed to produce incomplete descriptions using trivial sentence fragments that almost certainly match some parts of any set of references. This is the reason why the standard SCST implementation includes the special End-of-Sequence token, abbreviated as `<Eos>`, in the definition of the n-grams space. With this precaution, the reward function encourages a correct sentence termination leveraging the fact that the *tf-idf* of the `<Eos>` token out-weights those of function words.

CIDEr-D: 125.8		CIDEr-D: 130.1
A herd of sheep are standing in a field	->	A herd of sheep standing in a field with a
A group of people standing under an umbrella	->	A group of people standing under an umbrella in the
A group of people riding on the back of elephants	->	A group of people riding on elephants in a
A dog standing next to a fence with a stuffed animal	->	A dog is standing next to a fence with a stuffed animal
A bunch of bananas in a box on a table	->	A bunch of bananas in a box with a
A man sitting on a motorcycle	->	A man sitting on a motorcycle in front of a
A person in a boat in the water	->	A boat in the water with mountains in the
An elephant walking down a dirt road	->	An elephant walking down a dirt road in a
A small bird sitting on top of a rock	->	A bird sitting on top of a rock in the
A brown cow laying on the side of a motorcycle	->	A brown cow laying on the side of a mtorcycle
A group of people standing at a market with fruit	->	A woman standing in a market with fruit on
A person walking down a street with a clock station	->	A person walking in an ally with a clock on

Fig. 1. Captions generated by the same model (the Transformer [29]) trained with different implementations of SCST on the MS-COCO [17] data set. (Left) The model is optimized by the standard SCST and achieves 125.8 CIDEr-D on the validation set. (Right) The model is optimized by an implementation of SCST in which the `<Eos>` token is omitted and achieves 130.1 CIDEr-D on the validation set.

2.3 The Problem of the <Eos> omission

The inclusion or exclusion of the `<Eos>` token in the SCST algorithm represents a small and easy-to-overlook detail that significantly impacts a captioning system's performance. In case the `<Eos>` token is omitted, the descriptions generated by the network are often terminated by trivial sentence fragments such as "`and a`", "`in the`", "`on top of`" and "`in front of`" (more examples in Fig. 1).

However, despite the presence of artifacts, they achieve superior performances on popular benchmarks compared to the correct ones (Fig. 1). In particular, the number of additional points yielded by the artifacts can even be greater than the range of values in which different models developed around the same period typically compete. Therefore, the Image Captioning research field is currently suffering from a lack of transparency and, in some cases lack of awareness over the importance of the <Eos> token in the SCST. The problem can be described from multiple perspectives:

- If details over the <Eos> token in the SCST implementation are unavailable, omitted, or simply overlooked, it becomes difficult to compare models in the literature fairly.
- Researchers that are aware of the issue are given the difficult choice between less competitive results and poorly formulated outputs.
- Finally, researchers that are not aware of the issue (especially the newcomers in the field of Image Captioning) are indirectly encouraged to adopt the implementations that generate compromised sentences because of their superior performances.

3 <Eos> Omission Impact Analysis

3.1 Experimental Setup

For the qualitative and quantitative analysis of artifacts we implement the Transformer [29] with 3 layers, d_{model}=512 and d_{ff}=2048, trained on the COCO 2014 [17] data set using the Karpathy split [11]. The Faster-RCNN backbone provided by [1] is adopted. The learning procedure consists of a first training step on Cross Entropy loss for 8 epochs followed by the CIDEr-D optimization for 20 epochs. The following configurations are adopted:

1. batch size of 48, a learning rate of 2e-4 annealed by 0.8 every 2 epochs and warm-up of 10000 in case of Cross Entropy Loss;
2. batch size of 48, a learning rate of 1e-4 annealed by 0.8 every 2 epochs during the SCST.

Optimization details are provided only for the sake of reproducibility since the artifacts discussed in this work arise regardless of the architecture and optimization details. For the ensemble results, 4 model instances are generated with the aforementioned method differing only in the initialization seed. In the experiments, for each seed, the SCST in the Standard and No<Eos> configurations optimize the same pre-trained model.

3.2 Artifacts Analysis

The <Eos> token can be omitted in two aspects of SCST:

1. during the reward computation;
2. during the initialization of *tf-idfs*;

which leads to 4 implementation[1] instances in case sampled descriptions are tokenized consistently with respect to the ground-truths. Table 1 reports the impact of each configuration over the final descriptions.

Table 1. Impact of the `<Eos>` token in SCST over the final CIDEr-D score and outputs. "*tf-idf* Init." refers to the ground truth sentences involved in the calculation of document frequencies, and "Predictions" refers to the sampled predictions and respective references.

	tf-idf Init. w/`<Eos>`	*tf-idf* Init. w/o `<Eos>`
Predictions w/`<Eos>`	baseline score no artifacts	lower score with artifacts
Predictions w/o `<Eos>`	lower score with artifacts	higher score with artifacts

Two cases are the focus of this work since most popular implementations fall into the (*tf-idf* Init. w/ `<Eos>`, Prediction w/ `<Eos>`) and (*tf-idf* Init. w/o `<Eos>`, Prediction w/o `<Eos>`) configuration referred as "Standard" and "`No<Eos>`" respectively throughout the rest of this work.

In the `No<Eos>` configuration, results are affected by 8 classes of artifacts depending on how sequences are terminated, with the last token belonging to A={"`in`", "`a`", "`of`","`the`", "`with`", "`on`", "`and`" "`*`"}, where "`*`" represents all the possible remaining wrong cases. While all elements in the set A are just simplifications of longer trivial fragments such as "`and a`", "`in a`", "`with a`" and "`in front of`", the case of "`on`" may seem acceptable but the token is often part of uncommon formulations such as "`a beach with a surfboard on`" and "`a street with a bus on`". Nevertheless, "`on`" represents only a small fraction of all instances, which mostly end with the "`a`" token instead (see Fig. 2c). No artifacts were found in the case of Standard configuration.

Figure 2a showcases the number of artifacts converging to 50% of the whole testing set as the number of epochs increases. Thus, both correct and compromised sentences are produced by the `<Eos>` omission, which means the network learns to inject the fragments following a non-trivial and unpredictable criteria for each sequence.

Figure 2b and Table 2 showcase that a single model trained with SCST in the `No<Eos>` configuration consistently outperforms the standard one across all seeds, often by a large margin, with a maximum gain of +2.8 and +4.3 CIDEr-D in the offline test and validation set respectively. Whereas, by removing the artifacts from the latter predictions we observed the opposite trend with a maximum performance decrease of –2.3 and –2.0. Therefore, the increase in score is mostly due to the artifacts and the `<Eos>` omission poses an obstacle to the generation

[1] code available at: https://github.com/jchenghu/captioning_eos.

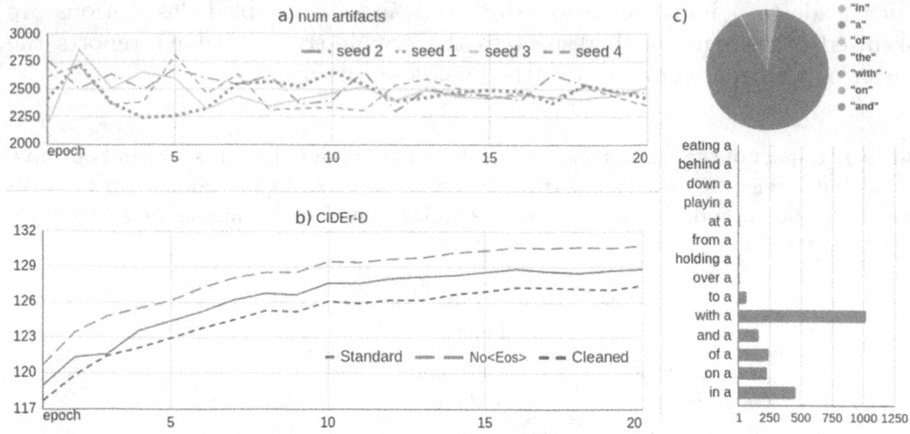

Fig. 2. a) The number of artifacts in the No<Eos> configuration on 5000 test set predictions. b) Average CIDEr-D score of 4 training instances (different seeds) in the Standard and No<Eos> configuration, "Cleaned" denotes the No<Eos> performance in case artifacts are removed before the evaluation. c) Artifacts distribution. Sequences terminated by "a" account for 89.8% of all cases (top). Histogram of sequences terminated by "a" (bottom).

Table 2. Performance comparison the CIDEr-D optimization in Standard and No<Eos> training. "Cleaned" refers to the No<Eos> results but artifacts are removed prior to the evaluation. $\sum$ refers to the ensemble of the four models, ε represents the percentage of artifacts and δ denotes the score difference between the No<Eos> configuration and the Standard one.

	Karpathy test split			Karpathy validation split		
	Standard	No<Eos> (ε)/δ	Cleaned/δ	Standard	No<Eos> (ε)/δ	Cleaned/δ
Seed 1	128.4	131.2 (48.3%)/+2.8	127.8/−0.6	125.8	130.1 (47.5%)/+4.3	126.4/+0.6
Seed 2	129.0	130.9 (49.3%)/+1.9	127.4/−1.6	127.0	129.9 (48.1%)/+2.9	126.2/−0.8
Seed 3	129.0	131.0 (50.3%)/+2.0	127.5/−1.5	127.2	129.3 (47.6%)/+2.1	125.7/−1.5
Seed 4	129.1	130.7 (50.4%)/+1.6	126.8/−2.3	128.0	130.0 (50.6%)/+2.0	126.0/−2.0
Avg	128.9	130.9 (49.6%)/+2.0	127.3/−1.1	126.9	129.8 (48.6%)/+2.8	126.0/−0.9
$\sum$	133.0	134.9 (50.2%)/+1.9	131.2/−1.8	131.8	133.8 (49.5%)/+2.0	129.8/−2.0

of semantically meaningful content. Similar behaviour is observed for ensemble performances (referred as $\sum$).

3.3 Literature Classification

We sample recent works in the research literature and classify each of them according to the way SCST is implemented. In Sect. 3.2 we observed that only half of the evaluated sentences are compromised, which means that if a paper provides only a few correct captioning examples, it is not enough to determine whether the <Eos> token was omitted or not. Because of that, the classification is made through code inspection. The classes and the respective criteria are defined as follows:

- Standard: <Eos> token is included in both SCST initialization and reward computation or complete results on either test or validation set are provided;
- No<Eos>: <Eos> token is omitted in both initialization and reward computation;
- Unknown: the code was not found or it was not available at the time this work was completed.

Table 3 showcases that only 12 of 25 works are confirmed to follow the Standard implementation, 8 fall in the No<Eos> category and 5 are unknown. The State-of-the-art architectures in 2019 [8] and 2020 [23] achieved 129.6 and 131.4 CIDEr-D scores respectively, which showcases the gradual improvement process

Table 3. SCST classification of recent Image Captioning works and their respective performances on the MS-COCO 2014 task. The offline case reports the CIDEr-D score of a single model in contrast to the online evaluation server results where an ensemble is adopted instead with some exceptions denoted with "*".

Year	Work	Offline	Online	SCST	Code inspection[a] (commit)
2018	GCN-LSTM [35]	127.6	–	Unknown	Code not found/available
2018	Up-Down [1]	120.1	120.5	Standard	peteanderson80/bottom-up-attention (514e561)
2019	HAN [32]	121.7	118.2	Unknown	Code not found/available
2019	LBPF [26]	127.6	–	Unknown	Code not found/available
2019	RDN [12]	117.3	125.2	Unknown	Code not found/available
2019	SGAE [34]	127.8	–	Standard	yangxuntu/SGAE (af88115)
2019	Obj.Rel.Transf. [6]	128.3	–	Standard	yahoo/object_relation_transformer (6cf5bd8)
2019	AoANet [8]	129.8	129.6	Standard	husthuaan/AoANet (94ffe17)
2020	Ruotian Luo [20]	129.6	–	Standard	ruotianluo/self-critical.pytorch (be1a526)
2020	M^2 [5]	131.2	132.1	No<Eos>	aimagelab/meshed-memory-transformer (e0fe3fa)
2020	X-Transformer [23]	132.8	133.5	Standard	JDAI-CV/image-captioning (d39126d)
2020	Unified VLP [38]	129.3	–	Standard	LuoweiZhou/VLP (74c4d85)
2021	GET [9]	131.6	132.5	No<Eos>	luo3300612/image-captioning-DLCT (575b4dd)
2021	DLCT [21]	133.8	135.4	No<Eos>	luo3300612/image-captioning-DLCT (575b4dd)
2021	RSTNet [37]	135.6	134.0	No<Eos>	zhangxuying1004/RSTNet (e60715f)
2022	PureT [33]	138.2	138.3	Standard	232525/PureT (8dc9911)
2022	ExpansionNet [7]	140.4	140.8	Standard	jchenghu/ExpansionNet_v2 (365d130)
2022	BLIP [15]	136.7	–	Standard	salesforce/BLIP (3a29b74)
2022	CaMEL[4]	138.9	140.0	No<Eos>	aimagelab/camel (67cb062)
2022	GRIT [22]	144.2	143.8	No<Eos>	davidnvq/grit (32afb7e)
2022	S^2 [36]	133.5	135.0	No<Eos>	zchoi/S2-Transformer (c584e4)
2022	OFA [31]	154.9	149.6*	Standard	OFA-Sys/OFA (1809b55)
2022	ER-SAN [14]	135.3	–	Standard	CrossmodalGroup/ER-SAN (e80128d)
2022	CIIC [18]	133.1	129.2*	Unknown	Code not found/available
2022	Xmodal-Ctx [13]	139.9	–	No<Eos>	GT-RIPL/Xmodal-Ctx (d927eec)

[a]Prefix https://github.com/

of the research activity and provides an example of the magnitude of improvements over the years. Unfortunately, such a difference in performance can be lower than the additional score yielded by artifacts (see Sect. 3.2). For instance, if AoANet adopted the No<Eos> configuration, its score would have been comparable to the State-of-the-art performances of the following year (X-Transformer) (see Table 4).

The amount of No<Eos> implementations in the last years confirms the phenomena described in Sect. 2.3.

Table 4. CIDEr-D performance increase, denoted by δ, observed in open source projects when the SCST configuration is changed from Standard into No<Eos> mode. Training details can be found in the respective works or repositories.

Model	RL epochs	Standard	No<Eos>	Ensemble	Set	δ
AoANet [8]	25	127.6	131.0	✗	test	+3.4
		126.2	130.3		val	+4.1
X-Transformer [23]	20	131.8	133.5	✗	test	+1.7
		130.1	132.3		val	+2.2
ExpansionNet [7]	12	143.7	145.3	✓	test	+1.6
		143.0	145.7		val	+2.7

4 SacreEOS

4.1 SacreEOS Signature

The lack of transparency and awareness over the <Eos> token in SCST originates from an easy-to-overlook implementation detail. Therefore, the natural solution is to disseminate awareness of the issue. To achieve this goal we introduce SacreEOS, a Python library whose main functionality consists of the generation of signatures that uniquely identify the key aspects of the SCST implementation. In particular, how the <Eos> token is handled. The sharing of the SacreEOS signature accomplishes three objectives:

1. it increases transparency and eases the comparison of models;
2. it informs the reader about the presence or absence of artifacts (those related to the <Eos> omission) in the results;
3. last but not least, it spreads awareness of the problem.

We believe this is especially useful in cases of works that do not release the code to the public.

Established researchers and existing implementations can manually generate the signature using the SacreEOS command line interface. The tool simply asks a few questions regarding the technical aspects of SCST, therefore it does not require any code integration. For new projects instead, SacreEOS consists of an SCST implementation helper, in this case, the signature is provided automatically.

Format and signature examples are the following:

Format:

```
<scst config>_<Init>+<metric[args]>+<base[args]>+<Version>
```

Examples:

```
STANDARD_w/oInit+Cider-D[n5,s6.0]+average[nspi5]+1.0.0

NO<EOS>MODE_wInit+Cider-D[n4,s6.0]+greedy[nspi5]+1.0.0

NO<EOS>MODE_w/oInit+BLEU[n4]+average[nspi5]+1.0.0
```

4.2 Implementation Helper and Limitations of the Approach

In addition to the functionality of signature generation, the SacreEOS library optionally provides helpful classes to ease the implementation of SCST in future projects. In particular, it covers the following aspects:

- *SCST class selection.* Given the number of established works implemented in both Standard and `No<Eos>` configurations, it is out of the scope of this paper to decide which one is the "correct" one (the library provides no default option in this regard). However, the tool helps the user to make informed decisions. Classes are currently defined by the reward metric, the reward base and whether the `<Eos>` token is included or omitted in both initialization and reward computation.
- *SCST initialization.* The library initializes the *tf-idfs* for the reward computation and performs input checks according to the selected class.
- *SCST reward computation.* The library currently supports the following reward functions CIDEr, CIDEr-D, CIDEr-R and BLEU. Results are consistent with the official repositories[5]. Each function is implemented in both Python and C, users can optionally enable the latter version to increase efficiency.
- *Signature generation.* In this case the SacreEOS signature is automatically determined by the class selection and does not require user intervention.

The library includes an intricate collection of assertions and input checks on all implementation levels, taylored to each specific class. Nevertheless, the SacreEOS does not prevent misreporting. In case the signature is manually generated, it relies on the user to provide the correct data.

5 Related Works

The work of [27] mentioned the role of the End-of-Sequence token. However, it only provided a few qualitative examples and did not report numerical details.

[5] CIDEr, CIDEr-D, BLEU: github.com/vrama91/cider
CIDEr-R: github.com/gabrielsantosrv/coco-caption.

Several works in the past focused on improving the evaluation of Image Captioning systems but they mostly proposed alternatives to the CIDEr metric, such as TIGEr [10], SPIDEr [19], and CIDEr-R [28]. None of them addressed the issue discussed in this work.

The main inspiration of SacreEOS is SacreBLEU [25], in the field of Machine Translation, where ambiguities can arise from different tokenization and detokenization choices that ultimately affect the BLEU score [24].

6 Conclusion

Our work discussed the role of `<Eos>` in the Self-Critical Sequence Training and how the lack of transparency and awareness over its function pose an obstacle to the scientific progress in the Image Captioning field. We described the source of the problem from a qualitative and quantitative perspective. We classified recent works in the scientific literature according to the SCST configuration to showcase the pervasiveness and the importance of the matter. Finally, we proposed a possible solution that consists of sharing a unique signature with the help of a Python library called SacreEOS, to enable fair model comparisons and spread awareness regarding the issue.

References

1. Anderson, P., et al.: Bottom-up and top-down attention for image captioning and visual question answering. In: Proceedings of the IEEE Conference on Computer Vision and Pattern Recognition, pp. 6077–6086 (2018)
2. Anderson, P., et al.: SPICE: semantic propositional image caption evaluation. In: Leibe, B., Matas, J., Sebe, N., Welling, M. (eds.) ECCV 2016. LNCS, vol. 9909, pp. 382–398. Springer, Cham (2016). https://doi.org/10.1007/978-3-319-46454-1_24
3. Banerjee, S., Lavie, A.: METEOR: an automatic metric for MT evaluation with improved correlation with human judgments. In: Proceedings of the ACL Workshop on Intrinsic and Extrinsic Evaluation Measures for Machine Translation and/or Summarization, pp. 65–72 (2005)
4. Barraco, M., et al.: CaMEL: mean teacher learning for image captioning. arXiv preprint arXiv:2202.10492 (2022)
5. Cornia, M., et al.: Meshed-memory transformer for image captioning. In: Proceedings of the IEEE/CVF Conference on Computer Vision and Pattern Recognition, pp. 10578–10587 (2020)
6. Herdade, S., et al.: Image captioning: transforming objects into words. Adv. Neural Inf. Process. Syst. **32**, 1–11 (2019)
7. Hu, J.C., Cavicchioli, R., Capotondi, A.: ExpansionNet v2: block static expansion in fast end to end training for image captioning. arXiv preprint arXiv:2208.06551 (2022)
8. Huang, L., et al.: Attention on attention for image captioning. In: Proceedings of the IEEE International Conference on Computer Vision, pp. 4634–4643 (2019)
9. Ji, J., et al.: Improving image captioning by leveraging intra-and inter-layer global representation in transformer network. In: Proceedings of AAAI Conference on Artifical Intelligence, vol. 35, pp. 1655–1663 (2021)

10. Jiang, M., et al.: TIGEr: text-to-image grounding for image caption evaluation. arXiv preprint arXiv:1909.02050 (2019)
11. Karpathy, A., Fei-Fei, L.: Deep visual-semantic alignments for generating image descriptions. In: Proceedings of the IEEE Conference on Computer Vision and Pattern Recognition, pp. 3128–3137 (2015)
12. Ke, L., et al.: Reflective decoding network for image captioning. In: Proceedings of the IEEE/CVF International Conference on Computer Vision, pp. 8888–8897 (2019)
13. Kuo, C.-W., Kira, Z.: Beyond a pre-trained object detector: cross-modal textual and visual context for image captioning. In: Proceedings of the IEEE/CVF Conference on Computer Vision and Pattern Recognition, pp. 17969–17979 (2022)
14. Li, J., et al.: ER-SAN: enhanced-adaptive relation self-attention network for image captioning. In: De Raedt, L. (ed.) Proceedings of the Thirty-First International Joint Conference on Artificial Intelligence, Main Track, IJCAI-2022, pp. 1081–1087. International Joint Conferences on Artificial Intelligence Organization (2022)
15. Li, J., et al.: Blip: bootstrapping language-image pre-training for unified vision-language understanding and generation. arXiv preprint arXiv:2201.12086 (2022b)
16. Lin, C.Y.: Rouge: a package for automatic evaluation of summaries. In: Text Summarization Branches Out, pp. 74–81 (2004)
17. Lin, T.-Y., et al.: Microsoft COCO: common objects in context. In: Fleet, D., Pajdla, T., Schiele, B., Tuytelaars, T. (eds.) ECCV 2014. LNCS, vol. 8693, pp. 740–755. Springer, Cham (2014). https://doi.org/10.1007/978-3-319-10602-1_48
18. Liu, B., et al.: Show, deconfound and tell: Image captioning with causal inference. In: Proceedings of the IEEE/CVF Conference on Computer Vision and Pattern Recognition, pp. 18041–18050 (2022)
19. Liu, S., et al.: Improved image captioning via policy gradient optimization of spider. In: Proceedings of the IEEE International Conference on Computer Vision, pp. 873–881 (2017)
20. Luo, R.: A better variant of self-critical sequence training. arXiv preprint arXiv:2003.09971 (2020)
21. Luo, Y., et al.: Dual-level collaborative transformer for image captioning. In: Proceedings of the AAAI Conference on Artificial Intelligence, vol. 35, pp. 2286–2293 (2021)
22. Nguyen, V.Q., Suganuma, M., Okatani, T.: Grit: faster and better image captioning transformer using dual visual features. arXiv preprint arXiv:2207.09666 (2022)
23. Pan, Y., et al.: X-linear attention networks for image captioning. In: Proceedings of the IEEE/CVF Conference on Computer Vision and Pattern Recognition, pp. 10971–10980 (2020)
24. Papineni, K., et al.: Bleu: a method for automatic evaluation of machine translation. In: Proceedings of the 40th Annual Meeting of the Association for Computational Linguistics, pp. 311–318 (2002)
25. Post, M.: A call for clarity in reporting bleu scores. arXiv preprint arXiv:1804.08771 (2018)
26. Qin, Y., et al.: Look back and predict forward in image captioning. In: Proceedings of the IEEE/CVF Conference on Computer Vision and Pattern Recognition, pp. 8367–8375 (2019)
27. Rennie, S.J., et al.: Self-critical sequence training for image captioning. In: Proceedings of the IEEE Conference on Computer Vision and Pattern Recognition, pp. 7008–7024 (2017)
28. dos Santos, G.Q., Colombini, E.L., Avila, S.: Cider-r: robust consensus-based image description evaluation. arXiv preprint arXiv:2109.13701 (2021)

29. Vaswani, A., et al.: Attention is all you need. Adv. Neural Inf. Process. Syst. **30**, 5998–6008 (2017)
30. Vedantam, R., Zitnick, C.L., Parikh, D.: Cider: consensus-based image description evaluation. In: Proceedings of the IEEE Conference on Computer Vision and Pattern Recognition, pp. 4566–4575 (2015)
31. Wang, P., et al.: Ofa: unifying architectures, tasks, and modalities through a simple sequence-to-sequence learning framework. In: International Conference on Machine Learning, pp. 23318–23340. PMLR (2022a)
32. Wang, W., Chen, Z., Hu, H.: Hierarchical attention network for image captioning. In: Proceedings of the AAAI Conference on Artificial Intelligence, vol. 33, pp. 8957–8964 (2019)
33. Wang, Y., Xu, J., Sun, Y.: End-to-end transformer based model for image captioning. arXiv preprint arXiv:2203.15350 (2022b)
34. Yang, X., et al.: Auto-encoding scene graphs for image captioning. In: Proceedings of the IEEE/CVF Conference on Computer Vision and Pattern Recognition, pp. 10685–10694 (2019)
35. Yao, T., et al.: Exploring visual relationship for image captioning. In: Proceedings of the European Conference on Computer Vision (ECCV), pp. 684–699 (2018)
36. Zeng, P., et al.: S2 transformer for image captioning. In: De Raedt, L. (ed.) Proceedings of the Thirty-First International Joint Conference on Artificial Intelligence, IJCAI-2022, Main Track. International Joint Conferences on Artificial Intelligence Organization, pp. 1608–1614 (2022)
37. Zhang, X., et al.: Rstnet: captioning with adaptive attention on visual and non-visual words. In: Proceedings of the IEEE/CVF Conference on Computer Vision and Pattern Recognition, pp. 15465–15474 (2021)
38. Zhou, L., et al.: Unified vision-language pre-training for image captioning and vqa. In: Proceedings of the AAAI Conference on Artificial Intelligence, vol. 34, pp. 13041–13049 (2020)

Shallow Camera Pipeline for Night Photography Enhancement

Simone Zini(✉), Claudio Rota, Marco Buzzelli, Simone Bianco, and Raimondo Schettini

Department of Informatics, Systems and Communication, University of Milano–Bicocca, Milan, Italy
{simone.zini,marco.buzzelli,simone.bianco,raimondo.schettini}@unimib.it, c.rota30@campus.unimib.it

Abstract. Enhancing night photography images is a challenging task that requires advanced processing techniques. While CNN-based methods have shown promising results, their high computational requirements and limited interpretability can pose challenges. To address these limitations, we propose a camera pipeline for rendering visually pleasing photographs in low-light conditions. Our approach is characterized by a shallow structure, explainable steps, and a low parameter count, resulting in computationally efficient processing. We compared the proposed pipeline with recent CNN-based state-of-the-art approaches for low-light image enhancement, showing that our approach produces more aesthetically pleasing results. The psycho-visual comparisons conducted in this work show how our proposed solution is preferred with respect to the other methods (in about 44% of the cases our solution has been chosen, compared to only about 15% of the cases for the state-of-the-art best method).

Keywords: Night photography enhancement · Low-light image enhancement · Psycho-visual image quality assessment

1 Introduction

A digital camera processing pipeline is a series of steps that a digital camera performs to process the raw data captured by its image sensor into a final image. The pipeline is responsible for applying various adjustments to the image, such as corrections for lens distortion, white balancing, noise reduction, sharpening, and color enhancement. Although camera manufacturers may use varying processing algorithms and stages in their pipelines, these basic steps are commonly involved in most digital camera processing pipelines [9]. The parameters of the single processing modules are usually optimized by manufacturer for daylight or flashlight illuminated scenes. The problem addressed in this paper is the enhancement of night scenes when the rendering intent is not only the visibility of spatial details, but also to keep the naturalness of the scene depicted and possibly to improve

G. L. Foresti et al. (Eds.): ICIAP 2023, LNCS 14233, pp. 51–61, 2023.
https://doi.org/10.1007/978-3-031-43148-7_5

the aesthetic of the photos. From a technical point of view, when processing nighttime images multiple challenges occur in comparison to the processing of daytime scenes. Nighttime scenes are typically much darker: this can result in images with a low signal-to-noise ratio, making it difficult to extract useful information from the image. Color casts are generally present, due to the sometimes different artificial lighting sources in the scenes. Also, due to the extended exposure time necessary for shooting in dark scenarios, motion blur and sensor noise are likely to occur in the final results. To address these problems, recent methods that leverage deep neural networks to enhance low-light images have been proposed, achieving remarkable results [7,8,11,19,22]. Zhang *et al.* [22] designed a Convolutional Neural Network (CNN) that decomposes images into two components responsible for light adjustment and degradation removal, respectively, and is trained with paired images shot under different exposure conditions. Jiang *et al.* [8] proposed to enhance low-light images using Generative Adversarial Networks (GANs), exploiting unpaired data for the training of the model. Yang *et al.* [19] proposed a semi-supervised model that integrates CNNs and GANs to enhance low-light images in two stages: the first one learns a coarse-to-fine band representation and infers different band signals jointly, while the second one recomposes the band representation using adversarial learning. Recently, Guo *et al.* [7] proposed zero-shot learning methods to eliminate the requirement of paired and unpaired data, which was later improved by Li *et al.* [11]. However, these approaches suffer from several problems, such as over-enhancement, color distortion, and loss of details. Moreover, their heavy computational complexity, memory consumption, and energy requirements may not always meet the constraints for onboard deployment in digital cameras.

In this paper, we introduce a camera pipeline for the rendering of visually pleasing photographs in low-light conditions, containing several algorithms that address the challenges presented by low-light images and characterized by a shallow structure and by a low parameter count. At a time when the resolution of most imaging problems is delegated to the direct or indirect use of neural networks, we propose a "traditional" processing pipeline, whose main modules are designed on the basis of our knowledge of the mechanisms of human vision, and on the basis of our knowledge of the main limitations of traditional imaging devices. The few parameters of the different modules are heuristically set by the authors according to their personal preferences [3], without any reference to existing datasets of low-light images corrected by human experts or automatic approaches. This low parametric dependency means that our solution is flexible, as it can be potentially tuned to match individual users' preferences and to different sensors.

To prove the effectiveness of our method we adopted the dataset used in the NTIRE2022 Night Photography Rendering challenge [6]. Psycho-visual experimental results involving real user evaluations show that our solution produces more pleasing results with respect to several CNN-based state-of-the-art methods for low-light image enhancement.

2 Proposed Method

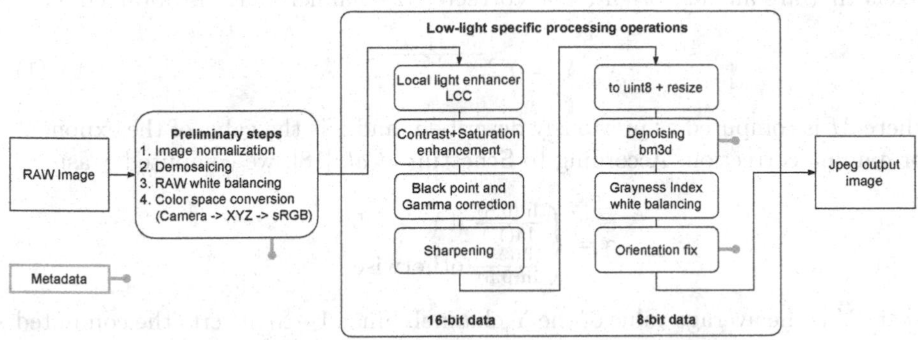

Fig. 1. Overview of the complete proposed pipeline. The entire pipeline can be divided in two parts: preliminary data preparation steps and low-light processing steps. Metadata extra information is exploited in the steps marked with the orange dot. (Color figure online)

The scheme of the proposed solution is depicted in Fig. 1. Our pipeline can be divided into two parts: the preliminary steps, which are the basic stages of a typical camera processing pipeline, and the low-light specific part, which instead contains steps to specifically handle night images. We refer multiple times within our pipeline to image metadata, which are indicated in the following using *italic text*.

2.1 Preliminary Steps

The first part of our pipeline is made of four steps working in the RAW domain. The first step is image normalization: the *black_level* as provided in the image metadata is subtracted, and the image values are rescaled so that the *white_level* is set to one. The demosaicing operation converts the single-channel RAW image into the three-channel RGB image using the appropriate color filter array pattern (*cfa_array_pattern*). Then, a preliminary automatic white balance step is performed using the Gray World algorithm [4], in order to provide a first approximate correction of the image cast. Finally, a color transformation step converts the image from the camera-specific color space to XYZ (obtained as the inverse of *color_matrix_1*) and finally to the sRGB color space.

2.2 Low-Light Specific Processing Operations

The second part of our pipeline has been specifically designed to handle images taken by night in low-light conditions.

The first step of this second part is the use of the Local Contrast Correction (LCC) algorithm by Moroney [15]. Here the local correction is performed on the Y channel of the YCbCr color space using a pixel-wise gamma correction, whose values are determined using a mask M obtained by blurring the luminance channel Y with a Gaussian filter in order to brighten dark areas and to not clip pixels that are already bright. The corrected $\hat{Y}$ channel image is obtained as

$$\hat{Y} = Y^{\gamma^{\frac{0.5-(1-M)}{0.5}}}, \tag{1}$$

where M is computed as previously described, and γ is the value of the exponent for gamma correction. According to Schettini *et al.* [18], we computed γ as:

$$\gamma = \begin{cases} \frac{\ln(0.5)}{\ln(\bar{Y})} & \text{if } \bar{Y} \geq 0.5 \\ \frac{\ln(\bar{Y})}{\ln(0.5)} & \text{otherwise} \end{cases}, \tag{2}$$

where $\bar{Y}$ is the average value of the Y channel. Since $1-M$ inverts the computed mask, bright areas are darkened by a gamma value lower than 1, and dark areas are brightened by a gamma value greater than 1.

The application of LCC tends to reduce the overall contrast and saturation, as noted by Schettini *et al.* [18]. Therefore, as subsequent steps, we perform contrast and saturation enhancement.

The contrast enhancement step adaptively stretches and clips the image histogram based on how the distribution of dark pixels changes before and after the contrast correction of LCC. Every histogram computed has 256 bins. The histogram range used for stretching and clipping is defined as follows: let any given pixel be "dark" if, in the YCbCr color space, its Y value is lower than 0.14 and its chroma radius, as defined in [18], is lower than 0.07. The lower range for histogram stretching is defined by the number of dark pixels after the application of LCC. If there is at least one pixel, the lower range is given by the difference of the bins corresponding to 30% of dark pixels in the cumulative histogram of $\hat{Y}$ and Y, which represent the output and input of LCC in Eq. 1. If there are no dark pixels, the lower range corresponds to the 2nd percentile value of the $\hat{Y}$ histogram. Concerning the "bright" pixels, the upper range for histogram stretching always corresponds to the 98th percentile value of the $\hat{Y}$ histogram. For both ranges, the maximum number of bins to clip is 50. Using the determined range, the image histogram is stretched and the histogram bins that fall outside are clipped.

For the saturation enhancement step, we correct each RGB channel as suggested by Sakaue *et al.* [17]:

$$\hat{C} = 0.5 \times \frac{\hat{Y}}{Y} \times (C + Y) + C - Y, \tag{3}$$

where C stands for each RGB channel, $\hat{C}$ is the corresponding output channel, $\hat{Y}$ and Y are the output and input Y channels used in Eq. 1.

After contrast and saturation enhancement, a black point correction step is performed in order to restore the natural aesthetics of night images, since LCC adjusts local statistics but produces an overall washed-out result. This operation

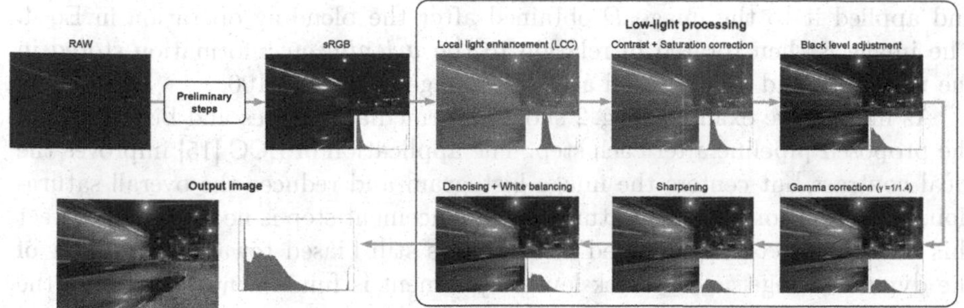

Fig. 2. Step-by-step results of the proposed pipeline. Along with images, we also reported histograms to show how the global pixel distribution changes.

is performed by clipping to zero all pixels below the 20th percentile value of the value channel V in the HSV color space. After this operation, a global gamma correction is performed with a gamma value set to $\frac{1}{1.4}$, followed by a sharpening operation using unsharp masking.

The image is then converted to 8-bit encoding, resized to match the predefined output size (imposed by the challenge to be 1300×866 for landscape orientation and 866×1300 for portrait one), and processed with the Block-Matching and 3-D Filtering (BM3D) denoising algorithm [5] to remove noise introduced by the poor light conditions typical of night scenes. Here the *noise_profile* value from the image metadata is used to determine the strength of the denoising operation, which is controlled by BM3D through a parameter σ that encodes an estimate of the noise standard deviation, used internally to control the parameters of the method. According to the distribution of the *noise_profile* values in the training data, we defined three classes representing different noise intensities, and we empirically assigned a σ value (0.2, 0.6 and 0.8) to each class. Since noise is more visible in dark regions rather than in bright regions and BM3D removes part of the high-frequency information, we performed a blending operation in RGB using a mask generated by blurring the luminance channel Y of the original noisy image in the YCbCr color space with a Gaussian filter. The final denoised image $\hat{D}$ is computed as

$$\hat{D} = I_{BM3D} \times (1 - mask \times u) + I \times (mask \times u), \tag{4}$$

where I_{BM3D} is the image denoised with BM3D, I is the original noisy image and the u parameter, empirically set to 0.6, controls the denoising effect in bright areas.

A second automatic white balance step is performed, this time on non-linear processed RGB data, in order to reduce color casts in those scenarios where the initial Gray World approach may have failed. Here the Grayness Index (GI) algorithm [16] is used. GI is very sensitive to noise, hence we estimated the image illuminant on the image I_{BM3D}, then we normalized it by its maximum value and applied it to the image $\hat{D}$ obtained after the blending operation in Eq. 4. The image is then rotated in relation to the *orientation* information stored in the metadata and finally saved as JPEG image at quality 100.

As illustrative example, Fig. 2 shows intermediate images and histograms of the proposed pipeline after each step. The application of LCC [15] improves the local contrast but centers the image histogram and reduces the overall saturation, hence the contrast and saturation enhancement step is necessary to correct this behavior. Yet, the obtained histogram is still biased towards the center of the dynamic range, and a black level adjustment is fundamental to restore the natural anesthetic of the image. Here a gamma correction can increase the overall brightness. Since BM3D [5] effectively removes noise but also part of the details, a preliminary sharpening operation that strengthens high frequencies helps preventing this problem.

3 Experiments

3.1 Dataset

We adopted the dataset used in the NTIRE2022 Night Photography Rendering challenge [6], which provides 250 RAW-RGB images of night scenes captured using a Canon EOS 600D device and encoded in 16-bit PNG files. Each RAW image has a resolution of 3646 × 5202 pixels. Image metadata are also available in JSON format. Due to the nature of the challenge, ground truth images are not available. According to the challenge organization, 50 images are provided as train set, 50 as the first validation set, 50 as the second validation set, and the remaining 100 as the final validation set (among these 100, only 50 were selected for the final evaluation). Since our solution does not need a training procedure, we used the train set to empirically select the few parameters required by our pipeline and used all validation sets to validate the results.

3.2 Results and Discussion

We evaluated our pipeline by comparing it with a subset of state-of-the-art approaches for low-light image enhancement and with other solutions that participated in the NTIRE2022 Night Photography Rendering challenge [6] using psycho-visual comparisons and Mean Opinion Score (MOS).

We selected eight recent state-of-the-art approaches for low-light image enhancement and performed a psycho-visual evaluation test using the same 50 images from the validation set, processed by the selected methods. More precisely, we selected DRBN [19], Kind [22] and Kind++ [21], TBEFN [14], EnlightenGAN [8], ExCNet [20], Zero-DCE [7] and Zero-DCE++ [11]. Since these methods expect images to be in sRGB color space, we used the preliminary steps described in Sect. 2 to convert the RAW images into sRGB images and applied these enhancement methods to them. The resulting images have been obtained using the LLIE platform [10]. The comparison has been performed by

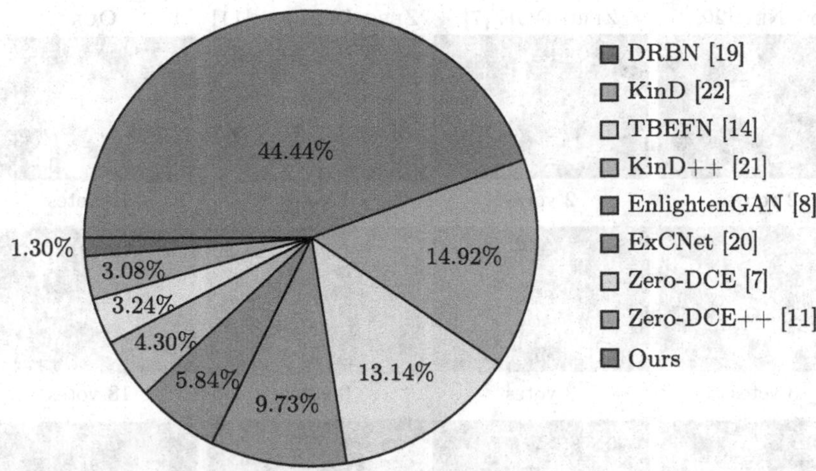

Fig. 3. Pie chart reporting the distribution of the results of the psycho-visual test, in terms of number of preferences, obtained comparing our proposed method with other state-of-the-art low-light image enhancement approaches.

a total of 31 users. Each user involved in the evaluation was shown a 3×3 grid containing the same image enhanced by the nine methods and was asked to click on the preferred one. This process was repeated for each of the 50 images. Grid composition and image order were randomly generated. The evaluation was done on monitors between 24 and 27 in. under controlled lighting conditions, and the images were shown on black background.

Figure 3 shows the pie chart with the results. As can be seen, in almost 45% of the cases the proposed approach is preferred with respect to the other ones. Zero-DCE [7] and Zero-DCE++ [11] obtained almost the same number of votes, followed by ExCNet [20]. From this first analysis, it is easy to notice how the proposed approach leads to more appreciated images with respect to the other methods. In order to provide a visual comparison for the readers, some results of the proposed pipeline are shown in Fig. 4. We also reported the same images corrected using different state-of-the-art deep learning-based low-light image enhancement approaches. In this figure are reported four different cases: the first two are cases in which our approach received the highest consensus in terms of user votes, while the last two are a mid-case scenario and a worst-case scenario, respectively. We can observe how our solution is better at removing noise, increasing sharpness, reducing color cast and preserving the mood that is typical of night scenes. This produces more pleasing results, as also confirmed by the vote distribution in Fig. 3. It is worth noting that the number of votes received by our solution is considerably higher than the votes received by other methods when it obtained the highest score. Instead, when other methods were preferred to ours, their vote counts are comparable to the number of votes received by our solution.

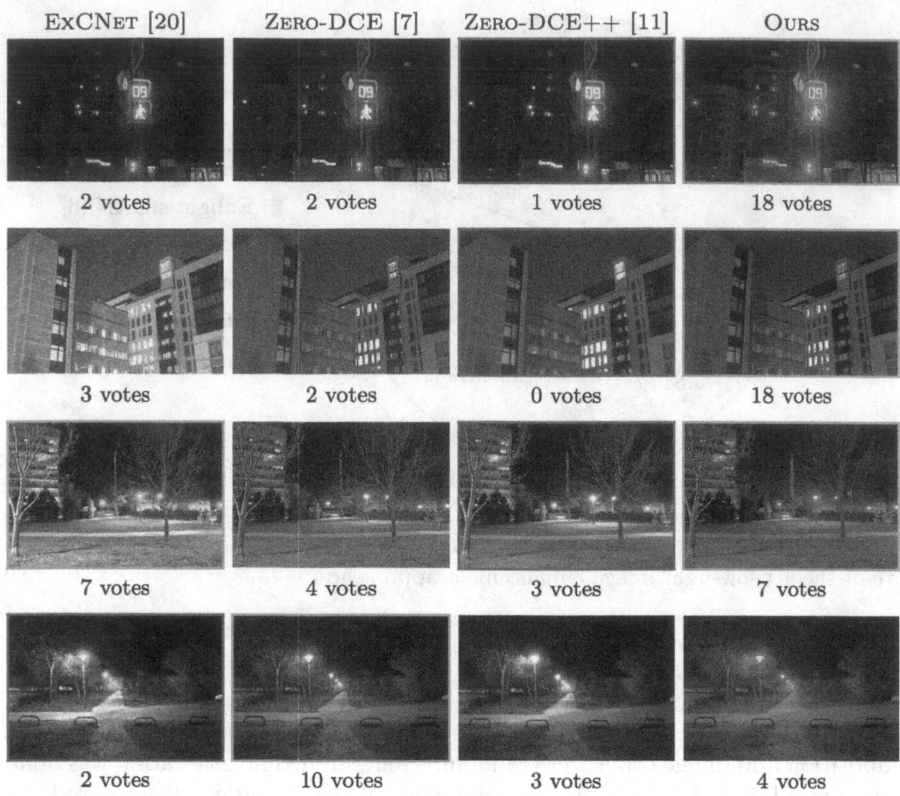

Fig. 4. Visual comparison between the proposed method and three highest score state-of-the-art approaches. Our solution produces sharper results, better reduces noise and color cast, and better maintains the mood of night photographs while the others tend to over-saturate colors and light casts. For each row, images framed in green are the ones with the highest score while the ones framed in red are the ones with the worst score. (Color figure online)

For what concerns the NTIRE2022 challenge, MOS results are obtained through visual comparison on the Yandex Toloka platform. Here every submission, consisting of 50 images of the final validation set, was included in 3250 comparisons. The results of the final leaderboard are reported in Table 1. As shown, our pipeline won the fifth place in the challenge obtaining 1935 votes. Note that our solution received only 112 fewer votes than the second winning solution (i.e. about 5% fewer votes) that uses different neural models for most of the operations in its pipeline [12].

In Table 1 we also add a further column, named significance score. First of all, for each method we compute the 95% confidence interval of the Score using the Binomial test. The significance score for each solution corresponds to the number

Table 1. Final leaderboard of the NTIRE2022 Night Photography Rendering challenge [6]. Every submission (50 images) was included in 3250 comparisons using the Yandex Toloka platform. Our team is highlighted in bold.

Rank	Team	Score	Votes	Sign. Score
1	MIALGO	0.8009	2603	12
2	Sorashiro	0.6298	2047	11
3	Feedback	0.6089	1979	11
4	OzU-VVGL	0.6045	1964	11
5	**IVLTeam**	**0.5955**	**1935**	**10**
6	NoahTCV	0.5742	1866	9
7	NTU607QCO	0.4798	1559	6
8	Winter	0.4631	1505	6
9	Sigma_WHU	0.4411	1433	5
10	Namecantbenull	0.3965	1288	3
11	BISPL	0.3683	1197	3
12	Baseline	0.2734	888	1
13	Low Light Hypnotize	0.0182	59	0

of solutions with respect to which it is statistically better or equivalent, i.e. the number of confidence intervals that are lower or overlap with the current one. The significance score highlights how the result achieved by the first solution [13] is statistically better than all the others, while the solutions ranked from the second position to the fourth one are actually statistically equivalent and therefore rank in the second place. They are followed by our solution, which ranks in the third place and is statistically better than all the remaining solutions. The results have been additionally evaluated by a professional photographer, who awarded our solution with the sixth place in the final leaderboard [6].

4 Conclusions

We have proposed a low-complexity handcrafted camera pipeline for the rendering of visually pleasing night photographs. Our solution includes several processing steps that address the challenges presented by low-light images and depends on a small number of free parameters, which we empirically set according to our personal preferences. However, the optimal parameters could be easily found with optimization methods if one had a suitable training set (whose cardinality however should not be so high as in the case of neural networks).

The effectiveness of our pipeline was validated through experiments involving real users, which demonstrated that our method produces more visually appealing results than other state-of-the-art methods for low-light image enhancement. These results have been further evaluated in the context of the NTIRE2022

Night Image Rendering challenge [6], also demonstrating that traditional imaging pipelines can compete with modern deep learning-based methods.

An interesting next step could be the parametrization and optimization of the proposed pipeline, adopting unsupervised training solutions to model user preferences [23]. Another promising research direction is that of exploiting saliency [1] to perform spatially varying enhancement or to exploit no reference image aesthetic metrics [2] to drive model parameters selection.

References

1. Bianco, S., Buzzelli, M., Ciocca, G., Schettini, R.: Neural architecture search for image saliency fusion. Inform. Fusion **57**, 89–101 (2020)
2. Bianco, S., Celona, L., Napoletano, P., Schettini, R.: Predicting image aesthetics with deep learning. In: Blanc-Talon, J., Distante, C., Philips, W., Popescu, D., Scheunders, P. (eds.) ACIVS 2016. LNCS, vol. 10016, pp. 117–125. Springer, Cham (2016). https://doi.org/10.1007/978-3-319-48680-2_11
3. Bianco, S., Cusano, C., Piccoli, F., Schettini, R.: Personalized image enhancement using neural spline color transforms. IEEE Trans. Image Process. **29**, 6223–6236 (2020)
4. Buchsbaum, G.: A spatial processor model for object colour perception. J. Franklin Inst. **310**(1), 1–26 (1980)
5. Dabov, K., Foi, A., Katkovnik, V., Egiazarian, K.: Image denoising by sparse 3-d transform-domain collaborative filtering. IEEE Trans. Image Process. **16**(8), 2080–2095 (2007)
6. Ershov, E., et al.: NTIRE 2022 challenge on night photography rendering. In: Proceedings of the IEEE/CVF Conference on Computer Vision and Pattern Recognition Workshops (2022)
7. Guo, C., et al.: Zero-reference deep curve estimation for low-light image enhancement. In: Proceedings of the IEEE/CVF Conference on Computer Vision and Pattern Recognition, pp. 1780–1789 (2020)
8. Jiang, Y., et al.: Enlightengan: deep light enhancement without paired supervision. IEEE Trans. Image Process. **30**, 2340–2349 (2021)
9. Karaimer, H.C., Brown, M.S.: A software platform for manipulating the camera imaging pipeline. In: Leibe, B., Matas, J., Sebe, N., Welling, M. (eds.) ECCV 2016. LNCS, vol. 9905, pp. 429–444. Springer, Cham (2016). https://doi.org/10.1007/978-3-319-46448-0_26
10. Li, C., Guo, C., Han, L., Jiang, J., Cheng, M.M., Gu, J., Loy, C.C.: Low-light image and video enhancement using deep learning: A survey. IEEE Trans. Pattern Anal. Mach. Intell. **44**(12), 9396–9416 (2021)
11. Li, C., Guo, C., Loy, C.C.: Learning to enhance low-light image via zero-reference deep curve estimation. IEEE Trans. Pattern Anal. Mach. Intell. **44**(8), 4225–4238 (2021)
12. Li, Z., Yi, S., Ma, Z.: Rendering nighttime image via cascaded color and brightness compensation. In: Proceedings of the IEEE/CVF Conference on Computer Vision and Pattern Recognition, pp. 897–905 (2022)
13. Liu, S., et al.: Deep-flexisp: a three-stage framework for night photography rendering. In: Proceedings of the IEEE/CVF Conference on Computer Vision and Pattern Recognition, pp. 1211–1220 (2022)

14. Lu, K., Zhang, L.: Tbefn: A two-branch exposure-fusion network for low-light image enhancement. IEEE Trans. Multimedia **23**, 4093–4105 (2020)
15. Moroney, N.: Local color correction using non-linear masking. In: Color and Imaging Conference, vol. 2000, pp. 108–111. Society for Imaging Science and Technology (2000)
16. Qian, Y., Kamarainen, J.K., Nikkanen, J., Matas, J.: On finding gray pixels. In: Proceedings of the IEEE/CVF Conference on Computer Vision and Pattern Recognition, pp. 8062–8070 (2019)
17. Sakaue, S., Nakayama, M., Tamura, A., Maruno, S.: Adaptive gamma processing of the video cameras for the expansion of the dynamic range. IEEE Trans. Consum. Electron. **41**(3), 555–562 (1995)
18. Schettini, R., Gasparini, F., Corchs, S., Marini, F., Capra, A., Castorina, A.: Contrast image correction method. J. Electron. Imaging **19**(2), 023005 (2010)
19. Yang, W., Wang, S., Fang, Y., Wang, Y., Liu, J.: From fidelity to perceptual quality: A semi-supervised approach for low-light image enhancement. In: Proceedings of the IEEE/CVF Conference on Computer Vision and Pattern Recognition, pp. 3063–3072 (2020)
20. Zhang, L., Zhang, L., Liu, X., Shen, Y., Zhang, S., Zhao, S.: Zero-shot restoration of back-lit images using deep internal learning. In: Proceedings of the 27th ACM International Conference on Multimedia, pp. 1623–1631 (2019)
21. Zhang, Y., Guo, X., Ma, J., Liu, W., Zhang, J.: Beyond brightening low-light images. Int. J. Comput. Vision **129**, 1013–1037 (2021)
22. Zhang, Y., Zhang, J., Guo, X.: Kindling the darkness: A practical low-light image enhancer. In: Proceedings of the 27th ACM International Conference on Multimedia, pp. 1632–1640 (2019)
23. Zini, S., Buzzelli, M., Bianco, S., Schettini, R.: A framework for contrast enhancement algorithms optimization. In: 2022 IEEE International Conference on Image Processing (ICIP), pp. 1431–1435. IEEE (2022)

GCK-Maps: A Scene Unbiased Representation for Efficient Human Action Recognition

Elena Nicora, Vito Paolo Pastore(✉), and Nicoletta Noceti

MaLGa - DIBRIS, University of Genoa, Genoa, Italy
elena.nicora@dibris.unige.it,
{vito.paolo.pastore,nicoletta.noceti}@unige.it

Abstract. Human action recognition from visual data is a popular topic in Computer Vision, applied in a wide range of domains. State-of-the-art solutions often include deep-learning approaches based on RGB videos and pre-computed optical flow maps. Recently, 3D Gray-Code Kernels projections have been assessed as an alternative way of representing motion, being able to efficiently capture space-time structures. In this work, we investigate the use of GCK pooling maps, which we called GCK-Maps, as input for addressing Human Action Recognition with CNNs. We provide an experimental comparison with RGB and optical flow in terms of accuracy, efficiency, and scene-bias dependency. Our results show that GCK-Maps generally represent a valuable alternative to optical flow and RGB frames, with a significant reduction of the computational burden.

Keywords: Action Recognition · Gray-Code Kernels · Motion Representation

1 Introduction

Human action recognition (HAR) from visual data is paramount in a broad range of applications, including robotics [34], assisted living, and well-being estimation [12], health care and rehabilitation [19], surveillance [25] entertainment [14]. A recent review on the topic can be found in [31].

In the last years, HAR has been commonly addressed using deep architectures, and in particular Convolutional Neural Networks (CNNs) [28], either in an end-to-end manner or as a representation learning approach. Despite the extensive application, deep learning models are notoriously data-hungry [1], usually requiring huge amounts of training data and thus resulting in a strong demand for computational resources and long training time. In the case of videos – the common input to address HAR – the employment of 3D CNNs, able to capture spatiotemporal structures in the data, further affects the computational load, as the number of parameters to be learned increases with the complexity of the model.

As a consequence, any help to accelerate the training process may be beneficial for the overall procedure. Attention mechanisms have been used to guide

E. Nicora and V. P. Pastore—These authors contributed equally to this work.

G. L. Foresti et al. (Eds.): ICIAP 2023, LNCS 14233, pp. 62–73, 2023.
https://doi.org/10.1007/978-3-031-43148-7_6

network learning. In [17] the authors propose VideoLSTM, which incorporates motion-based attention for action localization. The work in [10] introduces a novel learnable pooling layer that improves HAR performance. AttCell is a new neural cell proposed in [15], that acts at the level of features learning. An alternative approach consists in providing the network with a pre-processed input, in which parts of potential relevance for the downstream task have been emphasized. In this way, the learning method is guided to immediately see and learn more quickly the core information for the task of interest. On this line, common approaches make use of Optical Flow (OF), widely exploited for HAR, either as single input [33,35] or in addition to the RGB channel in two-stream architectures [6,29]. In the same spirit in [3] pose heat-map sequences are encoded in a comprehensive video representation for action recognition.

In this paper, we reason on the influence of the type of input for addressing HAR with CNNs on small-scale datasets with different characteristics. We compare classical RGB data, Optical Flow maps, and the Gray-Code Kernels Pooling Maps (GCK-Maps), a novel two-channel motion descriptor derived by 3D Gray-Code Kernels (GCKs) projections, recently proposed in [23,24]. Overall, we provide an experimental comparison between these three different representations for HAR in terms of accuracy, computational efficiency, and dependence on scene bias. Since the small size of the dataset we employed prevents us from training from scratch our networks, we also implicitly reason on the usability of the inputs in a transfer learning domain, when models pre-trained on ImageNet are employed. Our contributions can be referred to the following scientific questions

- How the scene bias is captured by the different inputs? What is the impact on the results?
- What is the computational load of the different solutions, also in reference to the obtained accuracy?
- Under what conditions the maps based on GCKs can be used as a suitable alternative to the more popular Optical Flow?

In our analysis, we found that, as one could imagine, the raw RGB input is more prone to rely on scene bias, while GCK-Maps are a suitable alternative to OF and RGB, providing higher or comparable results with an inferior computational load.

The remainder of the paper is organized as follows: in Sect. 2 we describe the related works for HAR; in Sect. 3 we provide details about the GCKs pooling maps, while in Sect. 4 we present and discuss the obtained results.

2 Related Works

In classical approaches to motion recognition, the representation phase was usually decoupled from the actual learning [26]. Optical Flow (OF) was a widely used solution for the representation step, employed for motion detection [36], tracking [35], or action recognition [9]. Among other existing approaches, we mention the ones employing motion history images [2], HOG descriptors [8], or shape context [37], just to name some. With the advent of deep architectures, HAR has been most often addressed with end-to-end pipelines. Recurrent Neural Networks and Long-Short Term Memory Networks are naturally able to model temporal data, and they have been employed for HAR for instance in [29,30]. However, a more widely adopted deep solution for HAR is represented by 3D CNNs, which can be seen as an extension of standard 2D CNNs capable of directly extracting spatiotemporal features from the input. One of the first 3D-CNN models applied for human action recognition is C3D [32], where the authors showed the benefit of using 3D kernels as opposed to 2D kernels. A few years later, Inflated 3D ConvNet (I3D) [6] extended C3D by inflating 2D kernels from an ImageNet pre-trained 2D CNN in order to improve convergence and training time. It adopted a transfer-learning framework, reducing the number of required training data, and enabling the analysis of small-scale datasets. In I3D, OF was used in conjunction with RGB data in a two-stream architecture exploiting the complementarity of the two. On the same line in [29] dense OF is used together with RGB frames in a two-stream neural network.

The interesting analysis in [27] highlights how the OF remains, to date, an essential tool for action recognition, and that its potential can be further exploited in a number of directions still largely unexplored. However, its computation is often expensive in terms of time and resources. Very recently, GCKs have been proposed as a more efficient, although less precise, alternative to OF for motion detection tasks [24]. More in general, they have been rarely applied to video data, for instance in [20] for motion estimation, and in [21] for foreground segmentation. Differently from these approaches, here our aim is to evaluate their potential as attention mechanisms for action recognition tasks.

We finally briefly mention works specifically focused on investigating the role of scene bias for action recognition, that if on the one hand can lead to poor generalization, on the other has been poorly investigated within action recognition tasks. The work in [7] proposes a loss function specifically tuned to mitigate scene bias of CNNs; in [16] the authors investigate approaches for assembling datasets without bias.

In the described context of HAR, our contribution focuses on comparing three different inputs in terms of accuracy, efficiency, and dependence on scene bias. We tackle small-scale datasets with a transfer-learning framework, compensating for the limited amount of available data. The aim of our analysis is to identify situations where the GCK-Maps may be a valuable and efficient alternative to RGB and OF.

3 GCK-Maps: A Motion Descriptor Based on 3D Gray-Code Kernels

The GCKs [5] are a family of filters that provide an efficient projection framework for filtering images and videos. A nice property is that they provide the result of a classical convolution in constant time. It has been experimentally

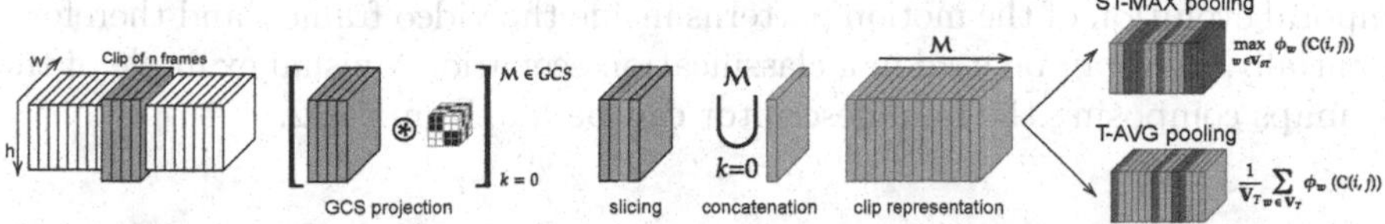

Fig. 1. Visual representation of the video projection scheme based on 3D Gray-Code Kernels computation: for each clip of size n, $\mathcal{M}$ 3D projections are computed, with $\mathcal{M}$ corresponding to the number of GCKs belonging to the family. After that, the central slice of each 3D projection is selected and concatenated to the others, forming a clip representation of size $\mathcal{M}$. Pooling maps are obtained by computing the max and the mean of two subsets of projections. (Color figure online)

assessed that they can be profitably used as a starting point for motion detection and segmentation [23,24], but they also showed an interesting descriptive potential that makes them ideal candidates to be used in HAR pipelines. Indeed, as with the Optical Flow, the maps naturally incorporate spatiotemporal information, together with direction and motion development cues, and considering their computational efficiency, they may represent a more convenient alternative. Indeed, if on the one hand OF algorithms have been studied for decades and their usefulness to derive motion representations HAR is undoubted, on the other they are often obtained with heavier computations (a rough comparison of OF algorithms' computational complexity can be found in [4]). The efficiency of GCKs is related to a cascade filtering approach: given a bank of filters, applying them in a sequence defined according to specific rules (called Gray-Code Sequence) allows to compute successive convolutions of an image with only two operations per pixel regardless of filter or image size. We refer the interested reader to [5,24] for a more in-depth analysis. In [24] the original work has been extended to tackle spatiotemporal data. The result is a filtering scheme, sketched in Fig. 1, able to very efficiently map the raw video data into a different space where space-time structures are effectively captured.

More specifically, the filtering is applied to overlapping clips of frames and it produces a 3D projection (the result of the convolution) for every kernel belonging to the family ($\mathcal{M}$). The central slice of the projection incorporates the highest amount of time information and is thus selected to be provided to the next step of the pipeline. All the slices are concatenated and the final representation is further compacted by pooling the values together.

By construction, a set of 3D GCKs includes filters that focus on spatial, temporal, and spatiotemporal changes within the given clip of frames. Clip descriptors that have been exploited in this paper were obtained considering the subsets of kernels that include temporal information: ST-MAX, obtained by computing the max pooling of the spatiotemporal projections at each position (i,j), and T-AVG, obtained by calculating the mean of the temporal projections at each position (i,j). These maps have been empirically proven to be effective in the representation of salient motion [24] providing cues on location, direction, and temporal evolution of the motion patterns inside the video frames, and therefore, potentially, they can be used in a classification scenario. A visual example of the two maps composing the final descriptor can be found in Fig. 2.

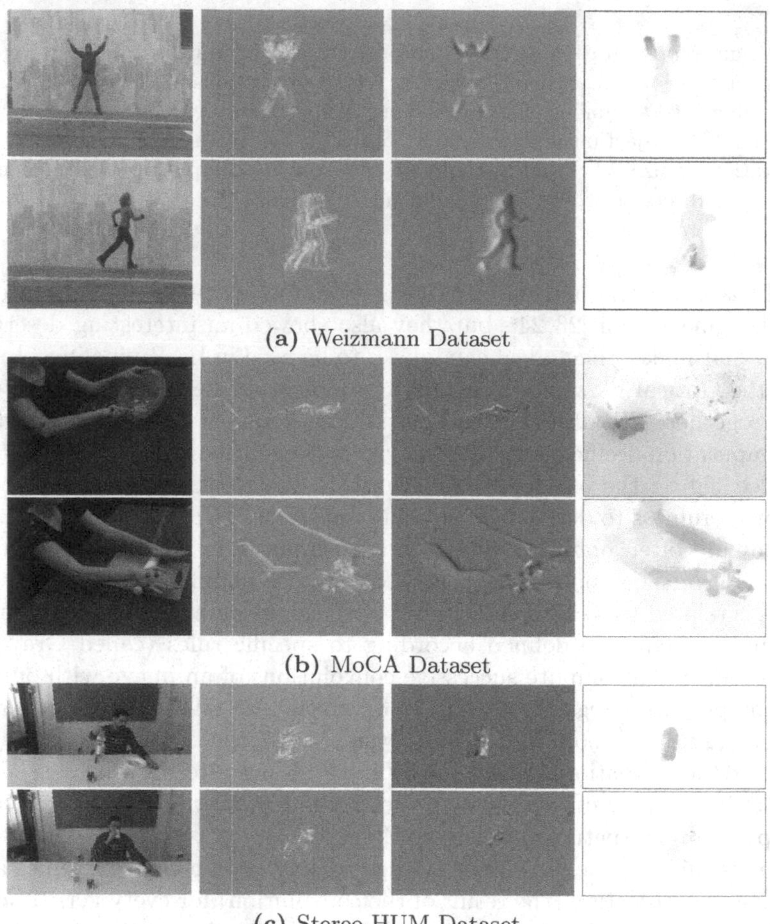

(a) Weizmann Dataset

(b) MoCA Dataset

(c) Stereo-HUM Dataset

Fig. 2. Examples of input considered in our experimental protocol from the three datasets included: RGB frames, followed by ST-MAX and T-AVG (the two channels of GCK-Maps) and siftFlow maps.

4 Experiments

In this section, we investigate the potentiality of GCK-Maps as spatiotemporal descriptors. As specified in the Introduction, we are particularly interested in the influence of different network inputs for what concerns accuracy and efficiency (in terms of training/convergence time). For these experiments, we chose a 2D densely connected CNN [13] and the Inception3D network [6]. The comparison also enables us to draw conclusions about the presence of scene bias in the datasets and the relative importance of appearance and motion information in the classification task. We employ RGB video frames, GCK pooling maps, and OF maps as input. Besides that, we maintained the same network parameters throughout all the computations for a more objective and fair comparison, starting from the number of frames (total and per batch) and size of the 2D input in terms of rows × columns.

In the following, we briefly describe the datasets used in our experiments, then we provide implementation details for reproducibility. Finally, we present and discuss the obtained results.

4.1 Datasets

In order to properly monitor the behavior of the different inputs, we resorted to action recognition datasets smaller than benchmarks widely used nowadays. All the datasets include sequences acquired with a fixed camera and static background; a single moving subject is present in the scene. The datasets we employed are the following:

- **Weizmann Dataset** [11] (Fig. 2a). It is a classical public dataset including 10 different full-body actions performed by 9 subjects. Motion patterns are limited to one or two main directions: subjects either move on a single spot (*jumping, waving*) or transit towards the east or towards the west (*walking, trotting, running*). A little bit more complex are limb movements that can undergo different patterns within the same sequence (*i.e. jumping jacks*).
- **MoCA dataset** [22] (Fig. 2b). The *Kinematic and Multi-View Visual Streams of Fine-Grained Cooking Actions* dataset consists of 20 upper-body actions composed by different motion patterns, i.e. circular and "back and forth" actions that develop spatially on different planes, movements formed by more than one gesture separated by pauses (*eating, transporting objects*), manipulations, that hence involve the use of tools (a pan for flipping pancakes, a pestel, a knife), but also fine-grained actions, in which only a hand (*beating eggs*) or just the fingers of the subject (*opening/closing a bottle*) are moving. Every action sequence is composed of multiple repetitions of the same movement ending with several instances.
- **Stereo-HUM** (Fig. 2c). The *Stereo Human Upper body Motion for interaction* dataset has been acquired in-house[1] to study human-object interactions in a controlled environment. Among the goals of the datasets, we mention the analysis of the role of object properties in manipulation tasks, the influence of the environment and its elements on movements, and the study of human attention for action anticipation purposes. It consists of 10 upper-body actions performed by 16 subjects. The scene is composed of a table

[1] The dataset will be soon released.

on which are placed a number of tools. Differently from the MoCA dataset, here each video depicts a single action instance (*sanitize hands, play with Rubik's cube, touch objects*), with the exception of actions that are repetitive by nature, like *eating* and *opening/closing a bottle.*

4.2 Implementation Details

We considered two widely employed architectures for 2D and 3D image data: the Inception3D [6], a classical action recognition model, limited to the stream pre-trained on ImageNet, and the DenseNet [13]. The latter, usually employed for image classification, has been included to help us draw conclusions on the amount of scene bias carried inherently by the different representations. With action recognition tasks ideally relying on motion information only, we can consider the use of a 2D architecture as a sanity check, in order to understand in which cases appearance plays a more significant role. A sketch of the different pipelines can be found in Fig. 3. The Inception3D model takes in input 3D samples composed of 10 frames (Fig. 3a) in mini-batches of 5 random samples, giving a single action prediction for each sample. As for the DenseNet, the batch size has been chosen in order to be in line with the analogous 3D experiment (i.e. mini-batch of 5 samples of width 10 for the I3D and a single batch of 50 frames for the DenseNet). Among all the alternatives that the literature offers, the flow maps employed come from a SIFT flow algorithm, a sparse Optical Flow technique that exploits SIFT features [18] and represents a good compromise between the quality of the final maps and the algorithm's computational requirements[2]. We resized input images to roughly have similar rows $\times$ columns ratio for all three datasets.

The classification protocol for the Weizmann dataset is based on a leave-one-out approach where the test set is composed, at each computation, by the subset of actions of a single actor. The validation set, instead, is composed each time by the sequences of one random subject among the remaining others. As for the MoCA and the Stereo-HUM datasets, instead, every action comes with a pair of sequences, one for training and one for testing. The number of training/validation/test samples is different for all the datasets (hence the differences in training time, Table 2).

We set a limit of 30 epochs and introduced an early stopping optimization technique that had the aim of avoiding overfitting while not compromising the accuracy of the model.

4.3 Results

First, we exploited our customized implementation of the i3D model to perform action recognition on the three datasets described in Sect. 4.1. The three datasets differ in terms of size and impact of the scene bias on the recognition of the performed action. The Weizmann dataset is the smallest one included in our experiments, with no scene bias. The MoCA dataset has longer videos

[2] Implementation available at http://people.csail.mit.edu/celiu/OpticalFlow/.

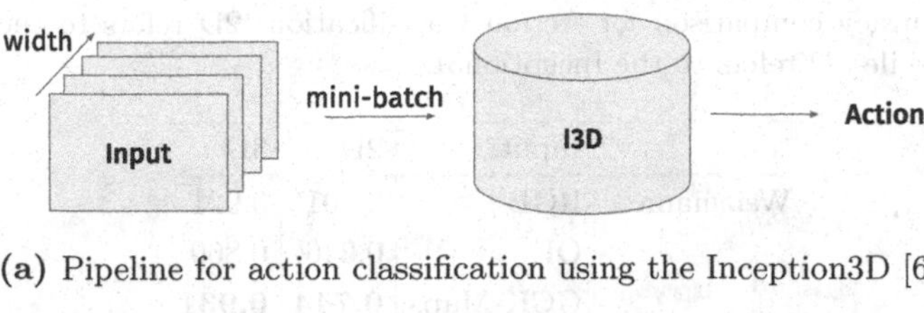

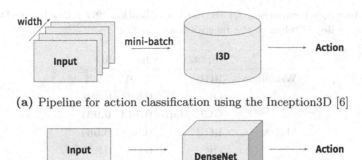

(a) Pipeline for action classification using the Inception3D [6]

(b) Pipeline for action classification using the DenseNet [13]

Fig. 3. A sketch of the two main network configurations considered in this experiment: input can either be RGB frames (3 channels), GCKs pooling maps (2 channels) or OF maps (2 channels). The DenseNet batch size has been chosen in order to be in line with the same 3D experiment.

with a strong scene bias, represented by the kitchen utensils characterizing each cooking action. The Stereo-HUM is the largest one in terms of the number of available frames, with scene bias only present for specific actions. Table 1 summarizes the obtained results in terms of test accuracy. As we can see, on the Weizmann dataset the GCK-Maps provide the best representation, with an accuracy of 0.931 and an improvement of 7% with respect to the OF. In this work, we evaluated the impact of scene bias considering a 2D CNN for the task of action recognition. In this way, we removed the explicit time information from our inputs. As a consequence, the RGB only relies on appearance cues, while the GCK-Maps and OF still possess the temporal information intrinsic in their computation. For the 2D analysis of the Weizmann dataset, the GCK-Maps outperform RGB and OF. Our hypothesis is that the GCK-Maps still maintain sufficient temporal cues in each map, differently from the other inputs. In the MOCA dataset, as expected for the high scene bias, the RGB outperforms GCK-Maps and OF, with an improvement of 12% in test accuracy. The high scene bias is confirmed by the 2D analysis, where the RGB shows only a small drop of $\approx 7\%$ with respect to the 3D model. In this case, the OF shows higher test accuracy for the 2D analysis, suggesting that it encodes more scene bias with respect to the GCK-Maps.

In the Stereo-HUM dataset, all the inputs perform similarly when using the 3D CNN. The OF provides the best input in the 2D analysis, with a difference of $\approx 11\%$ with respect to RGB and GCK-Maps. This result suggests that the appearance and temporal cues in a single frame of OF may be the most complete when datasets have an intermediate level of scene bias and enough data to learn from. Then, we evaluated the three different inputs in terms of computational efficiency. We measure the efficiency of each input considering the

Table 1. Accuracy comparison for Action Classification: 2D refers to the DenseNet architecture while 3D refers to the Inception3D.

	Input	2D	3D
Weizmann	RGB	0.501	0.921
	OF	0.246	0.860
	GCK-Maps	**0.744**	**0.931**
MoCA	RGB	**0.898**	**0.961**
	OF	0.871	0.837
	GCK-Maps	0.720	0.828
Stereo-HUM	RGB	0.629	0.633
	OF	**0.736**	**0.659**
	GCK-Maps	0.621	0.651

Table 2. A comparison between training time and the number of epochs needed by the network to converge using three different data sources with the Inception3D CNN. About the MoCa, to ensure balance in the number of training samples, and thus to avoid some sort of bias towards actions represented by a higher number of samples, we considered a fixed amount of frames for each action.

		Training Time (sec/epoch)	Convergence (avg #epochs)	Training samples
Weizmann	RGB	73.67	**5**	4000
	OF	62.15	12	
	GCK-Maps	**61.01**	7	
MoCA	RGB	105.59	5	6000
	OF	99.69	5	
	GCK-Maps	**65.69**	**5**	
Stereo-HUM	RGB	535.23	22	32000
	OF	440.67	19	
	GCK-Maps	**437.65**	**16**	

average time needed to perform one epoch of training, and the average number of epochs to reach convergence, using the early stopping criterium described in Subsect. 4.2. As we can see in Table 2, the GCK-Maps require the lowest training time for epoch, with a minimum difference of 10 seconds for the smallest dataset (Weizmann) and a maximum difference of 100 seconds for the biggest one (Stereo-HUM) with respect to the RGB. The average difference in training time with respect to the OF is ≈ 14 s. Moreover, it is worth noticing that computing the OF requires significantly higher computational resources than the GCK-Maps [24], further proving the efficiency of this motion representation. Finally, excluding the Weizmann dataset, the GCK-Maps require the same number or fewer epochs to converge with respect to the RGB and the OF. We expect such differences to become even more significant and impactful in the case of large-scale HAR datasets.

5 Discussion

In this paper, we compared the use of different inputs for 2D and 3D CNNs for action recognition on small-scale datasets with different characteristics. We considered raw RGB data, OF maps, and GCK-Maps, a new two-channel motion descriptor derived by 3D GCKs. Our experimental analysis was particularly focused on accuracy, computational efficiency, and dependence on scene bias. We summarize the main observations in the following, with reference to the scientific questions we reported in the Introduction.

- Overall, raw RGB is more prone to scene bias, as expected. On average, GCK-Maps showed less dependence on scene bias with respect to OF maps. We hypothesize this is mainly due to the different levels of precision of the two maps: OF provides more detailed maps than GCKs, able to capture more scene cues than the latter.
- GCK-Maps provide, in general, a more efficient solution, with shorter training time. This is coupled with recognition performances that are comparable to or higher than the ones obtained with OF. Additionally, GCK-Maps' computation requires significantly fewer resources.
- GCK-Maps generally represent a valuable alternative to OF in small-scale scenarios. Moreover, when the dataset is not significantly affected by scene bias, GCK-Maps show equal or higher accuracy than raw RGB.

Finally, our comparative analysis highlighted the complementarity between the investigated inputs. Thus, we are currently working on implementing solutions for multi-stream architectures involving the use of GCK-Maps.

Acknowledgments. This work has been supported by AFOSR with the grant n. FA8655-20-1-7035. VPP was supported by FSE REACT-EU-PON 2014-2020, DM 1062/2021.

References

1. Adadi, A.: A survey on data-efficient algorithms in big data era. Jour. Big Data **8**(1) (2021)
2. Ahad, M.A.R., Tan, J.K., Kim, H., Ishikawa, S.: Motion history image: its variants and applications. Mach. Vis. Appl. **23**(2), 255–281 (2012)
3. Asghari-Esfeden, S., Sznaier, M., Camps, O.: Dynamic motion representation for human action recognition. In: IEEE WACV, pp. 557–566 (2020)
4. Baker, S., Scharstein, D., Lewis, J., Roth, S., Black, M.J., Szeliski, R.: A database and evaluation methodology for optical flow. IJCV **92**, 1–31 (2011)
5. Ben-Artzi, G., Hel-Or, H., Hel-Or, Y.: IEEE PAMI **29**(3), 382–393 (2007)
6. Carreira, J., Zisserman, A.: Quo vadis, action recognition? a new model and the kinetics dataset. In: IEEE CVPR, pp. 6299–6308 (2017)
7. Choi, J., Gao, C., Messou, J.C., Huang, J.B.: Why can't i dance in the mall? learning to mitigate scene bias in action recognition. In: Advances in Neural Information Processing Systems 32 (2019)

8. Dalal, N., Triggs, B.: Histograms of oriented gradients for human detection. In: IEEE CVPR, vol. 1, pp. 886–893 (2005)
9. Gehrig, D., Kuehne, H., Woerner, A., Schultz, T.: Hmm-based human motion recognition with optical flow data. In: IEEE-RAS Humanoids, pp. 425–430 (2009)
10. Girdhar, R., Ramanan, D.: Attentional pooling for action recognition. In: Advances in Neural Information Processing Systems 30 (2017)
11. Gorelick, L., Blank, M., Shechtman, E., Irani, M., Basri, R.: Actions as space-time shapes. IEEE PAMI **29**(12), 2247–2253 (2007)
12. Grossi, G., Lanzarotti, R., Napoletano, P., Noceti, N., Odone, F.: Positive technology for elderly well-being: a review. PR Lett. **137**, 61–70 (2020)
13. Huang, G., Liu, Z., Van Der Maaten, L., Weinberger, K.Q.: Densely connected convolutional networks. In: IEEE CVPR, pp. 4700–4708 (2017)
14. Kong, Y., Fu, Y.: Human action recognition and prediction: a survey. IJCV **130**(5), 1366–1401 (2022)
15. Li, D., Yao, T., Duan, L., Mei, T., Rui, Y.: Unified spatio-temporal attention networks for action recognition in videos. IEEE Trans. Multimedia **21**(2), 416–428 (2018)
16. Li, Y., Li, Y., Vasconcelos, N.: RESOUND: towards action recognition without representation Bias. In: Ferrari, V., Hebert, M., Sminchisescu, C., Weiss, Y. (eds.) ECCV 2018. LNCS, vol. 11210, pp. 520–535. Springer, Cham (2018). https://doi.org/10.1007/978-3-030-01231-1_32
17. Li, Z., Gavrilyuk, K., Gavves, E., Jain, M., Snoek, C.G.: Videolstm convolves, attends and flows for action recognition. In: CVIU, vol. 166, pp. 41-50 (2018)
18. Lowe, D.G.: Object recognition from local scale-invariant features. In: IEEE ICCV, vol. 2, pp. 1150–1157 (1999)
19. Moro, M., et al.: A markerless pipeline to analyze spontaneous movements of preterm infants. Comput. Methods Programs Biomed. **226**, 107119 (2022)
20. Moshe, Y., Hel-Or, H.: Video block motion estimation based on gray-code kernels. IEEE TIP **18**(10), 2243–2254 (2009)
21. Moshe, Y., Hel-Or, H., Hel-Or, Y.: Foreground detection using spatiotemporal projection kernels. In: IEEE CVPR, pp. 3210–3217 (2012)
22. Nicora, E., Goyal, G., Noceti, N., Vignolo, A., Sciutti, A., Odone, F.: The moca dataset, kinematic and multi-view visual streams of fine-grained cooking actions. Scientific Data **7**(1), 1–15 (2020)
23. Nicora, E., Noceti, N.: Exploring the use of efficient projection kernels for motion saliency estimation. In: ICIAP, pp. 158–169 (2022)
24. Nicora, E., Noceti, N.: On the use of efficient projection kernels for motion-based visual saliency estimation. Front. Comput. Sci. **4** (2022)
25. Noceti, N., Odone, F.: Learning common behaviors from large sets of unlabeled temporal series. ImaVis **30**(11), 875–895 (2012)
26. Poppe, R.: A survey on vision-based human action recognition. ImaVis **28**(6), 976–990 (2010)
27. Sevilla-Lara, L., Liao, Y., Güney, F., Jampani, V., Geiger, A., Black, M.J.: On the integration of optical flow and action recognition. In: GCPR, pp. 281–297 (2019)
28. Shekokar, R.U., Kale, S.N.: Deep learning for human action recognition. In: 2021 6th International Conference for Convergence in Technology (I2CT), pp. 1–5 (2021)
29. Simonyan, K., Zisserman, A.: Two-stream convolutional networks for action recognition in videos. In: Advances in Neural Information Processing Systems 27 (2014)
30. Sun, L., Jia, K., Chen, K., Yeung, D.Y., Shi, B.E., Savarese, S.: Lattice long short-term memory for human action recognition. In: IEEE ICCV, pp. 2147–2156 (2017)

31. Sun, Z., Ke, Q., Rahmani, H., Bennamoun, M., Wang, G., Liu, J.: Human action recognition from various data modalities: a review. In: IEEE PAMI, pp. 1–20 (2022)
32. Tran, D., Bourdev, L., Fergus, R., Torresani, L., Paluri, M.: Learning spatiotemporal features with 3d convolutional networks. In: IEEE ICCV, pp. 4489–4497 (2015)
33. Tu, Z., et al.: Multi-stream cnn: Learning representations based on human-related regions for action recognition. PR **79**, 32–43 (2018)
34. Vignolo, A., Noceti, N., Rea, F., Sciutti, A., Odone, F., Sandini, G.: Detecting biological motion for human-robot interaction: a link between perception and action. Front. Robotics AI, 14 (2017)
35. Wang, H., Schmid, C.: Action recognition with improved trajectories. In: IEEE ICCV, pp. 3551–3558 (2013)
36. Yao, R., Lin, G., Xia, S., Zhao, J., Zhou, Y.: Video object segmentation and tracking: A survey. ACM TIST **11**(4), 1–47 (2020)
37. Zhang, Z., Hu, Y., Chan, S., Chia, L.-T.: Motion context: a new representation for human action recognition. In: Forsyth, D., Torr, P., Zisserman, A. (eds.) ECCV 2008. LNCS, vol. 5305, pp. 817–829. Springer, Heidelberg (2008). https://doi.org/10.1007/978-3-540-88693-8_60

Autism Spectrum Disorder Identification from Visual Exploration of Images

Marco Bolpagni[1] and Francesco Setti[2](✉)

[1] Department of Computer Science, University of Verona, Verona, Italy
[2] Department of Engineering for Innovation Medicine, University of Verona, Verona, Italy
francesco.setti@univr.it

Abstract. Autism Spectrum Disorder (ASD) affects 1 in 77 children in Italy, but the diagnostic process is slow and costly. As autistic individuals exhibit different gaze patterns from healthy controls in visual exploration of images and semantic interpretation, these are promising biomarkers to exploit in diagnosis. This study aims at developing a model to assist in the diagnosis of ASD using gaze data when static images are presented to the subjects. We first propose a set of features, each one motivated by psychological studies and findings. Then we apply a feature selection mechanism based on Boruta algorithm with SHAP values. Finally we use CatBoost to perform binary classification, and a strategy to optimize model hyperparameters using a multivariate Tree Parzen Estimator. We validated our model on the popular Saliency4ASD dataset, outperforming state of the art models tested with the same protocol by more than 3% in accuracy. We also provide an in-depth analysis of the feature importance and we show how these results are in line with the psychological literature.

Keywords: Autism Spectrum Disorder · Eye tracking · CatBoost

1 Introduction

Autism Spectrum Disorder (ASD) is a neurodevelopmental disorder characterized by impairment in social interaction and communication. Common symptoms associated with ASD include verbal and nonverbal expression difficulties, limited interest in social games and activities, repetitive and stereotyped behaviors, difficulties in adapting to changes in routine or environment, and difficulties in understanding others' thoughts and emotions [13]. While it was previously believed that individuals with autism also had intellectual deficits, literature has disproved this myth, as many individuals with high functioning autism have intellectual abilities equal to –or superior to– their peers and are able to perform normal professional activities or excel in academics [22]. Additionally, autism is a heterogeneous disorder that presents in many forms. Until 2013, when the Diagnostic and Statistical Manual of Mental Disorders (DSM-5) [2] came out, there were up to five independent autism-related conditions (Autistic Disorder, Asperger Syndrome, Rett Syndrome, Pervasive Developmental Disorder

G. L. Foresti et al. (Eds.): ICIAP 2023, LNCS 14233, pp. 74–86, 2023.
https://doi.org/10.1007/978-3-031-43148-7_7

Not Otherwise Specified, and Childhood Disintegrative Disorder). New research brought all these profiles into a single diagnostic category that best represents the underlying characteristics common to all of them. Today, the severity of the disorder is no longer evaluated through assignment to a specific diagnostic subtype, rather it is evaluated based on the degree of support an individual with ASD requires to manage the daily challenges of life. The causes of autism have not been identified yet, but researchers are largely in agreement that it is a multifactorial disorder, meaning that the cause is the result of the interaction between genes, environmental factors, and other biological variables such as metabolic and immune system anomalies [35]. With regard to the incidence of the disorder epidemiological studies have reported a generalized increase in the prevalence of ASD that might also be influenced by the progress in diagnosis and greater awareness of the what autism is in the general population. In Italy, it is estimated that 1 in 77 children (age 7 to 9) have an autism spectrum disorder, with a higher prevalence in males: males are 4.4 times more likely to be autistic than females. ASD diagnosis usually occurs during childhood, however, in cases where autism presents in a mild form, the person may never receive a diagnosis or only receive it in adulthood. The guidelines of the Italian National Institute of Health (ISS) for the diagnosis of ASD in developmental age provide for a comprehensive clinical evaluation carried out by a multidisciplinary team (child psychiatrist, psychologist, neuropsychomotor therapist for developmental age, speech therapist, and educator) using standardized assessment tools such as ADI-R (Autism Diagnostic Interview. Revised), CARS (Childhood Autism Rating Scale), and ADOS (Autism Diagnostic Observation Schedule). In terms of adult diagnosis, the situation is more complex and no guidelines are available. Despite progress in managing the diagnostic process, obtaining a diagnosis can take on average 3.5 years from the onset of symptoms to official diagnosis [8]. This aspect is particularly critical as early diagnosis allows for the implementation of more effective therapies and subsequently improves the individual's quality of life. In recent years scholars focused on the development of Artificial Intelligence (AI) systems to support the diagnostic process. These AI systems are based on the use of different biomarkers (genetic profile, voice, gaze, movement, brain activity and structure) to identify individuals potentially affected by autism. However, models developed so far are still at their infancy and they require further validation before being used in real contexts. In particular, it is necessary to identify economically sustainable systems that can support the diagnosis of all categories of individuals, from children to adults, from mild autism cases to more severe conditions. In light of these needs, this study will focus on the development of a gaze-based model capable of automatically classify individuals with ASD using eye-tracking data.

2 Related Work

Recently, *gaze data*, commonly obtained through the use of eye trackers, are receiving major attention from researchers and are providing promising results in supporting diagnosis of ASD. Eye tracking devices allow to monitor eye movements to assess how a subject explores the images presented to him/her and what details he/she dwells on. This approach is based on evidence gathered over the past decades of research in psychology. Studies showed substantial differences in ASD subjects with respect to both visual saliency (related to content) and visual exploration patterns (related to cognition) when compared to the typical development (TD) control subjects. When it comes to visual saliency, ASD individuals show a general increase in pixel-level saliency at the expense of the semantic-level [34], reduced attentional bias toward threat-related scenes, reduced saliency for faces, and more attention to non-social objects [29]. When looking at images with people, ASD individuals have longer latency in orienting the attention to people in the scene, and look less to faces and more at the hands [19]. In terms of visual exploration patterns ASD individuals show decreased saccade duration [5], a smaller overall number of fixations, a stronger image center bias regardless of object distribution [34], difficulties in disengagement and shifting from a region of interest [10], and stagnant exploration in general [25]. Based on such differences, researchers use eye tracking data to explore the discriminative power of gaze patterns in face images and natural scene images.

Early studies used a bag-of-words approach to extract features from face images and trained a Support Vector Machine (SVM) to classify ASD vs. TD individuals [21]. While the results were promising, the sample size was relatively small, and the findings were not generalizable enough for thorough evaluation. Later studies focused on the discriminative power of gaze in natural scene images. For instance, researchers conducted experiments on high-functioning autistic adults and predicted the individuals' visual saliency maps [34]. The ground-truth human fixation maps were created from the gaze data, and a linear SVM classifier was used to predict fixation allocation. This model demonstrated the potential of using gaze patterns to identify individuals with ASD. Similarly, [16] proposed a modified SALICON network [15] composed of two parallel VGG-16 networks adding a convolutional layer to predict the Difference of Fixation (DoF) map and a sigmoid activation to produce a binary classification output.

In 2019, the Saliency4ASD challenge provided a dataset consisting of 300 images and associated gaze data of 28 children [12]. This competition had the double effect of fostering research in the field and providing a standardized benchmark for evaluating performances of classification methods. Five teams participated in the challenge developing relevant models based on scanpath, saliency, and face-driven features. Techniques used ranged from decision trees [4,31] to artificial neural networks, either dense [36] or convolutional and recurrent [32].

Recently, researchers are investigating the application of eye tracking for ASD classification in virtual reality (VR) environments. Traditional methods like naive bayes, random forests, k-Nearest Neighbors, decision trees and gradient boosting are benchmarked in [20], while SVM is employed in [1].

To conclude, gaze is a solid biomarker for the classification of ASD as demonstrated by [24], where three types of features are extracted from scanpath data: namely, content features, fixation features, and bias features. Seven classifiers

–and some variations– are tested in binary classification task, with an accuracy of over 68% on Saliency4ASD dataset.

3 Method

Our method is based on the extraction of a large set of well motivated gaze features. We then provide automatic feature selection using BorutaSHAP algorithm to remove noisy features and reduce risk of overfitting the model. Finally, we train a CatBoost classifier for reconginzing ASD and TD subjects. We also perform automatic tuning of model hyperparameters via multivariate Tree Parzen Estimation.

3.1 Features

All the features we considered have a solid theoretical foundation in psychological literature. According to Treisman's Feature Integration Theory (FIT) [33], in a visual search task the visual system first performs a preattentive analysis of the image, *i.e.* a fast scan where information are gathered, then focus its attention on the most salient regions of the image for high level processing of the content semantics. Psychological studies demonstrate that ASD individuals present impairments in both phases –namely, *visual exploration* and *semantics*– and thus we propose a set of features covering both of them. The complete list of features is reported in Table 1.

Visual Exploration Features. Literature highlighted significant differences regarding image exploration in individuals with ASD related to attention allocation processes, anomalies in oculomotor abilities, and more generally to visual-motor integration [3]. We first included two popular features already proved significant in recent literature: *number of gaze points* [34] and *time spent on image* [7].

A recent study identified a significant linear relationship between the fixation count and visual complexity of an image [14]. We predicted *image complexity* by fine tuning a MobileNetV2 model on the SAVOIAS [28] dataset, adding a dense layer at the tail of the network to output a complexity score.

In terms of saccades and fixation patterns, relevant differences between ASD and TD subjects is identified in [5]. Such difference is relevant because vision is suppressed during a saccade and new information is only acquired during fixations [6]. We computed the *percentage of saccades and fixations* of each scanpath using a simple velocity-threshold fixation identification (I-VT) algorithm. Indeed, the velocity profiles eye movements show essentially two distributions of velocities: low velocities for fixations, and high velocities for saccades. This aspect of saccadic eye movements makes velocity-based discrimination fairly straightforward and robust. Motor coordination deficits in gaze patterns due to poor use of visual feedback during motor learning tasks can also impact the saccadic and fixation velocities [17,30]. Thus, we computed four features representing the *average and standard deviation of both saccadic and fixation velocity*.

As for the type of fixations, previous studies suggested that both increased perseverative and regressive fixations can be a marker of autism [25]. Perserverative fixations are fixations occurred in succession towards the same region (*e.g.* the same object) and are thought identify attentional stickiness or mental disengagement, both appearing to be closely related to the ASD's cognitive style that struggle to orient attention to new stimuli [10]. Regressive fixations occur when a participant returns his/her gaze to a specific region that was already been explored. ASD subjects may revisit more previously viewed areas due to their inability to process information contained in that portion of the image [25]. In order to synthesize information about these type of fixations we computed both the *percentage of perseverative and regressive fixations.* A point of the scanpath is considered a perseverative fixation if it is located within a 85×85px area having the previous point as its center. Regressive fixations are points lying within an area of the same size centered in any of the previously visited fixation points (excluding the previous case).

Finally, geometric features are considered to account for the tendency of ASD subjects towards hypoarousal [9] and a strong center bias [34]. As for the former, we define the *image coverage* as the ratio between the union of squared boxes of 85 pixels centered in the each point of the scanpath and the area of the image itself. For the latter, we introduce a set of features focusing on the center of the image (*i.e.* a circle with radius of 192px). We compute the *percentage of points in image center* and the *percentage of points on ROI within image center*, *i.e.* the percentage of points lying within the image center and belonging to a semantic element of interest. The introduction of this feature allows us to identify whether a person is looking at a salient part of the image positioned in the center or he/she is in a situation of attentional disengagement and is looking, for example, at the background. Following the same rationale, we compute the *percentage of time spent in image center* and the *percentage of time spent on ROI within image center.* To further characterize gaze patterns, we compute the *average distance of points in the image center*, the *average distance of points on ROI in image center*, and the *average distance of points not on ROI in image center.*

Semantic Features. High level semantic features are meaningful objects or categories recognized by the humans, such categories include objects, people, animals and text. ASD individuals struggle to orient their attention to stimuli depicting humans or human interactions [11] and when they do, they focus on different parts of the body (such as hands) [19]. Although high functioning individuals present similar visual patterns to the TD counterparts, they show some differences in the order or amount of time spent looking at social areas of the image [34], leading to the conclusion that the temporal dimension will also play a fundamental role and should be considered when extracting features.

Table 1. List of features extracted from scanpaths of eye tracking data. Each feature is motivated by a literature study that proved its relevance for the diagnostic task.

Visual Exploration	Semantics
Time spent on image [7]	*presence of people, objects, background*
Number of gaze points [34]	*% of points on people, objects, background*
Image complexity [14]	*% of time on people, objects, background*
Percentage of saccades and fixations [5]	*latency people, objects, background*
Average and standard deviation of	*presence of body parts*
saccadic and fixation velocity [17,30]	*% of points on body parts*
% of perseverative fixations [25]	*% of time spent on body parts*
% of regressive fixations [25]	*latency on body parts*
Image coverage [9]	*presence of text*
% of points in image center	*% of points on text*
% of points on ROI within image center	*% of time spent on text*
% of time spent on image center	*latency of text*
% of time spent on ROI in image center	
avg. distance of points in image center	
avg. distance of points on ROI	
in image center	
avg. distance of points not on ROI	
in image center	

To identify the areas of stimuli that contain semantic information, *i.e.* high-level information that influences visual saliency such as objects or people, we first apply semantic segmentation using Detectron2 library[1]. Following, we use Bodypix 2.0 model[2] to further segment people and identify body parts (head, arms, hands, torso, legs, feet). Lastly, we use EasyOCR[3] to identify regions containing text.

Four features were computed for each segmentation class (*i.e.* for Detectron2: person, object, background; for Bodypix: head, arms, hands, torso, legs, feet; for EasyOCR: text): the *presence of class instances*, a binary value indicating wheter one instance of that class is represented in the image; the *percentage of gaze points* falling in that specific class; the *percentage of time spent* on a specific class; the *latency* of the specific class, defined as the timestamp of the first time the subject watched that specific class.

Feature Selection. In order to avoid overfitting and problems related to noisy features we apply feature selection before training the classifier. We use Boru-

[1] https://github.com/facebookresearch/detectron2.
[2] https://github.com/tensorflow/tfjs-models/tree/master/body-segmentation.
[3] https://github.com/JaidedAI/EasyOCR.

taSHAP[4], an algorithm that compares the importance of each feature in the original dataset to the importance of randomized versions of the same dataset, and selects features that are consistently more important than their randomized counterparts. BorutaSHAP extends Boruta [18] by integrating SHAP values, a model-agnostic measure of feature importance that overcomes limitations of Gini impurity used by the original algorithm.

3.2 CatBoost Classifier

CatBoost [27] is a type of gradient boosting method that is particularly well-suited for use in medical applications. This is a specific implementation of gradient boosting that uses decision trees as its base model and fashions a smart boosting strategy called ordered boosting. We chose CatBoost for two main reasons: (1) it allows an effective modelling of all the subgroups corresponding to the different phenotypes found in autism disorder; and (2) it provides an easy interpretation of results, which is always desirable in medical applications.

We tuned hyperparameters of the model by adopting a bayesian approach. We use a Sequential Model-Based Optimization (SMBO) algorithm with Multivariate Tree Parzen Estimator (TPE). It works by constructing a probability distribution over the hyperparameters of the black-box function, evaluating the function at the most promising points, and updating the distribution based on the results to guide the search towards better-performing regions of the hyperparameter space. Unlike independent TPE that estimates the densities of good/bad hyperparameters using the product of univariate Parzen window estimators, Multivariate TPE models them by a multivariate approach that captures the inter-dependencies among parameters. This is particularly relevant in tree based models as some parameters (*e.g.* the depth of trees) might be related to others (*e.g.* the number of leaf nodes).

4 Experiments

To validate our model we used the Saliency4ASD challenge dataset [12]. This dataset consists of scanpath data (with information on both position of gaze points and their duration) related to 28 children (14 ASD and 14 TD) with an average age of 8 years. Subjects looked at set of 300 natural images (in total) containing everyday life scenes that include people, animals, and objects. Some of them are static scenes while others depict complex social interactions. Eye tracking was performed using a Tobii T120 tracker, images were displayed for 3s on a 17-inch monitor at a distance between 50 and 80cm. In our evaluation we used 240 images for training and validating the model and 60 images for testing.

In addition to the analysis of standard classification metrics –*i.e.* accuracy, precision, recall, F_1-score, AUC-ROC– we analyzed the results of our method from an explainable AI (XAI) perspective by evaluating the impact of individual

[4] https://github.com/Ekeany/Boruta-Shap.

features through SHapley Additive exPlanations (SHAP) values. In machine learning SHAP values are used to assess features contribution to the outcome. To this aim, we adopted the approach proposed in [23] which consists in fitting a TreeExplainer and computing global and local information about the predictions.

We report binary classification results for our method and two state of the art competitors in Table 2. Our classifier outperforms competitors in discriminating between ASD individuals and TD controls reaching an accuracy of 71.77% on the test set. Our classifier is also well balanced in terms of precision and recall, with a discrepancy of only 2%.

Table 2. Model performance on test set compared to other models that used the same dataset and similar train-test split.

Model	Accuracy	Precision	Recall	F1-score	AUC-ROC
Wu *et al.* [36]	65.41%	–	0.66	–	–
Mazumdar *et al.* [24]	68.50%	–	–	–	–
Proposed model	71.77%	0.74	0.76	0.75	0.71

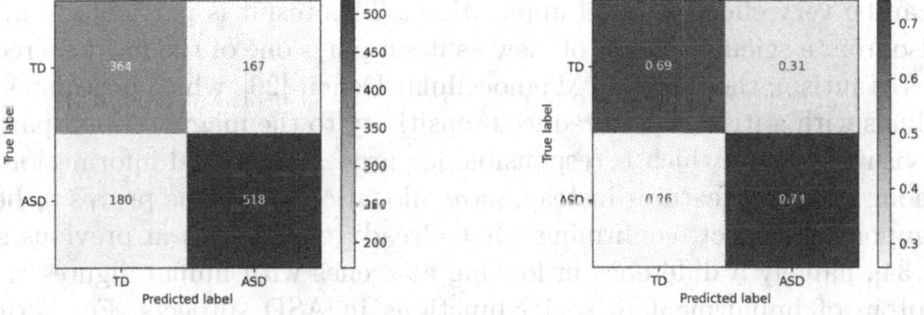

Fig. 1. Test set confusion matrix (Raw and Normalized data).

An in-depth analysis of the confusion matrices in Fig. 1 reveals that the model performs better in classifying ASD individuals than TD controls, in fact 74% of the total ASD individuals were correctly classified compared to 69% of healthy controls (TD). This is particularly interesting as it suggests that ASD individuals might exhibit distinctive visual patterns, whereas healthy controls adopt more heterogeneous behaviors when exploring a scene.

From an XAI perspective, results on both global and local information are considered. We should note that global information give hints about the feature importance while local information focus on how the features impact the prediction of the model, thus suggesting feature level patterns. Plots related to global and local information analysis are reported in Fig. 2 and 3 respectively.

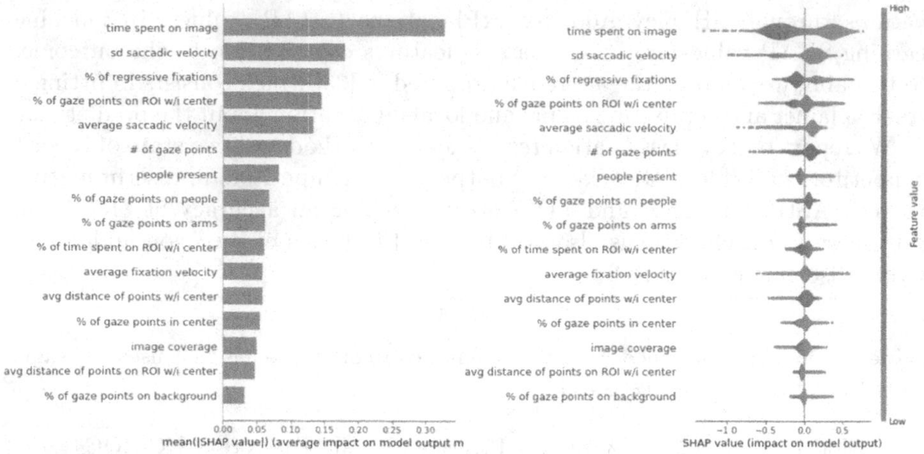

Fig. 2. Global feature importance

Fig. 3. Local explanation summary.

Global feature importance plot highlights that visual exploration features are more relevant for classification than the semantic ones suggesting that impairments in attention allocation, anomalies in oculomotor abilities, and more generally in visual-motor integration are prevalent patterns in ASD and that such patterns can be very effective in AI applications. This result is particularly important also from a scientific point of view as it supports one of the most accredited theories of autism: the theory of Magnocellular Deficit [26], which postulates that individuals with autism have a reduced sensitivity to the magnocellular pathway in the visual system, which is responsible for processing visual information.

Among semantic features instead, gaze allocation on people proves to be the most important aspect, confirming what already emerged from previous studies [29, 34], namely a difference in looking at scenes with human figures can be a symptom of impairment of social functions in ASD subjects. Furthermore, the preference for specific body parts distant from the face (*e.g.* arms) among the relevant features suggests that even when individuals with autism spectrum disorder (ASD) look at a person, they do so differently from TD individuals, prioritizing "less social" body parts. In this analysis, we should also consider features that were excluded by the feature selection algorithm, specifically semantic features which were almost entirely dropped. Considering body parts, only the percentage of gaze points on the arms was kept, while all other features were excluded. Another aspect that did not emerge is a difference in gaze patterns related to text and objects, *i.e.* salient areas with low social valence that might be preferred by ASDs as safe areas to look at. Finally, latency was not found to be relevant for any of the semantic features, unlike what recorded in [34].

Information about the time spent on image suggest that longer times have a high positive contribution to ASD prediction. This finding is compatible with anomaly in attention allocation and processing typical of ASD individuals [25] and this phenomenon is further confirmed by the fact that a high number of regressive fixation and high fixation velocity have a high positive contribution to ASD prediction. The combination of increased regressive fixations and higher fixation velocity suggests also that information about the areas the subject visited have not been properly processed, potentially because very fast fixations are close to saccades and might not consent information acquisition. The information provided by the model also confirms a static exploration style: an individual is more likely to be autistic if he/she shows low values of coverage and decreased saccadic velocity.

Another interesting aspect emerges from local information summary of center bias features. According to the violinplot, a high proportion of gaze points in the center of the image have positive contribution to ASD prediction, thus suggesting that this is common in autistic individuals. Furthermore, if a semantic content is present in the center of the image, the model considers a positive contribution to ASD prediction when less time is spent at looking at such ROI and when average distance of gaze points on the ROI is higher. This complex constellation of measures suggests that ASD subjects prefer central areas of the image, but if they encounter a semantic instance within that area they reorient the gaze in order to avoid exposure to social contents, or if not possible they focus on peripheral area of such content.

Finally, the model confirms what hypothesized analyzing global feature importance in respect of semantic features. No clear difference in the proportion of gaze point on people emerged, rather a difference in how ASDs look at people; according to the model a high proportion of gaze points on arms have a positive contribution to ASD prediction. This is an strong evidence of the preference for less social body parts such as arms when it comes to individuals with autism. Furthermore, a preference for background in autistic individuals is emerged (high proportion of gaze points on background contributes to ASD prediction). A qualitative example of these considerations is shown in Fig. 4.

Fig. 4. Example of difference in the scanpath of TD (left) and ASD (right) individuals regarding fixations on body parts (head, torso, and arms).

5 Conclusions

This research aimed to develop a model to assist in the diagnosis of Autism Spectrum Disorder (ASD) using gaze data. We developed a machine learning model with the goal of improving classification performance with respect to state-of-the-art approaches, identifying the most significant features among a vast pool of psychologically grounded candidates, and understanding how the values of these features influence prediction. We used the BorutaSHAP feature selection algorithm to select the most promising features and trained the model by a CatBoost classifier due to its high effectiveness and interpretability. After training the model, we fitted a TreeExplainer to extract the global feature importance and the local explanation summary. The former provided insights into the importance of features, while the latter illustrated how the values of individual features contributed to predict one class over the other. In terms of performance, our model achieved an outstanding accuracy of 71.77% on Saliency4ASD dataset. Based on these promising results, it is worth extending the analysis to a more exhaustive dataset with more subjects and a higher variability in terms of gender and age.

References

1. Alcañiz, M., et al.: Eye gaze as a biomarker in the recognition of autism spectrum disorder using virtual reality and machine learning: A proof of concept for diagnosis. Autism Res. **15**(1), 131–145 (2022)
2. American Psychiatric Association: Diagnostic and statistical manual of mental disorders: DSM-5. American psychiatric association Washington, DC (2013)
3. Apicella, F., Costanzo, V., Purpura, G.: Are early visual behavior impairments involved in the onset of autism spectrum disorders? insights for early diagnosis and intervention. Eur. J. Pediatr. **179**, 225–234 (2020)
4. Arru, G., Mazumdar, P., Battisti, F.: Exploiting visual behaviour for autism spectrum disorder identification. In: ICME Workshops (2019)
5. Bast, N., et al.: Saccade dysmetria indicates attenuated visual exploration in autism spectrum disorder. J. Child Psychol. Psychiatry **62**(2), 149–159 (2021)
6. Binda, P., Morrone, M.C.: Vision during saccadic eye movements. Ann. Rev. Vis. Sci. **4**, 193–213 (2018)
7. Chita-Tegmark, M.: Social attention in ASD: a review and meta-analysis of eye-tracking studies. Res. Dev. Disabil. **48**, 79–93 (2016)
8. Crane, L., Chester, J.W., Goddard, L., Henry, L.A., Hill, E.: Experiences of autism diagnosis: A survey of over 1000 parents in the united kingdom. Autism **20**(2), 153–162 (2016)
9. Cuve, H.C., Gao, Y., Fuse, A.: Is it avoidance or hypoarousal? a systematic review of emotion recognition, eye-tracking, and psychophysiological studies in young adults with autism spectrum conditions. Res. Autism Spectrum Disorders **55**, 1–13 (2018)
10. Dawson, G., Meltzoff, A.N., Osterling, J., Rinaldi, J., Brown, E.: Children with autism fail to orient to naturally occurring social stimuli. J. Autism Dev. Disord. **28**(6), 479–485 (1998)
11. Frazier, T.W., et al.: A meta-analysis of gaze differences to social and nonsocial information between individuals with and without autism. J. Am. Acad. Child Adolesc. Psychiatr. **56**(7), 546–555 (2017)

12. Gutiérrez, J., Che, Z., Zhai, G., Le Callet, P.: Saliency4asd: Challenge, dataset and tools for visual attention modeling for autism spectrum disorder. Signal Process. Image Commun. **92**, 116092 (2021)
13. Hodges, H., Fealko, C., Soares, N.: Autism spectrum disorder: definition, epidemiology, causes, and clinical evaluation. Translational Pediatrics **9**(1), S55 (2020)
14. Hu, R., Weng, M., Zhang, L., Li, X.: Art image complexity measurement based on visual cognition: evidence from eye-tracking metrics. In: AHFE (2021)
15. Huang, X., Shen, C., Boix, X., Zhao, Q.: SALICON: reducing the semantic gap in saliency prediction by adapting deep neural networks. In: ICCV (2015)
16. Jiang, M., Zhao, Q.: Learning visual attention to identify people with autism spectrum disorder. In: ICCV (2017)
17. Johnson, B.P., Rinehart, N.J., White, O., Millist, L., Fielding, J.: Saccade adaptation in autism and asperger's disorder. Neuroscience **243**, 76–87 (2013)
18. Kursa, M.B., Jankowski, A., Rudnicki, W.R.: Boruta-a system for feature selection. Fund. Inform. **101**(4), 271–285 (2010)
19. Kwon, M.K., Moore, A., Barnes, C.C., Cha, D., Pierce, K.: Typical levels of eye-region fixation in toddlers with autism spectrum disorder across multiple contexts. J. Am. Acad. Child Adolesc. Psychiatr. **58**(10), 1004–1015 (2019)
20. Lin, Y., Gu, Y., Xu, Y., Hou, S., Ding, R., Ni, S.: Autistic spectrum traits detection and early screening: A machine learning based eye movement study. J. Child Adolesc. Psychiatr. Nurs. **35**(1), 83–92 (2022)
21. Liu, W., Li, M., Yi, L.: Identifying children with autism spectrum disorder based on their face processing abnormality: A machine learning framework. Autism Res. **9**(8), 888–898 (2016)
22. Lorenz, T., Heinitz, K.: Aspergers-different, not less: Occupational strengths and job interests of individuals with asperger's syndrome. PLoS ONE **9**(6), e100358 (2014)
23. Lundberg, S.M., et al.: From local explanations to global understanding with explainable Ai for trees. Nature Mach. Intell. **2**(1), 56–67 (2020)
24. Mazumdar, P., Arru, G., Battisti, F.: Early detection of children with autism spectrum disorder based on visual exploration of images. Signal Process: Image Commun. **94**, 116184 (2021)
25. Nayar, K., Shic, F., Winston, M., Losh, M.: A constellation of eye-tracking measures reveals social attention differences in ASD and the broad autism phenotype. Molecular Autism **13**(1), 1–23 (2022)
26. Plaisted Grant, K., Davis, G.: Perception and apperception in autism: rejecting the inverse assumption. Philos. Trans. Royal Soc. B: Biolog. Sci. **364**(1522), 1393–1398 (2009)
27. Prokhorenkova, L., Gusev, G., Vorobev, A., Dorogush, A.V., Gulin, A.: Catboost: unbiased boosting with categorical features. In: Advances in Neural Information Processing Systems 31 (2018)
28. Saraee, E., Jalal, M., Betke, M.: Savoias: A diverse, multi-category visual complexity dataset. arXiv preprint arXiv:1810.01771 (2018)
29. Sasson, N.J., Elison, J.T., Turner-Brown, L.M., Dichter, G.S., Bodfish, J.W.: Brief report: circumscribed attention in young children with autism. J. Autism Dev. Disord. **41**(2), 242–247 (2011)
30. Schmitt, L.M., Cook, E.H., Sweeney, J.A., Mosconi, M.W.: Saccadic eye movement abnormalities in autism spectrum disorder indicate dysfunctions in cerebellum and brainstem. Molecular Autism **5**(1), 1–13 (2014)
31. Startsev, M., Dorr, M.: Classifying autism spectrum disorder based on scanpaths and saliency. In: ICME Workshops (2019)

32. Tao, Y., Shyu, M.L.: SP-ASDNet: CNN-LSTM based ASD classification model using observer scanpaths. In: ICME Workshops (2019)
33. Treisman, A.M., Gelade, G.: A feature-integration theory of attention. Cogn. Psychol. **12**(1), 97–136 (1980)
34. Wang, S., et al.: Atypical visual saliency in autism spectrum disorder quantified through model-based eye tracking. Neuron **88**(3), 604–616 (2015)
35. Waye, M.M.Y., Cheng, H.Y.: Genetics and epigenetics of autism: A review. Psychiatry Clin. Neurosci. **72**(4), 228–244 (2018)
36. Wu, C., Liaqat, S., Cheung, S.c., Chuah, C.N., Ozonoff, S.: Predicting autism diagnosis using image with fixations and synthetic saccade patterns. In: ICME Workshops (2019)

Target-Driven One-Shot Unsupervised Domain Adaptation

Julio Ivan Davila Carrazco[1,3](✉), Suvarna Kishorkumar Kadam[1], Pietro Morerio[1], Alessio Del Bue[1], and Vittorio Murino[1,2,4]

[1] Pattern Analysis and Computer Vision (PAVIS), Italian Institute of Technology, Genoa, Italy
{julio.davila,suvarna.kadam,pietro.morerio,alessio.delbue, vittorio.murino}@iit.it

[2] Department of Computer Science and Technology, Bioengineering, Robotics and Systems Engineering, University of Genova, Genoa, Italy

[3] Department of Marine, Electrical, Electronic and Telecommunications Engineering, University of Genoa, Genoa, Italy

[4] Department of Computer Science, University of Verona, Verona, Italy

Abstract. In this paper, we introduce a novel framework for the challenging problem of One-Shot Unsupervised Domain Adaptation (OS-UDA), which aims to adapt to a target domain with only a single unlabeled target sample. Unlike existing approaches that rely on large labeled source and unlabeled target data, our Target-Driven One-Shot UDA (TOS-UDA) approach employs a learnable augmentation strategy guided by the target sample's style to align the source distribution with the target distribution. Our method consists of three modules: an augmentation module, a style alignment module, and a classifier. Unlike existing methods, our augmentation module allows for strong transformations of the source samples, and the style of the single target sample available is exploited to guide the augmentation by ensuring perceptual similarity. Furthermore, our approach integrates augmentation with style alignment, eliminating the need for separate pre-training on additional datasets. Our method outperforms or performs comparably to existing OS-UDA methods on the Digits and DomainNet benchmarks.

Keywords: Unsupervised domain adaptation · Data augmentation · One-Shot

1 Introduction

Training deep learning models typically requires the availability of a large amount of data, which also represents a core problem for applications in which such information is not easily accessible. Hence, there is a need for methods to effectively learn from either fewer or alternative data that is easily accessible. The situation becomes even more challenging when we have to deal with different "domains", i.e., when data distributions in training (source domain) and test

G. L. Foresti et al. (Eds.): ICIAP 2023, LNCS 14233, pp. 87–99, 2023.
https://doi.org/10.1007/978-3-031-43148-7_8

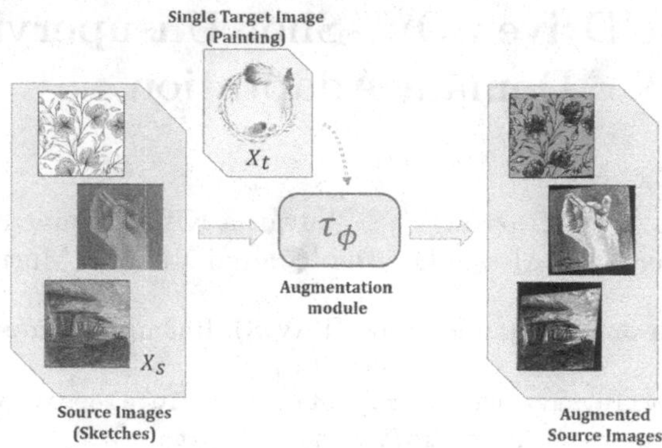

Fig. 1. Target-driven augmentation of source samples with TOS-UDA. τ is a learnable augmentation function with parameters ϕ that transforms the source images into target-like images (Sketches to Painting in this case).

(target domain) are different. The term 'domain' in this context is not clearly defined, and could roughly be identified with "a set of features", such as color or texture, characterizing samples of a specific dataset. The typical task here is to properly learn from the *labeled* source data and the *unlabeled* target data, i.e., how to transfer knowledge from source to target, the latter being also our test scenario. This is called unsupervised domain adaptation (UDA). A much less investigated, but more challenging task is to explore what happens when the source and/or target data are just a few samples, or even just a single data item. In these cases, we do not have enough statistics to capture the data variability, and then to properly generalize to new data. Hence, we might resort to methods to generate new or surrogate data from the few samples available in order to execute proper training. These UDA scenarios are named Few-Shot UDA (FS-UDA) [22,23] or, to an extreme case, One-Shot UDA (OS-UDA) [7,11]. FS-UDA denotes the presence of a few (labeled) samples in the source domain only, while there is enough unlabeled data in the target domain. OS-UDA instead assumes the presence of only one (unlabeled) sample in the target domain. FS-UDA and OS-UDA are difficult to learn tasks due to insufficient training data. Existing methods for solving FS-UDA, OS-UDA are quite limited. To the best of our knowledge, there is just one OS-UDA method, ASM [11] that performs UDA for classification and segmentation tasks.

In this paper, we introduce a new OS-UDA method where a single target sample is available to align the source domain with the target domain. We design a novel augmentation method implementing gradual transformations of the source images driven by the similarity with target's style. Our augmentation strategy achieves diverse, but controlled augmentations using a custom adversarial loss that rewards stronger augmentations but, at the same time, also penalizes too large deviations from the target's style. These diverse augmentations finally support robust adaptation to the target domain. In summary, our main contributions are as follows.

- We present a novel approach to tackle OS-UDA task, addressing the extreme case of target domain adaptation when only one unlabeled target sample is available. Our method performs target-driven augmentation of source images in an adversarial manner, and is the first end-to-end OS-UDA method that does not require specially pretrained style transfer modules.
- We guide source image augmentations with a style alignment module that controls the augmentation parameters while enforcing the style similarity between source and target domains.
- We report comparable, and in few cases, state-of-the-art results for OS-UDA on Digits and DomainNet benchmarks.

The rest of this paper is structured as follows: Sect. 2 reviews the literature on UDA, FS-UDA, OS-UDA, and augmentation methods. Section 3 presents our proposed framework and training strategy. Section 4 outlines the benchmarks, baselines, experimental results, and ablation study. Finally, Sect. 5 summarizes our approach and provides concluding remarks.

2 Related Work

Unsupervised domain adaptation (UDA) methods leverage the available data for learning and can face different scenarios where either the source, target or both domains have scarce data. Data augmentation is often the *de facto* approach to address data scarcity, as it employs diverse transformation strategies to increase the variability of available data. Our proposed approach is based on a novel adversarial augmentation strategy for UDA, and therefore we briefly discuss related work in both areas.

2.1 Data Augmentation

Image augmentation methods are broadly categorized as model-free, model-based, and optimization-based [21]. The model-free methods use classic image processing to alter geometric or color information. Geometric transformations [18] are often applied to a single image by flipping, rotating, or cropping it. Model-based image augmentation methods pre-train generative models such as GANs [6,12] and variants, to generate images. The distribution of generated images is expected to be similar to the original dataset. The optimization-based methods implement trainable augmentation strategies [3,4,17] that learn the transformations, and are usually preferred as they do not need to be configured for specific datasets. While data augmentation mostly improves model generalization, it can result in transformations that are too extreme [17]. Our proposed approach is optimization-based that uses the single target image's style to control the source image augmentations. Our novel target-driven augmentation strategy is inspired by the fact that we can separate the style from the content of the available single target image, and use it to adjust the transformations to be applied to source domain images.

2.2 Unsupervised Domain Adaptation

Most common UDA settings transfer knowledge from a label-rich source domain to unlabeled target domain, where the source and target domains share the same label space (i.e., same classes). Many UDA methods try to find a shared feature space where the overlap (confusion) between the source and target distribution is maximum. In this way, for any given sample projected in such space, it is often difficult to discriminate if it belongs to the source or target domain. UDA methods e.g. [2,19,20] assume that only unlabeled target samples are available. Some UDA methods focus on expanding the available target data to train their models, e.g., Volpi et al. [19] trained a domain-invariant feature extractor by applying feature augmentation using GANs. Minghao et al. [20] introduced Domain Mixup (DM-ADA), which uses a GAN approach and mixup [24] to learn a more continuous domain-invariant latent space. Moreover, Few-Shot Unsupervised Domain Adaption (FS-UDA) operates in a UDA setting where unlabeled target images are accessible for training and the number of labeled samples in the source domain is small [22,23]. One-shot UDA is a more constrained UDA scenario that assumes the presence of *only one* (unlabeled) sample in the target domain. The literature focusing on solving OS-UDA is quite limited. Luo et al. [11] presented Adversarial Style Mining (ASM) that applies style transfer to generate new samples of the target domain. ASM applies domain randomization to generalize better for OS-UDA, and needs an external dataset (WikiArts) to pre-train its style transfer model. Our approach is distinct as we do not need any external dataset to pre-train style transfer module. In our work, we exploit the perceptual similarity in style that exists between the source and target, to learn useful transformations while augmenting source samples. Unlike ASM [11] the augmentations generated by our method are affected solely by target image's style.

3 Method

Adapting a model by exploiting a *single* target sample is a quite challenging task. Our method tackles it by learning transformations for source domain samples to mimic the style of the target domain. A classifier trained with such augmented data is expected to generalize better to the target distribution. Notably, the problem here is how to adapt to the target distribution when the only sample available does not constitute sufficient statistics to generate a reliable distribution. So, our envisioned solution is to implement adversarial augmentation to learn robust transformations. To this end, we introduce a novel "style alignment" module to guide and control the augmentations using a single target sample. This guiding mechanism is not limited by the source distribution. Consequently, the augmented samples will be able to resemble the target distribution. In consequence, a classifier trained on augmented samples will be able to classify correctly samples from the target domain.

Our proposed model is composed of three modules: an augmentation module, a style alignment module, and a classifier. The augmentation module tries to

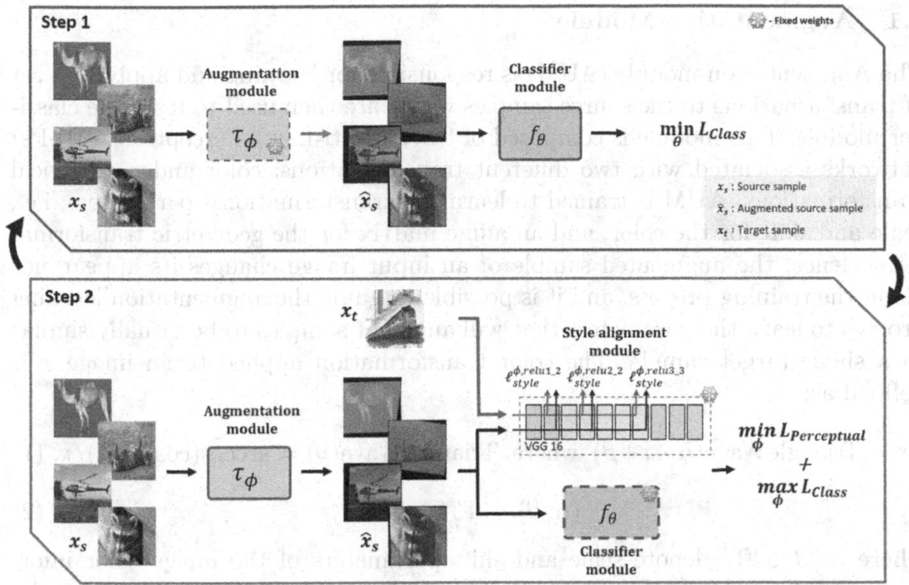

Fig. 2. Training of Target-Driven OS-UDA that alternatively updates the Classifier f_θ (Step-1) and Augmentation module T_ϕ (Step-2). Step 1 is repeated n times before step 2, where n is a hyperparameter.

learn the parameters of a set of transformations (e.g., color shift, geometrical transformation) that are applied to the source samples (see Fig. 2, Step 1). The style alignment module helps to drive the augmentation learning process by evaluating the perceptual similarities in style between the augmented source and the single target sample (see Fig. 2, Step 2). To train this architecture, we adopt a two-step strategy that alternates between updating the classifier and the augmentation modules (see Fig. 2). In the first step, the classifier is trained using augmented samples and a classification loss. In the second step, the augmentation module is trained by using the style alignment module and the classifier in an adversarial manner. The two-step strategy was firstly used by TeachAugment [17]. Unlike TeachAugment method, which also implements an adversarial augmentation mechanism, our method does not need a separate teacher model for controlling the augmentations. The use of a teacher limits the range of possible augmentations. This is because the augmented samples are bounded to be correctly classified by the teacher. Therefore, the distribution of the augmented samples needs to be similar to the source distribution. Our style alignment module allows better control for driving the augmentations towards a specific target. The architecture's modules and the two-step strategy will be explained further in the next subsections.

3.1 Augmentation Module

The Augmentation module (AUM) is responsible for learning and applying a set of transformations to the source samples which are then used to train the classifier module. This module is composed of three Multi-Layer Perceptrons (MLPs) networks associated with two different transformations: color and geometrical transformations. AUM is trained to learn the transformations' parameters, i.e., scale and shift for the color, and an affine matrix for the geometric transformation. Hence, the augmented sample of an input image changes its appearance along the training process, and it is possible to guide the augmentation learning process to learn the parameters that will augment samples to be visually similar to a single target sample. The color transformation applied to an image x is defined as:

$$\hat{x} = \text{TriangleWave}(\alpha \odot x + \beta) \text{ where } \text{TriangleWave}(p) = \arccos(\cos(p \cdot \pi))/\pi \quad (1)$$

$$(\alpha, \beta) = \mathrm{T}_{\phi}^{c}(x, z, c), \quad (2)$$

where α, $\beta \in \mathbb{R}^3$ denote scale and shift parameters of the image color information generated by the augmentation module sub-network T_{ϕ}^{c}. The TriangleWave function transforms the input variable p into a triangular waveform in the range $[0, 1]$, and $\odot$ denotes the element-wise multiplication. The sub-network T_{ϕ}^{c} takes as input, an image x, random noise $z \sim \mathcal{N}(0, I_N)$, where $\mathcal{N}(0, I_N)$ is N-dimensional unit Gaussian distribution, and c that represents one-hot vector encoding for class label that serves as a context for the module (see Fig. 3, Color transformation). The geometrical module applies an affine transformation to augment the input data. The transformation is defined as:

$$\hat{x} = \text{Affine}(x, A + I) \text{ , where } A = \mathrm{T}_{\phi}^{g}(z, c) \quad (3)$$

where Affine$(\cdot)$ denotes an affine operation applied to x, A denotes an affine 2×3 matrix generated by the augmentation module T_{ϕ}^{g}, $\mathrm{T}_{\phi}^{g}(\cdot)$ is the augmentation module sub-network which is a MLP that receives as inputs, a noise vector $z \sim \mathcal{N}(0, I_N)$ and c, which is a one-hot encoding of the sample class, and generates the parameters of the affine transformation (see Fig. 3, Geometrical transformation).

3.2 Style Alignment Module

The purpose of the Style Alignment Module (SAM) is to drive the learning process of the augmentation module. For this module, a VGG-16 [16] pretrained on ImageNet is used. This module leverages the style transfer methodology by exploiting a style loss. Johnson et al. [9] introduced the perceptual loss to replace the traditional pixel-by-pixel loss that was used for style transfer. The perceptual loss compares a target sample, a source sample, and the augmented source sample to calculate the perceptual differences between them. This loss focuses on two aspects of the input image, its style and content. The style refers to the texture information (e.g., color, texture, common patterns, etc.) present in the image.

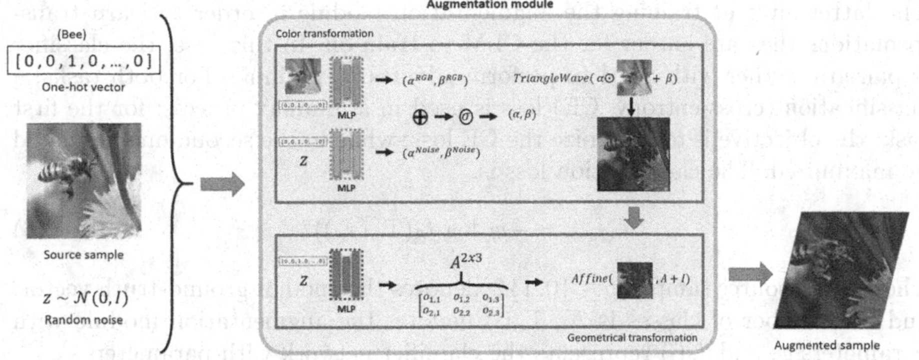

Fig. 3. Transformations performed within augmentation module: a) Color, b) Geometrical transformation. Color transformations are applied first, followed by geometrical transformation.

The style loss is calculated as the squared Frobenius norm between the Gram matrices of the augmented source and the target sample. The Gram matrix serves to extract the style representation via the correlations between different filters of the extracted feature maps [5] at several network layers. The Gram matrix is defined as:

$$G_j(x) = \frac{1}{C_j H_j W_j} \sum_{h=1}^{H_j} \sum_{w=1}^{W_j} \mathcal{H}_j(x) \mathcal{H}_j^T(x) \tag{4}$$

where x is an input image, $\mathcal{H}_j$ represents the feature map from the j-layer of the SAM; C_j, H_j, and W_j denote the number of channels, the height, and the width of $\mathcal{H}_j$, respectively. The style loss is then defined as:

$$L_{style} = \sum_j \ell_j^{style}(\mathrm{T}_\phi(x_s), x_t) \quad \text{and} \quad \ell_j^{style}(\hat{x_s}, x_t) = \|G_j(\hat{x_s}) - G_j(x_t)\|_F^2 \tag{5}$$

where x_t and $\hat{x_s}$ represent the target sample and the augmented source sample, respectively. $\mathrm{T}_\phi(\cdot)$ represents the AUM with parameters ϕ, and j is the indexes of the selected layers of the SAM related to the style loss terms (more details in Sect. 4). The layers selection is based on the well-known fact that lower layers preserve the texture information better[5]. The total style loss L_{style} is the sum of the individual style losses of the selected layers of the SAM.

3.3 Classifier Module

The classifier module (CLM) has a twofold task, namely, to classify augmented samples and to collaborate with the SAM for training the AUM. The former task focuses on learning how to correctly classify the augmented source samples.

The latter aims at training the augmentation module in order to learn transformations that are harder for the CLM to train on. In this case, the classifier is paired together with SAM to perform adversarial training. For both tasks, a classification (cross-entropy, CE) loss is used in a minmax process: for the first task, the objective is to minimize the CE loss, while in the second one, it should be maximized. The classification loss is:

$$L_{class} = -y_k \log f_\theta(\mathrm{T}_\phi(x_s)) \tag{6}$$

where x is a source sample, $y \in \{0,1\}^K$ denotes the one-hot ground-truth vector, and the number of classes is K, $\mathrm{T}_\phi(\cdot)$ denotes the augmentation module with parameters ϕ, and $f_\theta(\cdot)$ represents the classifier network with parameters θ.

3.4 Two-Step Training Process

Our training method optimizes the augmentation and classifier modules end-to-end with an alternate 2-step strategy (refer to Fig. 2). In Step 1, we train the CLM, while AUM is frozen. The input for CLM is the augmented source image, $\hat{x}$, passed through the augmentation module, while SAM is not used. To optimize CLM, the classification loss (Eq. 6) is minimized (supervised learning). In Step 2, AUM is then optimized. Its purpose is to learn a transformation to apply onto the source samples. In this step, the classifier aims at maximizing the classification loss on the augmented samples. By doing so, AUM will learn stronger transformations for the source samples, but this may not be sufficient to perform domain adaptation on the target. That is why our the style alignment module is introduced so as to improve the classification on the target domain. In our case, SAM drives the learning process by imposing a style loss (Eq. 5). As a result, the learned transformations will generate samples that resemble the style of the target domain. Finally, the augmentation module parameters ϕ are updated by adversarial training by maximizing the classification loss (Eq. 6) and minimizing the style loss (Eq. 5).

4 Experiments

4.1 Benchmarks

To evaluate our approach, we make use of two well-known DA benchmarks: Digits and DomainNet [14]. **Digits.** It is composed of three datasets: MNIST [10], USPS [8], and SVHN [13] datasets. They represent a collection of images of digits from 0 to 9. The evaluation is done by testing the model's performance on M→S, U→S, and M→U tasks. **DomainNet.** We follow the setup used in [15], in which only four domains (Real (R), Clipart (C), Painting (P), and Sketch (S)) are used, with 126 classes only. The evaluated tasks are: R→P, R→C, R→S, P→C, P→S, C→S, and S→P. In Fig. 4, sample images belonging to the aforementioned benchmarks are presented.

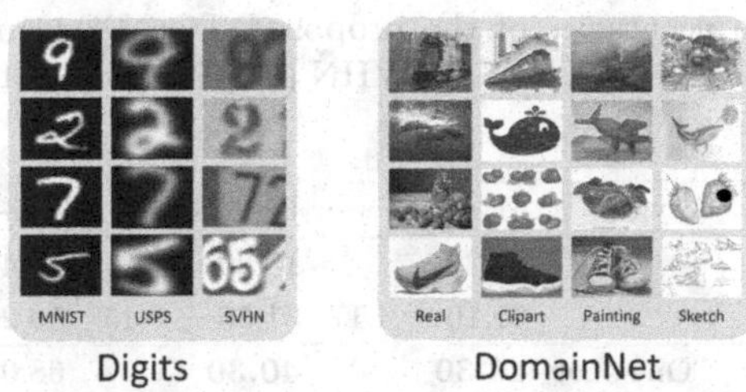

Fig. 4. Domain adaptation benchmarks used for evaluation of TOS-UDA: Digits with three domains (MNIST, USPS, and SVHN) and DomainNet with four domains (Real, Clipart, Painting, and Sketch).

4.2 Baseline and Comparative Methods

A number of baselines have been tested in order to validate the most important components of our approach and their contributions. **Source Only (SO).** It is a model trained using only the labeled source images with no adaptation. **Adversarial Style Mining (ASM).** It is a model that also focuses on OS-UDA. Similarly to us, ASM applies style transfer to generate target data for training. But their process requires pre-training a style generator module in an external dataset [11]. This is the only work with which we can directly compare to since it assumes the same starting hypotheses of our work.

4.3 Setup

To train our model, we used two classifier models. For Digits, the classifier consists of 2 blocks of a convolutional layer and a max pooling operation layer followed by three fully connected layers [19]. For DomainNet, the classifier network is a ResNet-101 pretrained on the source dataset. The augmentation module consists of three multi-layer perceptron (MLP) networks. For the style alignment module, a VGG-16 [16] pre-trained on ImageNet was used. In VGG-16, only the features maps from the layers *relu*1_2, *relu*2_2, *relu*3_3, and *relu*4_3 were used as we only apply the loss for the style in our experiments.

4.4 Results

In Table 1, the results on Digits are presented. The table shows the average classification accuracy and the related standard deviation, as we ran each experiment five times. Although our main objective is UDA with just one-shot, we also carried out experiments with more target samples (32). We randomly selected these samples at the start of the experiment. To calculate the style loss, each target sample was randomly paired with a sample from the current batch. For MNIST to SVHN (M→S) and USPS to SVHN (U→S), our approach obtains comparable results with respect to the state-of-the-art with differences lower than 1.0%. Although these results do not represent an improvement over the

Table 1. Classification accuracies of the proposed TOS-UDA method on three DA tasks: MNIST to SVHN (M→S), USPS to SVHN (U→S), and MNIST to USP (M→U).

Model	Type	M → S	U → S	M → U	Average
Source only	-	20.30	15.30	65.40	33.67
TeachAugment [17]	–	31.10 ± 4.17	31.56 ± 2.15	50.26 ± 2.16	37.64
ASM [11]	One-shot	**46.30**	**40.30**	68.00	51.53
OST [1]	One-shot	42.50	34.00	74.80	50.43
Our Model	One-shot	45.03 ± 1.86	36.67 ± 2.51	79.45 ± 6.73	53.72
Our Model (32 Targets)	Few-shot	45.88 ± 3.62	**39.96** ± 3.63	**79.97** ± 4.27	**55.27**

Table 2. Classification accuracies of the proposed TOS-UDA method on DomainNet across seven DA task focusing on four domains: Real (R), Clipart (C), Painting (P), and Sketch (S).

Model	Type	R → C	R → P	R → S	P → C	P → R	C → S	S → P	Average
Source only	–	56.59 ± 0.79	56.79 ± 0.50	46.25 ± 0.86	**55.55** ± 0.83	**66.20** ± 0.72	52.07 ± 1.01	44.81 ± 1.59	54.04
TeachAugment	–	53.84 ± 0.56	56.70 ± 0.59	46.70 ± 1.34	50.40 ± 1.27	58.64 ± 0.68	50.52 ± 0.09	44.89 ± 0.83	51.67
ASM	One-shot	39.74 ± 0.56	46.39 ± 1.53	31.37 ± 5.51	4.31 ± 0.60	5.87 ± 2.33	37.12 ± 1.12	19.67 ± 2.99	26.35
Our Model	One-shot	**58.11** ± 0.38	**58.57** ± 0.20	**49.87** ± 0.97	54.24 ± 0.62	62.72 ± 0.32	**52.88** ± 0.25	**47.94** ± 1.12	**54.90**

state of the art, they demonstrate that our method can achieve similar results as ASM without the need of external datasets and pretraining. For MNIST to USPS (M→U), our approach in both versions (1 and 32 targets) performs better than the baseline 74.80% with an increase of +4.66% and +5.17% for the 1-target and the 32-targets versions respectively.

In Table 2, we present results for DomainNet. We compare our model against three baselines: Source only (SO), TeachAugment and Adversarial Style Mining (ASM). We generate the results for these three baselines given that to the best of our knowledge is the first time that the DomainNet benchmark is used for OS-UDA. We report the accuracy and standard deviation for each task and their average across all tasks. Compared to ASM, our approach obtains higher accuracy in all seven DA tasks, setting the SoTA for five of these tasks. For P→C and P→R, our approach performs poorly by having results similar to the SO baseline. This indicates that for some domains, the augmentation module may struggle to learn transformations for the domain complex style from just 1 target.

4.5 Ablation Analysis

We conducted ablations to analyze the contribution of the augmentation and style alignment modules. Specifically, we removed the loss for style alignment and the adversarial loss component in two separate experiments. Table 3 shows the performance of our model compared to a pre-trained SO model and our method with excluded losses. Our method outperformed the SO model even with only one of the losses guiding the augmentations. However, the best results

Table 3. Ablation analysis of the proposed TOS-UDA approach on Digits benchmark

Model Setup	M → S	U → S	M → U
Source only	20.30	15.30	65.40
Our w/o style. loss	22.43 ± 1.71	28.45 ± 1.02	74.74 ± 3.01
Our w/o classif. loss	20.55 ± 1.28	20.93 ± 2.93	71.22 ± 10.48
Our method	45.03 ± 1.86	36.67 ± 2.51	79.45 ± 6.73

were achieved when both losses were used, confirming the contribution of both modules.

5 Conclusions

In this paper, we present our novel Target-Driven One-Shot Unsupervised Domain Adaptation (TOS-UDA) approach. Our method focuses on solving OS-UDA, which is the problem of performing domain adaptation with only one unlabeled target sample. Our method gradually augments the source samples to match them in style with the available target. To guide the augmentation process, adversarial training was used to encourage the augmentation module to learn diverse augmentations, while reaching a style similarity to the target sample. In this way, our augmentation module can learn effective transformations for OS-UDA. Further, our proposed training pipeline is simpler and end-to-end as compared to existing OS-UDA method such as ASM [11]. We tested our approach for image classification tasks in two well-known domain adaptation benchmarks: Digits and DomainNet. We demonstrated that our method performs better that the selected baselines in almost all the DA tasks. In the future, we will focus on extending the augmentation module to perform more specialized augmentations as the basic geometric and color augmentations may not be sufficient to simulate the style complexity of certain domains.

References

1. Benaim, S., Wolf, L.: One-shot unsupervised cross domain translation. In: Advances in Neural Information Processing Systems 31 (2018)
2. Chen, M., Zhao, S., Liu, H., Cai, D.: Adversarial-learned loss for domain adaptation. In: Proceedings of the AAAI Conference on Artificial Intelligence, vol. 34, pp. 3521–3528 (2020)
3. Cubuk, E.D., Zoph, B., Mane, D., Vasudevan, V., Le, Q.V.: Autoaugment: Learning augmentation strategies from data. In: Proceedings of the IEEE/CVF Conference on Computer Vision and Pattern Recognition, pp. 113–123 (2019)
4. Cubuk, E.D., Zoph, B., Shlens, J., Le, Q.V.: Randaugment: practical automated data augmentation with a reduced search space. In: Proceedings of the IEEE/CVF Conference on Computer Vision and Pattern Recognition Workshops, pp. 702–703 (2020)

5. Gatys, L.A., Ecker, A.S., Bethge, M.: Image style transfer using convolutional neural networks. In: Proceedings of the IEEE Conference on Computer Vision and Pattern Recognition, pp. 2414–2423 (2016)
6. Goodfellow, I., et al.: Generative adversarial networks. Commun. ACM **63**(11), 139–144 (2020)
7. Gu, M., Vesal, S., Kosti, R., Maier, A.: Few-shot unsupervised domain adaptation for multi-modal cardiac image segmentation. In: Bildverarbeitung für die Medizin 2022. I, pp. 20–25. Springer, Wiesbaden (2022). https://doi.org/10.1007/978-3-658-36932-3_5
8. Hull, J.: A database for handwritten text recognition research. IEEE Trans. Pattern Anal. Mach. Intell. **16**(5), 550–554 (1994). https://doi.org/10.1109/34.291440
9. Johnson, J., Alahi, A., Fei-Fei, L.: Perceptual losses for real-time style transfer and super-resolution. In: Leibe, B., Matas, J., Sebe, N., Welling, M. (eds.) ECCV 2016. LNCS, vol. 9906, pp. 694–711. Springer, Cham (2016). https://doi.org/10.1007/978-3-319-46475-6_43
10. LeCun, Y., Cortes, C.: MNIST handwritten digit database (2010). https://yann.lecun.com/exdb/mnist/
11. Luo, Y., Liu, P., Guan, T., Yu, J., Yang, Y.: Adversarial style mining for one-shot unsupervised domain adaptation. In: Advances in Neural Information Processing Systems 33, pp. 20612–20623 (2020)
12. Mirza, M., Xu, B., Warde-Farley, D., Ozair, S., Courville, A., Bengio, Y., Goodfellow, I.J., Pouget-Abadie, J.: Generative adversarial nets. In: Proceedings of the Advances in Neural Information Processing Systems, vol. 27, pp. 2672–2680 (2014)
13. Netzer, Y., Wang, T., Coates, A., Bissacco, A., Wu, B., Ng, A.Y.: Reading digits in natural images with unsupervised feature learning (2011)
14. Peng, X., Bai, Q., Xia, X., Huang, Z., Saenko, K., Wang, B.: Moment matching for multi-source domain adaptation. In: Proceedings of the IEEE/CVF International Conference on Computer Vision, pp. 1406–1415 (2019)
15. Saito, K., Kim, D., Sclaroff, S., Darrell, T., Saenko, K.: Semi-supervised domain adaptation via minimax entropy. In: Proceedings of the IEEE/CVF International Conference on Computer Vision (ICCV) (October 2019)
16. Simonyan, K., Zisserman, A.: Very deep convolutional networks for large-scale image recognition. arXiv preprint arXiv:1409.1556 (2014)
17. Suzuki, T.: Teachaugment: Data augmentation optimization using teacher knowledge. In: Proceedings of the IEEE/CVF Conference on Computer Vision and Pattern Recognition (CVPR), pp. 10904–10914 (June 2022)
18. Taylor, L., Nitschke, G.: Improving deep learning with generic data augmentation. In: 2018 IEEE Symposium Series on Computational Intelligence (SSCI), pp. 1542–1547. IEEE (2018)
19. Volpi, R., Morerio, P., Savarese, S., Murino, V.: Adversarial feature augmentation for unsupervised domain adaptation. In: Proceedings of the IEEE Conference on Computer Vision and Pattern Recognition, pp. 5495–5504 (2018)
20. Xu, M., et al.: Adversarial domain adaptation with domain mixup. In: Proceedings of the AAAI Conference on Artificial Intelligence, vol. 34, pp. 6502–6509 (2020)
21. Xu, M., Yoon, S., Fuentes, A., Park, D.S.: A comprehensive survey of image augmentation techniques for deep learning. arXiv preprint arXiv:2205.01491 (2022)
22. Yang, W., Yang, C., Huang, S., Wang, L., Yang, M.: Few-shot unsupervised domain adaptation via meta learning. In: 2022 IEEE International Conference on Multimedia and Expo (ICME), pp. 1–6 (2022). https://doi.org/10.1109/ICME52920.2022.9859804

23. Yue, X., et al.: Prototypical cross-domain self-supervised learning for few-shot unsupervised domain adaptation. In: Proceedings of the IEEE/CVF Conference on Computer Vision and Pattern Recognition, pp. 13834–13844 (2021)
24. Zhang, H., Cisse, M., Dauphin, Y.N., Lopez-Paz, D.: mixup: Beyond empirical risk minimization. arXiv preprint arXiv:1710.09412 (2017)

Combining Identity Features and Artifact Analysis for Differential Morphing Attack Detection

Nicolò Di Domenico, Guido Borghi(✉), Annalisa Franco, and Davide Maltoni

Dipartimento di Informatica - Scienza e Ingegneria (DISI), University of Bologna, 47521 Cesena, Italy
{nicolo.didomenico,guido.borghi,annalisa.franco,davide.maltoni}@unibo.it

Abstract. Due to the importance of the Morphing Attack, the development of new and accurate Morphing Attack Detection (MAD) systems is urgently needed by private and public institutions. In this context, D-MAD methods, *i.e.* detectors fed with a trusted live image and a probe tend to show better performance with respect to S-MAD approaches, that are based on a single input image. However, D-MAD methods usually leverage the identity of the two input face images only, and then present two main drawbacks: they lose performance when the two subjects look alike, and they do not consider potential artifacts left by the morphing procedure (which are instead typically exploited by S-MAD approaches). Therefore, in this paper, we investigate the combined use of D-MAD and S-MAD to improve detection performance through the fusion of the features produced by these two MAD approaches.

Keywords: Morphing Attack · Morphing Attack Detection · Differential MAD (D-MAD) · Single image MAD (S-MAD) · Feature Fusion

1 Introduction

Through an image morphing algorithm, it is possible to merge two images into one. In particular, this process can be applied to face images to create an intermediate one which includes facial characteristics of the two contributing subjects. A *Morphing Attack* [10] employs the aforementioned process to break the unique link between an official document and its owner: specifically, a subject with no criminal records (*accomplice*) can apply for a passport using a morphed mugshot picture to conceal the identity of a *criminal*. Indeed, several studies [29,32] have shown the effectiveness of this attack, capable of fooling both the human control (*e.g.* a police officer) and the current commercial Facial Recognition Systems.

In particular, the morphing attack poses a significant security threat to Automated Border Control (ABC) gates located at international airports. These systems are designed to automatically verify the facial image stored in the electronic Machine Readable Travel Document (eMRTD) against a live image captured at the gate. Indeed, the presence of a morphed face can effectively bypass these

G. L. Foresti et al. (Eds.): ICIAP 2023, LNCS 14233, pp. 100–111, 2023.
https://doi.org/10.1007/978-3-031-43148-7_9

security checks, allowing both the criminal and the accomplice to pass through the gate. Therefore, it is essential to develop robust and efficient *Morphing Attack Detection* (MAD) algorithms [26] capable of detecting the presence of a morphed face automatically, not only when the document is used by the criminal – *i.e.* the primary task – but also by the accomplice.

Recently, several MAD methods have been proposed in the literature. Generally, these algorithms are classified into two families of approaches [3]: *Single image MAD* (S-MAD) and *Differential MAD* (D-MAD). S-MAD methods receive as input a single image, they examine only the potentially morphed mugshot picture and mainly rely on the potential traces (*e.g.* artifacts) left by the morphing process [2]. Differently, D-MAD methods compare the potentially morphed image against a trusted one and then their main hypothetical usage is at the airport gates, in which the document image is compared with the live-captured one. Two examples of input couples of a D-MAD system are reported in Fig. 1.

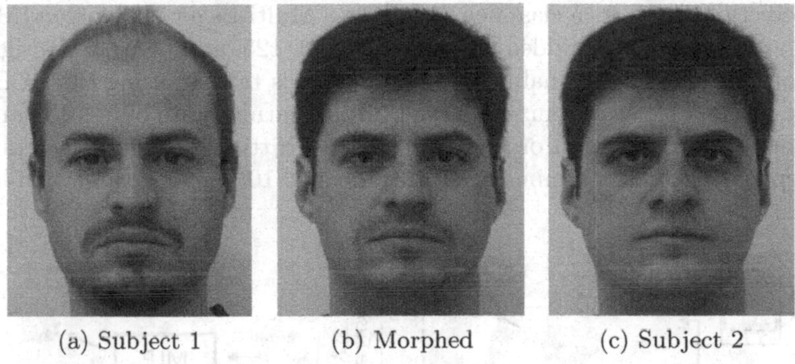

(a) Subject 1 (b) Morphed (c) Subject 2

Fig. 1. Input example of a D-MAD system, consisting of a morphed image (center) and a live acquisition depicting the criminal (Subject 1) or the accomplice (Subject 2).

From a general point of view, D-MAD methods exhibit greater performance than S-MAD methods in detecting morphed mugshot photos [32]. Unfortunately, since D-MAD systems are mainly based on the comparison of the two face identities provided in input, their efficacy is worsened with input images are similar (*i.e.* the morphed image is created from look-alike subjects or the morphing factor privileges one of the contributing subjects). Therefore, we focus this work on the development of D-MAD methods that exploit also artifact-related information, in order to improve their performance with similar identities or even extend the use of D-MAD systems to the document enrollment procedure (in which the ID image is very similar to the applicant to fool the human examiner).

Starting from the observation that D-MAD methods usually tend not to consider the presence of artifacts left by the morphing procedure, we explore different strategies for combining S-MAD and D-MAD features: in particular, we investigate the performance of the SoA D-MAD approach [31] by introducing an S-MAD module that operates only on the suspected morphed image. The underlying idea is that the S-MAD module can improve the final accuracy since it can detect visible or invisible artifacts produced by the morphing process that are normally overlooked by a more traditional D-MAD algorithm. Experimental results reveal that the proposed method improves the accuracy, especially in detecting morphed images in couples in which the accomplice is present (*i.e.* when the identity features extracted from the two images are very similar).

2 Proposed Method

The proposed method, depicted in Fig. 2, mainly consists of two different modules, *i.e.* S-MAD and D-MAD: the first module is responsible for the extraction of feature from the potentially morphed image, while the second one extracts features from the same image and the live probe. These features are then merged – through a feature fusion procedure investigated in the following sections – to create the input for the final classifier that produces in the output the final morphing detection score. As a classifier, we adopt a Multi-Layer Perceptron (MLP), with an architecture of 3 hidden layers of size 250, 125, and 64 with the sigmoid activation function in the final neuron. The MLP is trained using the BCE loss function, Adam [14] as optimizer with an initial learning rate of $5 \cdot 10^{-4}$ and an early stopping procedure in order to prevent overfitting, stopping the training after 5 epochs without a minimum improvement of 10^{-3} in the validation loss.

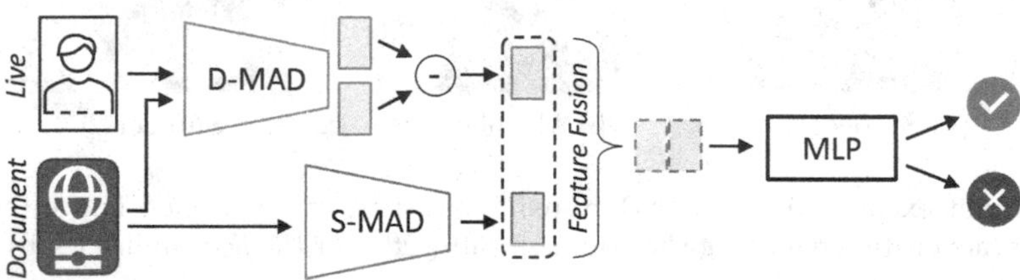

Fig. 2. Overview of the proposed method. As shown, S-MAD and D-MAD modules extract different features that are fused together and used by the final MLP classifier.

2.1 S-MAD Module

The S-MAD module consists of a backbone, specifically we adopt an Inception-Resnet V1 [33] architecture, pre-trained on the VGG-Face2 [5] dataset. In particular, the model is fine-tuned on several morphing datasets obtained with various morphing algorithms (described in Sect. 3.1), producing images ranging from low to medium quality. Moreover, we run a supplemental fine-tuning process to improve the algorithm's performance on heavily compressed, ICAO-compliant [35], JPEG images. A 512 dimensional feature vector is finally obtained removing the last fully connected layer of the architecture exploited for the classification task. For the training procedure, we adopt the Stochastic Gradient Descent (SGD) with a learning rate of 10^{-3} and the early-stopping procedure exploited for the training of the whole method. No momentum decay is exploited. This module is developed leveraging the Revelio framework[1].

2.2 D-MAD Module

As the D-MAD module, we take inspiration from the solution proposed in [31] that can be regarded as the current state of the art, as also shown in the results published on the FVC-onGoing platform [1].

Specifically, we use a ResNet-50 [12] network trained for the face recognition task [23] through the ArcFace loss [7], to extract the embeddings of the facial input images. Since the input is represented by two images, this module outputs two different embeddings of size 512 that are combined through a subtraction, and then the final feature is represented by a single embedding with the same size of 512. Authors show that the ArcFace loss function produces robust embeddings since it tends to maximize the geodesic distance between different identities and that the produced embeddings contain therefore information exclusively related to the input face identity. Our implementation is based on the publicly available Deepface[2] framework.

3 Experimental Validation

3.1 Datasets

For our experimental evaluation, we employ several publicly available datasets, with varying quality levels, briefly described and discussed in the following.

- *Progressive Morphing Database* (PMDB) [11]: it is a collection of 1108 morphed images generated by applying a public morphing algorithm to AR [18], FRGC [22], and Color Feret [21] datasets. The dataset contains 280 subjects, divided into 134 males and 146 females. No manual retouching procedures have been applied to enhance the visual quality of the images. As a result, the images may contain artifacts such as blurred areas or ghosts.
- *Idiap Morph* [27,28]: it is a collection of five datasets created using different morphing algorithms (OpenCV [17], FaceMorpher [24], StyleGAN [13], WebMorph [6], and AMSL [20]), created starting from face images from the Feret [21], FRGC [22], and Face Research Lab London Set (FRLL) [6] datasets. The visual quality of the morphed images generated with the OpenCV and FaceMorpher morphing algorithms is negatively affected by artifacts in both the background and foreground. Morphed faces generated with the StyleGAN algorithm exhibit typical GAN-related textures [37]. The AMSL morphing algorithm is used to generate 2175 morphed images from 102 adult faces: these images are compressed to a maximum size of 15 kB, simulating the process to embed an image into the chip of the eMRTD.

[1] https://miatbiolab.csr.unibo.it/revelio-framework.

[2] https://github.com/serengil/deepface.

- *MorphDB* [11]: it is a dataset of 100 morphed images generated using the Sqirlz Morph 2.1 [36] algorithm, applied to images from the Color Feret [21] and FRGC [22] datasets. This dataset is composed of 50 male and 50 female subjects. As all images are manually retouched, their visual quality is excellent. This element makes this dataset particularly challenging, although the limited number of images may make it unsuitable for conducting an extensive performance review of a MAD algorithm.
- *FEI* [34]: it is a dataset generated using the images contained in the *FEI Face Database*, which includes 200 subjects, equally split between male and female. All faces are mainly represented by subjects between 19 and 40 years old with distinct appearances, hairstyles, and accessories. This dataset contains 6000 morphed images obtained with three different morphing algorithms, namely FaceFusion [9], UTW [25], and NTNU [25], employing two different morphing factors (0.3 and 0.5).

3.2 Experimental Protocol

Experimental results are split into three distinct scenarios, according to the identity of the trusted live image: i) *Criminal*: contains bona fide attempts (*i.e.* the document image is not morphed) and morphed attempts where the document image is morphed and the live image belongs to the criminal subject; ii) *Accomplice*: contains bona fide and morphed attempts in which the live image comes from the accomplice. iii) *Both*: this scenario contains all the couples belonging to the criminal and accomplice ones.

A sample for each kind of couple can be found in Fig. 1.

In all the following results, we focus on the performance obtained in the *Accomplice* scenario that represents the practical case in which the accomplice presents the morphed image for the enrollment procedure. As mentioned, due to the greater similarity between the subjects present in both pictures, these experiments are generally considered more challenging than the previously mentioned one (*Criminal*) for D-MAD methods.

3.3 Metrics

To evaluate and compare MAD systems, there are several metrics commonly used for assessing their performance [30]: *Bona Fide Presentation Classification Error Rate* (BPCER), which represents the proportion of bona fide images incorrectly classified as morphed, and *Attack Presentation Classification Error Rate* (APCER), which represents the proportion of morphed images incorrectly labeled as bona fide. They are formulated as follows:

$$\text{BPCER}(\tau) = \frac{1}{N}\sum_{i=1}^{N} H(b_i - \tau), \quad \text{APCER}(\tau) = 1 - \left[\frac{1}{M}\sum_{i=1}^{M} H(m_i - \tau)\right] \tag{1}$$

In both definitions, τ is the score threshold on which b_i, m_i, the detection scores, are compared; $H(x) = \{1 \text{ if } x > 0, 0 \text{ otherwise}\}$ is defined as a step function. Typically, the BPCER is measured with respect to a given APCER value, *i.e.* $B_{0.1}$, $B_{0.05}$ and $B_{0.01}$, representing the lowest BPCER with APCER $\leq 10\%$, $\leq 5\%$, $\leq 1\%$, respectively. Ideally, a MAD algorithm employed in a real-world setting would need to operate at a low APCER (*i.e.* letting almost no criminals through) of around 0.1%, while maintaining an acceptable corresponding BPCER (*i.e.* generating few false positives) of around 1%.

The *Equal Error Rate* (EER), *i.e.* the error rate for which both BPCER and APCER are equal, is usually reported as a single value.

4 Experimental Results

4.1 S-MAD and D-MAD Module Assessment

Firstly, we assess the performance of S-MAD and D-MAD modules separately. For the S-MAD module, we test the same network described in Sect. 2.1, while in the D-MAD module we add the MLP architecture described in Sect. 2.2 that acts as a classifier. In this manner, we aim to understand the detection capabilities of each module, and these results offer a useful baseline to better analyze the performance of the proposed method in the following experiments.

Results are reported in Table 1. As expected, the metrics obtained through the S-MAD module are identical regardless of the type of couple, as in both cases only the same suspected morphed images are used. Besides, the D-MAD module provides considerably better performance than the S-MAD one when the live image contains the criminal, indicating that a trusted, live-capture image proves to be effective in tackling the task by comparing the two input identities. Moreover, we observe that not only the performance gap between S-MAD and D-MAD is nearly canceled when the accomplice is present in the live-capture image, but also the D-MAD performance is significantly worsened (about +10% in EER): we prove that the greater similarity between the identities makes the classification task more challenging.

Table 1. Morphing detection scores obtained on the FEI test set. Results are reported in terms of Equal Error Rate (EER), the lowest BPCER related to APCER $\leq 10\%$, $\leq 5\%$, and $\leq 1\%$, respectively.

Module	Accomplice			Criminal			Both		
	EER	$\mathbf{B_{0.05}}$	$\mathbf{B_{0.01}}$	**EER**	$\mathbf{B_{0.05}}$	$\mathbf{B_{0.01}}$	**EER**	$\mathbf{B_{0.05}}$	$\mathbf{B_{0.01}}$
S-MAD	.186	.360	.515	.186	.360	.515	.186	.360	.515
D-MAD	.180	.470	.827	.085	.147	.447	.141	.343	.767

4.2 Investigation on Feature Fusion

Previous results suggest the opportunity of exploring the combination of S-MAD and D-MAD classifiers so that the overall performance of the system does not only rely on the identity present in the live image. Therefore, we test the proposed method (see Sect. 2), merging the input features through different approaches described as follows. In addition, we also test our method by replacing the MLP with an SVM classifier, trained using an RBF kernel with a $C = 3$ regularization factor and a γ kernel coefficient which is inversely proportional to the variance of the training data received in input. This choice has been driven by the use of SVM in many morphing-related works in the literature that suggest the importance of this type of classifier in the MAD field.

Firstly, we start our investigation with a simple concatenation for the feature fusion produced by the S-MAD and D-MAD modules. Results are reported in the first line of Table 2 and show that the concatenation (indicated with the letter C) provides results that are similar to those obtained with the S-MAD module in the accomplice scenario. This behavior suggests that S-MAD features have a strong impact on the classification, outweighing the features provided by D-MAD and thus negating the benefits they bring in comparing the identities.

Table 2. Morphing detection scores obtained on the FEI test set across different classifiers and feature fusion techniques. C stands for concatenation, MM for Min-Max, and MV for Mean-Variance (see Sect. 4.2).

Fusion	Class.	Accomplice			Criminal			Overall		
		EER	$\mathbf{B}_{0.05}$	$\mathbf{B}_{0.01}$	**EER**	$\mathbf{B}_{0.05}$	$\mathbf{B}_{0.01}$	**EER**	$\mathbf{B}_{0.05}$	$\mathbf{B}_{0.01}$
C	MLP	.168	.317	.510	.168	.345	.520	.168	.330	.515
	SVM	.175	.248	**.458**	.175	.275	.460	.175	.265	.458
MM	MLP	**.132**	**.245**	.463	.138	.265	.463	.135	.260	.463
	SVM	.140	.278	.475	.160	.320	.475	.147	.295	.475
MV	MLP	.175	.338	.543	.185	.398	.563	.181	.365	.555
	SVM	.195	.317	.495	.257	.423	.530	.225	.387	.522

To further investigate our hypothesis, we run a *t-distributed Stochastic Neighbor Embedding* (t-SNE) [16] dimensionality reduction on the input features, divided both by source (*i.e.* D-MAD or S-MAD) and by class (*i.e.* bona fide or morphed). The resulting plot, shown in Fig. 3, highlights how the features can easily be separated by their respective source, suggesting that they may occupy different portions of the feature space. However, there is no clear separation between bona fide and morphx ed feature vectors; this reinforces the hypothesis that the classifier could be prioritizing the S-MAD features while disregarding those generated by the D-MAD module. Moreover, to test how the high dimensionality of the two concatenated vectors might affect the system's performance, we employ the PCA algorithm [19], which indicates that the optimal intrinsic feature dimensionality is only one less than the original. Thus, we infer that all features of both vectors may be required to achieve the best results and that other fusion methods must be investigated.

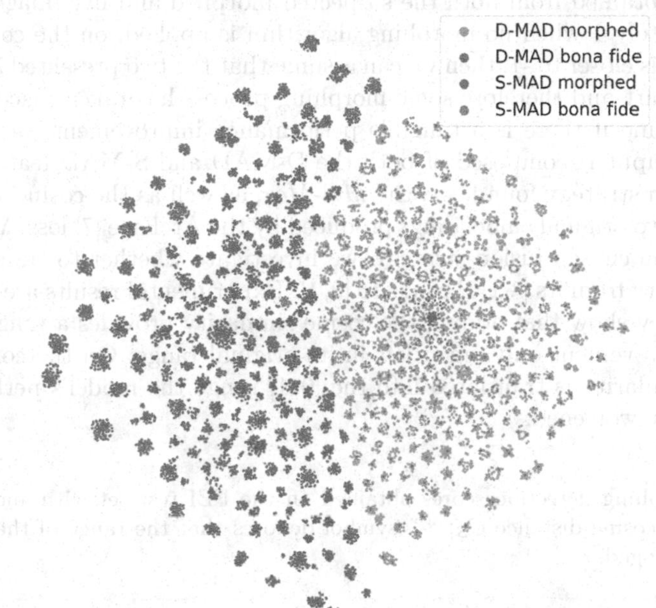

Fig. 3. Visualization of the *t-distributed Stochastic Neighbor Embedding* (t-SNE) [16] of D-MAD and S-MAD feature vectors divided by ground truth (best on screen).

Therefore, we investigate two additional fusion strategies: i) *Min-Max* (MM): before concatenating the two feature vectors, they are separately rescaled to have each component in the $[0, 1]$ range; ii) *Mean-Variance* (MV): before concatenating the two feature vectors, they are separately rescaled to have each component with mean value $\mu = 0$ and variance $\sigma = 1$.

Results of the above-mentioned experiments are reported in Table 2. We observe the performance gap that was previously found between MLPs and SVMs is not present when the two features are merged together; indeed, the former almost always outperforms the latter. Besides, the Mean-Variance strategy provides unsatisfactory results, which are worse than the simple concatenation strategy. An ex-post numerical analysis on the normalized feature vectors used for training shows that, even when each component is rescaled to have $\mu = 0$ and $\sigma = 1$, the D-MAD and S-MAD features still show significant differences in range. This could be a possible explanation for the great performance of the Min-Max fusion strategy, thus proving that translating the two feature vectors to the same numeric range helps improve the model's performance.

In addition to this investigation, we also test the performance impact of including the cosine similarity (C_S) [15] between the two D-MAD feature vectors. The underlying idea comes from the fact that, as explained in [7], the embeddings produced by the model are optimized so that the geodesic angle between each identity is maximized. Therefore, the cosine similarity between the embeddings obtained from both the suspected morphed and live images should be approximately 1 when no morphing algorithm is applied; on the contrary, if the similarity is closer to -1, then we can assume that the two presented identities are too far apart and therefore some morphing process has taken place.

To determine if there is a tangible performance improvement, we train an MLP whose input is composed of both the D-MAD and S-MAD features with the best fusion strategy found, *i.e.* the *Min-Max*, as well as the cosine similarity between the two original embeddings produced by the ArcFace [7] loss. Moreover, inspired by the chosen fusion strategy, we investigate whether to translate the cosine similarity from its $[-1, 1]$ range to $[0, 1]$. Experimental results are reported in Table 3: they show that adding the cosine similarity provides a tangible performance improvement only when left in its original range. On the contrary, if the cosine similarity is translated into the $[0, 1]$ range the model's performance is considerably worsened.

Table 3. Morphing detection scores obtained on the FEI test set with and without employing the cosine distance C_S. "-" symbol denotes that the range of the distance has kept unchanged.

C_S **range**	**Accomplice**			**Criminal**			**Overall**		
	EER	$\mathbf{B}_{0.05}$	$\mathbf{B}_{0.01}$	**EER**	$\mathbf{B}_{0.05}$	$\mathbf{B}_{0.01}$	**EER**	$\mathbf{B}_{0.05}$	$\mathbf{B}_{0.01}$
-	.132	.245	**.463**	.138	.265	.463	.135	.260	.463
$[-1, 1]$	**.125**	**.237**	.468	.125	.235	.440	.125	.235	.445
$[0, 1]$	.132	**.237**	.465	.141	.290	.470	.136	.270	.468

4.3 Comparison with the State of the Art

Finally, we test the performance of our proposed method against the current D-MAD literature methods. We compare our proposed algorithm (in particular, we select the best configuration obtained, *i.e.* the Min-Max feature fusion and the cosine similarity in the $[-1, 1]$ range) against the methods proposed in [4,31]. Experimental results are reported in Table 4.

It is worth noting that the proposed method overcomes both competitors when the identity in the live image belongs to the accomplice, consistently in all metrics reported, suggesting that S-MAD features can effectively improve the performance of D-MAD methods. However, when the criminal is present, the proposed algorithm has still room for improvement indicating the need to investigate further feature fusion methods and to develop specific MAD techniques to differently address image pairs with criminals and accomplices.

Table 4. Morphing detection scores obtained on the FEI test set through the proposed methods with respect to the current literature solutions.

Method	Accomplice			Criminal			Overall		
	EER	$\mathbf{B}_{0.05}$	$\mathbf{B}_{0.01}$	EER	$\mathbf{B}_{0.05}$	$\mathbf{B}_{0.01}$	EER	$\mathbf{B}_{0.05}$	$\mathbf{B}_{0.01}$
[31]	.175	.475	.780	.066	.085	.310	.129	.343	.690
[4]	.153	.345	.563	.060	.095	.370	.115	.257	.515
Ours	**.125**	**.237**	**.468**	.125	.235	.440	.125	.235	.445

Lastly, we also test the proposed method through the FVC-onGoing platform [8] on the sequestered DMAD-SOTAMD_D-1.0 benchmark, even though in this dataset the available morphed images are only compared with the criminal or bonafide subjects. We obtain an EER of about 10%, a worse result with respect to the method [31] that is SotA in couples with criminal (with an EER = 4.5%), but also a better result with respect to the algorithm proposed in [11] (EER = 14%), showing that machine learning-based techniques yield overall better results and represent promising solutions. These results suggest the need for the development of a strategy able to select the best MAD algorithm: specifically, this system should be able to detect whether the criminal or the accomplice is present in the live image, and then either use, for instance, the standalone state-of-the-art D-MAD algorithm (*e.g.* [31]) or the method proposed in this paper. In this manner, we may be able to overcome the limitations of both systems, which respectively underperform when the live image contains the accomplice or the criminal.

5 Conclusions and Future Works

In this paper, we have investigated the fusion of S-MAD and D-MAD features, to create a system that is capable of detecting morphed images in the challenging scenario in which the accomplice is used for comparison. Experimental results reveal that effectively combining the two kinds of embeddings is not a trivial task. Specifically, we demonstrate that the features produced by S-MAD and D-MAD methods occupy different regions of the feature space, and their normalization in a predefined range improves the model's overall effectiveness. Moreover, we show that including a further feature represented by the cosine distance between the two embeddings produced by the D-MAD feature extractor improves the algorithm's performance. As a future work, we plan to improve the overall performance by developing a model able to preliminarily discriminate between couples with the accomplice or the criminal, thus enabling the selective use of a specific method to address the D-MAD task.

Acknowledgment. This work is part of the iMARS project. The project received funding from the European Union's Horizon 2020 research and innovation program under Grant Agreement No. 883356. Disclaimer: this text reflects only the author's views, and the Commission is not liable for any use that may be made of the information contained therein.

References

1. Biolab: FVC-onGoing. https://biolab.csr.unibo.it/fvcongoing/
2. Borghi, G., Franco, A., Graffieti, G., Maltoni, D.: Automated artifact retouching in morphed images with attention maps. IEEE Access **9**, 136561–136579 (2021)
3. Borghi, G., Graffieti, G., Franco, A., Maltoni, D.: Incremental training of face morphing detectors. In: 2022 26th International Conference on Pattern Recognition (ICPR), pp. 914–921. IEEE (2022)
4. Borghi, G., Pancisi, E., Ferrara, M., Maltoni, D.: A double siamese framework for differential morphing attack detection. Sensors **21**(10), 3466 (2021)
5. Cao, Q., Shen, L., Xie, W., Parkhi, O.M., Zisserman, A.: Vggface2: A dataset for recognising faces across pose and age. In: 13th IEEE International Conference on Automatic Face & Gesture Recognition, FG 2018, Xi'an, China, May 15–19 (2018)
6. DeBruine, L., Jones, B.: Face research lab London set. Psychol. Methodol. Des, Anal (2017)
7. Deng, J., Guo, J., Xue, N., Zafeiriou, S.: Arcface: Additive angular margin loss for deep face recognition. In: IEEE Conference on Computer Vision and Pattern Recognition, CVPR 2019, Long Beach, CA, USA, June 16–20, 2019 (2019)
8. Dorizzi, B., et al.: Fingerprint and on-line signature verification competitions at ICB 2009. In: Advances in Biometrics, Third International Conference, ICB 2009, Alghero, Italy, June 2–5, 2009. Proceedings. Lecture Notes in Computer Science, vol. 5558 (2009)
9. FaceFusion: Facefusion. https://www.wearemoment.com/FaceFusion/
10. Ferrara, M., Franco, A., Maltoni, D.: The magic passport. In: IEEE International Joint Conference on Biometrics, Clearwater, IJCB 2014, FL, USA, September 29 - October 2, 2014 (2014)
11. Ferrara, M., Franco, A., Maltoni, D.: Face demorphing. IEEE Trans. Inf. Forensics Secur. **13**(4) (2018)
12. He, K., Zhang, X., Ren, S., Sun, J.: Deep residual learning for image recognition. In: Proceedings of the IEEE Conference on Computer Vision and Pattern Recognition, pp. 770–778 (2016)
13. Karras, T., Laine, S., Aittala, M., Hellsten, J., Lehtinen, J., Aila, T.: Analyzing and improving the image quality of StyleGAN. In: Proceedings of the IEEE/CVF Conference on Computer Vision and Pattern Recognition, pp. 8110–8119 (2020)
14. Kingma, D.P., Ba, J.: Adam: A method for stochastic optimization. In: Bengio, Y., LeCun, Y. (eds.) 3rd International Conference on Learning Representations, ICLR 2015, San Diego, CA, USA, May 7–9, 2015, Conference Track Proceedings (2015). https://arxiv.org/abs/1412.6980
15. Li, B., Han, L.: Distance weighted cosine similarity measure for text classification. In: Yin, H., et al. (eds.) IDEAL 2013. LNCS, vol. 8206, pp. 611–618. Springer, Heidelberg (2013). https://doi.org/10.1007/978-3-642-41278-3_74
16. Van der Maaten, L., Hinton, G.: Visualizing data using t-SNE. J. Mach. Learn. Res. **9**(11) (2008)
17. Mallick, S.: Face morph using OpenCV - C++ / Python. https://learnopencv.com/face-morph-using-opencv-cpp-python/
18. Martinez, A., Benavente, R.: The AR face database: Cvc technical report, 24 (1998)
19. Minka, T.: Automatic choice of dimensionality for PCA. In: Advances in Neural Information Processing Systems, vol. 13 (2000)
20. Neubert, T., Makrushin, A., Hildebrandt, M., Kraetzer, C., Dittmann, J.: Extended stirtrace benchmarking of biometric and forensic qualities of morphed face images. IET Biometrics **7**(4), 325–332 (2018)

21. Phillips, P.J., Wechsler, H., Huang, J., Rauss, P.J.: The FERET database and evaluation procedure for face-recognition algorithms. Image Vision Comput. **16**(5) (1998)
22. Phillips, P.J., et al.: Overview of the face recognition grand challenge. In: 2005 IEEE Computer Society Conference on Computer Vision and Pattern Recognition (CVPR'05). vol. 1. IEEE (2005)
23. Pini, S., Borghi, G., Vezzani, R., Maltoni, D., Cucchiara, R.: A systematic comparison of depth map representations for face recognition. Sensors **21**(3), 944 (2021)
24. Quek, A.: FaceMorpher morphing algorithm. https://github.com/alyssaq/face_morpher
25. Raja, K., et al.: Morphing attack detection-database, evaluation platform, and benchmarking. IEEE Trans. Inf. Forensics Secur. **16**, 4336–4351 (2020)
26. Raja, K.B., et al.: Morphing attack detection-database, evaluation platform, and benchmarking. IEEE Trans. Inf. Forensics Secur. **16**, 4336–4351 (2021)
27. Sarkar, E., Korshunov, P., Colbois, L., Marcel, S.: Vulnerability analysis of face morphing attacks from landmarks and generative adversarial networks. arXiv preprint arXiv:2012.05344 (2020)
28. Sarkar, E., Korshunov, P., Colbois, L., Marcel, S.: Are gan-based morphs threatening face recognition? In: ICASSP 2022–2022 IEEE International Conference on Acoustics, Speech and Signal Processing (ICASSP), pp. 2959–2963. IEEE (2022)
29. Scherhag, U., Debiasi, L., Rathgeb, C., Busch, C., Uhl, A.: Detection of face morphing attacks based on PRNU analysis. IEEE Trans. Biom. Behav. Identity Sci. **1**(4) (2019)
30. Scherhag, U., Rathgeb, C., Merkle, J., Breithaupt, R., Busch, C.: Face recognition systems under morphing attacks: a survey. IEEE Access **7**, 23012–23026 (2019)
31. Scherhag, U., Rathgeb, C., Merkle, J., Busch, C.: Deep face representations for differential morphing attack detection. IEEE Trans. Inf. Forensics Secur. **15** (2020)
32. Scherhag, U., et al.: Biometric systems under morphing attacks: Assessment of morphing techniques and vulnerability reporting. In: International Conference of the Biometrics Special Interest Group, BIOSIG 2017, Darmstadt, Germany, September 20–22, 2017. LNI, vol. P-270 (2017)
33. Szegedy, C., et al.: Going deeper with convolutions. In: IEEE Conference on Computer Vision and Pattern Recognition, CVPR 2015, Boston, MA, USA, June 7–12, 2015 (2015)
34. Thomaz, C., Giraldi, G.: A new ranking method for principal components analysis and its application to face image analysis. Image and Vision Comput. **28**(6), 902–913 (2010)
35. Wolf, A.: ICAO: Portrait quality (reference facial images for MRTD), version 1.0. standard. International Civil Aviation Organization (2018)
36. xiberpix: Sqirlz morphing algorithm. https://sqirlz-morph.it.uptodown.com/windows
37. Zhang, X., Karaman, S., Chang, S.F.: Detecting and simulating artifacts in GAN fake images. In: 2019 IEEE International Workshop on Information Forensics and Security (WIFS), pp. 1–6. IEEE (2019)

SynthCap: Augmenting Transformers with Synthetic Data for Image Captioning

Davide Caffagni[1], Manuele Barraco[1], Marcella Cornia[1(✉)], Lorenzo Baraldi[1], and Rita Cucchiara[1,2]

[1] University of Modena and Reggio Emilia, Modena, Italy
{davide.caffagni,manuele.barraco,marcella.cornia,lorenzo.baraldi, rita.cucchiara}@unimore.it
[2] IIT-CNR, Pisa, Italy

Abstract. Image captioning is a challenging task that combines Computer Vision and Natural Language Processing to generate descriptive and accurate textual descriptions for input images. Research efforts in this field mainly focus on developing novel architectural components to extend image captioning models and using large-scale image-text datasets crawled from the web to boost final performance. In this work, we explore an alternative to web-crawled data and augment the training dataset with synthetic images generated by a latent diffusion model. In particular, we propose a simple yet effective synthetic data augmentation framework that is capable of significantly improving the quality of captions generated by a standard Transformer-based model, leading to competitive results on the COCO dataset.

Keywords: Image Captioning · Synthetic Data · Vision-and-Language

1 Introduction

Image captioning is a complex task that involves the description of an image in natural language, posing challenges at the intersection of Computer Vision and Natural Language Processing fields. The most promising solutions to tackle the task are represented by deep learning-based captioning architectures which have become the de facto standard for the task [46]. Despite achieving state-of-the-art results, it is becoming difficult to further improve their performance, primarily because of the struggles in finding datasets containing a satisfactory amount of image-caption pairs. To overcome this issue, the predominant approach in the field is to train captioning networks [13,20,51,58] on large-scale datasets collected from the web [42,44], usually downloading an image along with the description provided in its "alt" tag. As a matter of fact, there is no surprise in witnessing more and more advanced deep learning-based models being trained on web-collected data, especially after the spread of large-scale language models [10, 59] and cross-modal architectures [36]. The knowledge found on the web, indeed, excels for size and variety, stimulating the robustness and sensibility of deep

G. L. Foresti et al. (Eds.): ICIAP 2023, LNCS 14233, pp. 112–123, 2023.
https://doi.org/10.1007/978-3-031-43148-7_10

learning models to long-tail concepts. However, its quality and ethics might be questionable, especially for image captioning which requires proper alignment between visual and textual contents. Although there are successful attempts to refine or distinguish web-based information [13,24], it is unfeasible to completely filter out wrong and noisy data when its extent grows too much.

Synthetic data seems an appealing alternative to match the scaling requirements of modern neural networks while attenuating the drawbacks of web-crawled data. In fact, synthetic data can be produced on-demand, are virtually infinite, and their annotations are in most cases at no cost. Moreover, from an ethical perspective, they usually offer better control over biases than their web counterparts. While the usage of synthetically generated data has led to promising results in various Computer Vision tasks [1,5,9,11,16], limited research efforts have been done in the context of image captioning.

Motivated by the recent advancements in Generative AI, in this work we explore the usage of synthetic images to boost the performance of captioning architectures. In particular, we leverage the well-known Stable Diffusion model [39] to generate synthetic images associated with human-annotated textual sentences and employ these newly generated data to augment the most widely used dataset in the image captioning field (*i.e.* COCO [28]). From a technical point of view, we introduce a simple yet effective framework to employ synthetic data that probabilistically replace real pictures with fake ones and apply it to a standard Transformer-based architecture [48]. To validate our proposal, we conduct extensive experiments to evaluate whether synthetic images can be leveraged to improve the quality of generated captions. Experimental results on the popular COCO dataset [28] demonstrate the effectiveness of our solution, which achieves better results than a baseline model without synthetic data augmentation and competitive performance compared to previously proposed approaches. We believe that our analysis can serve as a starting point for employing synthetically generated images as an effective data augmentation strategy in the field of image captioning and other vision-and-language tasks.

2 Related Work

Image Captioning. Early deep learning-based image captioning models were based on a basic encoder-decoder scheme, with the use of RNNs and LSTMs as popular choices for the text generation part along with CNNs to encode the visual content [22,38,50]. Following these initial attempts, subsequent techniques have steadily advanced both the image encoding and language generation stages. Regarding the image encoding, remarkable progress has been achieved through the introduction of additive attention mechanisms to incorporate spatial knowledge, first from a grid of CNN features [55] and later utilizing image regions extracted from pre-trained object detectors [4], eventually considering their semantic and spatial relationships encoded by graph neural networks [56,57]. Nowadays, Transformer-based architectures [48], initially designed for machine translation and language comprehension purposes and then employed in a variety of tasks [15,35,47], have been adopted in the domain of image captioning as well.

These models are commonly used both in the visual encoding stage [12,21,29,53] and as language models [14,17,30,60], also leading to the design of effective variants of the self-attention operator [14,21,33].

Recent advancements have been obtained by large-scale vision-and-language pre-training which usually employs noisy image-text pairs to increase the number of training samples, thus further enhancing the performance of fully-attentive image captioning models [13,20,51,58]. Effective alternatives also involve the use of visual features from large-scale cross-modal architectures [7,8,45] like CLIP [36]. These multimodal architectures also allow for the enrichment of predicted textual sentences employing retrieval components, that can be added to the captioning model, and external knowledge from which to extract additional information to improve the final performance [26,32,41].

Synthetic Data. To the best of our knowledge, there is a limited amount of works that explore the usage of synthetic data in image captioning. In particular, Hossain *et al.* [19] introduced artificial images into a captioning system, by creating new pictures thanks to generative adversarial networks. More recently, Xiao *et al.* [54] leveraged a latent diffusion model [39] to augment the training dataset, also employing paraphrasing sentences to pair with the generated pictures. However, they only achieved promising results when using limited training instances or when switching to an unpaired image captioning setting. Concurrently, Li *et al.* [25] proposed to employ fake images as a replacement for difficult samples to finetune a large-scale vision-and-language model for captioning. In this work, we stick with the same latent diffusion model to generate fake images (*i.e.* Stable Diffusion [39]), but we do not require any additional textual data outside of captions from the COCO dataset, demonstrating the effectiveness of synthetic data augmentation for the standard image captioning task.

3 Proposed Method

In this section, we introduce SynthCap, a novel image captioning architecture trained with the proposed synthetic augmentation strategy. Figure 1 shows an overview of our complete model.

3.1 Model Architecture

Visual Encoder. Our architecture is based on a fully-attentive Transformer network that takes as input visual features extracted from a pre-trained visual encoder. For the latter, we leverage the image encoder of a pre-trained CLIP-based model [36] and we freeze its weights throughout all the experiments. Specifically, we opt for the CLIP ViT-L/14 version which is based on the Vision Transformer (ViT) backbone [15].

Transformer Model. Our language model is a standard encoder-decoder Transformer network [48]. Each encoder layer is made of a self-attention block followed by a feed-forward layer. The former refines the supplied visual tokens via bi-directional self-attention. The latter operates on single tokens with two

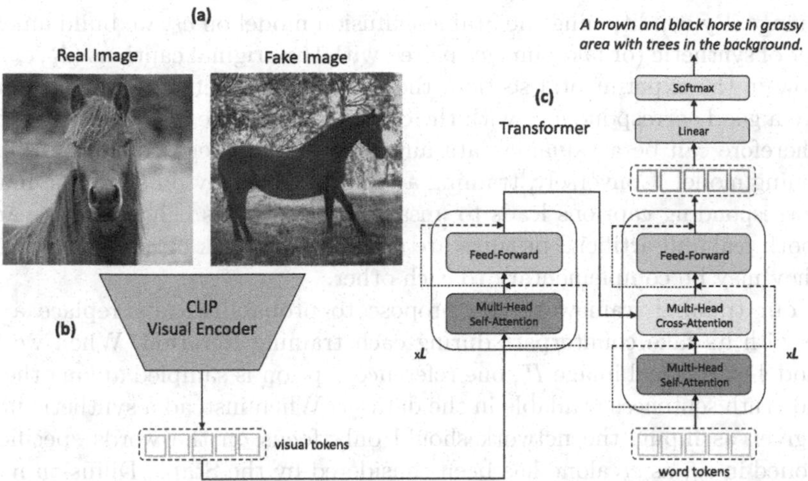

Fig. 1. Overview of the proposed method: (a) we select either a real or a synthetic image, according to a λ_s weight; (b) the CLIP-based visual encoder converts the input image into a sequence of visual tokens; (c) the encoder-decoder Transformer network generates the caption grounded on the visual token.

dense layers, featuring a GELU non-linearity in between. The output of each block is summed along with its input through a residual connection and then normalized. The decoder network shows a similar architecture to the encoder, but it comprises a cross-attention block interposed between the self-attention and feed-forward block. This additional component is critical, as here occurs the cross-modal integration between visual and textual modalities. In detail, the tokens representing the partial caption generated by the decoder up to time t act as queries, that attend the visual tokens from the encoder, *i.e.* keys and values. Unlike the encoder self-attention block, the decoder self-attention requires a causal mask to prevent tokens from attending to the future. Specifically, masking is implemented by artificially zeroing the entries of the self-attention matrix with row-column indexes $(i, j)_{\forall j>i}$. The output of the decoder is a token sequence $\widetilde{\mathbf{x}} = \{\widetilde{x}_t\}_{t=1,\ldots,N}$ whose length is equal to the input. To select the next word $\widetilde{x}_{t+1}$, we sample from a probability distribution over all the possible words in the reference vocabulary, obtained by feeding $\widetilde{x}_t$ to a linear and a softmax layer. At inference time, the decoder works in an auto-regressive manner, meaning that the token produced at time t will be included in the input for time $t+1$.

3.2 Synthetic Data Augmentation

Our goal is to probe whether synthetic images can be a valuable source of information to train captioning algorithms. We leverage Stable Diffusion [39] to generate fake images to extend the training set of the COCO dataset [22], which is originally composed of more than half a million image-caption pairs (I^r, c_k), with $k = 1, 2, 3, 4, 5$, *i.e.* there are five different reference descriptions available for each image. By conditioning the Stable Diffusion model on c_k, we build an extra dataset of synthetic (or fake) images paired with the original captions (I_k^s, c_k). As we show in the experimental section, the synthetically generated images prove to have a good correspondence with the captions they have been generated from and therefore can be a valuable data augmentation strategy to train an image captioning model. Conversely, training a model exclusively on synthetic images and corresponding captions leads to unsatisfactory results. Therefore, we argue that both real and artificial pictures are useful for the task of image captioning, and they may be complementary to each other.

In our training framework, we propose to probabilistically replace a real image with its fake counterpart during each training iteration. When we feed the model with a real image I^r, one reference caption is sampled among the five ground-truth sentences available in the dataset. When instead a synthetic image I_k^s is given as input, the network should only focus on the words specifically mentioned in c_k, as c_k alone has been considered by the Stable Diffusion model when generating I_k^s. Formally, given a caption c_k, we build an image-text pair (I, c_k), in which the visual component is chosen as follows:

$$I = \begin{cases} I_k^s & \text{if } \epsilon < \lambda_s \\ I^r & \text{otherwise,} \end{cases} \tag{1}$$

where λ_s is a hyperparameter controlling the probability of using synthetic data at each training iteration and $\epsilon \sim U(0,1)$. When we set $\lambda_s = 0$, the training set is the original one without any synthetic data augmentation, while when $\lambda_s = 1$ the training set is composed only of fake images and corresponding textual sentences. Note that, regardless of λ_s, the amount of processed samples per epoch remains the same as in the original training process.

Training Procedure. We adhere to the two-phase training typically used in image captioning [46] which consists of a pre-training step with cross-entropy loss followed by a finetuning phase based on the self-critical sequence training (SCST) proposed in [38], which optimizes the captioning model with reinforcement learning using the CIDEr metric [49] as a reward.

During SCST optimization, the baseline reward is chosen as the average score over all the sequences sampled using beam search within the same beam, following [14]. According to this setup, whenever we require a synthetic image to replace its associated real one, we opt to randomly draw from the five available fake images. Formally, $I_k^s \sim \{I_1^s, I_2^s, I_3^s, I_4^s, I_5^s\}$. Note that, although for each k, the synthetic image I_k^s has been created from a single description c_k, the CIDEr metric still measures the consensus of the captions generated by our model among all five reference captions $c_{k=1,\ldots,5}$.

4 Experimental Evaluation

4.1 Implementation Details

Dataset and Evaluation Metrics. We evaluate our proposal on the Microsoft COCO dataset [28], using the standard Karpathy splits [22]. We report the results according to evaluation metrics typically used for image captioning: BLEU [34], METEOR [6], ROUGE [27], CIDEr [49], and SPICE [3].

Architecture. Before being fed to the CLIP visual encoder, each input image undergoes a pre-processing pipeline. The first step involves a resize to reduce the longer side length to a maximum of 224 pixels, keeping the original aspect ratio. It follows a center crop plus a channel-wise normalization. The resulting input is a tensor with shape $3 \times 224 \times 224$, from which the ViT-based CLIP encoder extracts a grid of 256×1024 features, *i.e.* the visual tokens. Our Transformer-based image captioning network comprises $L = 3$ layers in both the encoder and decoder, operating on a hidden size $d = 512$. We therefore apply a linear projection over the CLIP visual features to match this dimensionality. We employ multi-head attention with 8 different heads in each attentive layer, plus dropout with probability 0.1. To convert words into tokens, we leverage the same byte-pair encoding (BPE) tokenizer [43] used by the CLIP text module.

Training Details. During cross-entropy optimization, we stick with the setup suggested in [26] using a batch size of 32 and the learning rate scheduling strategy of [48] with warmup equal to 20,000 iterations. In the SCST phase, we use a batch size of 16, a constant learning rate of 10^{-6}, and apply beam search decoding with a beam size equal to 5. For both training phases, we employ Adam [23] as optimizer. All experiments have been carried out with mixed precision [31] and ZeRO memory offloading [37], using the Huggingface Transformers library [52].

Synthetic Data Generation. All synthetic images are generated following [2], by feeding Stable Diffusion with the reference captions from the COCO Karpathy training split using the standard prompt "*An image of*". As Stable Diffusion model, we employ the implementation provided by the Huggingface library[1].

4.2 Ablation Studies and Analysis

In this section, we conduct ablation studies to discuss the main design choice of our proposal and validate the proposed synthetic data augmentation strategy.

Overall Validation of Synthetic Images. We first validate the correspondence of generated synthetic images with associated textual sentences by computing the image-text similarity between cross-modal embeddings extracted from CLIP-based visual and textual backbones. As demonstrated in recent literature [18,40], this image-text similarity is effective for evaluating image captioning models. As shown in Table 1, on average, synthetic images seem to have a slightly higher affinity with their descriptions compared to the real ones. This suggests that they could be a valuable source of information to feed an image captioning model during training.

Percentage of Synthetic Data. In our framework, we control the probability to replace a real image with a synthetic one thanks to λ_s. Table 2 presents the results when varying this parameter in comparison with a baseline model trained without synthetic data. When $\lambda_s = 1.0$, we entirely rely on synthetic images and

[1] https://huggingface.co/CompVis/stable-diffusion-v1-4.

Table 1. CLIP-based image-text similarity scores for real and synthetic images and corresponding textual sentences.

	Mean	Median	Min	Max
Real images	0.256	0.257	0.004	**0.463**
Synthetic images	**0.263**	**0.262**	**0.098**	0.437

Table 2. Analysis using different percentages of synthetic data. Results are reported after cross-entropy pre-training.

Synth. Data	λ_s	B-1	B-4	M	R	C	S
✗	-	77.5	37.2	30.0	58.6	126.5	23.3
✓	0.1	77.3	37.1	30.3	58.8	127.2	23.5
✓	0.2	77.5	37.9	30.3	59.0	128.1	23.4
✓	0.3	77.7	37.7	30.3	59.1	127.7	23.5
✓	0.4	77.8	37.8	30.4	59.0	128.3	23.5
✓	0.5	77.7	37.6	30.3	58.9	**128.6**	23.4
✓	0.6	77.9	37.6	30.1	58.8	127.5	23.3
✓	0.7	77.4	37.0	30.0	58.7	126.5	23.4
✓	1.0	72.7	29.2	25.5	53.1	100.2	19.0

experience a consistent drop with respect to the baseline. This behavior can be due to the reality gap between real and synthetic images which prevents the model to generalize on real data when it is trained on synthetically generated samples only. This means that synthetic images, despite the advancements in Generative AI, are still far from exactly mimicking pictures from the natural distribution. On the other hand, all other models benefit from augmented training with synthetic images. In detail, we reach the highest CIDEr score when feeding the model with fake images half of the time (*i.e.* $\lambda_s = 0.5$), but we still observe improvements with up to 60% of synthetic images. The positive effects of synthetic data appear to worsen with $\lambda_s = 0.7$, even though the performance is still competitive against the baseline without synthetic data augmentation.

Effectiveness of Synthetic Data. To prove that the observed improvements truly come from using synthetic images to augment our training set, we repeat the setup explained in Sect. 3.2 but change the source of visual input for augmentation. Since a synthetic image is naturally similar to the original image, a reasonable comparison should rely on visually similar but real images. Thus, in this case, given an image I from the COCO dataset, we replace it with probability λ_s with I_k^r, that corresponds to a real image randomly selected among the top-k similar images with respect to I. In particular, following [41], we extract a feature vector for each image from a pre-trained CLIP model. Then, given an encoded query image, the k most similar ones are retrieved with $k = 1, 3, 5$, using the cosine similarity between pairs of feature vectors as a similarity measure. For this experiment, we employ $\lambda_s = 0.5$ that corresponds to the configuration leading to the highest CIDEr score in the previous analysis. According to the results reported in Table 3, we can notice that our synthetic data augmentation strategy achieves the best performance compared to both the baseline and the employed retrieval-based augmentation solution.

Table 3. Analysis using our best configuration (*i.e.* $\lambda_s = 0.5$), replacing synthetic images with real ones selected among the top-k similar images. Results are reported after cross-entropy pre-training.

	Synth. Data	B-1	B-4	M	R	C	S
Transformer	✗	77.5	37.2	30.0	58.6	126.5	23.3
Transformer (w/ similar images, $k = 1$)	✗	77.6	37.0	29.8	58.3	125.1	22.8
Transformer (w/ similar images, $k = 2$)	✗	76.7	37.0	30.0	58.5	125.5	23.1
Transformer (w/ similar images, $k = 3$)	✗	76.8	37.0	29.8	58.2	124.6	22.9
SynthCap	✓	**77.7**	**37.6**	**30.3**	**58.9**	**128.6**	**23.4**

Table 4. Comparison with the state of the art on the COCO Karpathy test.

	Cross-Entropy Loss						**CIDEr Optimization**					
	B-1	B-4	M	R	C	S	B-1	B-4	M	R	C	S
Up-Down [4]	77.2	36.2	27.0	56.4	113.5	20.3	79.8	36.3	27.7	56.9	120.1	21.4
GCN-LSTM [57]	77.3	36.8	27.9	57.0	116.3	20.9	80.9	38.3	28.6	58.5	128.7	22.1
SGAE [56]	77.6	36.9	27.7	57.2	116.7	20.9	81.0	39.0	28.4	58.9	129.1	22.2
AoANet [21]	77.4	37.2	28.4	57.5	119.8	21.3	80.2	38.9	29.2	58.8	129.8	22.4
$\mathcal{M}^2$ Transformer [14]	-	-	-	-	-	-	80.8	39.1	29.2	58.6	131.2	22.6
X-Transformer [33]	77.3	37.0	28.7	57.5	120.0	21.8	80.9	39.7	29.5	59.1	132.8	23.4
DLCT [30]	-	-	-	-	-	-	81.4	39.8	29.5	59.1	133.8	23.0
RSTNet [60]	-	-	-	-	-	-	81.8	40.1	29.8	59.5	135.6	23.3
DIFNet [53]	-	-	-	-	-	-	81.7	40.0	29.7	59.4	136.2	23.2
CaMEL [8]	78.3	39.1	29.4	58.5	125.7	22.2	82.8	41.3	30.2	60.1	140.6	23.9
COS-Net [26]	**79.2**	**39.2**	29.7	**58.9**	127.4	22.7	82.7	42.0	30.6	60.6	141.1	24.6
Transformer	77.5	37.2	30.0	58.6	126.5	23.3	82.9	42.2	30.7	60.9	141.9	24.6
SynthCap	77.7	37.6	**30.3**	**58.9**	**128.6**	**23.4**	**83.0**	**42.4**	**30.8**	**61.1**	**143.1**	**24.7**

4.3 Comparison to the State of the Art

We now test SynthCap against other state-of-the-art captioning models. In our analysis, we include earlier approaches featuring LSTM as language models and attention over image regions, like Up-Down [4], eventually boosted with graph-based encoding (GCN-LSTM [57] and SGAE [56]) or self-attention, such as

AoANet [21]. Further, we include more recent proposals that rely on the Transformer network, namely $\mathcal{M}^2$ Transformer [14], X-Transformer [33], DLCT [30], RSTNet [60], DIFNet [53], CaMEL [8], and COS-Net [26]. We report the results in Table 4. As it can be seen, SynthCap beats the baseline across all the metrics, in both the cross-entropy pre-training and CIDEr-based optimization stages. Compared to the other better-performing approaches, our framework achieves competitive results, while being based on a simple encoder-decoder Transformer model without any other specific architectural component.

To further confirm the effectiveness of our data augmentation strategy, we report the results on the COCO online test server in Table 5. Following previous literature, we leverage an ensemble of four models trained using different random seeds. Also in this setting, SynthCap achieves the best results according to all evaluation metrics. Finally, in Fig. 2, we show some qualitative results on sample images from the COCO dataset, comparing captions generated by our model with those generated by the baseline without synthetic data augmentation.

Table 5. Leaderboard of various methods on the online COCO test server.

	BLEU-1		BLEU-2		BLEU-3		BLEU-4		METEOR		ROUGE		CIDEr	
	c5	c40	c5	c40	c5	c40	c5	c40	c5	c40	c5	c40	c5	c40
Up-Down [4]	80.2	95.2	64.1	88.8	49.1	79.4	36.9	68.5	27.6	36.7	57.1	72.4	117.9	120.5
SGAE [56]	81.0	95.3	65.6	89.5	50.7	80.4	38.5	69.7	28.2	37.2	58.6	73.6	123.8	126.5
AoANet [21]	81.0	95.0	65.8	89.6	51.4	81.3	39.4	71.2	29.1	38.5	58.9	74.5	126.9	129.6
$\mathcal{M}^2$ Transformer [14]	81.6	96.0	66.4	90.8	51.8	82.7	39.7	72.8	29.4	39.0	59.2	74.8	129.3	132.1
X-Transformer [33]	81.9	95.7	66.9	90.5	52.4	82.5	40.3	72.4	29.6	39.2	59.5	75.0	131.1	133.5
RSTNet [60]	82.1	96.4	67.0	91.3	52.2	83.0	40.0	73.1	29.6	39.1	59.5	74.6	131.9	134.0
DLCT [30]	82.4	96.6	67.4	91.7	52.8	83.8	40.6	74.0	29.8	39.6	59.8	75.3	133.3	135.4
COS-Net [26]	83.3	96.8	68.6	92.3	54.2	84.5	42.0	74.7	30.4	40.1	60.6	76.4	136.7	138.3
CaMEL [8]	83.2	97.3	68.3	92.7	53.6	84.8	41.2	74.9	30.2	39.7	60.2	75.6	137.5	140.0
SynthCap	**83.7**	**97.6**	**69.2**	**93.5**	**54.9**	**86.3**	**42.8**	**77.1**	**30.9**	**41.3**	**61.4**	**77.7**	**140.1**	**142.6**

Transformer: *A woman wearing a bunch of bananas on her head.*
SynthCap: *A woman wearing a costume with a bunch of bananas on her head.*

Transformer: *A girl blowing out candles on a spoon.*
SynthCap: *A girl is sitting at a table with a birthday cake with a candle.*

Transformer: *A close up of a zebra behind a fence.*
SynthCap: *A zebra standing behind a chain link fence.*

Transformer: *An old truck sitting in a field of flowers.*
SynthCap: *An old rusty truck is parked in a field with yellow flowers.*

Fig. 2. Qualitative comparison between SynthCap and the baseline on sample images from the COCO dataset.

5 Conclusion

In this work, we propose a novel image captioning framework enhanced with a synthetic data augmentation strategy. In particular, we leverage the well-known Stable Diffusion model to generate additional images that can be effectively employed as additional training samples. The proposed strategy is widely usable, given the easy accessibility of advanced text-to-image generative models and their increasingly impressive results. Experimentally, the proposed solution is capable of boosting the performance of a standard Transformer-based model, working only at the data level and maintaining the exact same network.

Acknowledgements. This work has partially been supported by the European Commission under the PNRR-M4C2 (PE00000013) project "FAIR - Future Artificial Intelligence Research", by the Horizon Europe project "European Lighthouse on Safe and Secure AI (ELSA)" (HORIZON-CL4-2021-HUMAN-01-03), co-funded by the European Union, and by the PRIN project "CREATIVE: Cross-modal understanding and generation of Visual and textual content" (CUP B87G22000460001), co-funded by the Italian Ministry of University.

References

1. Allegretti, S., Bolelli, F., Cancilla, M., Pollastri, F., Canalini, L., Grana, C.: How does connected components labeling with decision trees perform on GPUs? In: CAIP (2019)
2. Amoroso, R., Morelli, D., Cornia, M., Baraldi, L., Del Bimbo, A., Cucchiara, R.: Parents and Children: Distinguishing Multimodal DeepFakes from Natural Images. arXiv preprint arXiv:2304.00500 (2023)
3. Anderson, P., Fernando, B., Johnson, M., Gould, S.: SPICE: Semantic Propositional Image Caption Evaluation. In: ECCV (2016)
4. Anderson, P., et al.: Bottom-up and top-down attention for image captioning and visual question answering. In: CVPR (2018)
5. Azizi, S., Kornblith, S., Saharia, C., Norouzi, M., Fleet, D.J.: Synthetic Data from Diffusion Models Improves ImageNet Classification. arXiv preprint arXiv:2304.08466 (2023)
6. Banerjee, S., Lavie, A.: METEOR: An automatic metric for MT evaluation with improved correlation with human judgments. In: ACL Workshops (2005)
7. Barraco, M., Cornia, M., Cascianelli, S., Baraldi, L., Cucchiara, R.: The Unreasonable effectiveness of CLIP features for image captioning: an experimental analysis. In: CVPR Workshops (2022)
8. Barraco, M., Stefanini, M., Cornia, M., Cascianelli, S., Baraldi, L., Cucchiara, R.: CaMEL: Mean Teacher Learning for Image Captioning. In: ICPR (2022)
9. Bolelli, F., Allegretti, S., Grana, C.: One DAG to rule them all. IEEE Trans. PAMI **44**(7), 3647–3658 (2021)
10. Brown, T., et al.: Language models are few-shot learners. In: NeurIPS (2020)
11. Chen, Y., Li, W., Chen, X., Gool, L.V.: Learning semantic segmentation from synthetic data: a geometrically guided input-output adaptation approach. In: CVPR (2019)
12. Cornia, M., Baraldi, L., Cucchiara, R.: Explaining transformer-based image captioning models: an empirical analysis. AI Commun. **35**(2), 111–129 (2022)

13. Cornia, M., Baraldi, L., Fiameni, G., Cucchiara, R.: Universal Captioner: Inducing Content-Style Separation in Vision-and-Language Model Training. arXiv preprint arXiv:2111.12727 (2022)
14. Cornia, M., Stefanini, M., Baraldi, L., Cucchiara, R.: Meshed-Memory Transformer for Image Captioning. In: CVPR (2020)
15. Dosovitskiy, A., et al.: An Image is Worth 16x16 Words: Transformers for Image Recognition at Scale. In: ICLR (2021)
16. Fabbri, M., et al.: MOTSynth: How Can Synthetic Data Help Pedestrian Detection and Tracking? In: ICCV (2021)
17. Herdade, S., Kappeler, A., Boakye, K., Soares, J.: Image Captioning: Transforming Objects into Words. In: NeurIPS (2019)
18. Hessel, J., Holtzman, A., Forbes, M., Bras, R.L., Choi, Y.: CLIPScore: A Reference-free Evaluation Metric for Image Captioning. In: EMNLP (2021)
19. Hossain, M.Z., Sohel, F., Shiratuddin, M.F., Laga, H., Bennamoun, M.: Text to image synthesis for improved image captioning. IEEE Access **9**, 64918–64928 (2021)
20. Hu, X., et al.: Scaling Up Vision-Language Pre-training for Image Captioning. In: CVPR (2022)
21. Huang, L., Wang, W., Chen, J., Wei, X.Y.: Attention on attention for image captioning. In: ICCV (2019)
22. Karpathy, A., Fei-Fei, L.: Deep visual-semantic alignments for generating image descriptions. In: CVPR (2015)
23. Kingma, D.P., Ba, J.: Adam: a method for stochastic optimization. In: ICLR (2015)
24. Li, J., Li, D., Xiong, C., Hoi, S.: BLIP: Bootstrapping Language-Image Pre-training for Unified Vision-Language Understanding and Generation. In: ICML (2022)
25. Li, W., Lotz, F.J., Qiu, C., Elliott, D.: Data curation for image captioning with text-to-image generative models. arXiv preprint arXiv:2305.03610 (2023)
26. Li, Y., Pan, Y., Yao, T., Mei, T.: Comprehending and ordering semantics for image captioning. In: CVPR (2022)
27. Lin, C.Y.: Rouge: A package for automatic evaluation of summaries. In: ACL Workshops (2004)
28. Lin, T.Y., et al.: Microsoft COCO: Common Objects in Context. In: ECCV (2014)
29. Liu, W., Chen, S., Guo, L., Zhu, X., Liu, J.: CPTR: Full Transformer Network for Image Captioning. arXiv preprint arXiv:2101.10804 (2021)
30. Luo, Y., et al.: Dual-Level Collaborative Transformer for Image Captioning. In: AAAI (2021)
31. Micikevicius, P., et al.: Mixed Precision Training. In: ICLR (2018)
32. Moratelli, N., Barraco, M., Morelli, D., Cornia, M., Baraldi, L., Cucchiara, R.: Fashion-oriented image captioning with external knowledge retrieval and fully attentive gates. Sensors **23**(3), 1286 (2023)
33. Pan, Y., Yao, T., Li, Y., Mei, T.: X-Linear Attention Networks for Image Captioning. In: CVPR (2020)
34. Papineni, K., Roukos, S., Ward, T., Zhu, W.J.: BLEU: a method for automatic evaluation of machine translation. In: ACL (2002)
35. Pipoli, V., Cappelli, M., Palladini, A., Peluso, C., Lovino, M., Ficarra, E.: Predicting gene expression levels from DNA sequences and post-transcriptional information with Transformers. Comput. Methods Prog. Biomed. **225**, 107035 (2022)
36. Radford, A., et al.: Learning Transferable Visual Models From Natural Language Supervision. In: ICML (2021)
37. Rajbhandari, S., Rasley, J., Ruwase, O., He, Y.: ZeRO: Memory optimizations Toward Training Trillion Parameter Models. In: SC (2020)

38. Rennie, S.J., Marcheret, E., Mroueh, Y., Ross, J., Goel, V.: Self-Critical Sequence Training for Image Captioning. In: CVPR (2017)
39. Rombach, R., Blattmann, A., Lorenz, D., Esser, P., Ommer, B.: High-resolution image synthesis with latent diffusion models. In: CVPR (2022)
40. Sarto, S., Barraco, M., Cornia, M., Baraldi, L., Cucchiara, R.: Positive-Augmented Contrastive Learning for Image and Video Captioning Evaluation. In: CVPR (2023)
41. Sarto, S., Cornia, M., Baraldi, L., Cucchiara, R.: Retrieval-augmented transformer for image captioning. In: CBMI (2022)
42. Schuhmann, et al.: LAION-5B: An open large-scale dataset for training next generation image-text models. In: NeurIPS (2022)
43. Sennrich, R., Haddow, B., Birch, A.: Neural machine translation of rare words with subword units. In: ACL (2016)
44. Sharma, P., Ding, N., Goodman, S., Soricut, R.: Conceptual Captions: A Cleaned, Hypernymed. ACL, Image Alt-text Dataset For Automatic Image Captioning. In (2018)
45. Shen, S., et al.: How much can CLIP benefit vision-and-language tasks? In: ICLR (2022)
46. Stefanini, M., Cornia, M., Baraldi, L., Cascianelli, S., Fiameni, G., Cucchiara, R.: From show to tell: a survey on deep learning-based image captioning. IEEE Trans. PAMI **45**(1), 539–559 (2022)
47. Stefanini, M., Lovino, M., Cucchiara, R., Ficarra, E.: Predicting gene and protein expression levels from DNA and protein sequences with Perceiver. Computer Methods and Programs in Biomedicine **234**, 107504 (2023)
48. Vaswani, A., et al.: Attention is all you need. In: NeurIPS (2017)
49. Vedantam, R., Lawrence Zitnick, C., Parikh, D.: CIDEr: Consensus-based Image Description Evaluation. In: CVPR (2015)
50. Vinyals, O., Toshev, A., Bengio, S., Erhan, D.: Show and tell: A neural image caption generator. In: CVPR (2015)
51. Wang, Z., Yu, J., Yu, A.W., Dai, Z., Tsvetkov, Y., Cao, Y.: SimVLM: Simple Visual Language Model Pretraining with Weak Supervision. In: ICLR (2022)
52. Wolf, T., et al.: Transformers: State-of-the-Art Natural Language Processing. In: EMNLP (2020)
53. Wu, M., et al.: DIFNet: Boosting Visual Information Flow for Image Captioning. In: CVPR (2022)
54. Xiao, C., Xu, S.X., Zhang, K.: Multimodal Data Augmentation for Image Captioning using Diffusion Models. arXiv preprint arXiv:2305.01855 (2023)
55. Xu, K., et al.: Show, attend and tell: Neural image caption generation with visual attention. In: ICML (2015)
56. Yang, X., Tang, K., Zhang, H., Cai, J.: Auto-encoding scene graphs for image captioning. In: CVPR (2019)
57. Yao, T., Pan, Y., Li, Y., Mei, T.: Exploring visual relationship for image captioning. In: ECCV (2018)
58. Zhang, P., et al.: VinVL: Revisiting visual representations in vision-language models. In: CVPR (2021)
59. Zhang, S., et al.: OPT: Open Pre-trained Transformer Language Models. arXiv preprint arXiv:2205.01068 (2022)
60. Zhang, X., et al.: RSTNet: Captioning with adaptive attention on visual and non-visual words. In: CVPR (2021)

An Effective CNN-Based Super Resolution Method for Video Coding

Jun Yin[1,2], Shuang Peng[1], Jucai Lin[1], Dong Jiang[1(✉)], and Cheng Fang[1]

[1] Zhejiang Dahua Technology Co., Ltd, Hangzhou, China
{yin_jun,peng_shuang,lin_jucai,jiang_dong,fang_cheng1}@dahuatech.com

[2] Zhejiang University, Hangzhou, China

Abstract. The reference picture resampling (RPR) technique in Versatile Video Coding allows the input video to be down-sampled before encoding, and then the decoded video will be up-sampled. Compared with the handcrafted up-sampling interpolation filter, the convolutional neural network based super resolution (SR) method has better restoration effect. In this paper, we introduce the RPR technique into the SR method, to effectively restore the details lost in the down-sampling and encoding process. Meanwhile, the normalized attention module in the network is designed to adaptively determine the weight of the extracted features. Moreover, the prediction picture generated during the codec process is fed into the network as side information, which can provide the directional and texture information of the original picture, and implicitly derive the lost details together with the reconstruction picture. Compared with VTM-11.0-NNVC, the experimental results show that the proposed method achieves 12.79% and 10.53% BD-rate reductions, under all intra and random access configurations, respectively.

Keywords: Super resolution · Convolutional neural network · Reference picture resampling · Versatile Video Coding

1 Introduction

Resampling-based video coding scheme is proposed to solve the problem of video transmission under the limited bandwidth environment [1], especially for high-resolution video content, such as 4K videos. In this scheme, the input video is down-sampled before encoding and the decoded video is up-sampled after decoding. In Versatile Video Coding (VVC) [2], a reference picture resampling (RPR) technique is used to support this scheme, which utilizes the handcrafted filter to achieve down-sampling and up-sampling.

Besides, the resampling scheme can bring some coding gains at the low bitrates, especially when combined with the convolutional neural network (CNN) based super resolution (SR) method. For example, Wei et al. [3] proposed an end-to-end resampling coding scheme based on CNN to improve the compression efficiency, and used the trained down-sampling and up-sampling networks as the pre-processing and post-processing of HEVC, respectively. Lin et al. [4] proposed a block-level CNN-based

G. L. Foresti et al. (Eds.): ICIAP 2023, LNCS 14233, pp. 124–134, 2023.
https://doi.org/10.1007/978-3-031-43148-7_11

up-sampling method, and adaptively selected the coding mode between full-resolution and low-resolution based on rate-distortion cost. Liu et al. [5] proposed a CNN-based residual SR method, which performed down-sampling and up-sampling in the residual domain and utilized the prediction signal to improve the restoration quality. Lin et al. [6] proposed a CNN-based SR method using decoded information, and the trained model was used to replace the RPR up-sampling filter in VVC, which effectively improved the performance of I-slice. Furthermore, the improved methods [7] are used for B-slice. Peng et al. [8] designed an advanced attention residual block to improve the performance of CNN-based up-sampling filter. Kotra et al. [9] retrained a simplified version of EDSR [10] and used it as the up-sampling filter in the RPR process.

Although the previous resampling-based video coding methods combined with CNN improve the compression efficiency, there is still room for improvements. In this paper, an effective CNN-based SR method is proposed. Firstly, the RPR technique is introduced into CNN, which enables the network to focus on restoring the residual details lost in the resampling and encoding process. Secondly, the importance of the extracted features is determined by the normalized attention module. Thirdly, the prediction picture generated during the codec process is input into the network as side information, which can provide the directional and texture characteristics of the picture to be restored. Moreover, the residual details can be implicitly derived from the reconstruction picture and the prediction picture via the network. Lastly, the quantization parameter (QP) is fed into the network to improve model generalization, so that the trained model can be used for all QPs rather than one model for each QP.

The remainder of this paper is organized as follows. Section 2 presents the proposed SR method. Section 3 provides the experimental results and analyses. Finally, Sect. 4 concludes this paper.

2 Proposed Method

The overall framework of the proposed CNN-based SR method for video coding is shown in Fig. 1. The original full-resolution video is firstly down-sampled by the RPR filter before encoding, and then the decoded low-resolution video is up-sampled by a specially designed CNN. The architecture of the CNN is described in Sects. 2.1 and 2.2, and the training and inference details are described in Sects. 2.3 and 2.4.

2.1 Architecture of CNN

The architecture of the designed CNN is shown in Fig. 2, which includes the RPR-based global skip connection branch and the CNN-based SR branch. In the skip branch, the low-resolution picture provides basic information for the full-resolution picture via RPR up-sampling, so that the other branch only needs to learn the remaining residual information. In the SR branch, an attention-based neural network is designed, which is based on EDSR but removes the scaling factor and introduces the attentional mechanism [11].

As shown in Fig. 2, the restored full-resolution picture H_{Rec} can be calculated as follows.

$$H_{Rec} = SKIP_{RPR} + SR_{CNN} \tag{1}$$

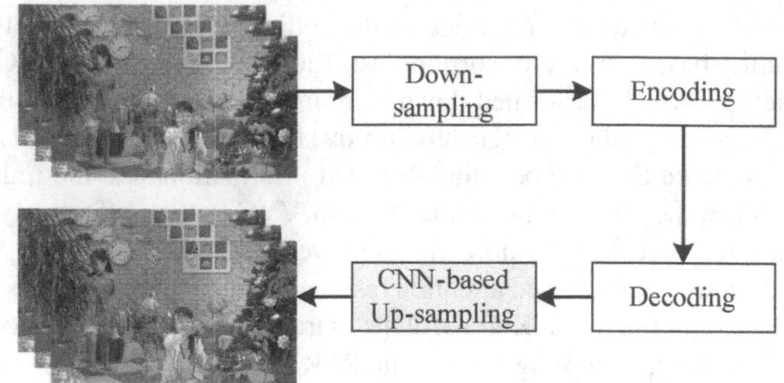

Fig. 1. The overall framework of the proposed method.

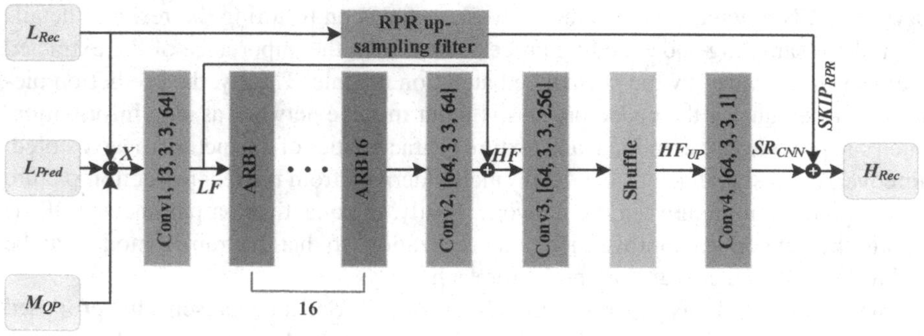

Fig. 2. The architecture of the designed CNN.

where $SKIP_{RPR}$ and SR_{CNN} are the output of the skip and SR branches, respectively. For the skip branch, the input low-resolution reconstruction picture is up-sampled by the RPR filter. For the SR branch, the input consists of three parts: the low-resolution reconstruction picture L_{Rec}, its corresponding low-resolution prediction picture L_{Pred}, and M_{QP}. M_{QP} is a matrix with the same shape as L_{Rec}, which is expanded by QP. The output of this branch is calculated as follows.

$$SR_{CNN} = UP_{CNN}\left(L_{Rec}, L_{Pred}, M_{QP}\right) \tag{2}$$

where $UP_{CNN}(\cdot)$ is the network of the SR branch, which contains 4 convolutional (Conv) layers, 16 attention residual blocks (ARBs), a concatenation layer, a shuffle layer, and a local skip connection from Conv1 to Conv2. The parameters of each convolutional layer are shown as $[C_{in}, kz, kz, C_{out}]$ in this figure, where C_{in}, kz and C_{out} represent the number of input channels, the kernel size and the number of output channels, respectively.

In the processing of the SR branch, all inputs are first concatenated as one input X, which can be described as the following equation.

$$X = \text{Cat}\left(L_{Rec}, L_{Pred}, M_{QP}\right) \tag{3}$$

Then, the convolutional layer Conv1 is used to extract the low-level feature LF from the input X, which is calculated as follows.

$$LF = Conv_1(X) \tag{4}$$

where $Conv_i\ (\cdot)$ represents the i-th convolutional layer. Further, LF is mapped to the high-level feature HF via the attentional residual block and the local skip connection, which is calculated as follows.

$$HF = LF + Conv_2(ARB_{16}(\ldots(ARB_2(ARB_1(LF)))\ldots)) \tag{5}$$

where $ARB_n(\cdot)$ represents the n-th ARB, and the specific structure of ARB is described in Sect. 2.3. Subsequently, a convolutional layer Conv3 and the Shuffle layer are used in sequence to expand the channel number of HF and obtain the full-resolution feature HF_{UP}, respectively, which is formulated as follows.

$$HF_{UP} = Shuffle(Conv_3(HF)) \tag{6}$$

Finally, HF_{UP} is mapped to SR_{CNN} via the convolutional layer Conv4.

2.2 Attention Residual Block

In this section, the structure of the ARB is given, as shown in Fig. 3. For each ARB, the combination of a convolutional layer, an activation layer and a convolutional layer is firstly used to extract the nonlinear features, and then the channel attention extraction module obtains the weight, as shown in the dashed line in the figure. Finally, the weighted features are added to the input through the skip connection to get the output features.

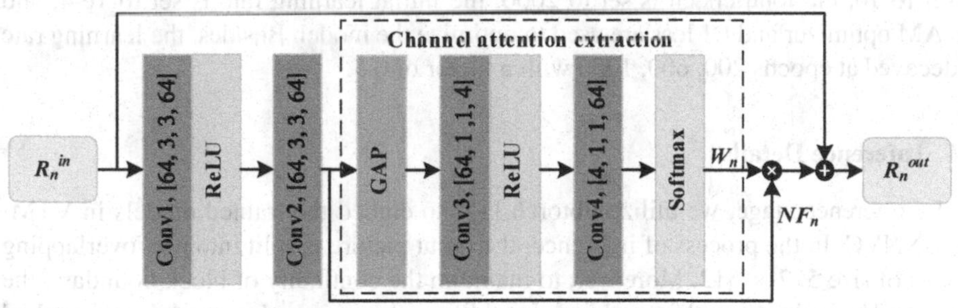

Fig. 3. The structure of the n-th ARB.

The process of the n-th ARB can be calculated as follows.

$$R_n^{out} = ARB_n\left(R_n^{in}\right) = R_n^{in} + W_n \odot NF_n \tag{7}$$

where $R_n{}^{out}$, $R_n{}^{in}$, W_n and NF_n are the output and the input, the weights and the nonlinear features, respectively, and $\odot$ is channel-wise multiplication. NF_n is calculated as follows.

$$NF_n = Conv_{n-2}(ReLU(Conv_{n-1}\left(R_n^{in}\right))) \tag{8}$$

where $Conv_{n\text{-}j}$ represents the j-th convolutional layer, and $ReLU$ is the Rectified Linear Unit. The calculation process for W_n is expressed as follows.

$$W_n = Softmax(Conv_{n-4}(ReLU(Conv_{n-3}(GAP(NF_n))))) \tag{9}$$

In this process, the global average pooling (GAP) layer firstly converts the dimension from $H \times W \times C$ to $1 \times 1 \times C$, where H, W, and C are the width, height, and number of channels, respectively. Then, convolutional layer and activation layer are used to extract the attention features. Finally, Softmax layer converts the attention features into the normalized attention weight.

2.3 Training Details

We use BVI-DVC [12] dataset to train the proposed SR method. The dataset contains 4 different spatial resolutions from 270p to 2160p, including 200 sequences for each resolution, and 64 frame pictures for each sequence. In this paper, only the 2160p video sequences are used, in which the first 190 sequences are used for training and the remaining 10 sequences are used for validation. The training data is down-sampled by 2x using the RPR filter, and then encoded by the neural network-based video coding reference software VTM-11.0-NNVC [13] with QPs equal to {22, 27, 32, 37, 42}, under all intra (AI) and random access (RA) configurations, respectively. It is noted that the model of I-slice and B-slice are trained separately due to their different encoding features.

The proposed method is implemented based on PyTorch [14] framework, and accelerated by 4 NVIDIA A40 GPUs. The input patch of the model with 128 × 128 size is randomly cropped from the raw picture. In additional, the patch is randomly flipped and rotated before being fed into the model. In the process of training, the batch size is set to 16, the total epoch is set to 2000, the initial learning rate is set to 1e-4, and ADAM optimizer and L1 loss are used to optimize the model. Besides, the learning rate is decayed at epochs 200, 600, 1000 with a factor of 0.5.

2.4 Inference Details

In the inference stage, we utilize libtorch [15] to embed the trained models in VTM-11.0-NNVC. In the process of inference, the input picture is split into non-overlapping blocks of size 512 × 512. Moreover, to maintain the continuity of block boundary, the adjacent 16 pixels are used for padding. In addition, the proposed method is only applied for the luma component.

3 Experimental Results

In this section, the experimental results of the proposed method are described and analyzed in detail. Firstly, the test conditions are given in Sect. 3.1, and then the experimental results and analysis are given in Sect. 3.2. Further, quality evaluations with subjective and objective are given in Sect. 3.3. Moreover, some ablation experiments are conducted to validate the effectiveness of the proposed method in Sect. 3.4.

3.1 Test conditions

All experiments are conducted under the common test conditions and evaluation procedures for neural network-based video coding technology [16]. Considering that resampling-based video coding scheme is mainly effective for high resolution content, the test sequences from the Class A1 and A2 are used for testing. The compression performance and the quality improvement are measured with BD-rate and BD-PSNR [17]. The anchor used for comparison is VTM-11.0-NNVC. In addition, there is an obvious the bitrates between the test with the low-resolution picture and the anchor with full-resolution picture in the resampling based video coding method. The QPs of the anchor is offset by 5 to reduce the bitrates difference, that is, the QPs of test and anchor are set to {22, 27, 32, 37, 42} and {27, 32, 37, 42, 47} respectively.

3.2 Results and Analysis

Table 1 and Table 2 show the test BD-rate results of the proposed method under AI and RA configurations, respectively. It can be found that the proposed method brings significant coding gains. The results show that the proposed method achieves 12.79% and 10.53% BD-rate reductions on average, under AI and RA configurations. Table 3 show the BD-PSNR results of the proposed method. The results show that the proposed method achieves 0.42 dB and 0.28 dB PSNR improvements on average, under AI and RA configurations. It can be seen that the coding gain is improved on all test sequences, which proves that the proposed method can effectively restore the full-resolution picture.

Table 1. BD-rate over VTM-11.0-NNVC under AI configuration (%)

Class	Sequence	AI			
		Lin [7]	Peng [8]	Kotra [9]	Proposed
Class A1	Tango2	-12.54	-12.69	-11.68	**-13.72**
	FoodMarket4	-6.85	-7.27	-6.31	**-8.43**
	Campfire	-19.70	-17.37	-17.88	**-18.48**
Class A2	CatRobot	-11.54	-11.14	-9.65	**-12.35**
	DaylightRoad2	-7.12	-6.10	-5.62	**-8.15**
	ParkRunning3	-14.16	-15.30	-13.65	**-15.63**
Average on Class A1		-13.03	-12.44	-11.96	**-13.54**
Average on Class A2		-10.94	-10.85	-9.64	**-12.04**
Average Overall		-11.99	-11.64	-10.80	**-12.79**
STD		4.79	4.41	4.63	**4.05**
TNP (Million)		5.51	4.07	**2.74**	2.89
MAC (KMAC/pixel)		764	508	**344**	361

Table 2. BD-rate over VTM-11.0-NNVC under RA configuration (%)

Class	Sequence	RA			
		Lin [7]	Peng [8]	Kotra [9]	Proposed
Class A1	Tango2	-11.84	-8.47	-10.32	**-13.02**
	FoodMarket4	-10.98	-7.43	-9.98	**-11.64**
	Campfire	**-18.75**	-8.47	-16.83	-17.99
Class A2	CatRobot	-5.73	-1.56	-3.36	**-5.74**
	DaylightRoad2	-0.79	3.60	2.35	**-1.47**
	ParkRunning3	-10.34	-8.96	-10.17	**-13.30**
Average on Class A1		-13.86	-8.12	-12.38	**-14.22**
Average on Class A2		-5.62	-2.31	-3.73	**-6.84**
Average Overall		-9.74	-5.22	-8.05	**-10.53**
STD		6.06	**5.12**	6.64	5.93
TNP (Million)		5.51	4.07	**2.74**	2.89
MAC (KMAC/pixel)		764	508	**344**	361

In addition, the results of comparison methods [7, 8] and [9] are also given in Table 1 and Table 2, and the best results for each row are identified in bold. The STD in the table is the standard deviation of the results of all test sequences, which is used to measure the stability of the method on different video content. The TNP and MAC in the table are the total number parameters and the number of multiply-accumulates, which reflect the memory consumption and the computational complexity. As the results show, overall the proposed method achieves the highest coding gains under AI and RA configurations. It can also be observed that the stability and the memory consumption of the proposed method is better than that of the comparison method. Although the STD of the proposed method is greater than that of method [8] under RA configuration, the performance of the proposed method is much higher than that of method [8].

Table 3. The BD-PSNR of the proposed method (dB).

	Tango2	Food-Market4	Campfire	CatRobot	Day-lightRoad2	Park-Running3	Average Overall
AI	0.38	0.35	0.34	0.47	0.20	0.78	0.42
RA	0.32	0.41	0.42	0.11	-0.04	0.46	0.28

Table 4. Results of the ablation experiments

	BD-rate (%)				BD-PSNR (dB)			
	proposed	w/o skip	w/o pred	w/o attn	proposed	w/o skip	w/o pred	w/o attn
AI	-12.79	-12.52	-11.83	-9.79	0.42	0.41	0.39	0.31
RA	-10.53	-3.62	-9.92	-8.58	0.28	0.03	0.26	0.21

Furtherly, we analyze the relationship between gain and computational complexity. The calculation for kilo-MAC (KMAC) per pixel is as follows.

$$\begin{aligned} \text{KMAC/pixel} = & \sum((kz_m)^2 \cdot (C_m^{in} \cdot W_m^{in} \cdot H_m^{in}) \\ & \cdot (C_m^{out} / (W_m^{out} \cdot H_m^{out})))/1000 \end{aligned} \quad (10)$$

where kz_m is the kernel size of the m-th convolutional layer, and $C_m{}^{in}$, $W_m{}^{in}$, $H_m{}^{in}$, $C_m{}^{out}$, $W_m{}^{out}$ and $H_m{}^{out}$ are the channel number, width and height for input and output of the m-th convolutional layer respectively.

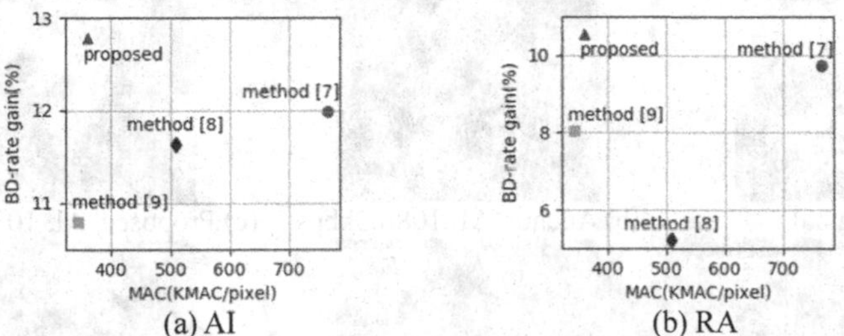

(a) AI (b) RA

Fig. 4. Gain vs. computational complexity under AI and RA configurations.

For a clear observation, gain versus computational complexity is given in Fig. 4. Compared with method [9], the proposed method brings significant gain with similar computational complexity. The proposed method has higher gain and lower computational complexity than method [7] and [8]. The results show that the proposed method achieves the best tradeoff between performance and complexity.

3.3 Quality Evaluations

For objective quality evaluation, the rate-distortion curves for each sequence are given under RA configuration, as shown in Fig. 5. Overall, the curves of the proposed method are located in the upper left position of the anchor. The results show that the proposed method can provide better quality at the same bitrates. It can be noticed that the proposed method has more significant quality improvements at the low bitrates than at the high bitrates. Moreover, some sequences show potential quality improvements at the high bitrates, such as Parkrunning3.

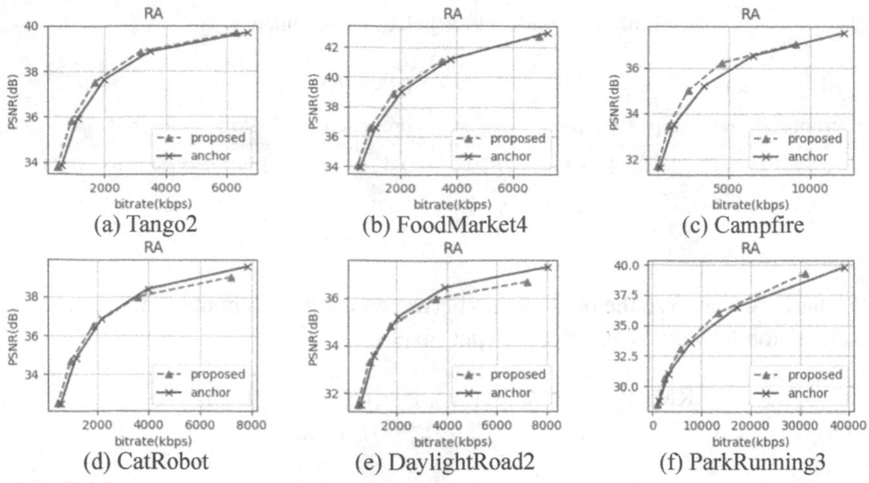

(a) Tango2 (b) FoodMarket4 (c) Campfire

(d) CatRobot (e) DaylightRoad2 (f) ParkRunning3

Fig. 5. The rate-distortion curves comparing the anchor with the proposed method.

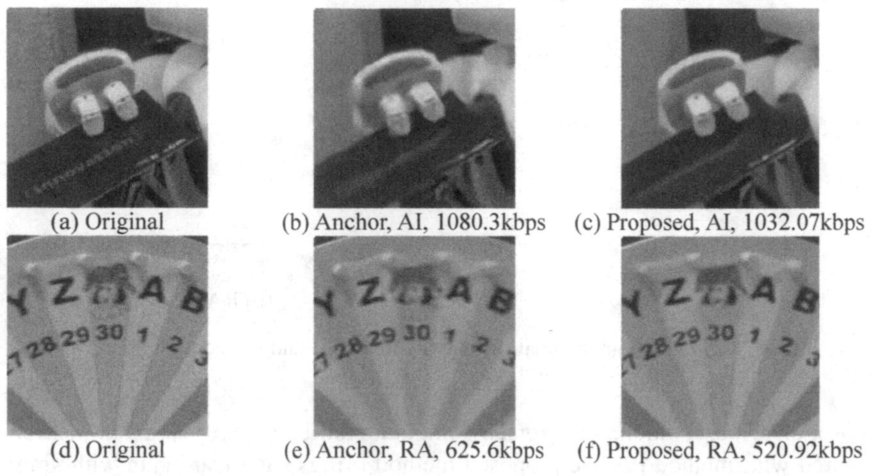

(a) Original (b) Anchor, AI, 1080.3kbps (c) Proposed, AI, 1032.07kbps

(d) Original (e) Anchor, RA, 625.6kbps (f) Proposed, RA, 520.92kbps

Fig. 6. The comparison of subjective quality between the anchor and the proposed method under AI and RA configurations.

For subjective quality evaluation, Fig. 6 shows the subjective results of the proposed method. Figure 6 (a) and (d) are the original pictures cropped from the first frame and the second frame of the sequence CatRobot with the size of 256×256. Figure 6 (b) and (e) are the corresponding pictures compressed by the anchor with the bitrates of 1080.3 kbps and 625.6 kbps under AI and RA configurations, and Fig. 6 (c) and (f) are the corresponding pictures using our method with the bitrates of 1032.07 kbps and 520.92 kbps under AI and RA configurations, respectively. It can be seen that compared with the anchor, the edge and the content of the compressed picture are clearer in our method, which means our method effectively reduce artifacts and blurring, and provides more visually pleasing results.

3.4 Ablation Experiments

To validate the effectiveness of the proposed method, some ablation experiments are investigated, and the results are shown in Table 4. The w/o skip denotes the proposed method removing the global skip connection, the w/o pred denotes removing the prediction picture, and the w/o attn denotes removing the attention extraction module from ARB. It can be observed that removing any one of these three elements results in a coding degradation. The coding gain decreases from 10.53% to 3.62% under RA configuration when the skip connection is removed, which indicates that the skip connection is the key to the performance improvement. The results of the w/o pred show a degradation from 12.79% to 11.83% and from 10.53% to 9.92% under AI and RA configurations, which indicates the importance of using the prediction picture as side information. Similarly, the results of w/o attn show the coding gain drops to 9.79% and 8.58% under AI and RA configurations, which indicates that the attention extraction module is very effective for the performance improvement. In addition, it can be seen that the results of BD-PSNR are consistent with those of BD-rate.

4 Conclusion

In this paper, we propose an effective CNN-based SR method with RPR-based global skip connection. And the attention mechanism is introduced into the network. Meanwhile, the side information generated during the codec process is utilized to further improve the coding gain. Experimental results show that the proposed method significantly improves the coding performance. Specifically, the proposed method achieves 12.79% and 10.53% BD-rate reductions, compared with the VTM-11.0-NNVC, under AI and RA configurations, respectively. Moreover, the ablation experiments are designed to validate the effectiveness of the proposed method.

References

1. Bruckstein, A.M., Elad, M., Kimmel, R.: Down-scaling for better transform compression. IEEE Trans. Image Process. **12**(9), 1132–1144 (2003)
2. Bross, B., Chen, J., Liu, S., Wang, Y.: Versatile video coding (draft10). In: 19th JVET Meeting, JVET-S2001, JVET Document, Teleconference (2020)
3. Wei, Y., Chen, L., Song, L.: Video compression based on jointly learned down-sampling and super-resolution networks. In: 2021 International Conference on Visual Communications and Image Processing, pp. 1–5 (2021)
4. Lin, J., Liu, D., Yang, H., Li, H., Wu, F.: Convolutional neural network-based block up-sampling for HEVC. IEEE Trans. Circuits Syst. Video Technol. **29**(12), 3701–3715 (2019)
5. Liu, K., Liu, D., Li, H., Wu, F.: Convolutional neural network-based residue super-resolution for video coding. In: 2018 IEEE Visual Communications and Image Processing, pp. 1–4 (2018)

6. Lin, C., Li, Y., Zhang, K., Zhang, Z., Zhang, L.: CNN-based super resolution for video coding using decoded information. In: 2021 International Conference on Visual Communications and Image Processing, pp. 1–5 (2021)
7. Lin, C., Li, Y., Zhang, K., Zhang, L.: EE1–2.2: CNN-based super resolution for video coding using decoded information. In: 24th JVET Meeting, JVET-X0064, JVET Document, Teleconference (2021)
8. Peng, S., Dong, J., Lin, J., Fang, C., Zhang, X.: AHG11: An improved cNN-based super resolution method. In: 25th JVET Meeting, JVET-Y0087, JVET Document, Teleconference (2022)
9. Kotra, A.M., Reuzé, K., Chen, J., Wang, H., Karczewicz, M., Li, J.: EE1–2.3: neural network-based super resolution. In: 23th JVET Meeting, JVET-W0105, JVET Document, Teleconference (2021)
10. Lim, B., Son, S., Kim, H., Nah, S., Lee, K.M.: Enhanced deep residual networks for single image super-resolution. In: 2017 IEEE Conference on Computer Vision and Pattern Recognition Workshops, pp. 1132–1140 (2017)
11. Hu, J., Shen, L., Sun, G.: Squeeze-and-excitation networks. In: 2018 IEEE Conference on Computer Vision and Pattern Recognition, pp. 7132–7141 (2018)
12. Ma, D., Zhang, F., Bull, D.: BVI-DVC: a training database for deep video compression. arXiv preprint arXiv:2003.13552 (2020)
13. Available: https://vcgit.hhi.fraunhofer.de/jvet/VVCSoftware_VTM/-/tags/VTM-11.0-NNVC
14. Paszke, A., et al: PyTorch: An Imperative Style, High-Performance Deep Learning Library. arXiv preprint arXiv:1912.01703 (2019)
15. Available: https://download.pytorch.org/libtorch/cpu/libtorch-cxx11-abi-shared-with-deps-1.13.0%2Bcpu.zip
16. Liu, A., Segall, A., Alshina, E., Liao, R.: JVET common test conditions and evaluation procedures for neural network-based video coding technology. In: 24th JVET meeting, JVET-X2016, JVET document, Teleconference (2021)
17. Bjontegaard, G.: Calcuation of average PSNR differences between rdcurves. In: 13th VCEG Meeting, VCEG-M33, VCEG Document, Austin (2001)

Medical Transformers for Boosting Automatic Grading of Colon Carcinoma in Histological Images

Pierluigi Carcagnì[1], Marco Leo[1(✉)], Luca Signore[2], and Cosimo Distante[1,2]

[1] CNR-ISASI, Ecotekne Campus via Monteroni, 73100 Lecce, Italy
marco.leo@cnr.it

[2] Università del Salento, Via Monteroni, 73100 Lecce, Italy

Abstract. Developing computer-aided approaches for cancer diagnosis and grading is receiving an uprising demand since this could take over intra- and inter-observer inconsistency, speed up the screening process, allow early diagnosis, and improve the accuracy and consistency of the treatment planning processes. The third most common cancer worldwide and the second most common in women is ColoRectal Cancer (CRC). Grading CRC is a key task in planning appropriate treatments and estimating the response to them. Automatic systems have the potential to speed up and make it more robust but, unfortunately, the most recent and promising machine learning techniques have not been applied for automatic CRC grading so far. For example, there is no work exploiting transformer networks, which outperform convolutional neural networks (CNN) and are replacing them in many applications, for CRC detection and grading at a large scale. To fill this gap, in this work, a transformer-based network endowed with an additional control mechanism in the self-attention module is exploited to understand discriminative regions in large histological images. These relevant regions have been used to train the most suited Convolutional Neural Network (as emerged from recent research findings) for the automatic grading of CRC. The experimental proofs on the largest publicly available CRC dataset demonstrated marked improvement with respect to the leading state-of-the-art approaches relying on CNN.

Keywords: Colon carcinoma · Artificial intelligence · Deep learning · Transformer networks · Histological diagnosis

1 Introduction

Colorectal carcinoma (CRC) is a well-characterized heterogeneous disease induced by different tumorigenic modifications in colon cells.

Developing computer-aided approaches for cancer diagnosis and grading is receiving an uprising demand since this could take over intra- and inter-observer inconsistency, speed up the screening process, favourite early diagnosis and then improve the accuracy and consistency of treatment planning processes [5,7,14].

G. L. Foresti et al. (Eds.): ICIAP 2023, LNCS 14233, pp. 135–146, 2023.
https://doi.org/10.1007/978-3-031-43148-7_12

In the last years, several works introduced modern machine learning strategies for making computer-aided approaches for cancer diagnosis and grading more and more reliable. Most of them rely on deep learning, and in particular on Convolutional Neural Networks. It is well known that CNN models often fail in global context modelling: due to the inherent locality of convolution operations, CNN usually shows limitations in clearly modelling dependencies. Since the boundary between multiple cells and organs is difficult to be distinguished on the image and then global information can help. To overcome this problem, researchers in machine learning have introduced transformer architectures [22] which exploit self-attention and even at the very first layer of information processing make connections between distant image locations [12]. In the field of medical image analysis, transformers have been successfully used in to full-stack clinical applications, including image synthesis/reconstruction, registration, segmentation, detection, and diagnosis [8]. On the other side, Transformers are hungry for annotated data which cannot easily be built, especially in clinical tasks where high-quality and consistent annotations require excellent skills, e.g. cancer grading. This is why, from the literature, it emerged that there are no works integrating transformers into algorithmic pipelines aimed at colon cancer grading. Existing work has demonstrated the effectiveness of this methodology in other related tasks, at first glandular regions segmentation [16]. The main contribution of this paper is to introduce a two-stage colon adenocarcinoma grading pipeline. The first stage aims at segmenting glandular regions whereas the second step is devoted to the grading of regions retained after segmentation. The second contribution is to merge the pros of CNN and Transformers architectures. Transformers are exploited for the segmentation step (for which a large amount of consistent annotated data are available) to precisely determine glandular boundaries to be supplied to the following multiclass grading problem lying on CNN exploiting local patterns of cells' configurations. Experiments on the largest, publicly available, CRC dataset demonstrated the effectiveness of the two-stage strategy and, besides, the actual contribution of transformers with respect to other segmentation strategies into the first stage. The rest of the paper is organized as follows: Sect. 2 describes related work whereas Sects. 3 and 4 introduce methods and data, respectively. Then Sect. 5 reports experimental results about grading colon carcinoma using transformers and convolutional networks. Finally, Sect. 6 concludes the paper.

2 Related Work

The focus of the paper is colon cancer grading. There are a few works specifically designed for colon cancer grading due to some challenges i.e.: extracting knowledge from limited datasets of labelled and biased data (the labelling task requires high clinical skills, and it is time-consuming); getting rid of intra-class variance and inter-class similarity derived from the continuum existing from the different grading levels.

Early attempts relied on handcrafted features and shallow classifiers [1,3,20]. Recently, deep learning-based approaches demonstrated their superiority in colon cancer grading. CNNs are generally used for representation learning from small image patches (e.g. 224 × 224) extracted from digital histology images due to computational and memory constraints. Some of them made use of an intermediate tissue classification [23]. The patch-level classification results have to be then combined to aggregate predictions and model the fact that not all patches will be discriminative [10]. In [4] the authors combined convolutional and recurrent architectures to train a deep network to predict colorectal cancer outcomes based on images of tumour tissue samples. Some contextual information can be embedded in the model in order to improve accuracy as proposed in [17] where a computational model, that utilises feature sharing across scales and learns dependencies between scales using long-short term memory (LSTM) unit, was tested. In [25] a novel cell-graph convolutional neural network (CGC-Net) that converts each large histology image into a graph, where each node is represented by a nucleus within the original image and cellular interactions are denoted as edges between these nodes according to node similarity, was proposed. More recently, a framework for context-aware learning of histology images has been proposed in [15]. The framework first learns the Local Representation by a CNN (LR-CNN) and then it aggregates the contextual information through a representation aggregation CNN (RA-CNN). Anyway, a recent study [11] compared several CNN architectures and it has demonstrated that models designed for the task of image classification can work better than domain-specific solutions due to the lack of data for a robust knowledge generalization in the target domain. In particular, it emerged that for CRC grading, EfficientNet-B1 and EfficientNet-B2 architectures [19] outperform all the approaches in the state of the art so far. Recent studies demonstrated that an attention mechanism, added in parallel to capture key features that facilitate network classification, can really help CNN in accomplishing knowledge extraction tasks from large histological images [13,24]. Finally, it has been proved that multi-step pipeline [6,9,24] may result more effective when analyzing histopathological images for highly complex tasks such as grading colon adenocarcinoma.

3 Methods

A schematic representation of the proposed pipeline exploiting a transformer architecture is reported in Fig. 1.

In this paper, a transformer-based model endowed with an additional control mechanism in the self-attention module is preliminarily exploited to understand discriminative regions in large histological images. All 165 provided images picturing stage T3 or T42 colorectal adenocarcinoma have been exploited to make the capability of generalization the largest as possible.

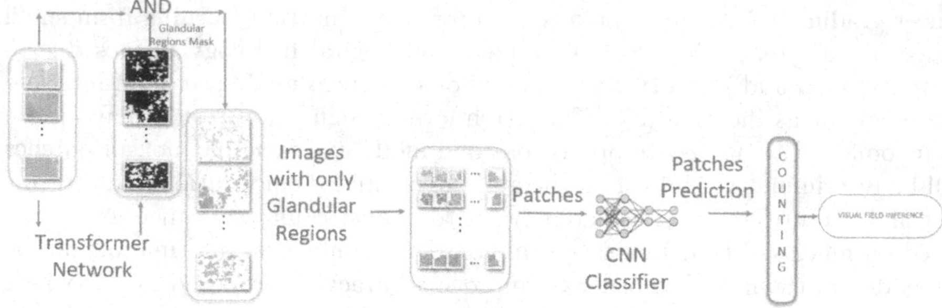

Fig. 1. A schematic representation of the proposed pipeline exploiting a transformer architecture. Through the transformer architecture trained on the GLAs dataset, glandular regions in each input image of the CRC dataset are extracted. Then non-overlapping patches of size 224 × 224 are extracted and given as input to CNN-based classifiers. Finally, the overall label of the input image is derived from counting the most predicted patch-level class (majority voting).

Visual fields[1] are given as input to the transformer network that combines local and global training to extract information from both the entire image and local patches in which finer details can be discovered [21]. It exploits a shallow global branch and a deep local branch. Each branch gets as inputs the feature maps extracted from an initial conv block with 3 convolution layers, followed by batch normalization and ReLU activation. The global branch has 2 blocks of encoding and 2 blocks of decoding. The local branch has 5 blocks of encoding and 5 blocks of decoding. The key idea is that, for colon carcinoma grading, the transformer architecture can preliminary help to understand which areas of the large-sized histological images could help the following CNN architectures to discriminate among carcinoma grades. This way models' training could be performed using fewer data and in a more effective way. The transformer was trained in order to learn how to distinguish glandular structures, which are currently considered one of the important biomarkers for tumour grade determination [2], from the rest of the visual field content. After training, it is able to provide, also on unseen visual fields, the corresponding binary masks that point out glandular regions. Non-overlapping patches of size 224 × 224 are then extracted on glandular regions and processed by CNN models obtaining as an outcome the predicted class for each patch. In particular EfficientNet architectures [19] have been exploited since they were proven to be the most suited for the CRC grading task. They use a compound coefficient to uniformly scale network width, depth, and resolution. Starting from a mobile-size baseline architecture created ad hoc (Efficient-B0) and following the proposed scaling strategy, a family of eight architectures has been built up with an increasing grade of complexity. Involved CNNs were modified to be adapted to a three classes inference

[1] with the term visual field, from now on, we refer to the total area acquired for each specimen.

problem. Finally, the overall label of the input image is derived from counting the most predicted patch-level class (majority voting).

4 Datasets

One of the most used datasets for adenocarcinoma grading tasks is CRC-Dataset [3]. It is comprised of visual fields extracted from 38 Hematoxylin and eosin-stained whole-slide images (often abbreviated as H&E stained WSIs) of colorectal cancer cases and consists of 139 visual fields with an average size of 4548×7520 pixels obtained at 20× magnification. These visual fields are classified into three different classes (normal, low grade, and high grade) based on the organization of glands in the visual fields by the expert pathologist. Recently, the CRC dataset has been extended with more visual fields extracted from another 68 H&E stained WSIs using the same criteria. This extended colorectal cancer dataset (from now on Extended CRC) consists of 300 visual fields with an average size of 5000×7300 pixels.

The extended CRC dataset was selected as a benchmark to test the different approaches investigated in Sect. 5. In it, there are 120 visual fields labelled as Normal, 120 labelled as Low Grade and 60 labelled as High Grade.

The GLAs dataset [18] consists of 165 images derived from 16 H&E stained histological sections of stage T3 or T42 colorectal adenocarcinoma. Each section belongs to a different patient, and sections were processed in the laboratory on different occasions. Thus, the dataset exhibits high inter-subject variability in both stain distribution and tissue architecture. The digitization of these histological sections into whole-slide images (WSIs) was accomplished using a Zeiss MIRAX MIDI Slide Scanner with a pixel resolution of 0.465μm. The WSIs were subsequently rescaled to a pixel resolution equivalent to 20× objective magnification. A total of 52 visual fields from both malignant and benign areas across the entire set of the WSIs were selected in order to cover as wide a variety of tissue architectures as possible. According to the overall glandular architecture, an expert pathologist (DRJS) then graded each visual field as either 'benign' or 'malignant'. The pathologist also delineated the boundary of each individual glandular object on that visual field. This manual annotation will be used as ground truth for automatic segmentation using the transformer network introduced in Sect. 3.

5 Experimental Results on the Extended CRC Dataset

In this section the experimental results, gathered by using the machine learning strategy described in Sect. 3 in order to analyze histological visual fields in the extended CRC dataset, are described.

The experimental results obtained by introducing a transformer network for segmenting glandular regions, as a preliminary step, are reported in the following. The starting point of this experimental phase consists in the results recently published in [11] which demonstrated that EfficientNet architectures, and in particular the ones named B0 and B1, were the most performing among the several tested in classify patches on the same CRC extended dataset. A transformer network was then added as a preliminary step for these CNN architectures as described in Sect. 3. The transformer was trained on images and binary masks provided with the GLAs dataset described in Sect. 4.

After training on the GLAs dataset, the learned configuration was exploited to a extract binary mask for the extended CRC dataset and then only patches corresponding to predicted glandular regions were used as input to the subsequent CNN-based colon carcinoma grading. For each visual field, non-overlapping patches of size 224×224 pixels were extracted and given as input to the subsequent training (batch size was set to 16) and classification steps. In all the experiments on CRC extended dataset, the same training, validation, and test splits as in [17] were used for a fair comparison with the existing methods. In particular, the data are separated into 56% for training, 14% for validation, and 30% for testing at the patient level.

For the same reason, binary and ternary classification performance are reported where binary classification means that examples with intermediate and high grades have been put together and considered a unique class against the class including only examples of lower-grade cancer. This could help to provide significant outcomes for the sub-task of differentiating between normal and tumour tissues. Two evaluation metrics were used: average accuracy and weighted accuracy. Moreover, ternary classification and binary classification have been performed. For binary classification, examples with intermediate and high grades have been put together and considered in that class.

In particular, for each fold j in range $[1, K]$ ($k = 3$ in the following experiments), *average accuracy* is computed as the average of:

$$acc_j = \frac{\sum_{i=1}^{c} TP_i}{\sum_{i=1}^{c} Ni} \tag{1}$$

Similarly, *weighted accuracy* is computed as the average of:

$$w\_acc_j = \frac{\sum_{i=1}^{C} \frac{TP_i}{N_i}}{C}. \tag{2}$$

where C indicated the number of classes (2 or 3), N_i is the number of elements in the class i and TP_i is the number of true positives for the class i.

Figure 2 reports an example of how the transformer architecture works. In Fig. 3a an original histological image containing a colon carcinoma of intermediate grade (grade 1) is shown. Figure 3b depicts the corresponding binary mask extracted by the transformer network that points out (white regions) glandular regions. The image obtained by a logical AND between the initial image and the mash is subsequently built and used as input for the classification step.

This way the transformer network mainly puts attention on regions relevant for grading, making it possible to discard those patches that just introduce noise in the learning process.

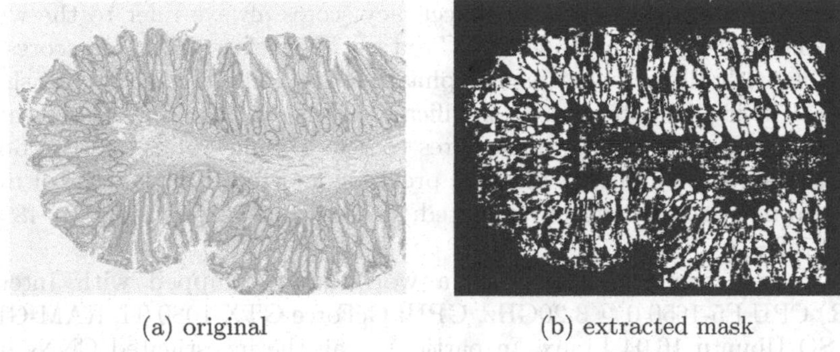

(a) original (b) extracted mask

Fig. 2. Glandular regions pointed out by the transformer network.

The transformer discarded 128914 patches, lying on masked areas of the original image. Summing up, without segmentation, the CNN-based classifiers received almost 190000 patches as input whereas after segmentation only 61025 patches (32%) were processed to make decisions about the grades of all the visual fields of the considered benchmark dataset.

It is worth noting that the number of visual fields to be classified (300 in the considered extended CRC dataset does not change but this just affects the number of patches that contributed to the final labelling of each of them. This way only patches providing an actual contribution to the classification step were retained.

Quantitative results are reported in Table 1 which reports grading results with and without exploiting transformer networks for pointing out discriminative regions.

Table 1. Results on the extended CRC dataset while integrating transformer networks for pointing out discriminative regions and CNN architectures for tumour detection and grading.

Model	Average (%) (Binary)	Weighted (%) (Binary)	Average (%) (3-classes)	Weighted (%) (3-classes)
EfficientNet-B1	95.64 ± 1.23	94.79 ± 1.15	85.89 ± 3.64	83.56 ± 3.39
EfficientNet-B2	96.99 ± 2.94	96.65 ± 3.11	87.58 ± 3.36	85.54 ± 2.21
T+EfficientNet-B1	**99.67 ± 0.47**	**99.72 ± 0.39**	89.58 ± 4.17	**87.50 ± 3.54**
T+EfficientNet-B2	98.66 ± 0.95	98.74 ± 0.91	**89.92 ± 2.50**	87.22 ± 2.08

The exploitation of the transformer network strengthened the CNN classification for all the models. The improvement in performance with respect to results gathered without transformer network was significant, on average of +3% in both binary and ternary problems.

It is useful to point out that the accuracy scores always refer to the whole set of visual fields in the extended CRC dataset. Therefore, gathered scores are comparable with all the experimental phases in [11]. The advantage of using a transformer is it puts attention to significant patches for the final classification goals, discarding those having no features to help to build robust CNN models. As a final result, the gathered f1-score, precision and recall in the case of using the transformer architecture are reported: f1-score=0.753, precision=0.748 and recall=0.752.

Experiments were performed on a workstation equipped with Intel(R) Xeon(R) CPU E5-1650 0 @ 3.20GHz, GPU: GeForce GTX 1080 Ti, RAM-GPU: 11GB, SO Ubuntu 16.04 Linux. In particular, all the investigated CNNs have been fine-tuned starting from the ImageNet pre-trained models provided with the reference implementations. Moreover, given the limited number of visual fields, data augmentation techniques were exploited, more precisely, horizontal and vertical flip, rotation by a random value sampled from the list [−90, −45, 45, 90], and shear on the x-axis by a random amount between −20 and 20 degrees.

Finally, the Stochastic Gradient Descent (SGD) optimizer was employed, with learning rate = 0.001, momentum = 0.9 and weight decay = 0.001, batch = 16 parameters, and an early stopping strategy of 10 epochs on the validation set with a max number of 100 training epochs were chosen.

For the transformer architecture, the training configuration specified in [21] was used, i.e. a batch size of 4, Adam optimizer and a learning rate of 0.001. Finally, the network was trained for 400 epochs.

5.1 Comparisons to Leading Approaches in the Literature

In this subsection, the results in Table 1 have been compared with the leading approaches in the literature for the task of colon carcinoma grading on the extended CRC dataset[2]. It is possible to observe that the proposed solutions outperformed most of the previous ones both for 2-classes and 3-classes classification tasks. In particular, for the 3-classes task, i.e. the most challenging and interesting for grading purposes, the introduced strategy based on the EfficientNet-B2 model achieved the best classification scores. The performance for the binary classification task was only slightly lower than the ones obtained by introducing contextual information as proposed in [15]. It is worth noting that in [15] the context-aware network is a complex architecture comprising different blocks. In particular, the contextual information is learned, by exploiting an attentive mechanism, starting from local features extracted by another CNN as well. Experiments carried out in this paper proved that an easier patch-based approach, if implemented by properly exploiting modern CNN architectures can be effective as well, and for the 3-class grading can work even better.

Compared architectures used different patch and batch sizes (for instance patches of 1792×1792 pixels and a batch of 64 images were used in [15]). Hence, it is possible to conclude that introduced models have been able to extract

[2] Data for previous works were taken from original papers.

Table 2. Comparisons of results obtained with strategies in this paper and those in the works in the literature.

Model	Average (%) (Binary)	Weighted (%) (Binary)	Average (%) (3-classes)	Weighted (%) (3-classes)
Proposed				
T+EfficientNet-B1	**99.67 ± 0.47**	**99.72 ± 0.39**	89.58 ± 4.17	87.50 ± 3.54
T+EfficientNet-B2	98.66 ± 0.95	98.74 ± 0.91	**89.92 ± 2.50**	**87.22 ± 2.08**
Previous works				
ResNet50 [15]	95.67 ± 2.05	95.69 ± 1.53	86.33 ± 0.94	80.56 ± 1.04
LR + LA-CNN [15]	97.67 ± 0.94	97.64 ± 0.79	86.67 ± 1.70	84.17 ± 84.17
CNN-LSTM [17]	95.33 ± 2.87	94.17 ± 0.79	82.33 ± 2.62	83.89 ± 2.08
CNN-SVM [10]	96.00 ± 0.82	96.39 ± 1.37	82.00 ± 1.63	76.67 ± 2.97
CNN-LR [10]	96.33 ± 1.70	96.39 ± 1.37	86.67 ± 1.25	82.50 ± 0.68
EfficientNet-B2 [11]	96.99 ± 2.94	96.65 ± 3.11	87.58 ± 3.36	85.54 ± 2.21
RegNetY-4.0GF [11]	95.64 ± 0.94	95.37 ± 1.52	84.55 ± 2.57	81.36 ± 1.43
RegNetY-6.4GF [11]	94.31 ± 2.48	94.26 ± 2.15	86.57 ± 2.12	83.58 ± 2.21

knowledge from a minor amount of data and to generalize it, making this way the effectiveness of the training process more independent from setup parameters.

5.2 Ablation Study

Finally, an ablation study to assess the contribution of transformer architecture is reported. In the same pipeline, a CNN-based segmentation model was used instead of transformer one in the first stage of the pipeline. In particular, the Faster Region-Based Convolutional Neural Network (Faster-RCNN, fRCNN) architecture for instance segmentation with a ResNet-101 feature extraction backbone was used inspired by the promising results gathered in [9]. As done for the transformer architecture it was trained on the GLAs dataset and tested on the extended CRC. Extracted patches were then split in folds and given in input to the EfficientNet-B1 and EfficientNet-B2 models.

Figure 3 shows the benefits of using transformers in the segmentation of the glandular regions. In particular, in Fig. 3a a portion of an image of the CRC dataset is reported whereas in Fig. 3b and 3c the corresponding glandular regions pointed out by the fRCNN and transformer network respectively are pointed out. The ability of the transformer to better follow glands contours and to discard unuseful regions can be trivially derived. This means that, by using the transformer to locate glands, the following CNN classifier is fed by more patches with actual informative contents for grading purposes. This can be quantitatively proved by comparing the classification results of the whole pipeline as reported in Table 3. Using fRCNN the accuracy in grading is quite similar to the one gathered without any preliminary segmentation, whereas by using the transformer it increases by almost 3%. This makes evident the importance of using an efficient segmentation scheme as a preliminary step of the multiclass classification. It can be also highlighted by a quantitative comparison in the visual field classification task as reported in Table 3.

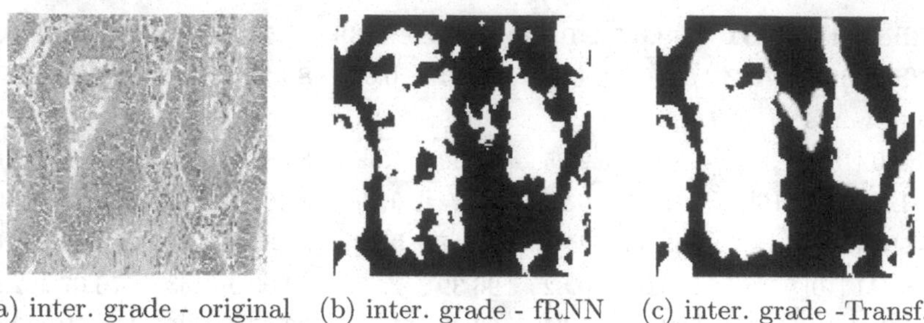

(a) inter. grade - original (b) inter. grade - fRNN (c) inter. grade -Transf.

Fig. 3. From left to right: a portion of an image from the CRC dataset, glandular regions are pointed out by the fRNN and transformer network respectively.

Table 3. Results on the extended CRC dataset while using fRCNN instead of transformer for pointing out discriminative regions.

Model	Average (%) (Binary)	Weighted (%) (Binary)	Average (%) (3-classes)	Weighted (%) (3-classes)
EfficientNet-B1	95.64 ± 1.23	94.79 ± 1.15	85.89 ± 3.64	83.56 ± 3.39
EfficientNet-B2	96.99 ± 2.94	96.65 ± 3.11	87.58 ± 3.36	85.54 ± 2.21
fRCNN+EfficientNet-B1	**96.14 ± 0.85**	**95.38 ± 1.41**	86.77 ± 2.25	**85.65 ± 1.45**
fRCNN+EfficientNet-B2	96.88 ± 1.15	97.08 ± 1.25	**88.10 ± 2.10**	86.44 ± 1.80
T+EfficientNet-B1	**99.67 ± 0.47**	**99.72 ± 0.39**	89.58 ± 4.17	**87.50 ± 3.54**
T+EfficientNet-B2	98.66 ± 0.95	98.74 ± 0.91	**89.92 ± 2.50**	87.22 ± 2.08

6 Conclusion

In this work, the most performing convolutional architectures in object classification tasks have been powered by introducing transformers to improve colon carcinoma grading in histological images. Exploiting embedded self-attention mechanism, only relevant visual information was retained and the results on the largest publicly available dataset demonstrated a substantial improvement with respect to leading state-of-the-art approaches. Besides, a speed-up in the training process was also achieved. This is an incredibly important result since the up top 3% of improvement in classification makes it possible to develop robustly and objective (reduced bias from different operator expertise and evaluation circumstances) computational tools to support healthcare professionals in the challenging task of carcinoma diagnosis and therapy planning. The main limitation of this work is the lack of testing in a clinical setting. This will be carried out in the next future. Future works will also deal with the use of different transformers architectures and CNN architectures also exploiting ensembles of networks to further improve classification accuracy.

Acknowledgment. This research was funded in part by Future Artificial Intelligence Research-FAIR CUP B53C220036 30006 grant number PE0000013.

References

1. Altunbay, D., Cigir, C., Sokmensuer, C., Gunduz-Demir, C.: Color graphs for automated cancer diagnosis and grading. IEEE Trans. Biomed. Eng. **57**(3), 665–674 (2009)
2. Awan, R., Al-Maadeed, S., Al-Saady, R., Bouridane, A.: Glandular structure-guided classification of microscopic colorectal images using deep learning. Comput. Electr. Eng. **85**, 106450 (2020)
3. Awan, R., et al.: Glandular morphometrics for objective grading of colorectal adenocarcinoma histology images. Sci. Reports **7**(1), 1–12 (2017)
4. Bychkov, D., et al.: Deep learning based tissue analysis predicts outcome in colorectal cancer. Sci. Reports **8**(1), 1–11 (2018)
5. Carcagnì, P., et al.: Classification of skin lesions by combining multilevel learnings in a DenseNet architecture. In: Ricci, E., Rota Bulò, S., Snoek, C., Lanz, O., Messelodi, S., Sebe, N. (eds.) ICIAP 2019. LNCS, vol. 11751, pp. 335–344. Springer, Cham (2019). https://doi.org/10.1007/978-3-030-30642-7_30
6. Chen, S., et al.: Automatic tumor grading on colorectal cancer whole-slide images: Semi-quantitative gland formation percentage and new indicator exploration. Front. Oncol. **12** (2022)
7. Das, A., Nair, M.S., Peter, S.D.: Computer-aided histopathological image analysis techniques for automated nuclear atypia scoring of breast cancer: a review. J. Digital Imag. **33**(5), 1091–1121 (2020)
8. He, K., et al.: Transformers in medical image analysis: a review. Intell. Med. **3**(1), 59–78 (2022)
9. Ho, C., et al.: A promising deep learning-assistive algorithm for histopathological screening of colorectal cancer. Sci. Reports **12**(1), 1–9 (2022)
10. Hou, L., Samaras, D., Kurc, T.M., Gao, Y., Davis, J.E., Saltz, J.H.: Patch-based convolutional neural network for whole slide tissue image classification. In: Proceedings of the IEEE Conference On Computer Vision and Pattern Recognition, pp. 2424–2433 (2016)
11. Leo, M., Carcagnì, P., Signore, L., Benincasa, G., Laukkanen, M.O., Distante, C.: Improving colon carcinoma grading by advanced cnn models. In: Image Analysis and Processing-ICIAP 2022: 21st International Conference, Lecce, Italy, May 23–27, 2022, Proceedings, Part I. pp. 233–244. Springer (2022)
12. Leo, M., Farinella, G.M.: Computer vision for assistive healthcare. Academic Press (2018)
13. Pei, Y., et al.: Colorectal tumor segmentation of CT scans based on a convolutional neural network with an attention mechanism. IEEE Access **8**, 64131–64138 (2020)

14. Saxena, S., Gyanchandani, M.: Machine learning methods for computer-aided breast cancer diagnosis using histopathology: a narrative review. J. Med. Imag. Radiation Sci. **51**(1), 182–193 (2020)
15. Shaban, M., et al.: Context-aware convolutional neural network for grading of colorectal cancer histology images. IEEE Trans. Med. Imaging **39**(7), 2395–2405 (2020). https://doi.org/10.1109/TMI.2020.2971006
16. Shamshad, F., et al.: Transformers in medical imaging: A survey. Med. Image Anal. **88**, 102802 (2023)
17. Sirinukunwattana, K., Alham, N.K., Verrill, C., Rittscher, J.: Improving whole slide segmentation through visual context - a systematic study. In: Frangi, A.F., Schnabel, J.A., Davatzikos, C., Alberola-López, C., Fichtinger, G. (eds.) MICCAI 2018. LNCS, vol. 11071, pp. 192–200. Springer, Cham (2018). https://doi.org/10.1007/978-3-030-00934-2_22
18. Sirinukunwattana, K., et al.: Gland segmentation in colon histology images: The glas challenge contest. Medical Image Analysis **35**, 489–502 (2017)
19. Tan, M., Le, Q.: Efficientnet: Rethinking model scaling for convolutional neural networks. In: International Conference on Machine Learning, pp. 6105–6114. PMLR (2019)
20. Tosun, A.B., Kandemir, M., Sokmensuer, C., Gunduz-Demir, C.: Object-oriented texture analysis for the unsupervised segmentation of biopsy images for cancer detection. Pattern Recogn. **42**(6), 1104–1112 (2009)
21. Valanarasu, J.M.J., Oza, P., Hacihaliloglu, I., Patel, V.M.: Medical transformer: Gated axial-attention for medical image segmentation. arXiv preprint arXiv:2102.10662 (2021)
22. Vaswani, A., et al.: Attention is all you need. In: Proceedings of the 31st International Conference on Neural Information Processing Systems, pp. 6000–6010 (2017)
23. Vuong, T.L.T., Lee, D., Kwak, J.T., Kim, K.: Multi-task deep learning for colon cancer grading. In: 2020 International Conference on Electronics, Information, and Communication (ICEIC), pp. 1–2. IEEE (2020)
24. Zhou, P., et al.: Hccanet: histopathological image grading of colorectal cancer using CNN based on multichannel fusion attention mechanism. Sci. Reports **12**(1), 15103 (2022)
25. Zhou, Y., Graham, S., Alemi Koohbanani, N., Shaban, M., Heng, P., Rajpoot, N.: CGC-net: Cell graph convolutional network for grading of colorectal cancer histology images. In: 2019 IEEE/CVF International Conference on Computer Vision Workshop (ICCVW), pp. 388–398 (2019). https://doi.org/10.1109/ICCVW.2019.00050

FERMOUTH: Facial Emotion Recognition from the MOUTH Region

Berardina De Carolis(✉), Nicola Macchiarulo, Giuseppe Palestra, Alberto Pio De Matteis, and Andrea Lippolis

Department of Computer Science, University of Bari, Bari, Italy
{berardinade.carolis,nicola.macchiarulo,giuseppe.palestra,
albertopiode.matteis,andrea.lippolis}@uniba.it

Abstract. People use various nonverbal communicative channels to convey emotions, among which facial expressions are considered the most important ones. Consequently, automatic Facial Expression Recognition (FER) is a crucial task for enhancing computers' perceptive abilities, particularly in human-computer interaction. Although state-of-the-art FER systems can identify emotions from the entire face, situations may arise where occlusions prevent the entire face from being visible. During the COVID-19 pandemic, many FER systems have been developed for recognizing emotions from the eye region due to the obligation to wear a mask. However, in many situations, the eyes may be covered, for instance, by sunglasses or virtual reality devices. In this paper, we faced the problem of developing a FER system that solely considers the mouth region and classifies emotions using only the lower part of the face. We tested the effectiveness of this FER system in recognizing emotions from the lower part of the face and compared the results to a FER system trained on the same datasets using the same approach on the entire face. As expected, emotions primarily associated with the mouth region (e.g., happiness, surprise) were recognized with minimal loss compared to the entire face. Nevertheless, even though most negative emotions were not accurately detected using only the mouth region, in cases where the face is partially covered, this area may still provide some information about the displayed emotion.

Keywords: Facial Emotion Recognition · CNN · Occlusion

1 Introduction

Humans employ different communication channels to express their emotions, such as facial expressions, voice intonation, gestures, and postures. The face is the main channel people use to decipher the feelings of others since it is the prime communicator of emotion [1,19]. Facial emotion recognition (FER) is a subfield of computer vision and pattern recognition that involves analyzing human facial expressions and recognizing emotional states. The ability to recognize emotions is fundamental to human communication and plays a critical role in our social

G. L. Foresti et al. (Eds.): ICIAP 2023, LNCS 14233, pp. 147–158, 2023.
https://doi.org/10.1007/978-3-031-43148-7_13

interactions, including non-verbal communication, empathy, and emotional regulation. Consequently, FER has become an active area of research in computer science, with significant advancements being made in recent years. FER has various applications, including in human-computer interaction, affective computing, and psychology. FER systems are used in numerous domains, such as intelligent tutoring systems [25], interactive game design [17], affective robots [2], driver fatigue monitoring [3], personalized services [23], and many others. Most existing research works are based on Ekman's theories, which suggest that there are six basic emotions universally recognized in all cultures: happiness, surprise, anger, sadness, fear, and disgust [8]. Despite the significant progress made in this field, various challenges remain, such as occlusions, individual differences, and variations in facial expressions across cultures. This has led to ongoing research to improve the accuracy and robustness of FER systems.

The problem of recognizing emotions in the presence of occlusion gained a lot of interest during the COVID-19 pandemic since the presence of the mask occludes a large part of the face, thus representing a big challenge for a FER system. Many FER systems were developed to address this issue. In our previous work [7], we addressed the issue to recognize emotions when the lower part of the face was covered by a mask with an approach similar to the one described in this paper. However, in many situations, the eyes may be covered by sunglasses or virtual reality devices in some interactions. To accomplish this, we developed a FER system that solely considers the mouth region and classifies emotions using only that portion of the face.

The mouth region is a very informative area for recognizing emotions, for instance, surprise and fear require the analysis of the mouth to be distinguished by humans. Another social study on the impact of face masks on emotion recognition and social judgments (perceived trustworthiness, likability, and closeness) was carried out in [13]. Such a study confirms that humans are less accurate in emotion recognition in the presence of masked faces. Nevertheless, the region of the eyes is very crucial in expressing and recognizing emotions and it may be informative enough to carry out a FER analysis.

Starting from this insight, we developed a FER system to automatically recognize emotions from the bottom part of the face, the mouth region based on deep learning models. The models exploit transfer learning, which uses general-purpose neural networks pre-trained on extensive datasets, later fine-tuned on the classification domain. In addition, the attention mechanism inside deep learning models has been used. We trained two models on two of the most used datasets in this domain: FER2013 [11] and RAF-DB [18]. In this way, besides understanding the extent to which a FER system can be effective in recognizing emotions from that region in the presence of occlusions, we could identify which emotions were confused with others when considering only the mouth region and observe whether the accuracy of the emotion recognition from the mouth was consistent across the three models or was depending on the dataset.

As a result, we found that, in general, the approach achieves a better performance in recognizing positive emotions than negative ones. In particular, the obtained results show that happiness and neutrality are the emotions that have minor loss compared to the accuracy of the models trained on the entire face. The negative emotions of Fear, Sadness and Anger have a more important loss compared to the accuracy of the models trained on the entire face. The paper is organized as follows. Section 2 describes research works addressing the same problem. Section 3 describes the pipeline designed and developed for our FER system. In Sect. 4 we show the experimental results. Finally, Sect. 5 draws the conclusions and future work.

2 Related Work

Recent research underlines that primary emotional states such as happiness, sadness, anger, disgust, fear, surprise can be recognized from facial expressions universally across cultures [8] with a sufficient accuracy [10]. FER systems are typically built using deep learning techniques that are trained on large datasets of facial images to recognize and classify different emotions. However, when a significant portion of the face is occluded, these systems face challenges in capturing the necessary facial cues and features to accurately determine emotions. The issue of facial occlusion in FER systems is currently an active area of research [15,26]. Researchers are exploring various solutions to improve the robustness of FER systems when faced with occlusion. One approach is to develop algorithms that can effectively handle partial or obstructed views of the face. By addressing this challenge, FER systems can maintain accuracy even when certain parts of the face are obscured. An interesting analysis that shows the need for improving the accuracy of FER systems in presence of occlusion and in the wild is illustrated in [21]. In this paper, the correlation between sports performance and facial expressions is investigated by demonstrating that there is a relationship between the signs of fatigue expressed by facial expressions and the performance of each runner. In addition, the study shows the great difficulty of applying FER models in a wild environment. Many researchers have concentrated their efforts on studying the recognition of emotions from the eye region in order to gain a better understanding of the impact of masks on human emotional communication. For instance in [7,12,22,24], the automatic FER in presence of masks is performed using Convolution Neural Networks (CNN) trained on eyes and forehead segments. On the opposite, there has been limited research on analyzing emotions solely from the mouth area.

The studies described in [4,9] explore emotion recognition based on the mouth. The researchers employ a transfer learning approach, which is an advanced machine learning technique that involves reusing a pre-existing model designed for one specific task as a starting point for developing a model for a different task. In this study, the accuracy of emotion recognition based on the mouth has also been compared to the corresponding full-face emotion recognition, revealing that the loss of precision is compensated by consistent performance in the domain of visual emotion recognition, emphasizing the importance of mouth detection in the overall process of emotion recognition. The analysis focused on the classification of emotions using only the mouth to study how much the mouth is involved in emotional expression and can provide accuracy in recognizing the entire face. Subsequently, training was carried out using 4 different CNNs (VGG16, Xception, Inception V3, and InceptionResNetV2), with InceptionResNetV2 being the best convolutional neural network based on the results obtained.

3 The FERMOUTH System

Generally, a Facial Expression Recognition system is composed of three main steps: preprocessing, feature extraction, and emotion recognition activity, which maps the extracted facial features into a label-mapping space that includes basic emotions plus neutral. To create a FER that analyzes only the portion of the face containing the mouth, the base pipeline is modified by adding an image cropping phase. After cropping the relevant part of the face, the emotion recognition model applies a CNN to classify the emotion as one of the following: happiness, surprise, anger, sadness, fear, disgust, and neutrality. The main stages of the proposed pipeline are detailed in the following sections.

3.1 Face Detection

The facial detection phase is first performed using the Python library detector Dlib [16]. The region encompassing the entire face is detected and cropped with dimensions of 224×224 to be compatible with the input size accepted by the Convolutional Neural Network (CNN).

3.2 Region-of-Interest: Lower Face Area

The next step is to crop the region of the face related to the mouth. For this step, we selected the region comprised from the chin and the tip of the nose. To this aim, it has been used an ensemble of regression trees in order to extract 68 face's landmark reference points $X = FL_i \quad \text{where} \quad i = 1, ..., 68$. Then, the Region-Of-Interest (ROI) has been selected through the rectangle that has its origin at the point x, y defined as:

$$x = FL_2.x, \qquad y = max(FL_2.y, FL_{14}.y)$$

and has width w and height h defined as:

$$w = FL_{14}.x, \qquad h = min(FL_6.y, FL_{10}.y)$$

3.3 Emotion Recognition

For emotion recognition, the developed models will only consider the portion of the face that includes the mouth. The image will be resized to 224×224 to be compatible with the model (Fig. 1).

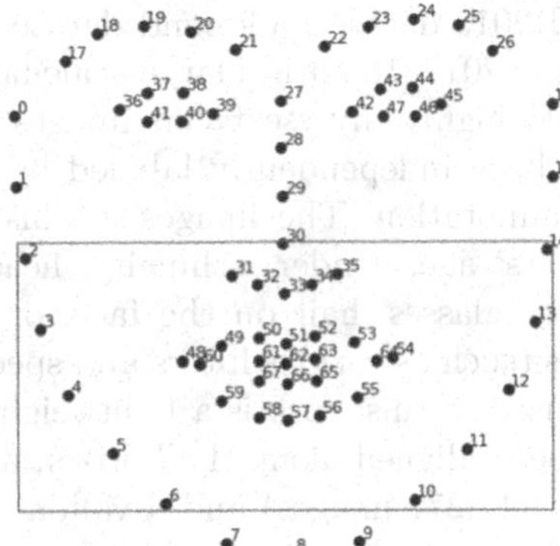

Fig. 1. The region of interest of the lower face area based on the 68 facial landmarks.

3.4 The Datasets

As mentioned in the introduction section, the experiment was carried out using two different datasets: FER2013 and RAF-DB, The FER2013 dataset was created in 2013 by the organizers of the "Challenges in Representation" competition [11]. The FER2013 dataset is a benchmark in facial expression recognition and contains a photographic set of faces in both posed and unposed positions, not generated by a bot. It is a widely used dataset in computer vision research, consisting of facial images labeled with basic emotions. Starting from this dataset, it has been created a reduced dataset, considering only those images where it was possible to detect facial landmarks. The resulting data set contains 33,095 grayscale images of faces resized to 48 × 48 pixels. Each image is annotated with one of the following labels: Angry, Disgust, Fear, Happy, Neutral, Sadness, and Surprise. The dataset is already divided into a training set (26,427 images), a validation set (3,335 images), and a test set (3,333 images) (Fig. 2).

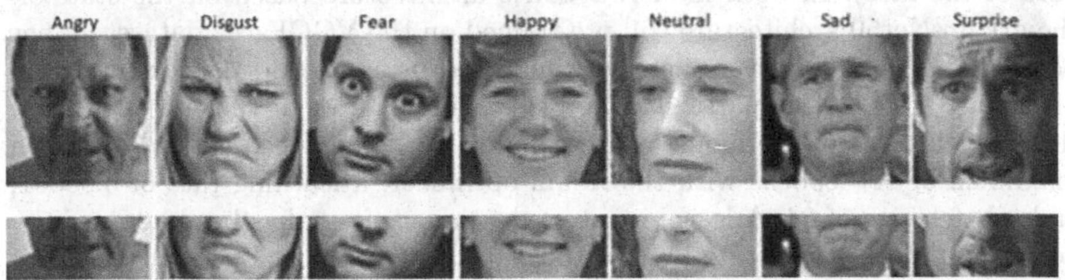

Fig. 2. Sample images from the FER 2013 dataset. The upper row contains images extracted from the FER2013 dataset. The lower row contains cropped images of the same emotion.

In addition to the FER2013 dataset, a second dataset has been used: RAF-DB. RAF-DB was created in 2017 [18], it is a large-scale facial expression dataset with approximately 30,000 highly diverse facial images downloaded from the internet. Each image has been independently labeled by around 40 annotators based on crowdsourcing annotation. The images in this database vary significantly in terms of subjects' age, gender, ethnicity, head poses, lighting conditions, occlusions (such as glasses, hair on the face, or self-occlusion), image post-processing operations (such as various filters and special effects), and more. The version of RAF-DB used in this work is a lightweight version consisting of 15,339 images that have been aligned along the horizontal axis of the eyes. It is divided into a training set (12,271 images) and a validation set (3,068 images). In this version, the images are annotated with 7 different emotion classes as in the previous dataset.

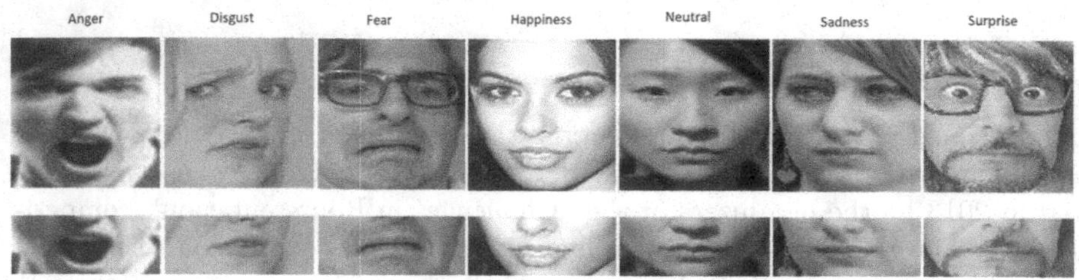

Fig. 3. Sample images from the RAF-DB dataset. The upper row contains images extracted from the RAF-DB dataset. The lower row contains cropped images of the same emotion.

3.5 The Models

For each dataset, two models have been trained: i) a model that analyzes the entire face; ii) and a model that analyzes only the region of interest (the lower part of the face). In both models a CNNs architecture based on the state-of-the-art ResNet50 architecture [14] pre-trained on the VGGFace2 dataset [5] has been used. The architecture has been improved using the Bottleneck Attention Module (BAM) [20], in particular, we placed three BAMs at the end of the first three bottlenecks of the models. The training phase has been carried out for a maximum of 100 epochs with the Adam optimizer, a learning rate of 0.00001, and a batch size of 64. A learning rate reduction strategy has been adopted, by decreasing the learning rate by a factor of 10 every 5 epochs without accuracy improvement. These values for hyperparameters were obtained empirically by carrying out several training runs with different values.

4 Experimental Results

The experiment consists in verifying the accuracy of each model (entire face vs. mouth-region) on the two datasets. Then, firstly we calculated the accuracy of the first model, which used the entire face. Then, the accuracy of the second model that used the lower part of the face, has been calculated. In the end, a comparison of the accuracy of the two models on the two datasets has been performed. The comparison aims to discover how much accuracy is lost, with the second model, in recognizing only the lower part of the face compared to a model that recognizes the entire face, to understand whether, in the presence of occlusions, the FER system can still be reliable in its task.

The following measures were computed to compare the FER models trained on the dataset of the mouth ROI region and on the dataset of entire faces:

- $Precision = \frac{TP}{TP+FP}$
- $Recall = \frac{TP}{TP+FN}$
- $F1 = \frac{2*(Recall*Precision)}{(Recall+Precision)}$
- $Accuracy = \frac{TP+TN}{TP+FP+FN+TN}$

TP denotes the number of true positives, TN indicates the number of false positives, the number of false positives is represented with FP and the number of false negatives is denoted with FN. These measures were used to calculate the confusion matrix. The comparative results of the models with attention mechanisms are summarized in Table 1. Both models trained on the mouth ROI have a loss in accuracy compared with those trained on the entire face, as expected.

Table 1. Validation set accuracy on FER2013 and RAF-DB datasets.

Dataset	Entire Face	Mouth-ROI
FER2013	72.99%	61.75%
RAF-DB	87.45%	70.99%

With the model related to the mouth trained on the FER2013 dataset, the accuracy of the "Happy" and "Surprise" emotions remains slightly lower in the mouth model but with a minimal loss compared to the others. Among the negative emotions, which are those which had a greater loss compared to the accuracy on the entire face (about 10% on average), Disgust is the one that is better recognized (Table 2).

From the analysis of data on model accuracy metrics on RAF-DB, it emerges that there is better accuracy compared to the FER-based model and, in the case of the entire face, there is better accuracy as more data on facial expressions are available. The trend of the comparison between the performances of the two models is similar those of the FER2013-based models. The negative valenced emotions present the major loss. However, it is possible to notice how "Disgust" has a very important loss in the model trained on the mouth, compared to the one on the entire face. Also in this case the positive emotions such as "Happy" and "Surprise", despite having lower precision in the case of the mouth, are the one that are better recognized only by the mouth-ROI (Table 3).

Table 2. Accuracy of the models (face and Mouth-ROI) on the FER2013 dataset.

Emotions	Precision		Recall		F1-score	
	Entire Face	*ROI-Mouth*	*Entire Face*	*ROI-Mouth*	*Entire Face*	*ROI-Mouth*
Anger	**0.66**	0.52	**0.63**	0.49	**0.65**	0.50
Disgust	0.67	**0.74**	**0.71**	0.48	**0.69**	0.58
Fear	**0.60**	0.47	**0.57**	0.44	**0.58**	0.46
Happiness	**0.90**	0.82	**0.92**	0.87	**0.91**	0.84
Neutral	**0.73**	0.58	**0.73**	0.59	**0.74**	0.59
Sadness	**0.59**	0.45	**0.61**	0.48	**0.60**	0.47
Surprise	**0.80**	0.73	**0.82**	0.71	**0.81**	0.72

Considering the confusion matrices in Fig. 4, and Fig. 5 we can notice that both models, just like humans do [6], confuse the negative-valenced emotions with each other. The *Happy* emotion is recognized with the highest accuracy than the others even only by the mouth ROI.

Table 3. Accuracy of the models (face and mouth-ROI) on the RAF-DB dataset.

Emotions	Precision		Recall		F1-score	
	Entire Face	*ROI-Mouth*	*Entire Face*	*ROI-Mouth*	*Entire Face*	*ROI-Mouth*
Anger	**0.86**	0.56	**0.80**	0.53	**0.83**	0.55
Disgust	**0.66**	0.39	**0.59**	0.28	**0.62**	0.33
Fear	**0.77**	0.50	**0.69**	0.42	**0.73**	0.46
Happiness	**0.95**	0.85	**0.95**	0.88	**0.95**	0.86
Neutral	**0.83**	0.67	**0.88**	0.73	**0.85**	0.70
Sadness	**0.83**	0.60	**0.87**	0.57	**0.85**	0.59
Surprise	**0.90**	0.65	**0.80**	0.62	**0.85**	0.63

4.1 Web Application

We deployed our FER pipeline in a real-time web application that is able to analyze new images, videos or real-time streaming and recognize the emotion expressed by the face or by the mouth ROI using the two proposed models. Figure 3 shows the application interface. On the left side, the original image or video is shown, while, on the right side, the entire, the lower part of the face, and the recognized emotions are shown. In the example, the happy emotion is correctly recognized by both models (Fig. 6).

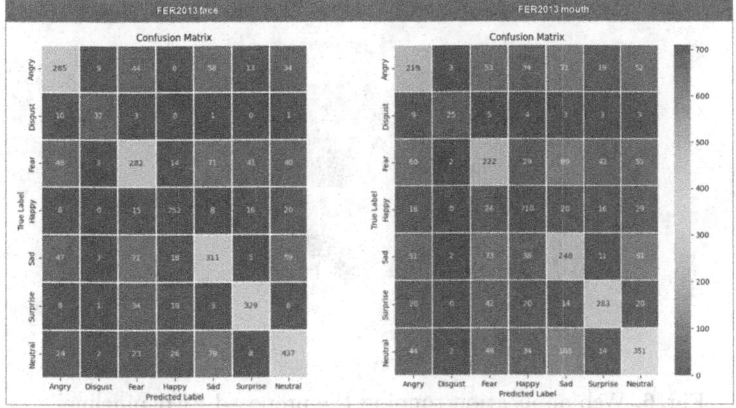

Fig. 4. Confusion matrices of the two models trained on the FER2013 dataset *Entire Face* vs. (*Mouth-ROI*).

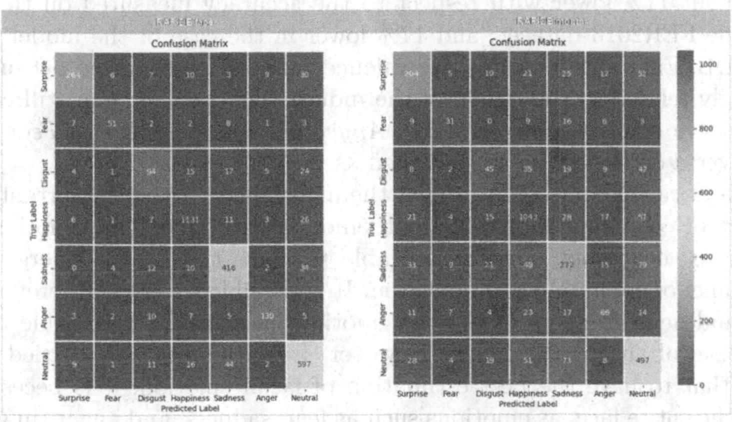

Fig. 5. Confusion matrices of the two models trained on RAF-DB *Entire Face* vs. (*Mouth-ROI*).

5 Conclusions and Future Work

Automatic facial expression recognition is an essential feature in human-computer interaction and human behavior understanding. It can be used in various scenarios in which the facial expression is impaired by an occlusion on the upper part of the face, such as in the case of sunglasses and virtual reality devices. In this study, we investigated how effective is the FER system in recognizing emotions by the mouth region and by comparing the same approach to the recognition of the same expressions on the entire face. The proposed pipeline is based on deep learning techniques and the adoption of a more sophisticated model improved the global accuracy in recognizing emotion even from only the

Fig. 6. Web application running the proposed FER pipeline.

upper part of the face. In general, the accuracy measured on the occluded faces is only about 11% lower with respect to the accuracy measured on the entire face on the FER2013 dataset, and 17% lower in the case of the model trained on RAF-DB. Emotions with positive valence are better recognized because they are strongly related to the region of the mouth. The FER system still confuses some negative emotions, for example, *Angry* is often classified as *Fear* or *Sad* and vice-versa or *Disgust* is misclassified as *Angry*.

The work carried out confirms that the mouth plays a very important role in the sphere of communication of human emotions. In particular, emotions such as happiness, neutrality, surprise are able to make the mouth emerge as the main channel of emotion communication. It is possible to assert, therefore, that positive and neutral emotions are the emotions most easily detectable through the presence of the mouth alone. However, from the analyses carried out, it emerged that to have the best estimation of facial emotion, it is necessary to examine the entire face, as emotions such as fear, sadness, and anger (in general, negative emotions) are more difficult to perceive through the mouth alone, as they can be more easily confused with the positive or neutral meanings of the same emotions. It should also be emphasized that in the field of deep learning, the recognition of facial emotions from the area of the mouth has been the subject of a limited number of detailed studies, therefore, it will be possible to achieve higher levels of accuracy in the future, while some emotions will remain difficult to interpret due to the objective lack of the upper part of the face.

Possible future developments will involve training the models on additional datasets, such as AffectNet, applying the concept of transfer learning as in this project. Furthermore, since we already trained a model only on the eye region [7], we will compare and evaluate which is the contribution of each part of the face in recognizing specific emotions and use these models in an ensemble. Another significant advancement in this research could be cross-dataset model learning, aimed at testing if performance improves when applied in real-world scenarios. In conclusion, the mouth can be considered a crucial area for human communication. However, it is important to note that certain emotions may involve the mouth to a lesser extent, which reduces the accuracy of emotion recognition when the facial region around the eyes, which plays a similarly important role, is not taken into account.

References

1. Encyclopedia of Human Behavior, 2nd edition. V. S. Ramachandran (2012)
2. Akbar, M.T., Ilmi, M.N., Rumayar, I.V., Moniaga, J., Chen, T.K., Chowanda, A.: Enhancing game experience with facial expression recognition as dynamic balancing. Procedia Computer Science 157, 388–395 (2019). doi: https://doi.org/10.1016/j.procs.2019.08.230,the 4th International Conference on Computer Science and Computational Intelligence (ICCSCI 2019) : Enabling Collaboration to Escalate Impact of Research Results for Society
3. Assari, M.A., Rahmati, M.: Driver drowsiness detection using face expression recognition. 2011 IEEE International Conference on Signal and Image Processing Applications (ICSIPA) pp. 337–341 (2011)
4. Biondi, G., Franzoni, V., Gervasi, O., Perri, D.: An approach for improving automatic mouth emotion recognition. In: Computational Science and Its Applications-ICCSA 2019: 19th International Conference, Saint Petersburg, Russia, July 1–4, 2019, Proceedings, Part I 19. pp. 649–664. Springer (2019)
5. Cao, Q., Shen, L., Xie, W., Parkhi, O.M., Zisserman, A.: Vggface2: A dataset for recognising faces across pose and age. 2018 13th IEEE International Conference on Automatic Face & Gesture Recognition (FG 2018) pp. 67–74 (2018)
6. Carbon, C.C.: Wearing face masks strongly confuses counterparts in reading emotions. Frontiers in Psychology **11**, 2526 (2020)
7. Castellano, G., De Carolis, B., Macchiarulo, N.: Automatic facial emotion recognition at the covid-19 pandemic time. Multimedia Tools and Applications **82**(9), 12751–12769 (2023)
8. Ekman, P.: Basic emotions. Handbook of cognition and emotion **98**(45–60), 16 (1999)
9. Franzoni, V., Biondi, G., Perri, D., Gervasi, O.: Enhancing mouth-based emotion recognition using transfer learning. Sensors **20**(18), 5222 (2020)
10. González-Lozoya, S.M., de la Calleja, J., Pellegrin, L., Escalante, H.J., Medina, M.A., Benitez-Ruiz, A.: Recognition of facial expressions based on cnn features. Multimedia tools and applications **79**, 13987–14007 (2020)
11. Goodfellow, I.J., Erhan, D., Carrier, P.L., Courville, A., Mirza, M., Hamner, B., Cukierski, W., Tang, Y., Thaler, D., Lee, D.H., Zhou, Y., Ramaiah, C., Feng, F., Li, R., Wang, X., Athanasakis, D., Shawe-Taylor, J., Milakov, M., Park, J., Ionescu, R., Popescu, M., Grozea, C., Bergstra, J., Xie, J., Romaszko, L., Xu, B., Chuang, Z., Bengio, Y.: Challenges in representation learning: A report on three machine learning contests (2013)
12. Greco, A., Saggese, A., Vento, M., Vigilante, V.: Performance assessment of face analysis algorithms with occluded faces. In: Del Bimbo, A., Cucchiara, R., Sclaroff, S., Farinella, G.M., Mei, T., Bertini, M., Escalante, H.J., Vezzani, R. (eds.) Pattern Recognition. ICPR International Workshops and Challenges. pp. 472–486. Springer International Publishing, Cham (2021)
13. Grundmann, F., Epstude, K., Scheibe, S.: Face masks reduce emotion-recognition accuracy and perceived closeness. PloS one **16**(4), e0249792 (2021)

14. He, K., Zhang, X., Ren, S., Sun, J.: Deep residual learning for image recognition. In: 2016 IEEE Conference on Computer Vision and Pattern Recognition (CVPR). pp. 770–778 (2016). DOI: 10.1109/CVPR.2016.90
15. Kim, G., Seong, S.H., Hong, S.S., Choi, E.: Impact of face masks and sunglasses on emotion recognition in south koreans. PLoS One **17**(2), e0263466 (2022)
16. King, D.E.: Dlib-ml: A machine learning toolkit. The Journal of Machine Learning Research **10**, 1755–1758 (2009)
17. Lankes, M., Riegler, S., Weiss, A., Mirlacher, T., Pirker, M., Tscheligi, M.: Facial expressions as game input with different emotional feedback conditions. In: Proceedings of the 2008 International Conference on Advances in Computer Entertainment Technology. p. 253–256. ACE '08, Association for Computing Machinery, New York, NY, USA (2008). DOI: 10.1145/1501750.1501809
18. Li, S., Deng, W., Du, J.: Reliable crowdsourcing and deep locality-preserving learning for expression recognition in the wild. In: 2017 IEEE Conference on Computer Vision and Pattern Recognition (CVPR). pp. 2584–2593. IEEE (2017)
19. Mehrabian, A.: Nonverbal Communication. Aldine-Atherton, New York (1972)
20. Park, J., Woo, S., Lee, J.Y., Kweon, I.S.: Bam: Bottleneck attention module. In: BMVC (2018)
21. Santana, O.J., Freire-Obregón, D., Hernández-Sosa, D., Lorenzo-Navarro, J., Sánchez-Nielsen, E., Castrillón-Santana, M.: Facial expression analysis in a wild sporting environment. Multimedia Tools and Applications **82**(8), 11395–11415 (2023)
22. Saxena, S., Tripathi, S., Sudarshan, T.: Deep facial emotion recognition system under facial mask occlusion. In: International Conference on Computer Vision and Image Processing. pp. 381–393. Springer (2020)
23. Tkalčič, M., Maleki, N., Pesek, M., Elahi, M., Ricci, F., Marolt, M.: Prediction of music pairwise preferences from facial expressions. In: Proceedings of the 24th International Conference on Intelligent User Interfaces. p. 150–159. IUI '19, Association for Computing Machinery, New York, NY, USA (2019). DOI: 10.1145/3301275.3302266
24. Yang, B., Wu, J., Hattori, G.: Facial expression recognition with the advent of human beings all behind face masks. MUM2020, Association for Computing Machinery, Essen, Germany (2020)
25. Zakka, B.E., Vadapalli, H.: Detecting learning affect in e-learning platform using facial emotion expression. In: Abraham, A., Jabbar, M.A., Tiwari, S., Jesus, I.M.S. (eds.) Proceedings of the 11th International Conference on Soft Computing and Pattern Recognition (SoCPaR 2019). pp. 217–225. Springer International Publishing, Cham (2021)
26. Zhang, L., Verma, B., Tjondronegoro, D., Chandran, V.: Facial expression analysis under partial occlusion: A survey. ACM Computing Surveys (CSUR) **51**(2), 1–49 (2018)

Consensus Ranking for Efficient Face Image Retrieval: A Novel Method for Maximising Precision and Recall

Anders Hast(✉)

Uppsala University, 751 05 Uppsala, Sweden
anders.hast@it.uu.se
http://www.andershast.com

Abstract. Efficient face image retrieval, i.e. searching for existing photographs of a person in unlabelled photo collections using a query photo, is evaluated for a novel method to find the top n results for Consensus Ranking. The approach aims to maximise precision and recall by using the retrieved photos, all ranked on similarity. The proposed method aims to retrieve all photos of the queried person while excluding images of other individuals. To achieve this, the method uses the top n results as temporary queries, recalculates similarities, and combines the obtained ranked lists to produce a better overall ranking. The method includes a novel and reliable procedure for selecting n, which is evaluated on two datasets, and considers the impact of age variation in the datasets.

1 Introduction

This paper demonstrates an efficient approach to searching for additional photographs of a specific person in large unlabelled photo collections, using a query photo as the reference. This process is generally known as face image retrieval (FIR). Foremost, a technique is suggested for selecting the top n items from the retrieved results to generate an enhanced ranked list. The approach is specifically designed for searching unlabelled databases and the objective is to maximise the number of photographs featuring the query face while minimising the chances of retrieving incorrect faces.

1.1 Face Recognition

The purpose of Face Recognition (FR) is to identify or verify the identity of an individual based on their facial features. This technology uses biometric algorithms to analyse unique characteristics of a person's face, such as the distance between the eyes, the shape of the nose and mouth, and the contours of the face. The process typically involves the following steps:

1. Face detection: The system uses computer vision techniques to locate and extract faces from images or video frames [27,30].
2. Face alignment: The system adjusts the position, rotation, and scale of the detected faces to align them with a standard reference frame [33,34].

G. L. Foresti et al. (Eds.): ICIAP 2023, LNCS 14233, pp. 159–170, 2023.
https://doi.org/10.1007/978-3-031-43148-7_14

3. Feature extraction: The system analyzes the facial features, such as the distance between the eyes, the shape of the nose and mouth, and the contours of the face, to create a unique facial signature, i.e. the facial feature vector (FFV) [5,16].
4. Matching/Verification: The system compares the extracted FFVs to a database of previously registered faces with their FFVs to find a match [4,6].

Image processing, computer vision and machine learning are essential components of several of these steps for FR. Image processing is used to detect and extract faces from images or video frames and to enhance their quality and resolution. Machine learning algorithms are then used to analyse the facial features and create a unique FFV for each individual. These algorithms can be trained on a large datasets of images to learn to recognise and differentiate between different faces. FR has many applications, including security, access control, and identification, but also entertainment to automatically tag people in photos and videos.

Face recognition algorithms can be based on different approaches, such as eigenfaces [26], or deep neural networks such as Convolutional Neural Networks (CNNs) [16,18], just to mention a few. During training, the CNN learns to recognise and differentiate between different facial features, such as the distance between the eyes, the shape of the nose, and the contours of the face. The output of the CNN is a high-dimensional FFV that represents the unique identity of the person.

1.2 Face Image Retrieval

Tang et al. [24] proposed a novel approach for FIR based on Deep Hashing, which learns feature representations of images, hashing functions, and classifiers in a joint manner. Hashing methods for FIR have gained a lot of attention and many studies have been published [11,12,25], just to mention a few.

Zaeemzadeh et al. [32] proposed a new face image retrieval framework that uses an adjustment vector to specify desired modifications to facial attributes and a preference vector to assign importance levels to different attributes. This approach allows users to retrieve images similar to a query image but with different desired attributes. Furthermore, Shi and Jain [22] proposed a semi-supervised framework to learn robust face representation that can generalize to unconstrained faces beyond the labeled training data.

1.3 Person Re-Identification

A similar problem arises in person re-identification (re-ID), which refers to the task of identifying the same individual across different non-overlapping camera views in a multi-camera surveillance system. It is a challenging problem due to variations in appearance caused by changes in illumination, viewpoint, occlusion, and other factors. The goal is to match a person of interest across different camera views in order to track their movements and activities throughout a surveillance network. Gupta et al. [7] proposed a method using residual networks and transfer learning for solving this task.

2 Background

The main objective of this approach is not only to enhance the search outcomes in facial image databases, but also to search for a particular face in large databases of untagged data, such as repositories of digitised facial photographs. This technique can be advantageous in scenarios where photos from multiple collections are available, but they have not yet been labeled. In such situations, the method of FIR can be employed to locate individuals in different collections.

2.1 Similarity Between FFVs

Typically, in both object-image retrieval and information retrieval, the results are ranked based on their relevance. In FIR, this process involves determining the similarity through pairwise comparisons, where the query image is compared against all retrieved face images using a selected similarity measure.

There are different ways to define such similarity, but here it was chosen to use the *cosine similarity* in the experiments, which is defined as

$$s_i := \frac{q \cdot w_i}{\|q\|\|w_i\|} \tag{1}$$

where q is the query FFV and w_i is the i-th FFV in the databases.

The name *cosine similarity*, comes from the fact that the dot product corresponds to the cosine of the angle between the two normalised vectors. This value can theoretically vary between -1 and 1, where 1 means that the face images themselves are the very same images. However, in practice the similarity measure often varies around 0.4–0.8 for the same person and falls below about 0.1 when two different persons are compared. It should be noted though that this threshold varies depending on the quality of the FFVs at hand. In this study, an optimal threshold τ for maximum efficiency will be discussed.

2.2 Consensus Ranking

In the work of Hast [9], a method called *consensus Ranking* (CR) for enhancing the quality of a ranked list is proposed. CR differs from re-ranking, which typically involves re-ranking the results using a *different* similarity computation method or even employing an alternative model for generating the feature vectors, i.e., FFVs.

CR involves selecting the top n items from the original ranked list, which is sorted on the similarities s to the query face q as in Eq. 1, and re-ranking them against one another. This process is iterated n times using the top n faces as queries $s_{1..n}$, resulting in n distinct rankings that are later combined to form a more comprehensive and improved ranking. In order to fuse these lists, the similarity measures for each instance in the lists are averaged. However, the challenge remains in defining the top n candidates and determining a suitable similarity threshold τ, which are highly dependent on the specific application. To address this, an automatic approach for selecting the top n candidates is proposed, and a reliable methodology for accomplishing this is explained in detail.

3 The Face Recognition Pipeline

In this work, one freely available pipeline for FR was used to produce FFVs called The *InsightFace* [10]. However, other approaches that could also have been used to produce FFVs are *DeepFace* [19–21], *CosFace* [28] *FaceNet* [18], *SphereFace* [13], and [23,29], just to mention a few.

The *InsightFace* [10] pipeline is an integrated Python library, mainly based on PyTorch [17] and MXNet [3]. *InsightFace* efficiently implements several state of the art algorithms for both face detection, face alignment and FR. It allows for automatic extraction of highly discriminative FFVs for each face, based on the Additive Angular Margin Loss (ArcFace) approach [5]. The model used in the analysis was $buffalo\_l$, which will produce 512 long FFVs.

4 Datasets

In the process of automated FR, images that were not correctly identified or contained multiple faces were removed as they presented problems. Additionally, duplicate face images, incorrectly labeled faces, and corrupt face feature vectors (FFVs) were also eliminated. Nonetheless, the following two datasets were used, which both covers several age spans per person.

4.1 AgeDB

The *AgeDB* dataset [14] contains a total of $16,516$ images. The images selected by requiring that each person included should have at least 30 face images covering at least three different age decades, i.e. (0–10, 11–20, 21–30, 31–40, 41–50, 51–60, 61–70 and 70+). A total of 9728 face images were extracted, yielding about 36 face images per person, on average.

4.2 CASIA

The original *CASIA-WebFace* [31] is quite large and contains about $500k$ images. However, the same selection and cleaning process resulted in a more feasible dataset containing 63609 face images, and more than 50 face images per person on average. Hence, it is larger than the *AgeDB* dataset.

5 Finding n

The main novelty is explained in this section. The new approach using CR with the proposed method of finding the top n FFVs will be referred to as *Optimised*. This will be compared to using no CR, which will be referred to as *Sorted* in the following sections, since it is just sorted on similarity s. Nonetheless, the study will also indicate for what thresholds the latter will work well and how it compares to the former. As will be shown the *Optimised* approach will outperform the *Sorted* approach.

When comparing the FFV of a query photograph with all other available FFVs and sorting the resulting similarities, one can observe a sharp increase in similarity that separates the photos into two distinct groups. The first group comprises FFVs of people who are highly unlikely to be the person being sought, while the second group consists of those who are very likely to be the same as the one in the query photo.

Figure 1 shows the sorted s and it exhibits a "knee", which refers to a point where the shape of the curve changes abruptly. This is often indicated by a change in the slope or concavity of the curve. In some cases, this knee point is associated with a significant change or transition in the underlying system that the curve represents.

When the second derivative of a function has a minimum, it indicates that the function is changing from a state of increasing concavity to a state of decreasing concavity. This is often associated with a change in the shape of the curve, and a potential knee point. Specifically, when the second derivative is positive, the curve is convex and when the second derivative is negative, the curve is concave. Thus, when the second derivative reaches its minimum, it indicates that the curve is transitioning from a more convex shape to a more concave shape, which could correspond to a knee point.

More formally, a curve is convex if and only if, for any two points on the curve, the line segment connecting them lies above the curve. In other words, a curve is convex if it "bulges out" towards the observer. On the other hand, a curve is concave if and only if, for any two points on the curve, the line segment connecting them lies below the curve.

The second derivative can be effectively computed by a double filtering approach which allows for smoothing of the curve in the same process, so that minor variations are even out [8]. In this case a combination of two Bézier filters were chosen, as explained in the same paper. It should also be mentioned that all the negative similarities where set to zero in order to avoid detecting the initial "knee" in the sorted list.

However, even if the value for n computed this way captures a majority of the persons searched for (of length d), it might also contain other persons faces, especially when d is small. It was empirically derived that computing n using $d/2$ generally yielded a much better and more reliable result.

6 Results

Several different measurements where used to evaluate the impact of using CR with the proposed method for finding n, i.e. the *Optimised* method, compared to a simple sorted list, (the *Sorted* method). Two measurements where computed on these two lists, namely Mean Average Precision and the Bhattacharyya Coefficient. Experiments where also done so that the threshold τ was varied while computing the so called F_1 *score*. This measurement and the two other measurements used are explained in the following subsections, together with the results.

6.1 Mean Average Precision

One way to determine the quality of a ranked list is to compute the mean Average Precision (mAP) [1]. This is a standard measurement used for retrieval systems. This is

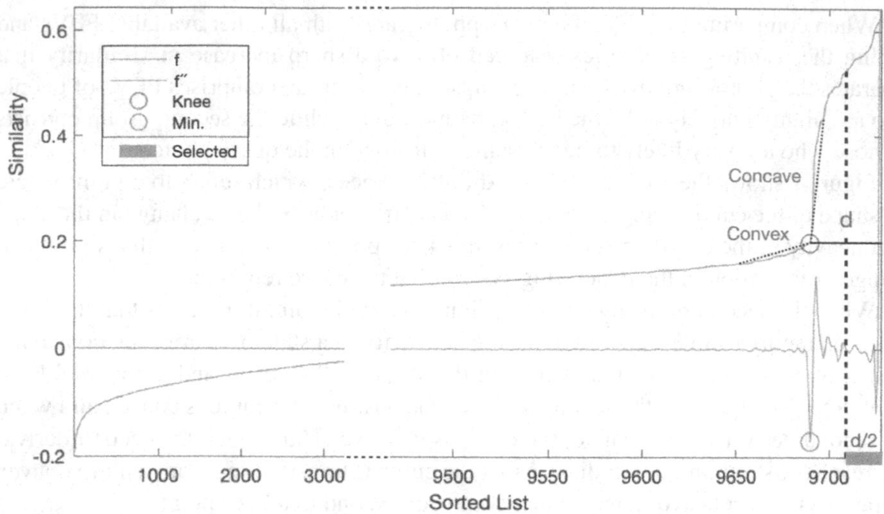

Fig. 1. Illustration of how n can be found by sorting the similarity s, finding the "knee" and then using the upper half in order to be safe not including false faces, especially when d is small.

done by computing all the similarities with respect to the FFV in question, compared to all other FFVs, and sorting them on the said similarity. Assume that there are m_k FFVs that are the same person as the query person q. Then the idea is that the Average Precision (AP) will be 1 if all the top n persons, where $n = m_k$, are the same person. However, if one of the instances in the top n ranked list is not the same person, then the AP will be lower, and so on. The mAP is defined as

$$\text{mAP} = \frac{1}{N}\sum_{i=1}^{N} AP_i \tag{2}$$

where $N = |s|$, and for each FFV

$$\text{AP}_\text{i} = \frac{\sum_{k=1}^{N} P@k \times r(k)}{m_k} \tag{3}$$

where $P@k$ is the precision for the N items in the sorted list, and $r(k)$ is a binary relevance function, indicating whether the k-th item in the list is a relevant object, i.e. belonging to the class in question.

6.2 F_1 Score

According to Opitz and Burst [15], the best choice to compute the F_1 score is the arithmetic mean over harmonic means, especially for imbalanced datasets. Hence, F_1 scores are computed for each person and then averaged via arithmetic mean, such that

$$\mathbb{F}_i = \frac{1}{N}\sum_{i=1}^{N} F_1 = \frac{1}{N}\sum_{i=1}^{N} \frac{2P_iR_i}{P_i + R_i} \tag{4}$$

where P_i and R_i are precision and recall respectively for each person i.

6.3 Bhattacharyya Measurements

The Bhattacharyya coefficient [2] is being used in numerous fields, such as statistics, image processing, and pattern recognition. It is a similarity measure between two probability distributions. Where 0 indicates no overlap between the distributions, and 1 indicates that the two distributions are identical. It is defined as the dot product of the normalised distributions, and is therefore capturing how much overlap there is between the distributions

$$BC = \sum_{i=1}^{N} \sqrt{P_i Q_i} \tag{5}$$

where P and Q are the distributions, of length N, of the intra- and interdistances respectively.

Here, one distribution of the similarity between images of the same person is computed, and one with similarities between all image pairs of different persons. Since CR performs a fusion of ranked lists, the similarities will be the average of n ranked lists. Hence the computed similarity for the *Optimised* approach will be different from the *Sorted* approach. Table 1 shows the results for both datasets, and it is clear that the *Optimised* approach is generally performing better. Keep in mind that mAP and the F1-score should be closer to 1, while the Bhattacharyya Coefficient should be closer to 0.

Figure 2 shows the results on the *AgeDB* dataset, and it is clear that the *Optimised* approach yields a more optimal Precision-Recall curve, but also a higher F_1 score. The shaded bar graph for the latter, depicts both the *mean value* for all persons in the datasets as well as the *variance*.

Moreover, the distribution graphs in the same figure, indicated that there is a smaller overlap for the *Optimised* approach, which means a smaller Bhattacharyya Coefficient. One can especially note that the interdistances becomes more packed around 0, while the intradistances are shifted towards higher values. In addition, a small bump to the right is visible that indicates that those cases are getting higher values than when using the *Sorted* approach. Hence, CR does have a positive impact on the results and the proposed method for finding n clearly improves the result.

The interesting thing about these two datasets are that they have a large age variation and is therefore rather challenging. Next, a series of experiments where done on the much larger *CASIA* dataset, where different age spans where extracted. Hence, the age variation will be smaller and it can be deduced how well the proposed method works for such cases.

Table 1. Comparison of mAP, F_1 score and Bhattacharyya Coefficient for the AgeDB and CASIA datasets.

Database	size	Sort. mAP	Opt. mAP	Sort. Thresh.	F_1	Optimised Thresh.	F_1	Sort. Dist.	Opt. Dist.
AgeDB	9728	0.9810	**0.9896**	0.3040	0.9482	0.2980	**0.9682**	0.0768	**0.0670**
CASIA	63609	0.9743	**0.9840**	0.3330	0.9479	0.3280	**0.9689**	0.0856	**0.0761**

Table 2 shows the results of this investigation. Interestingly, the *Optimised* approach always do better when it comes to the F_1 score, but not always for mAP. Furthermore, the Bhattacharyya coefficient is almost always smaller for the *Optimised* approach, which indicates it is generally better.

Table 2. Comparison of mAP, F_1 score and Bhattacharyya Coefficient for the AgeDB and CASIA datasets.

Age group	size	Sort. mAP	Opt. mAP	Sorted Thresh.	F_1	Optimised Thresh.	F_1	Sort. Dist.	Opt. Dist.
0–19	1337	**0.9664**	0.9637	0.3300	0.9456	0.3130	**0.9653**	0.1055	**0.0938**
20–24	3433	0.9715	**0.9723**	0.3320	0.9505	0.3210	**0.9704**	0.0886	**0.0745**
25–29	10487	0.9712	**0.9775**	0.3380	0.9494	0.3280	**0.9727**	0.0946	**0.0792**
30–34	12548	0.9778	**0.9825**	0.3410	0.9572	0.3340	**0.9768**	0.0840	**0.0726**
35–39	12548	0.9776	**0.9832**	0.3400	0.9617	0.3320	**0.9792**	0.0764	**0.0666**
40–44	7883	0.9812	**0.9847**	0.3360	0.9642	0.3250	**0.9810**	0.0716	**0.0619**
45–49	4679	0.9747	**0.9754**	0.3190	0.9625	0.3130	**0.9766**	0.0861	**0.0758**
50–54	1813	**0.9814**	0.9793	0.3180	0.9702	0.3100	**0.9813**	0.0755	**0.0612**
55–59	1873	0.9901	**0.9905**	0.3230	0.9828	0.3100	**0.9911**	0.0439	**0.0376**
60–69	1253	**0.9841**	0.9818	0.3270	0.9793	0.3060	**0.9880**	0.0526	**0.0486**
70–100	501	**0.9879**	0.9865	0.3000	0.9843	0.2970	**0.9899**	**0.0396**	0.0437

7 Discussion

As shown in the tables and the figures the *Optimised* approach improves the F_1 score for both datasets and all age spans. The Bhattacharyya coefficient indicates the same thing except for one age span. The deviation is probably due to the comparably small number of images in that age span, making the computation of the distributions less reliable. However, the fact that the mAP is not always better is less obvious. It must be stressed though, that it is generally better and that the deviating cases, once again are for the age spans with smaller sets. Nevertheless, the F_1 score depends on the threshold τ chosen that serves to optimise the Precision and Recall for all the different persons and is therefore a more interesting measurement. The mAP is independent of the threshold, but for the FIR being of practical use, a threshold τ needs to be defined.

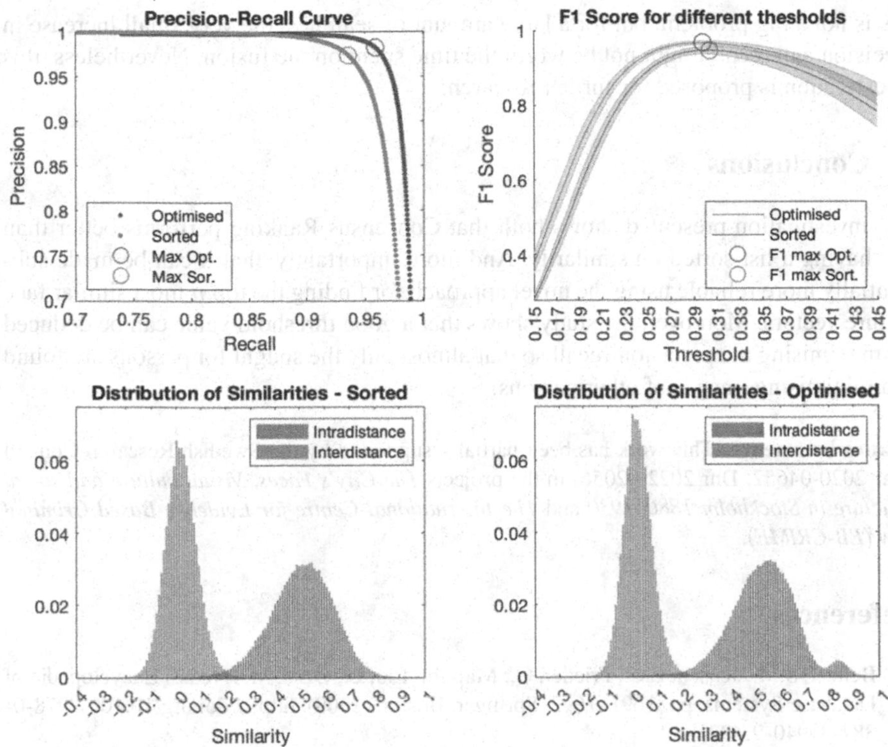

Fig. 2. Illustration of the difference between ranking using a sorted list and an optimised list. Both the precision-recall curve and the F_1 score is higher for the optimised approach. The distribution of similarities changes so that for the optimised approach the interdistances becomes more packed around 0 and the intradistances are shifted towards higher values, and a small bump indicates that those cases are getting higher values than when using sorting.

The novel thing proposed in this paper was to use CR with a robust way of finding the top n FFVs for the fusion performed in the CR. By smoothing the curve obtained from the sorted list of similarities, and in the same time computing the second derivative, any minor bumps are removed. Some FFVs of the sought person could be found below the "knee", but there is also a small probability to find the wrong person just above the "knee". Therefore, the distance d to the top was divided by 2 so the list would, to a much higher degree, contain the FFVs of the sought for persons. This will be especially important when d is small. However, when d is larger, the result of the CR is still reliable even if the list is contaminated with a few wrong persons. It was not investigated when this would be a problem and whether, d could been divided in any other way. It could also be different for different pipelines for computing the FFVs and therefore it is proposed for further research.

Nevertheless, it was experimentally concluded that a larger portion of d could be used when d itself was larger. However, then n becomes larger and the fusion takes more time, since more temporary queries needs to be performed. For a single search, this is not a big problem, but for a large amount of searches, the very small increase in precision and recall might not be worth the time spent on the fusion. Nevertheless, this investigation is proposed for further research.

8 Conclusions

The investigation presented shows both that Consensus Ranking performs better than just having a list sorted on similarity. And more importantly, that it can be made substantially more reliable using the novel approach for finding the top n most similar face feature vectors. Moreover, the study shows that a good threshold value can be deduced for maximising the precision-recall so that almost only the sought for persons are found and minimising images of other persons.

Acknowledgments. This work has been partially supported by the Swedish Research Council (Dnr 2020-04652; Dnr 2022-02056) in the projects *The City's Faces. Visual culture and social structure in Stockholm 1880–1930* and *The International Centre for Evidence-Based Criminal Law (EB-CRIME).*

References

1. Beitzel, S.M., Jensen, E.C., Frieder, O.: Map. In: Liu, L., Özsu, M.T. (eds.) Encyclopedia of Database Systems, pp. 1691–1692. Springer, Boston (2009). https://doi.org/10.1007/978-0-387-39940-9_492
2. Bhattacharyya, A.: On a measure of divergence between two statistical populations defined by their probability distribution. Bull. Calcutta Math. Soc. **35**, 99–110 (1943)
3. Chen, T., et al.: Mxnet: a flexible and efficient machine learning library for heterogeneous distributed systems. CoRR abs/1512.01274 (2015). http://arxiv.org/abs/1512.01274
4. Choy, C.B., Gwak, J.Y., Savarese, S., Chandraker, M.: Universal correspondence network. In: Proceedings of the 30th International Conference on Neural Information Processing Systems, NIPS 2016, pp. 2414–2422. Curran Associates Inc., Red Hook (2016)
5. Deng, J., Guo, J., Xue, N., Zafeiriou, S.: Arcface: additive angular margin loss for deep face recognition. In: Proceedings of the IEEE/CVF Conference on Computer Vision and Pattern Recognition, pp. 4690–4699 (2019)
6. Duchenne, O., Bach, F., Kweon, I.S., Ponce, J.: A tensor-based algorithm for high-order graph matching. IEEE Trans. Pattern Anal. Mach. Intell. **33**(12), 2383–2395 (2011). https://doi.org/10.1109/TPAMI.2011.110
7. Gupta, A., Pawade, P., Balakrishnan, R.: Deep residual network and transfer learning-based person re-identification. Intell. Syst. Appl. **16**, 200137 (2022). https://doi.org/10.1016/j.iswa.2022.200137. https://www.sciencedirect.com/science/article/pii/S2667305322000746
8. Hast, A.: Simple filter design for first and second order derivatives by a double filtering approach. Pattern Recogn. Lett. **42**, 65–71 (2014). https://doi.org/10.1016/j.patrec.2014.01.014. https://www.sciencedirect.com/science/article/pii/S0167865514000282
9. Hast, A.: Consensus ranking for increasing mean average precision in keyword spotting. In: Amelio, A., Borgefors, G., Hast, A. (eds.) Proceedings of 2nd International Workshop on Visual Pattern Extraction and Recognition for Cultural Heritage Understanding co-located with 16th Italian Research Conference on Digital Libraries (IRCDL 2020), Bari, Italy, 29 January 2020, CEUR Workshop Proceedings, vol. 2602, pp. 46–57. CEUR-WS.org (2020). https://ceur-ws.org/Vol-2602/paper4.pdf

10. InsightFace: Insightface (2023). https://insightface.ai. Accessed 30 Feb 2023
11. Jang, Y.K., Jeong, D., Lee, S.H., Cho, N.I.: Deep clustering and block hashing network for face image retrieval. In: Jawahar, C.V., Li, H., Mori, G., Schindler, K. (eds.) ACCV 2018. LNCS, vol. 11366, pp. 325–339. Springer, Cham (2019). https://doi.org/10.1007/978-3-030-20876-9_21
12. Lin, J., Li, Z., Tang, J.: Discriminative deep hashing for scalable face image retrieval. In: Proceedings of the Twenty-Sixth International Joint Conference on Artificial Intelligence, IJCAI-2017, pp. 2266–2272 (2017). https://doi.org/10.24963/ijcai.2017/315
13. Liu, W., Wen, Y., Yu, Z., Li, M., Raj, B., Song, L.: Sphereface: deep hypersphere embedding for face recognition. In: 2017 IEEE Conference on Computer Vision and Pattern Recognition (CVPR), pp. 6738–6746. IEEE Computer Society, Los Alamitos (2017). https://doi.org/10.1109/CVPR.2017.713
14. Moschoglou, S., Papaioannou, A., Sagonas, C., Deng, J., Kotsia, I., Zafeiriou, S.: Agedb: the first manually collected, in-the-wild age database. In: 2017 IEEE Conference on Computer Vision and Pattern Recognition Workshops (CVPRW), pp. 1997–2005 (2017). https://doi.org/10.1109/CVPRW.2017.250
15. Opitz, J., Burst, S.: Macro f1 and macro f1 (2021)
16. Parkhi, O.M., Vedaldi, A., Zisserman, A.: Deep face recognition. In: Xie, X., Jones, M.W., Tam, G.K.L. (eds.) Proceedings of the British Machine Vision Conference (BMVC), pp. 41.1–41.12. BMVA Press (2015). https://doi.org/10.5244/C.29.41
17. Paszke, A., et al.: PyTorch: an imperative style, high-performance deep learning library. Curran Associates Inc., Red Hook (2019)
18. Schroff, F., Kalenichenko, D., Philbin, J.: Facenet: a unified embedding for face recognition and clustering. In: 2015 IEEE Conference on Computer Vision and Pattern Recognition (CVPR), pp. 815–823 (2015). https://doi.org/10.1109/CVPR.2015.7298682
19. Serengil, S.I., Ozpinar, A.: Lightface: a hybrid deep face recognition framework. In: 2020 Innovations in Intelligent Systems and Applications Conference (ASYU), pp. 23–27. IEEE (2020). https://doi.org/10.1109/ASYU50717.2020.9259802
20. Serengil, S.I., Ozpinar, A.: Hyperextended lightface: a facial attribute analysis framework. In: 2021 International Conference on Engineering and Emerging Technologies (ICEET), pp. 1–4. IEEE (2021). https://doi.org/10.1109/ICEET53442.2021.9659697
21. Serengil, S.I., Ozpinar, A.: An evaluation of sql and nosql databases for facial recognition pipelines (2023). https://www.cambridge.org/engage/coe/article-details/63f3e5541d2d184063d4f569. https://doi.org/10.33774/coe-2023-18rcn, preprint
22. Shi, Y., Jain, A.K.: Boosting unconstrained face recognition with auxiliary unlabeled data. In: 2021 IEEE/CVF Conference on Computer Vision and Pattern Recognition Workshops (CVPRW), pp. 2789–2798. IEEE Computer Society, Los Alamitos (2021). https://doi.org/10.1109/CVPRW53098.2021.00314
23. Shi, Y., Jain, A.: Probabilistic face embeddings. In: 2019 IEEE/CVF International Conference on Computer Vision (ICCV), pp. 6901–6910 (2019). https://doi.org/10.1109/ICCV.2019.00700
24. Tang, J., Li, Z., Zhu, X.: Supervised deep hashing for scalable face image retrieval. Pattern Recogn. **75**(C), 25–32 (2018). https://doi.org/10.1016/j.patcog.2017.03.028
25. Tang, J., Lin, J., Li, Z., Yang, J.: Discriminative deep quantization hashing for face image retrieval. IEEE Trans. Neural Netw. Learn. Syst. **29**(12), 6154–6162 (2018). https://doi.org/10.1109/TNNLS.2018.2816743
26. Turk, M., Pentland, A.: Eigenfaces for recognition. J. Cogn. Neurosci. **3**(1), 71–86 (1991). https://doi.org/10.1162/jocn.1991.3.1.71
27. Viola, P., Jones, M.: Robust real-time face detection. In: Proceedings Eighth IEEE International Conference on Computer Vision, ICCV 2001, vol. 2, pp. 747–747 (2001). https://doi.org/10.1109/ICCV.2001.937709

28. Wang, H., et al.: Cosface: large margin cosine loss for deep face recognition. In: 2018 IEEE/CVF Conference on Computer Vision and Pattern Recognition, pp. 5265–5274 (2018). https://doi.org/10.1109/CVPR.2018.00552
29. Wen, Y., Zhang, K., Li, Z., Qiao, Yu.: A discriminative feature learning approach for deep face recognition. In: Leibe, B., Matas, J., Sebe, N., Welling, M. (eds.) ECCV 2016. LNCS, vol. 9911, pp. 499–515. Springer, Cham (2016). https://doi.org/10.1007/978-3-319-46478-7_31
30. Yang, M.H., Kriegman, D., Ahuja, N.: Detecting faces in images: a survey. IEEE Trans. Pattern Anal. Mach. Intell. **24**(1), 34–58 (2002). https://doi.org/10.1109/34.982883
31. Yi, D., Lei, Z., Liao, S., Li, S.Z.: Learning face representation from scratch (2014). https://doi.org/10.48550/ARXIV.1411.7923. https://arxiv.org/abs/1411.7923
32. Zaeemzadeh, A., et al.: Face image retrieval with attribute manipulation. In: 2021 IEEE/CVF International Conference on Computer Vision (ICCV), pp. 12096–12105 (2021). https://doi.org/10.1109/ICCV48922.2021.01190
33. Zhang, K., Zhang, Z., Li, Z., Qiao, Y.: Joint face detection and alignment using multitask cascaded convolutional networks. IEEE Signal Process. Lett. **23**(10), 1499–1503 (2016). https://doi.org/10.1109/LSP.2016.2603342
34. Zhang, Z., Luo, P., Loy, C.C., Tang, X.: Learning deep representation for face alignment with auxiliary attributes. IEEE Trans. Pattern Anal. Mach. Intell. **38**(5), 918–930 (2016). https://doi.org/10.1109/TPAMI.2015.2469286

Towards Explainable Navigation and Recounting

Samuele Poppi, Roberto Bigazzi, Niyati Rawal, Marcella Cornia(✉), Silvia Cascianelli, Lorenzo Baraldi, and Rita Cucchiara

University of Modena and Reggio Emilia, Modena, Italy
{samuele.poppi,roberto.bigazzi,niyati.rawal,marcella.cornia, silvia.cascianelli,lorenzo.baraldi,rita.cucchiara}@unimore.it

Abstract. Explainability and interpretability of deep neural networks have become of crucial importance over the years in Computer Vision, concurrently with the need to understand increasingly complex models. This necessity has fostered research on approaches that facilitate human comprehension of neural methods. In this work, we propose an explainable setting for visual navigation, in which an autonomous agent needs to explore an unseen indoor environment while portraying and explaining interesting scenes with natural language descriptions. We combine recent advances in ongoing research fields, employing an explainability method on images generated through agent-environment interaction. Our approach uses explainable maps to visualize model predictions and highlight the correlation between the observed entities and the generated words, to focus on prominent objects encountered during the environment exploration. The experimental section demonstrates that our approach can identify the regions of the images that the agent concentrates on to describe its point of view, improving explainability.

Keywords: Explainable AI · Visual Navigation · Image Captioning

1 Introduction

Recent advances in the field of Embodied AI aim to foster the next generation of autonomous and intelligent mobile agents and robots. Research in this field includes visual navigation [10,29], object-driven navigation [11], and the creation of new research platforms for simulation of embodied agents [33]. While this line has focused on providing mobile robots with perception and action capabilities, it is indubitable that future agents will need to interact seamlessly with human beings. In this regard, the research at the intersection of Computer Vision and NLP is of particular interest to the community, as it can provide robots with the ability to connect what they perceive with their linguistic abilities. Also, being able to describe what the robot sees can be a valuable means of explainability and bridge the gap between the black-box architecture and the user.

In this paper, we concentrate on a novel Embodied AI setting, which aims both to connect Vision and Language and provide explainability. In our setting, a robot navigates an unknown environment and has the ability to describe in

G. L. Foresti et al. (Eds.): ICIAP 2023, LNCS 14233, pp. 171–183, 2023.
https://doi.org/10.1007/978-3-031-43148-7_15

natural language what it sees. In particular, the agent needs to perceive the environment around itself, navigate it driven by an exploration goal, and describe salient objects and scenes in natural language. Beyond navigating the environment and translating visual cues in natural language, the agent also needs to identify appropriate moments to perform the explanation step, *i.e.*, it needs to employ an appropriate speaking policy. It is worthwhile to note that this setting poses challenges from different perspectives. Firstly, exploring an environment without any previous knowledge, nor a reference trajectory is, by itself, a significant challenge for Embodied AI. Secondly, while describing visual content in Natural Language has been previously addressed by the image captioning community [39], existing approaches have been applied to either natural or web-scale images – and applications of image captioning to images taken from the perspective of a mobile robot are still very limited in literature [12]. Moreover, an appropriate speaking policy needs to be defined to interconnect the navigation and captioning components of the approach and select appropriate moments for describing the visual perception. Lastly, for the generated descriptions to be a valuable explainable mean, their quality should be properly assessed, and their content properly aligned with the robot's actual perception.

In the following, we jointly address all the previously mentioned points. We devise an exploration strategy based on a surprisal reward [24], in which the agent is trained in a self-supervised manner to explore previously unknown environments. Further, we endow our agent with a Transformer-based captioner that can generate natural language descriptions. Lastly, we devise a component that can measure the explanation capability of the caption by aligning its content with what the visual encoder is attending from the input image. Experimentally, we assess the performance of the proposed approach on the Matterport3D dataset [9] by employing the Habitat simulation platform [33]. In particular, we analyze the navigation, captioning, and explainability capabilities of the model and demonstrate the appropriateness of the proposed solutions. Overall, our proposal makes a step forward in the direction of mobile agents that can be explainable by design and interact with human beings in natural language.

2 Related Work

As the goal of this work is to enable embodied agents to generate user-explainable descriptions of the perceived environment, this section describes state-of-the-art research in explainable AI, image captioning, and embodied exploration.

Explainable AI. Explainability properties are becoming an increasingly desired feature for deep learning models, especially when these models are employed by final users and not by experts. When dealing with models for Computer Vision applications, several explainability approaches have been developed, depending on the specific task to perform and the type of features used [13,16,20,23,26,27]. In this work, we borrow ideas from the image classification task, where the predominant approach in literature is that of building a saliency map out of the visual encoder, achieving some level of explainability by visualizing the most salient regions in the input image. Subsequently, an explanation map can be computed as the pixel-wise multiplication of the input image and the saliency map. Many different approaches have been proposed to obtain explanation maps, like visualization tools [37,45], gradient-based approaches [38,40], and Class Activation Mapping (CAM)-based methods [35,42,46]. In this work, we resort to a CAM-based approach [46] for providing explanation maps.

Image Captioning. Being a task at the intersection between vision and language, image captioning has benefited from the technical advancements in both these fields. The goal of the task is to generate a natural language description of a given image. To this end, the image must be properly represented. In most works, this has been done by employing convolutional neural networks to extract global or grid features [19,43], or image region features containing visual entities [1]. More recent approaches employ fully-attentive Transformer-like architectures [41] as visual encoders, which can also be applied directly to image patches [13]. The image representation is used to condition a language model that generates the caption. The language model can be implemented as a recurrent neural network [7,18,19] or Transformer-based fully-attentive models [14,15,32]. In this work, we develop a fully-attentive image captioning approach as our captioning module, as it is potentially suitable to be employed in a wide variety of real-world settings such as the embodied exploration one.

Embodied Exploration. Research in Embodied AI has witnessed increasing interest from the community, thanks to the development of photo-realistic 3D simulation environments [9,33] that bring exploration and navigation agents one step closer to real-world deployment. Those agents are commonly trained with reinforcement learning adopting a modular, hierarchical approach [10] where the agent learns to explore the environment by optimizing a self-supervision signal in the form of a reward function [4,10,29,30]. Once trained on the aforementioned simulators, the developed agents can be easily deployed on physical robotic platforms [6]. However, many state-of-the-art architectures are still considered black boxes, as their behavior lacks explainability [2]. In this respect, some attempts have been made to make the robot navigation and decision-making processes more interpretable for the end-user by letting it produce natural language descriptions of what it observes [3,5,12]. In this work, we consider a curiosity-driven exploration agent [30] and equip it with the ability to produce natural language descriptions of what it observes while navigating the environment, also exploiting explainable maps to enhance the interpretability of the descriptions.

3 Proposed Method

In our proposed approach, a deep reinforcement learning exploration agent navigates an unknown environment and collects interesting views according to a heuristic speaker policy. These images are then passed to an encoder-decoder

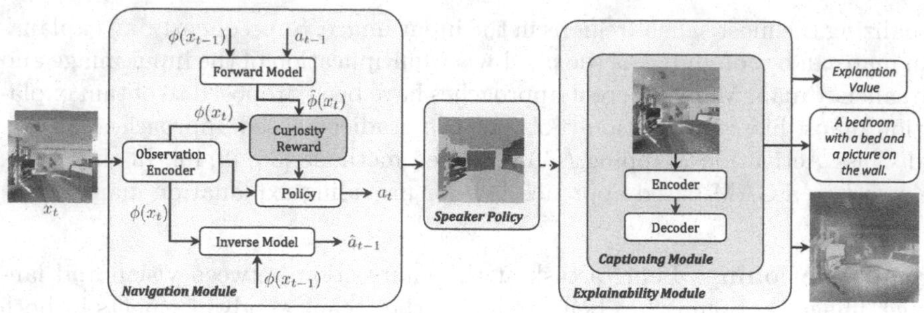

Fig. 1. Schema of the proposed "navigation and recounting" framework.

captioning model, combined with an explainability technique to provide a user-understandable interpretation of the environment internal representation of the agent. An overview of the proposed approach is depicted in Fig. 1.

3.1 Navigation Module

While moving inside the environment, our agent captures an RGB-D image from its current position x_t, which is encoded via a convolutional neural network ϕ. Note that, during training, the encoding network is maintained fixed for guaranteeing the stability of the resulting features during training, thus resulting in a more efficient training of our agent, as demonstrated in [8].

Our agent can move inside the environment by taking atomic actions at each timestep, $a_t \in \{\textit{turn left 15}^\circ,\ \textit{turn right 15}^\circ,\ \textit{move forward 0.25}\,\text{m}\}$. After the execution of each action, the agent captures a new observation x_{t+1}. From the observation x_t and action a_t, we can define the forward dynamics problem of predicting the next observation as $\hat{\phi}(x_{t+1}) = f(\phi(x_t), a_t; \theta_F)$, where $\hat{\phi}(x_{t+1})$ is the predicted visual embedding for x_{t+1} and f is the forward dynamics model, whose parameters are θ_F. Moreover, given two consecutive observations (x_t, x_{t+1}), we can define the inverse dynamics problem of predicting the action a_t performed between the two observations as $\hat{a}_t = g(\phi(x_t), \phi(x_{t+1}); \theta_I)$, where $\hat{a}_t$ is the predicted estimate for the action a_t and g is the inverse dynamics model, whose parameters are θ_I. The parameters θ_F and θ_I of the dynamics models are determined by minimizing respectively $\mathcal{L}_F$ and $\mathcal{L}_I$ losses:

$$\mathcal{L}_F = \frac{1}{2} \left\| \hat{\phi}(x_{t+1}) - \phi(x_{t+1}) \right\|_2^2, \quad \text{and} \quad \mathcal{L}_I = y_t \log \hat{a}_t, \tag{1}$$

where y_t is the one-hot representation of a_t.

Note that the actions the agent performs at each timestep are selected by a policy $\pi(\phi(x_t); \theta_\pi)$, that is trained to maximize the expected sum of a reward expressed as the discrepancy of the predictions of dynamics models and the actual observation, *i.e.*,

$$\max_{\theta_\pi} \mathbb{E}_{\pi(\phi(x_t);\theta_\pi)} \left[\sum_t r_t \right] \quad \text{s.t.} \quad r_t = \frac{\eta}{2} \left\| f(\phi(x_t), a_t) - \phi(x_{t+1}) \right\|_2^2 - p_t, \tag{2}$$

where η is a scaling factor and p_t is a penalty factor to prevent the agent from repeating the same action multiple consecutive times. In this work, p_t is set to 0 and then becomes equal to a constant value $\tilde{p}$ if the same action is repeated $\tilde{t}$ times. By combining the loss functions in Eq. 1 and 2, we obtain the overall optimization problem:

$$\min_{\theta_\pi, \theta_F, \theta_I} \left[-\lambda \mathbb{E}_{\pi(\phi(x_t);\theta_\pi)} \left[\sum_t r_t \right] + \beta \mathcal{L}_F + (1-\beta)\mathcal{L}_I \right] \tag{3}$$

where λ and β are regularization factors.

3.2 Object-Driven Speaker Policy

As the navigation proceeds, the relevant objects in the scene can be recognized from the observation x_t. Based on the analysis presented in [5], in this work, we adopt an object-driven speaker policy that elicits the description of the observation if a minimum number of salient objects are present in the scene. By adopting this policy, the captioning module produces descriptions when a sufficient number of characteristic objects are observed, which allow connoting, and thus, distinguishing, that view of the environment. In this work, we set to five the minimum number of objects in the scene for it to be described.

3.3 Captioning Module

The goal of the captioning module is that of modeling an autoregressive distribution probability $p(\boldsymbol{w}_t|\boldsymbol{w}_{\tau<t}, \mathbf{V})$, where $\mathbf{V}$ is an image captured from the agent and $\{\boldsymbol{w}_t\}_t$ is the sequence of words comprising the generated caption. This is usually achieved by training a language model conditioned on visual features to mimic ground-truth descriptions.

We represent each training image-caption pair as a pair of image and text $(\mathbf{V}, \mathbf{W})$, where $\mathbf{V}$ is encoded with a set of fixed-length visual descriptors. The text input is tokenized with lower-cased Byte Pair Encoding [36] with a vocabulary of 49,152 tokens. For multimodal fusion, we employ an encoder-decoder Transformer [41] architecture. Each layer of the encoder employs multi-head self-attention (MSA) and feed-forward layers, while each layer of the decoder employs multi-head self- and cross-attention (MSCA) and feed-forward layers. For enabling text generation, sequence-to-sequence attention masks are employed in each self-attention layer of the decoder. The visual descriptors $\mathbf{V} = \{\boldsymbol{v}_i\}_{i=1}^N$ are encoded via bi-directional attention in the encoder, while the token embeddings of the caption $\mathbf{W} = \{\boldsymbol{w}_i\}_{i=1}^L$ are inputs of the decoder, where N and L indicate the number of visual embeddings and caption tokens, respectively. The overall network operates according to the following schema:

$$\begin{aligned} &\text{encoder} \quad \tilde{\boldsymbol{v}}_i = \text{MSA}(\boldsymbol{v}_i, \mathbf{V}) \\ &\text{decoder} \quad \mathbf{O}_{\boldsymbol{w}_i} = \text{MSCA}(\boldsymbol{w}_i, \tilde{\mathbf{V}}, \{\boldsymbol{w}_t\}_{t=1}^i), \end{aligned} \tag{4}$$

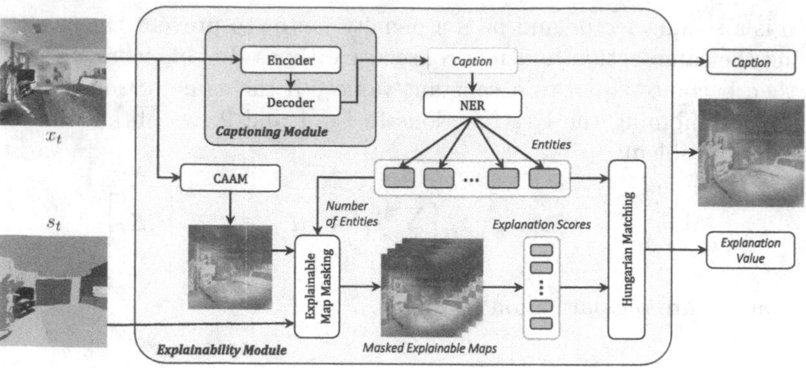

Fig. 2. Schema of the explainability module of our approach.

where $\mathbf{O}$ is the network output, $\text{MSA}(\boldsymbol{x}, \mathbf{Y})$ a self-attention with $\boldsymbol{x}$ mapped to query and $\mathbf{Y}$ mapped to key-values, and $\text{MSCA}(\boldsymbol{x}, \mathbf{Y}, \mathbf{Z})$ a self-attention with $\boldsymbol{x}$ as query and $\mathbf{Z}$ as key-values, followed by cross-attention with $\boldsymbol{x}$ as query and $\mathbf{Y}$ as key-values. We omit feed-forward layers and the dependency between consecutive layers for ease of notation. The captioning network is trained with a unidirectional language modeling loss based on cross-entropy.

Inference. Once the model is trained, at each time step t, it samples a token $\hat{\boldsymbol{w}}_t$ from the output probability distribution. This is then concatenated to previously predicted tokens to form a sequence $\{\hat{\boldsymbol{w}}_\tau\}_{\tau=1}^{t}$, which is employed as the input for the next iteration. Since the representation of output tokens does not depend on subsequent tokens, the past intermediate representations are kept in memory to avoid repeated computation and increase efficiency at prediction time.

Visual Features. To obtain the set of visual features $\mathbf{V}$ for an image, we employ a visual encoder pre-trained to match vision and language [28]. Compared to using features extracted from object detectors [1], our strategy is beneficial in terms of both computational efficiency feature quality. Specifically, we use one of the encoders proposed in CLIP [28], which can be either based on CNNs or Vision Transformers. In both cases, we employ the entire grid of features coming from the last visual encoder layer.

3.4 Captioning Assessment via Explanation Maps

As a measure of the explainability degree of the caption, given the output of the captioner, we align its content with portions of the image on which the visual encoder focuses (Fig. 2). Given an input image x, we firstly compute a Class-Agnostic Activation Map (CAAM) inspired by the Class Activation Mapping [46] literature. This is defined as a linear combination of the activation maps of the last layer of the encoder, weighted by their mean score, as follows:

$$\text{CAAM}(x) = \text{ReLU}\left(\sum\nolimits_{k=1}^{N_l} \alpha_k A_k\right), \tag{5}$$

where N_l denotes the number of activation maps, A_k is the k-th channel of the activation, and α_k are weight coefficients indicating the importance of each activation map, each of them defined as the average of A_k over the two spatial axes. A ReLU activation is employed to consider only the features that have a positive influence. Then, given a semantic segmentation map of the input image, *i.e.*, a set of binary masks, each associated to a semantic class, we average the values of CAAM(x) over each mask to obtain an "explanation score" for each semantic class. The explanation score of a semantic class indicates how much the network has focused on that semantic class. Finally, we align explanation scores with respect to concepts present in the caption. To this end, we firstly extract nouns from the generated caption by using Named Entity Recognition[1]. We then build a matching between the set of nouns in the caption and the set of semantic classes found in the segmentation map by using the Hungarian algorithm and GloVe embeddings [25] as similarity measure. With the obtained association between nouns mentioned in the caption and semantic concepts, we take a weighted average of explanation scores, according to the weights defined by the Hungarian matching for the optimal association. The final explanation value is a measure of how much the caption aligns with what the visual encoder is actually observing, and therefore it is an indirect score for the explainability power of the caption. In the following, we refer to this explanation value as EXPL-S.

4 Experiments

4.1 Exploration Experiments

The exploration capability of the agent influences the variability and information content of the observations extracted from the environment. We qualitatively validate the navigation module presenting the trajectories of the agent on some sample test episodes.

Implementation Details. In our experiments, we use Habitat simulator [33] with scenes from the Matterport3D dataset [9], for which dense semantic annotations are available with 40 different object categories. The observations for our navigation agent and the subsequent modules have a 640 × 480 resolution. Before being fed to the navigation module, the observations are resized to 84×84 grayscale images. Furthermore, instead of using a single observation for the current timestep, we use the last four observations stacked together for modeling temporal dependencies. Stacked grayscale observations are fed to the observation encoder to compute the features used by the dynamics models and the policy. The navigation module has been trained for 10K updates on the training set of Matterport3D dataset for a total of ≈ 1.3M frames, then, experiments are done on unseen environments of the test split. Episodes have a maximum length T of 500 and 1000 steps for the training and the testing phases, respectively.

[1] https://spacy.io/.

The policy and the dynamics models are trained using PPO [34] The parameters in Eq. 3 are set to $\lambda = 0.1$ and $\beta = 0.9$. The penalty introduced in Eq. 2 is triggered if the agent repeats the same action for $\tilde{t} = 5$ and assumes a value $p_t = \tilde{p} = 0.01$.

Fig. 3. Qualitative results of the agents' trajectories in sample exploration episodes.

Exploration Results. In Fig. 3 we show the agent's trajectory on the top-down environment map in some sample episodes. These results show that the agent is able to explore small environments completely while navigating efficiently for the duration of the time budget to optimize the area seen in larger environments.

4.2 Captioning and Explainability Experiments

In the considered setting, the explainability properties of the navigation agent depend on the quality of the captions produced by the captioning module and to their explainability power. In this section, we evaluate these two aspects.

Datasets. To assess the performance of the captioning and explainability modules, we first consider the COCO image captioning dataset [22]. Specifically, we follow the splits defined in [19], which consists of 113,287 images for training, 5,000 images for validation, and 5,000 for test. The images depict people and common-use objects and come with five ground-truth captions each. Moreover, for evaluating these modules on the Matterport3D dataset, we use the images gathered from the robot that are considered to be worth describing according to the speaker policy. In particular, these are the images in which the number of objects is greater than five, resulting in 16,828 images. Note that no ground-truth caption is available for these images, which are therefore used only in inference.

Implementation Details. We consider four variants of the captioning module, each of them based on a different CLIP-based visual encoder [28]. In particular, we consider three encoders based on CNNs (*i.e.*, CLIP-RN50, CLIP-RN50×4, and CLIP-RN50×16), and an encoder based on a Vision Transformer (*i.e.*, CLIP-ViT-B16). In all the variants, the decoder takes as inputs d-dimensional vectors with $d = 384$, and has $l = 3$ layers and $H = 6$ attention heads. To represent the position of the input words, we exploit the sinusoidal positional encoding as in [41]. We train the four variants of the captioning module by optimizing a standard cross-entropy loss with the LAMB optimizer [44]. We employ the learning rate scheduling strategy proposed in [41], which entails a warmup of 6,000 iterations whose resulting learning rate is multiplied by 5, and minibatch size equal to 1,080. We additionally fine-tune the models with the SCST strategy, by employing the Adam optimizer [21] and fixed learning rate 5×10^{-6}.

Table 1. Captioning results on the COCO Karpathy-test split.

	BLEU-4	METEOR	ROUGE	CIDEr	SPICE	CLIP-S
CLIP-RN50	36.9	28.1	57.5	125.2	21.5	0.738
CLIP-ViT-B16	38.6	29.2	58.8	132.5	23.0	0.749
CLIP-RN50×4	39.2	29.2	58.8	132.9	22.7	**0.753**
CLIP-RN50×16	**40.2**	**29.5**	**59.4**	**137.0**	**23.1**	0.750

Table 2. Captioning and explainability results on the images gathered from the agent in the Matterport3D dataset, and described according to the speaker policy.

	CLIP-S	$\text{Cov}_{>1\%}$	$\text{Cov}_{>3\%}$	$\text{Cov}_{>5\%}$	$\text{Cov}_{>10\%}$	EXPL-S
CLIP-RN50	0.697	0.492	0.569	0.625	0.721	**0.670**
CLIP-ViT-B16	**0.722**	0.521	0.591	0.651	0.735	0.694
CLIP-RN50×4	0.721	0.520	0.591	0.649	0.732	0.623
CLIP-RN50×16	0.719	**0.528**	**0.605**	**0.654**	**0.738**	0.554

Evaluation Setup. For evaluating the performance of the captioning module on the COCO dataset, we consider the standard image captioning metrics (*i.e.*, BLEU-4, METEOR, ROUGE, CIDEr, SPICE) [39] and, following recent advancements in the field [17,31], the CLIP-S [17] in its reference-free definition. As for the evaluation on Matterport3D, in which no ground-truth captions are available, the standard metrics mentioned above cannot be computed. For this reason, we consider a variant of the soft coverage score, computed between the set of nouns mentioned in the caption and the set of semantic classes in the image, as defined in [5]. In the considered variant, only objects whose area is greater than a threshold are included in the compared sets. Moreover, we compute the CLIP-S, as done for the images from COCO, and the EXPL-S proposed.

Captioning and Explainability Results. First, we evaluate the performance of the captioning module alone on the COCO dataset. The results of this analysis are reported in Table 1. From the table, it can be observed that the CLIP-RN50×16 variant is the best performing in terms of the classical paired metrics and the second-best in terms of the unpaired CLIP-S metric. In terms of this latter metric, CLIP-RN50×4 performs best. The performance of the CLIP-ViT-B16 is close to that of the mentioned variants. These results suggest that the considered captioners have the ability to generate meaningful and grammatically correct captions, and that larger visual encoders can generally improve caption quality and quantitative performance.

We also evaluate the considered variants for the captioning module on the images from the Matterport3D dataset. The results of this analysis are reported in Table 2. It emerges that, when applied to these images, CLIP-ViT-B16 is the best performing variant in terms of CLIP-S, while CLIP-RN50×4 is the

Fig. 4. Qualitative captioning and explainability results. For each image, we report the generated caption, the saliency map produced by CAAM, and the EXPL-S score.

second-best. The obtained scores are in line with those obtained on the COCO dataset, indicating the generalization capabilities of the considered captioners. In terms of the proposed EXPL-S, the variant with the highest explanation power is CLIP-ViT-B16. In terms of Coverage, the variant with CLIP-RN50×16 achieves the best result, while CLIP-ViT-B16 is the second best. This underlines that CLIP-RN50×16 can name more coherent semantic concepts, while CLIP-ViT-B16 is better terms of both caption quality and explainability power.

The high absolute values of EXPL-S achieved by most of the visual encoders outline that there is a significant agreement between nouns mentioned in the caption and regions attended by the visual encoders. This can also be observed in Fig. 4, where we show qualitative results on Matterport3D environments. For each sample, we report the RGB view of the agent, the generated caption, and the saliency map produced by CAAM, which is the basis for the computation of the EXPL-S score. As it can be noticed, the EXPL-S score quantifies the degree of alignment between the generated caption and the saliency map.

5 Conclusion

In this work, we have presented an embodied agent for exploration, whose internal representation of the environment can be interpreted even by non-expert users. This has been achieved by equipping the agent with the ability to produce a natural language description of the observed scene when the scene is deemed interesting according to a speaking policy. In addition, we have defined an explanation map-based score to measure the explainability power of the description produced, which matches the explanation map, the produced caption, and the observed scene. The experimental results have shown that the proposed approach is a viable solution to gain insights into the perception and navigation capabilities of embodied agents.

Acknowledgements. This work has been supported by the "Fit for Medical Robotics" (Fit4MedRob) project, funded by the Italian Ministry of University and

Research, and by the European Union's Horizon 2020 research and innovation programme under the Marie Skłodowska-Curie grant agreement No. 955778 for project "Personalized Robotics as Service Oriented Applications" (PERSEO).

References

1. Anderson, P., et al.: Bottom-up and top-down attention for image captioning and visual question answering. In: CVPR (2018)
2. Anjomshoae, S., Najjar, A., Calvaresi, D., Främling, K.: Explainable agents and robots: results from a systematic literature review. In: AAMAS (2019)
3. Bigazzi, R., Cornia, M., Cascianelli, S., Baraldi, L., Cucchiara, R.: Embodied agents for efficient exploration and smart scene description. In: ICRA (2023)
4. Bigazzi, R., Landi, F., Cascianelli, S., Baraldi, L., Cornia, M., Cucchiara, R.: Focus on impact: indoor exploration with intrinsic motivation. RA-L **7**(2), 2985–2992 (2022)
5. Bigazzi, R., Landi, F., Cornia, M., Cascianelli, S., Baraldi, L., Cucchiara, R.: Explore and explain: self-supervised navigation and recounting. In: ICPR (2020)
6. Bigazzi, R., Landi, F., Cornia, M., Cascianelli, S., Baraldi, L., Cucchiara, R.: Out of the box: embodied navigation in the real world. In: CAIP (2021)
7. Bolelli, F., Baraldi, L., Pollastri, F., Grana, C.: A hierarchical quasi-recurrent approach to video captioning. In: IPAS (2018)
8. Burda, Y., Edwards, H., Pathak, D., Storkey, A., Darrell, T., Efros, A.A.: Large-scale study of curiosity-driven learning. arXiv preprint arXiv:1808.04355 (2018)
9. Chang, A., et al.: Matterport3D: learning from RGB-D data in indoor environments. In: 3DV (2017)
10. Chaplot, D.S., Gandhi, D., Gupta, S., Gupta, A., Salakhutdinov, R.: Learning to explore using active neural SLAM. In: ICLR (2019)
11. Chaplot, D.S., Gandhi, D.P., Gupta, A., Salakhutdinov, R.R.: Object goal navigation using goal-oriented semantic exploration. In: NeurIPS (2020)
12. Cornia, M., Baraldi, L., Cucchiara, R.: SMArT: training shallow memory-aware transformers for robotic explainability. In: ICRA (2020)
13. Cornia, M., Baraldi, L., Cucchiara, R.: Explaining transformer-based image captioning models: an empirical analysis. AI Commun. **35**(2), 111–129 (2022)
14. Cornia, M., Baraldi, L., Fiameni, G., Cucchiara, R.: Universal captioner: inducing content-style separation in vision-and-language model training. arXiv preprint arXiv:2111.12727 (2022)
15. Cornia, M., Stefanini, M., Baraldi, L., Cucchiara, R.: Meshed-memory transformer for image captioning. In: CVPR (2020)
16. Hendricks, L.A., Akata, Z., Rohrbach, M., Donahue, J., Schiele, B., Darrell, T.: Generating visual explanations. In: Leibe, B., Matas, J., Sebe, N., Welling, M. (eds.) ECCV 2016. LNCS, vol. 9908, pp. 3–19. Springer, Cham (2016). https://doi.org/10.1007/978-3-319-46493-0_1
17. Hessel, J., Holtzman, A., Forbes, M., Bras, R.L., Choi, Y.: CLIPScore: a reference-free evaluation metric for image captioning. In: EMNLP (2021)
18. Huang, L., Wang, W., Chen, J., Wei, X.Y.: Attention on attention for image captioning. In: ICCV (2019)
19. Karpathy, A., Fei-Fei, L.: Deep visual-semantic alignments for generating image descriptions. In: CVPR (2015)

20. Kim, S.S., Meister, N., Ramaswamy, V.V., Fong, R., Russakovsky, O.: HIVE: evaluating the human interpretability of visual explanations. In: Brostow, G., Cissé, M., Farinella, G.M., Hassner, T. (eds.) ECCV 2016. LNCS, vol. 13672, pp. 280–298. Springer, Heidelberg (2022). https://doi.org/10.1007/978-3-031-19775-8_17
21. Kingma, D., Ba, J.: Adam: a method for stochastic optimization. In: ICLR (2015)
22. Lin, T.-Y., et al.: Microsoft COCO: common objects in context. In: Fleet, D., Pajdla, T., Schiele, B., Tuytelaars, T. (eds.) ECCV 2014. LNCS, vol. 8693, pp. 740–755. Springer, Cham (2014). https://doi.org/10.1007/978-3-319-10602-1_48
23. Lovino, M., Bontempo, G., Cirrincione, G., Ficarra, E.: Multi-omics classification on kidney samples exploiting uncertainty-aware models. In: ICIC (2020)
24. Pathak, D., Agrawal, P., Efros, A.A., Darrell, T.: Curiosity-driven exploration by self-supervised prediction. In: ICML (2017)
25. Pennington, J., Socher, R., Manning, C.D.: GloVe: global vectors for word representation. In: EMNLP (2014)
26. Poppi, S., Cornia, M., Baraldi, L., Cucchiara, R.: Revisiting the evaluation of class activation mapping for explainability: a novel metric and experimental analysis. In: CVPR Workshops (2021)
27. Poppi, S., Sarto, S., Cornia, M., Baraldi, L., Cucchiara, R.: Multi-class explainable unlearning for image classification via weight filtering. arXiv preprint arXiv:2304.02049 (2023)
28. Radford, A., et al.: Learning transferable visual models from natural language supervision. In: ICML (2021)
29. Ramakrishnan, S.K., Al-Halah, Z., Grauman, K.: Occupancy anticipation for efficient exploration and navigation. In: Vedaldi, A., Bischof, H., Brox, T., Frahm, J.-M. (eds.) ECCV 2020. LNCS, vol. 12350, pp. 400–418. Springer, Cham (2020). https://doi.org/10.1007/978-3-030-58558-7_24
30. Ramakrishnan, S.K., Jayaraman, D., Grauman, K.: An exploration of embodied visual exploration. IJCV **129**, 1616–1649 (2021)
31. Sarto, S., Barraco, M., Cornia, M., Baraldi, L., Cucchiara, R.: Positive-augmented contrastive learning for image and video captioning evaluation. In: CVPR (2023)
32. Sarto, S., Cornia, M., Baraldi, L., Cucchiara, R.: Retrieval-augmented transformer for image captioning. In: CBMI (2022)
33. Savva, M., et al.: Habitat: a platform for embodied AI research. In: ICCV (2019)
34. Schulman, J., Wolski, F., Dhariwal, P., Radford, A., Klimov, O.: Proximal policy optimization algorithms. arXiv preprint arXiv:1707.06347 (2017)
35. Selvaraju, R.R., Cogswell, M., Das, A., Vedantam, R., Parikh, D., Batra, D.: Grad-CAM: visual explanations from deep networks via gradient-based localization. In: ICCV (2017)
36. Sennrich, R., Haddow, B., Birch, A.: neural machine translation of rare words with subword units. In: ACL (2016)
37. Simonyan, K., Vedaldi, A., Zisserman, A.: Deep inside convolutional networks: visualising image classification models and saliency maps. arXiv preprint arXiv:1312.6034 (2013)
38. Springenberg, J.T., Dosovitskiy, A., Brox, T., Riedmiller, M.: Striving for simplicity: the all convolutional net. arXiv preprint arXiv:1412.6806 (2014)
39. Stefanini, M., Cornia, M., Baraldi, L., Cascianelli, S., Fiameni, G., Cucchiara, R.: From show to tell: a survey on deep learning-based image captioning. IEEE Trans. PAMI **45**(1), 539–559 (2022)
40. Sundararajan, M., Taly, A., Yan, Q.: Axiomatic attribution for deep networks. In: ICML (2017)

41. Vaswani, A., et al.: Attention is all you need. In: NeurIPS (2017)
42. Wang, H., et al.: Score-CAM: score-weighted visual explanations for convolutional neural networks. In: CVPR Workshops (2020)
43. Xu, K., et al.: Show, attend and tell: neural image caption generation with visual attention. In: ICML (2015)
44. You, Y., et al.: Large batch optimization for deep learning: training bert in 76 minutes. In: ICLR (2019)
45. Zeiler, M.D., Fergus, R.: Visualizing and understanding convolutional networks. In: Fleet, D., Pajdla, T., Schiele, B., Tuytelaars, T. (eds.) ECCV 2014. LNCS, vol. 8689, pp. 818–833. Springer, Cham (2014). https://doi.org/10.1007/978-3-319-10590-1_53
46. Zhou, B., Khosla, A., Lapedriza, A., Oliva, A., Torralba, A.: Learning deep features for discriminative localization. In: CVPR (2016)

Towards Facial Expression Robustness in Multi-scale Wild Environments

David Freire-Obregón(✉), Daniel Hernández-Sosa, Oliverio J. Santana, Javier Lorenzo-Navarro, and Modesto Castrillón-Santana

SIANI, Universidad de Las Palmas de Gran Canaria, Las Palmas de Gran Canaria, Spain
david.freire@ulpgc.es

Abstract. Facial expressions are dynamic processes that evolve over temporal segments, including onset, apex, offset, and neutral. However, previous works on automatic facial expression analysis have mainly focused on the recognition of discrete emotions, neglecting the continuous nature of these processes. Additionally, facial images captured from videos in the wild often have varying resolutions due to fixed-lens cameras. To address these problems, our objective is to develop a robust facial expression recognition classifier that provides good performance in such challenging environments. We evaluated several state-of-the-art models on labeled and unlabeled collections and analyzed their performance at different scales. To improve performance, we filtered the probabilities provided by each classifier and demonstrated that this improves decision-making consistency by more than 10%, leading to accuracy improvement. Finally, we combined the models' backbones into a temporal-sequence classifier, leveraging this consistency-performance trade-off and achieving an additional improvement of 9.6%.

Keywords: Facial expressions · Multi-scale resolution · Sequence-based approach

1 Introduction

Efficiently recognizing facial expressions in images hinges on a solid cultural comprehension of the human face and its constituents. In recent years, there have been notable advancements in Facial Expression Recognition (FER) owing to many studies tackling this issue. Most of these studies present approaches to identify facial expressions in a discrete domain, aiming to label a subject image with a facial expression designation [30]. Further work has pushed the FER problem based on psychological studies that suggest the existence of emotions in

This work is partially funded by the the Spanish Ministry of Science and Innovation under project PID2021-122402OB-C22, and by the ACIISI-Gobierno de Canarias and European FEDER funds under project, ProID2021010012, ULPGC Facilities Net, and Grant EIS 2021 04.

G. L. Foresti et al. (Eds.): ICIAP 2023, LNCS 14233, pp. 184–195, 2023.
https://doi.org/10.1007/978-3-031-43148-7_16

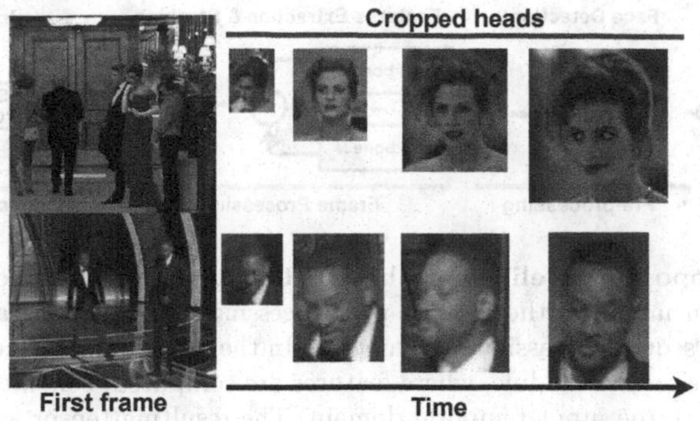

Fig. 1. Example of head size variation during a footage. The upper row belongs to a '90s movie sequence (Pretty Woman), whereas the lower row belongs to an unexpected situation during the 2022 Oscars ceremony (ABC Network). In both cases, it can be appreciated that head resolution increases in time as the subject moves towards the camera position.

the continuous domain, analyzing footage instead of images [24]. Therefore, FER can be tackled along with the facial-expression intensity, expanding the label complexity space from a fixed set of categories to a dynamic and time-dependent set of categories, i.e., happy+high-intensity, happy+medium-intensity, and so on [12].

Despite the promising strides in label complexity, benchmark evaluations FER on still images and videos remain unified, with a requirement for clear faces and minimal resolution variation throughout sequences. In this work, we take a step forward in separating these two common benchmark characteristics. In real-life situations, the acquisition device position usually remains fixed (or with a low range variability in the case of smartphones) during recording (see Fig. 1). Consequently, the recorded subject's face resolution is dynamic during the footage. Besides, Fig. 1 below row shows additional issues that may occur under unexpected situations such as motion blur or shadows appearance.

In this work, we have developed a comprehensive FER analysis considering state-of-the-art (SOTA) FER models and collections in the wild to benchmark those models. First, the considered face detector boosts multiple resolutions face acquisition. Then, several pre-trained models provide FER probabilities. Additionally, we carried out a performance analysis of the models, and the best performing classifiers are combined to focus not only on the accuracy but mainly on the model's consistency, that we define as the ability to make consistent emotion recognition across successive frames.

Our proposal was tested on various video collections in real-world settings. This included a FER-labeled movie collection consisting of 1156 video sequences,

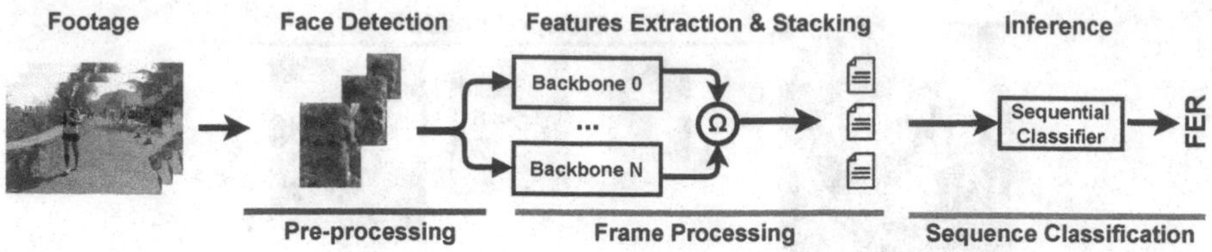

Fig. 2. The proposed pipeline for the FER system. The devised process comprises three main modules: the footage pre-processing module, the frame-processing module, and the sequence-classification module. In the first module, faces are detected and passed to the second module, where features are computed, concatenated (Ω) and stacked to preserve the input temporal domain. The resulting tensor acts as an input to the classifier, completing the FER process.

as well as three non-FER-related unlabeled collections containing a total of 3363 video sequences. The labeled collection is used to validate our FER results on the unlabelled datasets. The results are remarkable (more consistent predictions up to 15%), and they have also provided interesting insights. The first insight is the existence of a correlation between performance and consistency. Accuracy goes up as the model inference is more consistent between frames. More importantly, this work shows how facial resolution affects performance and consistency. Another insight is related to the importance of the FER models ensemble; combining the best models improves performance and consistency. Finally, we take advantage of the temporal coherence by building a sequence classifier on top of a pre-trained network to boost performance on video sequences in the wild, achieving a 56.2% accuracy. Our contributions can be summarized as follows: We conducted a systematic FER analysis on wild collections, utilizing several successful SOTA models, and all datasets used are publicly available. Through experimentation, we demonstrated that enhancing the classifier's consistency leads to improved accuracy. We calculated the median of the emotion's probability using a k-window size to provide more robust predictions. Additionally, we experimented using a soft-voting ensemble of the best-performing SOTA models, allowing multiple models to contribute to a prediction proportionately to their estimated performance. We further leverage current SOTA models in wild conditions by utilizing a temporal-sequence classifier instead of a dense-layer classifier. This approach is particularly useful in scenarios where the temporal dynamics of facial expressions play a crucial role, such as videos.

2 Related Work

Facial expression recognition (FER) plays a crucial role in interpersonal and human-computer interactions, as it helps the receiver interpret the intended meaning of a message, complementing other communication channels. FER also impacts any content, product, or service that is designed to elicit emotional arousal and facial responses, such as video games [1]. The goal of FER is to infer emotions from human-centered signals, including video, audio, and body pose. The universal emotions proposed by Ekman and Friesen [9] are widely adopted by the community, including sadness, happiness, fear, anger, surprise, and disgust. FER can be approached using two different perspectives based on the input format: static and dynamic methods.

Static FER Methods. Facial expression recognition (FER) on static images has been a long-standing challenge in facial biometrics, with two main approaches: geometric-based and appearance-based. Geometric-based methods use facial component shapes based on salient landmarks detected on faces [15,26], while appearance-based methods discretize video sequences by picking out keyframes [8]. Deep feature embedding approaches have been used in the last decade to learn discriminative features [5], with recent proposals successfully classifying footage keyframes [21,27]. However, our pipeline differs by using pre-trained networks to generate features to classify the entire sequence, rather than each frame separately, and assessing multi-scale FER consistency after feature extraction.

Dynamic FER Methods. These methods typically utilize the temporal aspect of facial expressions in a sequence of images and can be categorized into three main groups: geometric-based, appearance-based, and motion-based methods. Geometric-based methods use face landmarks to describe facial components and rely on the movement of these landmarks to infer facial expressions [24]. In contrast, the appearance-based methods extract texture features from facial images to identify expressions. Feng and Ren proposed a dynamic FER approach that utilizes a two-stream architecture with both spatial and temporal convolutional neural networks (CNN) and LBP-TOP features [11]. Motion-based methods aim to model the spatial-temporal evolution of facial expressions, with some techniques detecting both the expression and its intensity [12]. Deep learning techniques have also addressed the dynamic aspect of FER. For instance, Meng et al. proposed a method that highlights discriminative frames in an end-to-end framework [23], and Zhang et al. presented an end-to-end learning model for pose-invariant facial expression recognition that relies on geometry information [33]. However, our proposed approach processes all frames in the sequence without discrimination.

3 Description of the Proposal

3.1 The Proposed Architecture

This work proposes and evaluates a sequential training pipeline entailing three main modules: a face-detection module, a frame-processing module, and a time-sequence classification module, as shown in Fig. 2. This pipeline is a general scheme for the approach considered, where the backbones have been selected after ablation studies, depending on performance.

Face detection (FD) has been a broad research area for the last three decades. In the proposed scenario, FER requires a previous FD process.

The RetinaFace algorithm [6] in particular claims to provide SOTA results in the FDDB [18] and the WiderFace [31] datasets. We have adopted RetinaFace because we have found that it performs better than other face detectors like MTCNN [34] for datasets with large pose, illumination, and face resolution variations, especially for low-resolution images (+28% positive face detection rate).

Features Extraction and Stacking. The implemented encoding technique comprises of two steps that transform the input data into a feature vector. Firstly, the detected face is passed through several pre-trained backbones, each of which outputs a feature vector that is concatenated with the outputs of the other backbones. Secondly, the resulting feature vector is stacked through the temporal axis for each frame. This encoding approach effectively captures the temporal dynamics of facial expressions and results in a compact representation of the input data.

We have tested several pre-trained backbones, but only three of them are finally used in the proposed architecture. Savchenko [27] developed several models trained on the VGGFace2 dataset. They are based on CNNs lightweight architectures, such as MobileNet-V1 and EfficientNet. Unlike traditional architectures that scale a network's dimensions by increasing width, depth, and resolution, EfficientNet scales each dimension with a fixed set of scaling coefficients uniformly [28], configuring a family of eight (B0 to B7) models, each with a larger size. Practically, scaling individual dimensions improves model performance and preserves architectural growth. In this sense, we are using EfficientNet-B0 and EfficientNet-B2. On the other hand, MobileNet-V1 uses depthwise separable convolutions to reduce the model size and complexity [16], which is beneficial for mobile and embedded vision applications. Savchenko tested these pre-trained models on AFEW dataset [7] achieving a 55.3%, a 59.2%, and a 59% for MobileNet-V1, EfficientNet-B0, and EfficientNet-B2 respectively. The author considered a subset of enhanced facial frames provided by the AFEW dataset to achieve these remarkable results. We have replaced the pre-trained model's dense-layer classifiers with a unique time-sequence classifier.

Sequence Classification. The sequence of features provided by the backbones goes into a trainable classifier in the last step. It is composed of three sequentially-connected blocks. Each block is a sequential pipeline of a convolutional 1D layer (512 filters, with a kernel size of three), a batch normalization layer, and a ReLU activation. Dropout regularization is applied at the end of the second block. A final global average pooling layer followed by a softmax dense layer (seven units, one per emotion) provides the final output with the emotions probabilities. The network is fine-tuned with the training data of the AFEW dataset. The considered optimizer is Adam and the loss function is a sparse categorical cross entropy.

3.2 Consistency Assessment and Datasets

We have considered three approaches to evaluate prediction consistency: hard voting, a median-filter assessment, and a soft-voting ensemble. The former is

Fig. 3. Facial samples of the considered video datasets.

based on statistical evaluation, whereas the latter seeks to find a stronger correlation between prediction consistency and accuracy.

Hard Voting (Vanilla). Pre-trained classifiers are used without any training or architectural modification. The classifier generates a score at each time-step or frame. The classifier output is a set of seven probabilities, one per emotion. The frame prediction is the emotion with a higher probability. Then, the predicted class label for a particular video is the class label that represents the majority of the class labels predicted by each classifier.

Median-Filter Assessment (MFA). In this case, the probability of each emotion is computed considering a median filter of the k previous steps, which reduces the influence of spurious scores. Like in vanilla, the emotion of each frame is the one with the highest probability, and the predicted class label for a particular video is computed by a majority voting.

Soft-voting ensemble (SVE). In order to find the best combination of classifiers, we apply a soft-voting ensemble. During a grid search, specific weights are assigned to each classifier set of outputs. Consequently, the predicted emotion probabilities for each classifier are collected, multiplied by the classifier weight, and averaged. The final class label is then computed from the class label with the highest average probability.

The described soft-voting ensemble allows multiple models to contribute to a prediction in proportion to their trust or performance. The primary motivation is to exploit the dependence between the base learners. Therefore, the soft-voting output is a weighted sum of all considered models.

Table 1 displays the four collections considered in this study, all of which are wild datasets. The first one is FER-focused, while the remaining three are included due to their variability in facial resolution. The AFEW dataset [7] is the

Table 1. Face resolution (*pixels* = *width* × *height*) **after detecting with RetinaFace per timing quarter**. As can be appreciated, entries are sorted by dataset. We have divided each video into four parts. Precisely, each column represents a different video-sequence quarter.

Timing quarter	Q1	Q2	Q3	Q4
AFEW	35.2K ± 2.9K	35.3K ± 2.4K	35.7K ± 2.4K	35.8K ± 3.7K
AveRobot	13.3K ± 3K	14.8K ± 2.7K	17.5K ± 2.7K	19.2K ± 4.7K
Gotcha	6.3K ± 0.8K	7.1K ± 1K	10.1K ± 1.9K	15.7K ± 7.4K
TGC20ReId	0.3K ± 0.07K	0.5K ± 0.1K	1K ± 0.2K	1.9K ± 1.2K

only labeled collection analyzed and comprises video clips from movies and TV shows with spontaneous expressions, head poses, occlusions, and illuminations. AFEW 7.0 includes seven emotion labels and is divided into a train set and a validation set. While the test split is not publicly available, we fine-tuned our model on the training split and reported results on the validation set. The facial detection resolution barely varies in this collection (see Table 1), but we used it to guide the consistency-performance experiment.

The AveRobot dataset [22] is the second collection used in this work, consisting of 2664 video clips that present more challenges than AFEW. These challenges include lack of illumination in corridors, as shown in Fig. 3, and a wide range of identities, cameras, and locations within the same building. The videos in this collection depict a subject approaching the camera, making a short utterance, and then leaving. The AveRobot dataset has been recently used for research on multimodal robot-human interaction [14]. Table 1 demonstrates that the face resolution varies by 44% between the first and last timing quarters.

The third collection considered is the Gotcha dataset [13], and it consists of 6 clips per subject covering cooperative and non-cooperative modes, including indoor with artificial light, indoor without lights (flash camera on), and outdoor with sunlight. The dataset contains over 372 video clips (around 10 s each) where subjects are captured while they walk, exhibiting two different behaviors in front of the camera: avoiding the camera (non-cooperative) or just ignoring it (cooperative). A remarkable feature of this dataset is that the camera is not static. It moves horizontally and vertically while preserving the position, trying to capture a clear image of the individual. Table 1 shows that face resolution varies a 149% between the first and the last timing quarter.

Lastly, we have partially used the TGC20ReId dataset [25]. This collection contains runners of the TGC Classic who must cover 128 km in less than 30 h. Participants are recorded moving towards a fixed-lens camera until they surpass it. We have included 456 short video clips (around seven seconds each) in our study. In this dataset, faces may be affected by fatigue. Table 1 shows that face resolution varies more than a 600% between the first and the last timing quarter.

In addition to MobileNet-V1, EfficientNet-B0, and EfficientNet-B2, we also tested the pre-trained classifiers from Luan et al. [21] on their VEMO dataset.

These classifiers, based on backbones such as VGG19, Resnet34, and InceptionV3, were used in the Vanilla, MFA, and SVE experiments (Sect. 3.3). The top-performing backbones on the AFEW dataset, MobileNet-V1, EfficientNet-B0, and EfficientNet-B2, were selected for the proposed architecture experiment (see Sect. 3.4).

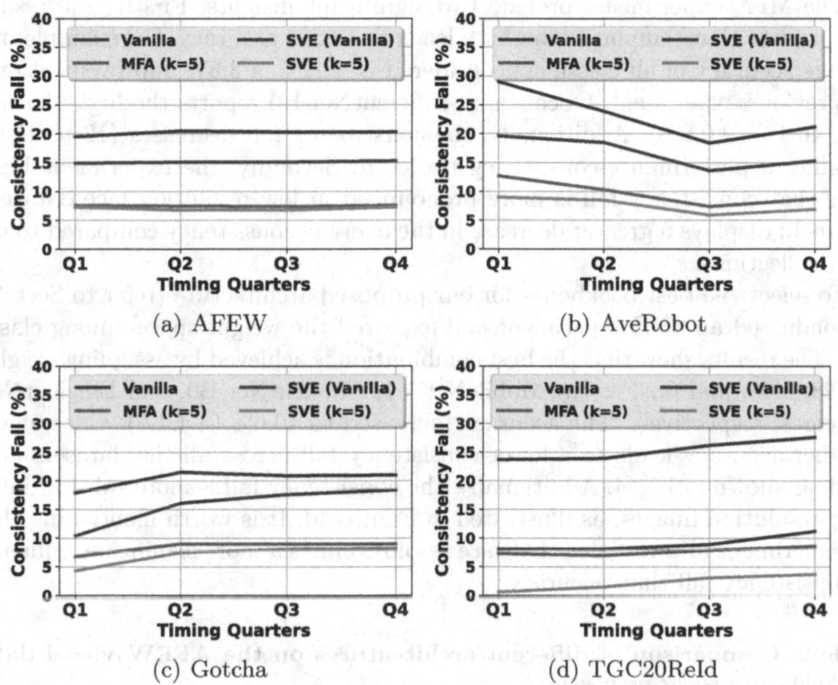

(a) AFEW (b) AveRobot

(c) Gotcha (d) TGC20ReId

Fig. 4. Consistency robustness assessment. The x-axis represents the timing quarter of the video clips, whereas the y-axis represents the model's consistency fall on average. Lower means a higher consistency.

3.3 Prediction Consistency-Performance Trade-Off

To evaluate the prediction consistency, we divided each video clip into four parts and analyzed them separately. The prediction consistency graph for each dataset is depicted in Fig. 4. The level of consistency is determined by the number of emotions that switch during the clip inference. Since the analyzed videos are relatively short (ranging from 7 to 10 s each), emotions are not expected to change significantly during the acquisition time.

The results of the Vanilla experiment indicate that this classifier exhibits the lowest consistency due to its nature. While it has shown excellent performance on static image datasets, it struggles in dynamic environments where a significant proportion of frames (20 to 30%) show an emotion switch from the previous frame. Figure 4b illustrates how low-light environments and low-resolution faces affect these predictions, while Fig. 4d confirms that low-resolution faces have a negative impact on decision-making. In particular, the TGC20ReId model performs poorly at the end of a clip due to more profile angles. Across all considered classifiers, the average accuracy is 39%, with the EfficientNet-B0 model achieving the highest accuracy of 45.5%.

The MFA experiments provide two significant insights. Firstly, the results demonstrate that temporal coherence leads to better accuracy. For example, the average accuracy of all classifiers considered is 42.2%, a 3.3% improvement over the previous experiment. Once again, EfficientNet-B0 reports the highest accuracy, 46.4% (+0.9%). Additionally, the consistency fall decreases (12 to 15%), revealing a performance-consistency trade-off. Secondly, the experiment highlights that consistency fall is more pronounced in low-resolution face datasets. Figure-4d displays a greater decrease in the average consistency compared to the other collections.

To select the best backbones for our proposed architecture (refer to Sect. 3), we conducted an SVE experiment and explored the weight space among classifiers. The results show that the best combination is achieved by assigning weights of 0.28, 0.49, and 0.23 to the MobileNet-V1, EfficientNet-B0, and EfficientNet-B2 scores, respectively. The accuracy increased to 46.6% (+4.4%). Once again, a higher accuracy leads to a lower consistency fall across all the datasets analyzed, as shown in Fig. 4. Additionally, the consistency fall is more substantial in lower-resolution images, as illustrated in Figure-4d. It is worth mentioning that the experiment also revealed that face resolution has a more significant influence on consistency fall than accuracy.

Table 2. Comparison of different architectures on the AFEW visual data. The bold entry is our proposal.

	Approach	Accuracy		Approach	Accuracy
Frame-based	HoloNet (2016) [32]	44.8%	Seq.-based	VGG16+SA+TP (2019) [2]	49.0%
	DSN HoloNet (2017) [17]	46.5%		Densenet-161 (2018) [20]	51.4%
	DSN VGG-Face (2018) [10]	48.0%		Variational LSTM (2019) [3]	51.4%
	FAN (2019) [23]	51.2%		Emotion-BEEU (2019) [19]	55.1%
	EfficientNet-B0 (2021) [27]	59.3%		**Our Model**	**56.2%**

3.4 Experimental Analysis

Table 2 provides a comparison of our approach with other state-of-the-art (SOTA) techniques from two perspectives: frame-based and sequence-based. The first set of columns presents the results for the AFEW dataset, while the second set shows the sequence-based approaches. The improvement rates in FER performance on the AFEW dataset have been slow over the years, with increases ranging from 1% to 2.5%. However, two significant breakthroughs can be observed.

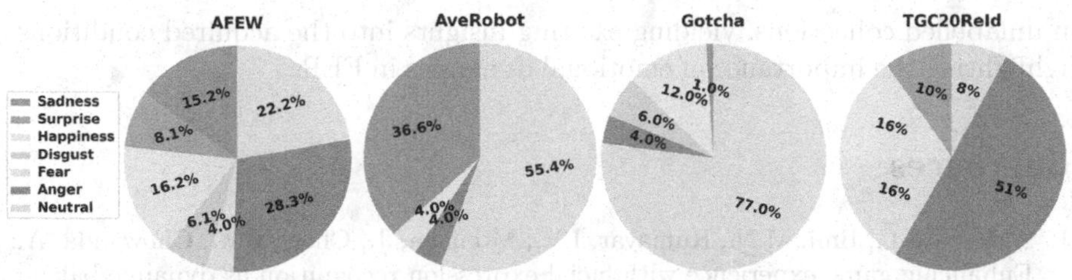

Fig. 5. Inference FER distribution of the considered collections.

In the sequence-based approaches, Kumar et al. achieved a performance of 55.1% (+3.7%) by considering specific facial regions for FER purposes [19]. In the frame-based approaches, Savchenko achieved a remarkable 8.1% improvement by using a subset of enhanced frames from the AFEW collection when testing his pre-trained models. In our work, we extend Savchenko's proposal by adapting his backbones to process high-variance resolution faces on video clips instead of a selection of enhanced frames. We fine-tuned the network using the AFEW training subset and achieved a significant accuracy of 56.2% (+1.1%).

The model's prediction distribution for each considered dataset is presented in Fig. 5. The AFEW dataset is noted for having no inductive bias, with a wide variety of predictions across different facial expressions. Tablc 2 displays the reported accuracy of the inference chart in bold. On the other hand, the AveRobot dataset reveals that participants may have experienced some stress due to the experiment's conditions, such as being recorded by multiple cameras and speaking in a non-native language, particularly when expressing neutral or sad emotions. Meanwhile, the Gotcha dataset features mostly neutral expressions, but some videos elicited happiness due to the presence of other participants during recording, resulting in a somewhat unusual situation. Lastly, the TGC20ReId dataset shows significant anger and happiness predictions in video sequences, possibly due to facial cues that are common markers for these emotions. For instance, studies have found that the activation frequency of the zygomatic major muscle, which triggers the lip corner pull movement, is remarkable in athletes undertaking great physical effort [29]. Additionally, humans tend to lower their brows under sun exposure, which can lead to cues that are associated with anger and fear [4].

4 Conclusions

FER has seen remarkable results on static images and indoor video sequences, but as newer devices produce more multimedia content, new approaches are needed for FER. This study explores extending existing models to the dynamic FER scope and evaluates consistency, considering how image resolution and accuracy affect it. Interestingly, a trade-off is observed between consistency and accuracy, where accuracy improvements lead to consistency falls. The AFEW dataset was used to guide the process, and the proposed deep learning pipeline outperforms state-of-the-art sequence-based approaches on this dataset. The FER model was also tested on unlabelled collections, yielding exciting insights into the acquired conditions, highlighting the importance of emotional dynamics in FER.

References

1. Akbar, M.T., Ilmi, M.N., Rumayar, I.V., Moniaga, J., Chen, T.K., Chowanda, A.: Enhancing game experience with facial expression recognition as dynamic balancing. Proc. Comput. Sci. **157**, 388–395 (2019)
2. Aminbeidokhti, M., Pedersoli, M., Cardinal, P., Granger, E.: Emotion recognition with spatial attention and temporal softmax pooling. In: Karray, F., Campilho, A., Yu, A. (eds.) Image Analysis and Recognition, pp. 323–331 (2019)
3. Baddar, W.J., Ro, Y.M.: Mode variational LSTM robust to unseen modes of variation: application to facial expression recognition. In: AAAI Conference on Artificial Intelligence. vol. 33, pp. 3215–3223 (2019)
4. Barrett, L.F., Adolphs, R., Marsella, S., Martinez, A.M., Pollak, S.D.: Emotional expressions reconsidered: challenges to inferring emotion from human facial movements. Psychol. Sci. Public Interest **20**(1), 1–68 (2019)
5. Bell, S., Bala, K.: Learning visual similarity for product design with convolutional neural networks. ACM Trans. Graph. **34**(4), 98:1–98:10 (2015)
6. Deng, J., Guo, J., Zhou, Y., Yu, J., Kotsia, I., Zafeiriou, S.: RetinaFace: Single-stage dense face localisation in the wild. CoRR abs/1905.00641 (2019)
7. Dhall, A.: EmotiW 2019: automatic emotion, engagement and cohesion prediction tasks. In: 2019 International Conference on Multimodal Interaction, pp. 546–550 (2019)
8. Dhall, A., Asthana, A., Goecke, R., Gedeon, T.: Emotion recognition using PHOG and LPQ features. In: 2011 IEEE International Conference on Automatic Face Gesture Recognition, pp. 878–883 (2011)
9. Ekman, P., Friesen, W.: Unmasking the Face: A Guide to Recognizing Emotions from Facial Expressions. Prentice Hall, Hoboken (1975)
10. Fan, Y., Lam, J.C.K., Li, V.O.K.: Video-based emotion recognition using deeply-supervised neural networks. In: 20th ACM International Conference on Multimodal Interaction, pp. 584–588 (2018)
11. Feng, D., Ren, F.: Dynamic facial expression recognition based on two-stream-CNN with LBP-TOP. In: 2018 5th IEEE International Conference on Cloud Computing and Intelligence Systems, pp. 355–359 (2018)
12. Freire-Obregón, D., Castrillón-Santana, M.: An evolutive approach for smile recognition in video sequences. Int. J. Pattern Recogn. Artif. Intell. **29**, 1550006 (2015)
13. Freire-Obregón, D., Castrillón-Santana, M., Barra, P., Bisogni, C., Nappi, M.: An attention recurrent model for human cooperation detection. Comput. Vis. Image Underst. **197–198**, 102991 (2020)
14. Freire-Obregón, D., Rosales-Santana, K., Marín-Reyes, P.A., Penate-Sanchez, A., Lorenzo-Navarro, J., Castrillón-Santana, M.: Improving user verification in human-robot interaction from audio or image inputs through sample quality assessment. Pattern Recogn. Lett. **149**, 179–184 (2021)
15. Happy, S.L., Routray, A.: Automatic facial expression recognition using features of salient facial patches. IEEE Trans. Affect. Comput. **6**(1), 1–12 (2015)
16. Howard, A.G., et al.: MobileNets: Efficient convolutional neural networks for mobile vision applications. CoRR abs/1704.04861 (2017)

17. Hu, P., Cai, D., Wang, S., Yao, A., Chen, Y.: Learning supervised scoring ensemble for emotion recognition in the wild. In: 19th ACM International Conference on Multimodal Interaction, pp. 553–560 (2017)
18. Jain, V., Learned-Miller., E.: FDDB: A benchmark for face detection in unconstrained settings. Tech. rep., University of Massachusetts, Amherst (2010)
19. Kumar, V., Rao, S., Yu, L.: Noisy student training using body language dataset improves facial expression recognition. In: Computer Vision - ECCV 2020 Workshops, pp. 756–773 (2020)
20. Liu, C., Tang, T., Lv, K., Wang, M.: Multi-feature based emotion recognition for video clips. In: 20th ACM International Conference on Multimodal Interaction, pp. 630–634 (2018)
21. Luan, P., Huynh, V., Tuan Anh, T.: Facial expression recognition using residual masking network. In: IEEE 25th International Conference on Pattern Recognition, pp. 4513–4519 (2020)
22. Marras, M., Marín-Reyes, P., Lorenzo-Navarro, J., Castrillón-Santana, M., Fenu, G.: AveROBOT: an audio-visual dataset for people re-identification and verification in human-robot interaction. In: Proceedings of the 8th International Conference on Pattern Recognition Applications and Methods, pp. 255–265 (2019)
23. Meng, D., Peng, X., Wang, K., Qiao, Y.: Frame attention networks for facial expression recognition in videos. In: 2019 IEEE International Conference on Image Processing, pp. 3866–3870 (2019)
24. Pantic, M., Patras, I.: Dynamics of facial expression: recognition of facial actions and their temporal segments from face profile image sequences. IEEE Trans. Syst. Man Cybern. **36**(2), 433–449 (2006)
25. Penate-Sanchez, A., Freire-Obregón, D., Lorenzo-Melián, A., Lorenzo-Navarro, J., Castrillón-Santana, M.: TGC20ReId: a dataset for sport event re-identification in the wild. Pattern Recog. Lett. **138**, 355–361 (2020)
26. Saeed, A., Al-Hamadi, A., Niese, R., Elzobi, M.: Effective geometric features for human emotion recognition. In: 2012 IEEE 11th International Conference on Signal Processing. vol. 1, pp. 623–627 (2012)
27. Savchenko, A.V.: Facial expression and attributes recognition based on multi-task learning of lightweight neural networks. In: 2021 IEEE 19th International Symposium on Intelligent Systems and Informatics, pp. 119–124 (2021)
28. Tan, M., Le, Q.V.: EfficientNet: Rethinking model scaling for convolutional neural networks. CoRR abs/1905.11946 (2019)
29. Uchida, M.C., et al.: Identification of muscle fatigue by tracking facial expressions. PLoS ONE **13**(12), e0208834 (2018)
30. Vyas, A.S., Prajapati, H.B., Dabhi, V.K.: Survey on face expression recognition using CNN. In: 2019 5th International Conference on Advanced Computing Communication Systems (ICACCS), pp. 102–106 (2019)
31. Yang, S., Luo, P., Loy, C.C., Tang, X.: WIDER FACE: a face detection benchmark. In: IEEE Conference on Computer Vision and Pattern Recognition, pp. 5525–5533. IEEE, Hawai, USA (2016)
32. Yao, A., Cai, D., Ping Hu, S.W., Sha, L., Chen, Y.: HoloNet: towards robust emotion recognition in the wild. In: 18th ACM International Conference on Multimodal Interaction, pp. 472–478 (2016)
33. Zhang, F., Zhang, T., Mao, Q., Xu, C.: Geometry guided pose-invariant facial expression recognition. IEEE Trans. Image Process. **29**, 4445–4460 (2020)
34. Zhang, N., Luo, J., Gao, W.: Research on face detection technology based on MTCNN. In: 2020 International Conference on Computer Network, Electronic and Automation, pp. 154–158 (2020)

Depth Camera Face Recognition by Normalized Fractal Encodings

Umberto Bilotti, Carmen Bisogni, Michele Nappi, and Chiara Pero(✉)

University of Salerno, 84084 Fisciano, Italy
{ubilotti,cbisogni,mnappi,cpero}@unisa.it

Abstract. Face recognition is a thriving topic in biometrics literature. In most cases, it is performed on RGB images using deep learning approaches. However, due to the wider scenarios in which face recognition can be applied, it is necessary to operate when illumination conditions are not in favor of the RGB image analysis. To overcome this problem, we present DLIF (Depth Landmark Identity by Fractal), a new depth face recognition method that uses fractal encoding to treat faces as depth features. A facial landmark predictor is developed to detect and extract the face, making DLIF completely in-depth. Comparisons of the fractal-encoded faces are obtained by the Canberra distance, which makes DLIF simple but effective. Statistical analysis is performed to improve DLIF, resulting in DLIF+, which takes into account the source of the image. DLIF and DLIF+ are tested on four challenging datasets and on a fifth dataset, named BIPS, with 72 subjects. The results obtained show that DLIF and DLIF+ are competitive with the state of the art and enhance the robustness of this method in relation to the number of subjects considered.

Keywords: face recognition · depth cameras · fractal encoding · facial keypoints

1 Introduction

The research advances in facial recognition, an extensively investigated field, yield striking performance in the RGB domain. At the same time, depth cameras and, specifically, depth maps have become of particular interest to the computer vision community. Depth images are useful in scenarios where RGB images may not work due to non-uniform illumination conditions, such as in the driver's scenario [6] or surveillance purposes [19]. Facial recognition algorithms that use depth images may not require RGB features, and some solutions avoid using RGB frames for face detection. Despite the increase in affordability of depth acquisition devices, there is still a lack of work in the field, and deep learning algorithms have limited training data due to the limited number of available datasets. On the other hand, traditional techniques can still be used on a limited number of subjects. In this article, we present "DLIF", the Depth Landmark Identity by Fractal method, which uses fractal encoding and a re-trained landmark predictor model to perform face recognition in depth, increasing both accuracy and efficiency. The main contributions of this work are:

G. L. Foresti et al. (Eds.): ICIAP 2023, LNCS 14233, pp. 196–208, 2023.
https://doi.org/10.1007/978-3-031-43148-7_17

– a depth landmark detector able to find facial landmarks on depth images regardless of the device used;
– a set of fractal-encoded facial features used for the first time on depth facial images;
– a completely depth face recognition method able to operate over extreme head poses, occlusions, and quality of images;
– an improved version of the presented algorithm that uses statistical analysis to speed up the process and improve accuracy.

The rest of the article is organized as follows. Section 2 presents the state-of-the-art in RGB-D and depth facial recognition methods. Section 3 describes each step of the proposed approach and the theoretical background; Sect. 4 illustrates the results of the experiments conducted and, finally, Sect. 5 concludes the article by summarizing the work done and the achievements.

2 Related Works

Starting from the works that use both RGB and depth features, we can find the work by [9]. They proposed a new dataset, in particular a video dataset of both RGB and depth images. Here, Kinect 1 and Kinect 2 are used to record 108 subjects, for whom quite all of their bodies are available. As a consequence, the face resolution results are quite small. The authors used LBP, TPLBP, and HOG as descriptors and VGGFace, RISE, mRISE, and a commercial algorithm to classify the subjects. To examine RGB-D face recognition under various conditions, [18] addressed the problem of low resolution by employing CLM-Z for 3D modeling, aligning the face images, and selecting the key points. To describe the features around landmarks, they used HOG and 3DLBP and classified them by SVMs. [25] used twin networks for 3D and 2D textures. The preprocessing of the images involves hole filling, data normalization, and cubic interpolation. The features are then extracted by a CNN for each kind of image and then fused. [8] used the Kinect Box One to perform their experiments. Also, in this case, the 3D data from the device was transformed into a 2D contour map. Their recognition method is similar to a fingerprint recognition technique when the contour lines are the discriminant between different subjects. Other works use RGB information only in the preprocessing step, as in the case of the algorithm proposed by [16], that performs face detection on the RGB image and, then, face segmentation on depth images. The face segmentation is helped by the annotated canonical face and permits excluding noise such as hair and background. Then, the 3D model is converted into a 2.5D object through an orthographic transformation. The recognition step is performed by searching for the elements stored in the model and comparing their keypoint distances with the keypoint distances of the input 2.5D model. The same authors presented two years later a dataset collected by using a Kinect sensor [17]. To enhance the contribution of depth images, we can find works that present their results using both RGB and depth and RGB-D images. In [1], the idea is to use depth features to improve RGB recognition. They performed a preprocessing step on depth images by using a zero-elimination median filter to remove holes and then using interpolation. The classification step is then performed by using Support Vector Machines (SVM) in which some faces are used as samples of each subject. [12] present a Deep Convolutional Neural Network (DCNN) to solve the depth face recognition problem. In particular, they used two different DCNN, one for RGB features and one for depth features. Then, the arrays obtained in the final layers are fused in an additional layer, followed by a 64-way softmax output. In some cases, like the one we present, only depth data is used since the algorithm is supposed to work in a scenario where no other data is available. The method in [22] is focused on the speed of 3D face recognition. In particular, they extract the features using an LBP algorithm, and then the features are classified using an SVM method.

3 Proposed Model

The main components of the proposed model DLIF can be seen in Fig. 1, where we illustrate the DLIF workflow. All the steps are presented in detail as follows.

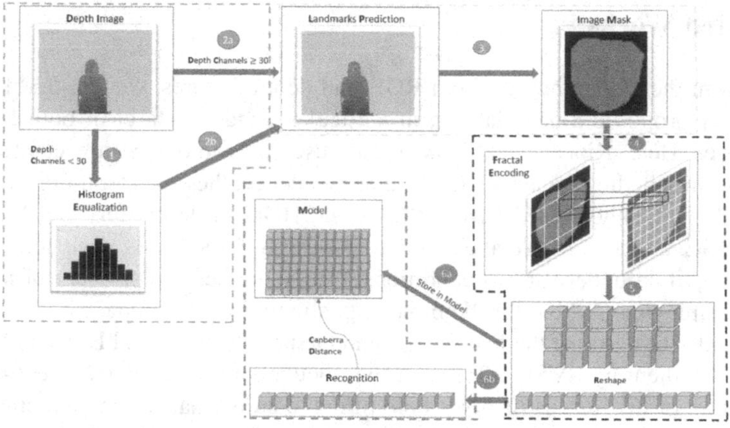

Fig. 1. DLIF Workflow. In green the histogram pre-processing, in yellow the face extraction, in black the face fractal encoding, and finally, in red the recognition step. (Color figure online)

Image Histogram. We compute the histogram for each image to perform histogram equalization and calibrate the contrast. We adopt normalization since the number of gray levels detected depends on the sensors used. Kinect 2 is used as a reference, and we observed that the number of gray levels is always higher than 30 in this case. If the number of gray levels is less than 30, we adjust the image contrast via histogram equalization to better distribute the intensities.

Face Extraction. To extract the facial part, we retrained a landmark predictor model to be used on depth images. The method is based on an ensemble of regression trees and is explained in detail in Sect. 4.2. After the depth landmarks are obtained, a facial mask is created. The landmarks that represent the border of the face are used to create a mask that contains the depth values in the inner points and 0 in the outer points. The mask is then cut by using the coordinates of the extreme non-zero points.

Fractal Encoding. The fractal encoding algorithm compresses image data by utilizing image self-similarity through contractive transformations. Image partitioning is done into domain and range blocks, with range blocks transformed using contractive transformations to map them into domain blocks, resulting in fractal elements. For the depth face, range blocks with dimension 4 and domain blocks with dimension 8 are used. Facial frames are resized to 128 × 128 pixels to ensure the same size for each encoding. The resulting fractal elements are matrices with 256 rows and 6 columns, containing the position of the matched domain block, applied transformations, contrast scale, and brightness shift. Each image has 1536 fractal elements in array format. More details are in Sect. 3.1.

Identity Estimation. The input fractal encoding element is then compared to the labeled fractal codes present in the model. In particular, for each of the fractal elements memorized in the set, the subject label is associated. The Canberra distance is used as a metric to recognize identity. Originally introduced by Lance and Williams [15], this metric computes the sum of absolute fractional distinctions in the coordinates of a pair of objects. The above-mentioned distance is formally defined as follows:

$$d_j = \sum_i \frac{|p_i - q_i|}{|p_i| + |q_i|} \tag{1}$$

where p_i and q_i represent, respectively, the input fractal encoding element and the template encoding (the j-th). It is proven that this distance is the most suitable for data from facial images encoded in fractal arrays [5]. In Fig. 2 we show an improved version of the recognition module. Here, the gray levels and the skewness of the histogram play a key role in harvesting the pool of candidates for the model, providing both a faster method and a more accurate result. We called the new workflow **DLIF+**. The details about the use of gray levels and skewness are described in Sect. 3.2. All the other modules of the workflow described above remain unchanged.

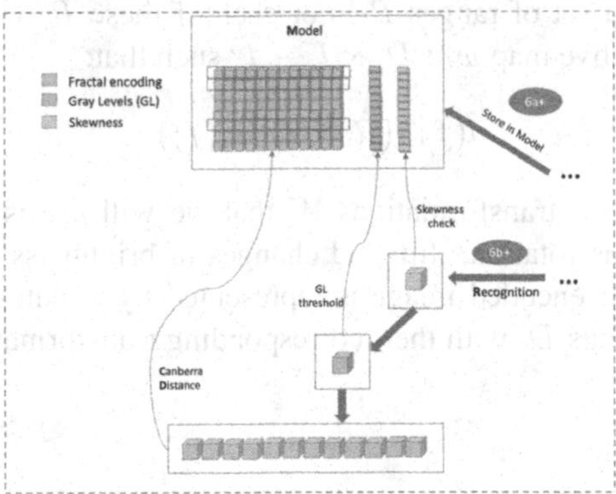

Fig. 2. DLIF+ Module. This module substitute the red one in DLIF. (Color figure online)

3.1 Fractal Encoding Details

The fractal encoding we obtain to implement our method is based on the concept of Iterated Function Systems (IFS) that we will introduce in this Section. First of all, we must define a Metric Space as a set X on which a real-valued distance $d : X \times X \rightarrow \mathbb{R}$ is defined. We assume this distance satisfies the metric properties. We then define a metric space X complete if every Chauchy sequence in X converges to a limit point in x, where a Cauchy sequence is defined as a sequence of points x_n such that for any $\epsilon > 0$, $d(x_m, x_n) < \epsilon$ for all $n, m > N$, with N integer. On a metric space X with d as a metric we can define a contractive map by the following. A map $w : X \rightarrow X$ is Lipschitz with Lipschitz factor s if there exists a positive real value s such that $d(w(x), w(y)) \leq sd(x, y)$ for every $x, y \in X$. If this Lipschitz constant s is lower than 1, then w is called a *contractive* map with contractivity s. It can be proven that a Lipschitz map is also continuous. In X, complete metric space, if we have $f : X \rightarrow X$ a contractive mapping, then, for the *Fixed-Point Theorem*, exists a unique point $x_f \in X$ called *the attractor* such that for any point $x \in X$, is $x_f = f(x_f) = \lim_{n\rightarrow\infty} f^n(x)$. As a consequence of this theorem, we have that $d(x, x_f) \leq \frac{1}{1-s} d(x, f(x))$. At this point, we can define an iterated function system as a collection of contractive maps $w_i : X \rightarrow X$ with $i = 1, ..., n$. Now, our goal is to find this set of maps that can be collected under a unique transformation $W = \cup w_i$ that is itself contractive. This map will satisfy the fixed point theorem, and we want its attractor to be close to the given image. In the case of images, our complete metric space is $F = \{f : I^2 \rightarrow \mathbb{R}\}$. All the previous assumptions can also be applied in this case. Given an image, find the map W such that its attractor $x_w = f$ is called the inverse problem and does not have an exact solution. However, it can be found a $f' \in F$ such that $d(f, f')$ is minimal when $f' = w_f$. For this reason, we define a set of domains D_i on which w_i are defined, and we search W such that:

$$f \approx W(f) = \cup_{i=1}^{n} w_i(f) \tag{2}$$

In practice, we have to cover f with part of itself, D_i, and the way in which this part covers f is defined by the w_i. In conclusion, the fractal encoding is thus obtained by partitioning I^2 in a set of ranges R_i. For each of these R_i we can find a domain $D_i \subset I^2$ and a contractive map $w_i : D_i \times I \rightarrow I^3$ such that:

$$d(f \cap (R_i \times I), w_i(f)) \tag{3}$$

is minimized. The set of transformations W that we will use is composed of affine transformations such as rotation, flip, and changes in brightness and contrast. At the end of the process, our encoded image is represented by a matrix that contains a set of ordered domain blocks D_i with their corresponding transformations w_i to cover the image.

3.2 DLIF+

In the improved method, two measures were used to reduce the search pool. Firstly, images belonging to the same subject have similar grayscale depth levels due to the facial traits of individuals. Secondly, statistical indices on gray levels were used to discriminate between datasets. These indices were estimated on a validation set by performing DLIF configuration on 20% of the elements in the model and obtaining the errors produced on this set. If images associated with each other to obtain the subject ID differ consistently in gray levels, they are likely to belong to different subjects. The gray levels of an image are formally defined as:

$$GL = \sum_i hist_i \qquad hist_i = \begin{cases} 0 & \textit{if} \quad freq_i = 0 \\ 1 & \textit{otherwise} \end{cases} \tag{4}$$

where $freq_i$ is the frequency of the i-th gray level in the image. As a consequence, we call $GL0$ the GL of the corrected classified images and $GL1$ the GL of the misclassified images. Then, we compute:

$$\Delta_1 = |GL1_g - GL1_p| \qquad \Delta_0 = |GL0_g - GL0_p| \tag{5}$$

for each image of the validation set, where ($GL0_g$,$GL1_g$) and ($GL0_p$, $GL1_p$) are the GL of the ground truth image and the GL of the image of the predicted subject, respectively. We then split the obtained Δ_0 and Δ_1 into classes based on their frequencies. If we consider, as an example, the BIWI dataset, the corrected images have a Δ_0 value lower than 10. Repeating the same on Δ_1, we observe that, on the other hand, the majority of errors have Δ_0 higher or equal to 10. For this reason, 10 is a good threshold for the Δ value. Together with thc fractal cncoding, we will also store the GL of the images in the model, which can be easily obtained by the histogram already computed in Step 1. When the algorithm computes the input image, if an image of the model has a Δ value higher than the GL threshold, we will discard it without performing further comparisons.

To set appropriate GL thresholds for different datasets, we used skewness as an additional index, specifically the Fisher-Pearson coefficient. Since this index needs to be independent of position and variability to compare different distributions, we introduced the concept of a normalized variable. Let x_i be a variable with a mean of μ and a standard deviation of σ. The normalized variable Z_i=$Z_i(x_i)$ is defined by this linear transformation: $Z_i = \frac{x_i - \mu}{\sigma}$. Then, we define the skewness index as the arithmetic mean of the third powers of the normalized variable, which is the GL distribution overlapped with a Gaussian, to study the difference between them: $\gamma_1 = \frac{1}{N}\sum_{i=1}^{N} Z_i^3$. The skewness value can be positive, zero, negative, or undefined. In particular, it will indicate where the distribution has the tail by values that can be:

- *negative:* the tail is longer on the left, the mass of the distribution is concentrated on the right, and the indices of position form the chain Mean < Median < Mode.
- *positive:* the tail is longer on the right, the mass of the distribution is concentrated on the left, and the indices of position form the chain Mode < Median < Mean.

We excluded from the evaluation of the skewness the frequencies of the value 0, which represents the black (e.g., the background) in the image. As demonstrated in experiments, by the skewness of an image, it is possible to distinguish between the dataset to which it belongs. This property will be used in BIPS to apply the right threshold to the input image. Once we obtained the skewness for each image, we considered the distribution of the skewness on the validation set to obtain the intervals that refer to each dataset.

4 Experiments

In this section, we will present the experiments we performed on both depth and RGB preprocessing. We then compared our results with DL methods over the same data, and we finally compared DLIF with the state-of-the-art. All the experiments are performed on a MacBook Pro 2,6 GHz Intel Core i7 6 core 16 GB 2667 MHz MHz DDR4 Intel UHD Graphics 630 1536 MB with Python 3.6.8.

4.1 Datasets

If RGB face recognition datasets are very popular in literature, the same cannot be said for depth images. RGB datasets present various challenges, such as occlusions, variations in pose, and illumination, that are not typically present in depth datasets. For this reason, we adapted existing depth image datasets to perform different tasks that take into account those challenges, and we adapted them for face recognition. By using ICT-3DHP, Biwi, and Pandora, we covered a wide range of head poses during testing. On the other side, by using Florence Superface, we also investigate the effects of different resolutions.

The *Biwi Kinect Head Pose dataset* [11] has been collected in 2011 by using a Kinect 1. The subjects are 20 and were recorded while turning their heads around freely, for a total of about 15K frames. Biwi also has a corresponding RGB image for each depth frame.

In 2012, the *ICT 3D Head Pose Database* [2] was recorded using the same device as Biwi. There are 10 participants who move their heads freely. The frames are about 14K, and an RGB reference is available for each frame.

The *Pandora dataset* [6] is the newest and biggest of the ones presented here, adapted for face recognition. Pandora was collected in 2017 with a Kinect 2 and has a total of more than 250K frames. The sequences are more varied since Pandora has 100 sequences for 22 subjects. This is because Pandora was originally developed to be used in a driving test scenario.

Fig. 3. Samples from the datasets. From left to right: Biwi, ICT-3DHP, Pandora, Superface. On the top 4 RGB-images and on the bottom 4 depth images of the same subject.

The *Florence Superface dataset* [4] includes the low-resolution and high-resolution 3D scans of 20 subjects. During the acquisitions, subjects move their faces around the yaw axis to an angle of approximately 60-70 °C.

Since each of these datasets contains a small number of subjects, we combined their images to obtain a new benchmark, namely *BIPS* (from **B**iwi, **I**CT-3DHP, **P**andora and **S**uperface). As a result, this new database contains 72 subjects, different devices used to collect data (Kinect 1 or Kinect 2), and a wide range of variations in pose, occlusions, and distance from the device.

4.2 Depth Face Detection

We adapted an RGB landmark predictor to work with depth images instead of RGB. We used the landmark detection process instead of facial detection because it adapts better to depth images. The predictor we used as a reference is by [14]. We retrained their dlib landmark predictor using the Biwi and ICT-3DHP datasets, as they are the only datasets that provided aligned RGB and depth images. We randomly selected 65668 images for training and 15092 images for testing. The training parameters were the same as in [14], and we evaluated the average distance between predicted landmarks and ground truth landmarks. The training error was 1.53, and the test error was 6.30. The predictor takes an average of 0.001s to predict the landmarks of a depth image, similar to the RGB predictor in [14]. Figure 4 shows an example of the prediction results for a Biwi image in the test set and a Pandora image, on which the predictor has never been trained.

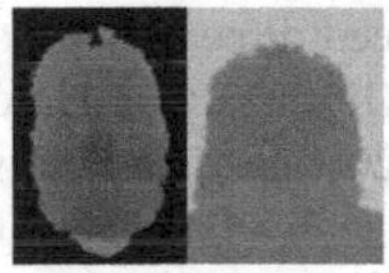

Fig. 4. Facial landmarks predicted on a Biwi image on the left and on a Pandora image on the right. We cropped the image of the face to enhance the results. The method works on the entire image.

The landmarks we will use are the border landmarks on the face. We do not perform prior face detection but rather landmark prediction directly. For this reason, once the landmarks are detected, we extract the face borders as the border landmarks that go from the eyebrows to the chin. At this point, we create a facial mask that contains only the depth information inside the face, as depicted in Fig. 1 step 3.

4.3 Depth Face Recognition

Once the facial mask is obtained, we can proceed using the steps described in Sect. 3. To ensure that the new landmark predictor does not negatively affect recognition, where possible, we tested the method using both RGB and depth landmarks to enhance the difference. For Pandora, we cannot use RGB landmarks in the preprocessing step because the original RGB and depth images are not aligned. For this reason, RGB landmarks cannot be used to extract the face from depth images. On the other hand, the quality of the depth images of Superface is so low (see Fig. 3) compared to the other datasets, that it is not possible to obtain depth landmarks on the latter, even if our predictor has been trained to work well in the case of different sensors (as in the Kinect 2 of Pandora). In this case, we will use RGB landmarks to detect the face, and then proceed by only using depth features. The ratio used to perform recognition is a classical one [13]: 80% of the images of each subject will compose the model, and the remaining 20% the test set. To ensure the depth landmarks do not negatively affect the accuracy, we compared the results obtained by using RGB and depth landmarks in the preprocessing. Those results were obtained by DLIF. On Biwi, the face recognition accuracy is 0.92 when using RGB landmarks and 0.9328 when using depth landmarks. This means that for Biwi, the accuracy obtained by using depth landmarks is even higher than RGB. We suppose that this is possible thanks to the high number of details in the facial parts of the depth image. For ICT-3DHP, we obtain an accuracy of 0.9342 for RGB and 0.9254 for depth landmarks. A small decrease, about 0.0088, that is more than tolerable if we consider that we used only depth images. For Pandora, as previously explained, we do not have RGB landmark results; however, considering that this dataset presents several occlusions and strong head pose variation, we can claim that the accuracy obtained of 0.9065 is more than acceptable. Finally, the worst result is from Superface, where we have 0.8303 of accuracy on RGB. Here, it is not possible to perform the depth landmark accuracy because of the missing ground truth. We can assume that this lower accuracy is due to the low detail level of the depth images since no other challenging scenario is in this dataset if we compare it to the others.

In the case of DLIF+, we proceed directly with depth landmarks, and we need to evaluate the previously introduced GL threshold. Here, the datasets have been split into 60% model, 20% validation, and 20% test, respectively. The obtained values of GL are 30 for ICT-3DHP, 10 for Biwi, 18 for Pandora, 10 for Superface and 20 for BIPS. In BIPS, we obtained a threshold that represents a value from images from different sources. As introduced in Sect. 3.2 we want to use the skewness to refine this result by distinguishing the source of the image. For this purpose, the skewnesses obtained on the validation set of each dataset have been seen as a distribution for which we calculate the mean μ and the standard deviation σ. The formula that allows us to obtain the interval of skewness that distinguishes between datasets is, thus $min = \mu - \sigma$ and $max = \mu + \sigma$. By representing those intervals, we obtain Fig. 5. Here we can notice that the skewness intervals of Biwi and Superface make them separable, whereas those intervals for Pandora and ICT-3DHP make them not separable. For this reason, since Biwi and Superface have the same GL threshold, if the skewness of the input image is between 4 and 6 or 14 and 17 (extreme excluded), we will apply 10 as the GL threshold. In other cases, we will apply the mean between the thresholds of ICT-3DHP and Pandora, obtaining 24.

The results obtained by applying DLIF+ with those GL thresholds and skewness values are shown in the following subsections.

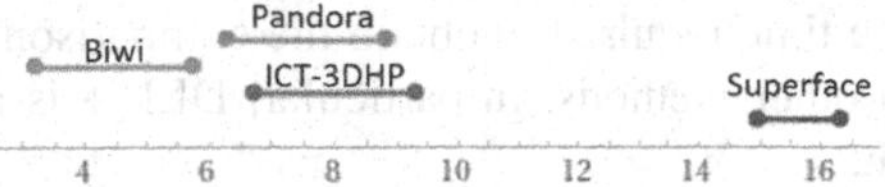

Fig. 5. The skewness intervals on the validation set images for the considered datasets.

4.4 Face Recognition Comparisons

Recently, DL-based approaches have gained promising results in the field of facial recognition. We evaluated the most popular architectures for face recognition, thus obtaining their performances in identification accuracy on the above-mentioned databases. The Deep Convolutional networks proposed are the following: FaceNet and FaceNet512 [21], OpenFace [3], DeepFace [23], DeepID [24], ArcFace [10], VGGFace [20] and VGGFace2 [7]. The resulting accuracy levels compared with our results are shown in Table 1. DLIF+ performs better than: 5 out of 7 Deep Networks (DN) over ICT-3DHP, with a difference of only 0.0148 from the best one; 3 out of 7 DN algorithms over Biwi, with a difference of only 0.0324 from the best one; 3 out of 7 DN algorithms over Pandora, with a difference of only 0.0316 from the best one; 1 out of 7 DN algorithms on Superface, with a difference of 0.0504 from the best one; 3 of 7 DN algorithms over BIPS, with a difference of only 0.02 from the best one.

Table 1. Comparisons of performance with different Deep Learning methods.

Model	ICT-3DHP	Biwi	Pandora	Superface	BIPS
VGGFace	0.8958	0.9568	0.9498	0.9310	0.9247
FaceNet	0.8996	0.9168	0.8897	0.9395	0.9062
Facenet512	0.8623	0.8622	0.8723	0.9063	0.8664
OpenFace	0.8225	0.8728	0.8074	0.8603	0.8292
DeepFace	0.9405	0.9706	0.9198	0.9701	0.9489
DeepID	0.9116	0.9649	0.9235	0.9735	0.9406
ArcFace	0.9436	0.9559	0.9451	0.9480	0.9475
DLIF	0.9253	0.9327	0.9065	0.8303	0.9098
DLIF+	0.9288	0.9382	0.9182	0.8976	0.9289

DLIF+ results are competitive with DL techniques, even if the time required to perform the comparisons of the subjects is significantly reduced compared to the time required for DL methods. In fact, we used the Python DeepFace library to implement the DL techniques, in which the time required to obtain the features from the model images is comparable to our model construction, but our subject identification step is faster. In Table 2 we can observe the time in seconds required to perform the prediction by using DLIF+ and the four methods that overcame DLIF+ on at least one dataset (VGGFace, DeepFace, DeepID and ArcFace). As can be appreciated, even if the accuracies are similar (see Table 1), the time required to obtain the comparison is significantly lower in DLIF+ compared to other methods. In particular, DLIF+ is also more stable with regard to dataset growth.

Table 2. Time (in seconds) of the best methods over the five datasets evaluated.

Model	ICT-3DHP	Biwi	Pandora	Superface	BIPS
VGGFace	10.3	4.9	6.7	1.3	24.2
DeepFace	14.1	6.2	9.2	6	34.4
DeepID	2	0.9	1.4	0.8	5
ArcFace	3.3	1.5	2.1	1.4	8.8
DLIF	0.6	0.3	0.5	0.3	1.5
DLIF+	**0.4**	**0.1**	**0.2**	**0.1**	**0.8**

At the state of the art, the datasets used are various, and we used for the first time the presented datasets. For this reason, we compared DLIF and DLIF+ with the literature by using the modality as a criteria, e.g., RGB-D, partial RGB-D (when the RGB is used only in the preprocessing step), depth only, and the stability with respect to the number of subjects in the dataset. As can be seen from Table 3, DLIF+ is one of the

Table 3. Comparisons with SOTA algorithms. 'Partial RGB-D' indicates methods that uses RGB images only in pre-processing.

Method	Modality	Subj.	Acc.
Chhokra et al. [9]	RGB-D	108	0.695
Kaashki et al. [18]	RGB-D	52	0.995
Kaashki et al. [18]	RGB-D	106	0.866
Cheng et al. [8]	RGB-D	20	0.925
Xu et al. [25]	RGB-D	123	0.942
Min et al. [16]	Partial RGB-D	20	0.979
Min et al. [17]	Depth	52	0.740
Ahmad et al. [1]	Depth	52	0.780
Feng et al. [12]	Depth	64	0.859
Shi et al. [22]	Depth	17	0.968
DLIF+ - ICT-3DHP	Depth	10	0.929
DLIF+ - Biwi	Depth	20	0.938
DLIF+ - Pandora	Depth	22	0.919
DLIF+ - Superface	Partial RGB-D	20	0.898
DLIF+ - BIPS	Depth	72	0.929

best methods that uses only depth images to perform recognition, surpassed only by [22] which, however, has a very small set of subjects (17).

5 Conclusions

We proposed a method for face recognition using only depth images, which is a challenging task due to the difficulty of performing preprocessing using depth images alone. Our method involves developing a depth landmark detector to locate and isolate the face, followed by histogram equalization to account for differences in depth levels. The features used for recognition are obtained through fractal encoding and compared using the Canberra distance. Our results are competitive with the state of the art and remain stable with an increasing number of subjects and variations in occlusions and head poses. In the future, we plan to combine our method with machine learning techniques like SVM to further improve the classification performance.

Acknowledgements. This work was partially supported by the project IDA included in the Spoke 2 - Misinformation and Fakes of the Research and Innovation Program PE00000014, "SEcurity and RIghts in the CyberSpace (SERICS)", under the National Recovery and Resilience Plan, Mission 4 "Education and Research" - Component 2 "From Research to Enterprise" - Investment 1.3, funded by the European Union - NextGenerationEU.

References

1. Ahmad, N., Ali, J., Khan, K., Naeem, M., Ali, U.: Robust multimodal face recognition with pre-processed kinect RGB-D images. J. Eng. Appl. Sci. **36**, 77–84 (2017)
2. Baltrušaitis, T., Robinson, P., Morency, L.: 3D constrained local model for rigid and non-rigid facial tracking. In: IEEE Conference on Computer Vision and Pattern Recognition, pp. 2610–2617 (2012)
3. Baltrušaitis, T., Robinson, P., Morency, L.P.: OpenFace: an open source facial behavior analysis toolkit. In: IEEE Winter Conference on Applications of Computer Vision, pp. 1–10 (2016)
4. Berretti, S., Del Bimbo, A., Pala, P.: Superfaces: a super-resolution model for 3D faces. In: Fusiello, A., Murino, V., Cucchiara, R. (eds.) ECCV 2012. LNCS, vol. 7583, pp. 73–82. Springer, Heidelberg (2012). https://doi.org/10.1007/978-3-642-33863-2_8
5. Bisogni, C., Nappi, M., Pero, C., Ricciardi, S.: PIFS scheme for head pose estimation aimed at faster face recognition. IEEE Trans. Biometrics, Behavior Identity Sci. **4**(2), 173–184 (2021)
6. Borghi, G., Venturelli, M., Vezzani, R., Cucchiara, R.: POSEidon: face-from-depth for driver pose estimation. In: IEEE Conference on Computer Vision and Pattern Recognition (CVPR), pp. 5494–5503 (2017)
7. Cao, Q., Shen, L., Xie, W., Parkhi, O.M., Zisserman, A.: VGGFace2: a dataset for recognising faces across pose and age. In: 13th IEEE International Conference on Automatic Face & Gesture Recognition (FG 2018), pp. 67–74 (2018)
8. Cheng, Z., Shi, T., Cui, W., Dong, Y., Fang, X.: 3D face recognition based on kinect depth data. In: 4th International Conference on Systems and Informatics, pp. 555–559 (2017)
9. Chhokra, P., Chowdhury, A., Goswami, G., Vatsa, M., Singh, R.: Unconstrained kinect video face database. Inf. Fusion **44**, 113–125 (2018)

10. Deng, J., Guo, J., Xue, N., Zafeiriou, S.: ArcFace: additive angular margin loss for deep face recognition. In: Proceedings of the IEEE/CVF Conference on Computer Vision and Pattern Recognition, pp. 4690–4699 (2019)
11. Fanelli, G., Weise, T., Gall, J., Van Gool, L.: Real time head pose estimation from consumer depth cameras. In: Mester, R., Felsberg, M. (eds.) DAGM 2011. LNCS, vol. 6835, pp. 101–110. Springer, Heidelberg (2011). https://doi.org/10.1007/978-3-642-23123-0_11
12. Feng, J., Guo, Q., Guan, Y., Wu, M., Zhang, X., Ti, C.: 3D face recognition method based on deep convolutional neural network. In: Panigrahi, B.K., Trivedi, M.C., Mishra, K.K., Tiwari, S., Singh, P.K. (eds.) Smart Innovations in Communication and Computational Sciences. AISC, vol. 670, pp. 123–130. Springer, Singapore (2019). https://doi.org/10.1007/978-981-10-8971-8_12
13. Gwyn, T., Roy, K., Atay, M.: Face recognition using popular deep net architectures: a brief comparative study. Future Internet **13**, 164 (2021)
14. Kazemi, V., Sullivan, J.: One millisecond face alignment with an ensemble of regression trees. In: IEEE Conference on Computer Vision and Pattern Recognition, pp. 1867–1874 (2014)
15. Lance, G.N., Williams, W.T.: Computer programs for hierarchical polythetic classification ("similarity analyses"). Comput. J. **9**(1), 60–64 (1966)
16. Min, R., Choi, J., Medioni, G., Dugelay, J.L.: Real-time 3D face identification from a depth camera. In: Proceedings of the 21st International Conference on Pattern Recognition (ICPR2012), pp. 1739–1742 (2012)
17. Min, R., Kose, N., Dugelay, J.L.: KinectFaceDB: a kinect database for face recognition. IEEE Trans. Syst. Man Cybern. Syst. **44**(11), 1534–1548 (2014)
18. Nourbakhsh Kaashki, N., Safabakhsh, R.: RGB-D face recognition under various conditions via 3D constrained local model. J. Vis. Commun. Image Represent. **52**, 66–85 (2018)
19. Pala, P., Seidenari, L., Berretti, S., Del Bimbo, A.: Enhanced skeleton and face 3D data for person re-identification from depth cameras. Comput. Graph. **79**, 69–80 (2019)
20. Parkhi, O.M., Vedaldi, A., Zisserman, A.: Deep face recognition. In: British Machine Vision Conference (2015)
21. Schroff, F., Kalenichenko, D., Philbin, J.: FaceNet: a unified embedding for face recognition and clustering. In: Proceedings of the IEEE Conference on Computer Vision and Pattern Recognition (CVPR) (June 2015)
22. Shi, L., Wang, X., Shen, Y.: Research on 3D face recognition method based on LBP and SVM. Optik **220**, 165157 (2020)
23. Taigman, Y., Yang, M., Ranzato, M., Wolf, L.: DeepFace: closing the gap to human-level performance in face verification. In: IEEE Conference on Computer Vision and Pattern Recognition, pp. 1701–1708 (2014)
24. Wong, S.Y., Yap, K.S., Zhai, Q., Li, X.: Realization of a hybrid locally connected extreme learning machine with deepID for face verification. IEEE Access **7**, 70447–70460 (2019)
25. Xu, K., Wang, X., Hu, Z., Zhang, Z.: 3D face recognition based on twin neural network combining deep map and texture. In: IEEE 19th International Conference on Communication Technology (ICCT), pp. 1665–1668 (2019)

Automatic Generation of Semantic Parts for Face Image Synthesis

Tomaso Fontanini(✉), Claudio Ferrari, Massimo Bertozzi, and Andrea Prati

IMP Lab, Department of Engineering and Architecture, University of Parma, Parma, Italy
{tomaso.fontanini,claudio.ferrari2,massimo.bertozzi, andrea.prati}@unipr.it

Abstract. Semantic image synthesis (SIS) refers to the problem of generating realistic imagery given a semantic segmentation mask that defines the spatial layout of object classes. Most of the approaches in the literature, other than the quality of the generated images, put effort in finding solutions to increase the generation diversity in terms of style *i.e.* texture. However, they all neglect a different feature, which is the possibility of manipulating the layout provided by the mask. Currently, the only way to do so is manually by means of graphical users interfaces. In this paper, we describe a network architecture to address the problem of automatically manipulating or generating the shape of object classes in semantic segmentation masks, with specific focus on human faces. Our proposed model allows embedding the mask class-wise into a latent space where each class embedding can be independently edited. Then, a bi-directional LSTM block and a convolutional decoder output a new, locally manipulated mask. We report quantitative and qualitative results on the CelebMask-HQ dataset, which show our model can both faithfully reconstruct and modify a segmentation mask at the class level. Also, we show our model can be put before a SIS generator, opening the way to a fully automatic generation control of both shape and texture. Code available at https://github.com/TFonta/Semantic-VAE.

Keywords: Image Synthesis · Variational Autoencoder · Face Editing

1 Introduction

The task of Semantic Image Synthesis (SIS) consists in generating a photo-realistic image given a semantic segmentation mask that defines the shape of objects. The mask is usually an image in which the pixel values define a specific semantic class (like eyes, skin, hair, *etc.* in the case of human face). This allows for accurately defining the spatial layout and shape of the generated images, while maintaining a high degree of freedom in terms of textures and colors. Indeed, those can be randomly generated [16] or by extracting a specific style from a reference image [7,9].

A nice feature of SIS methods is that the semantic mask can be manipulated to alter the shape of objects in the generated samples. However, currently this

G. L. Foresti et al. (Eds.): ICIAP 2023, LNCS 14233, pp. 209–221, 2023.
https://doi.org/10.1007/978-3-031-43148-7_18

is done manually by using custom painting software allowing the user to modify the shape of one or more mask parts. Attempts of performing automatic face shape parts manipulation have been done, yet with different techniques, such as by using a 3D deformable model of the face [3]. Whereas manual alteration of the semantic masks is fun, it turns out impractical when the objective is to modify the shape of a large number of images.

In the attempt of overcoming this limitation, in this paper we explore the problem of the automatic generation and manipulation of classes in segmentation masks, and propose a method that allows to generate and edit the shape of any number of parts. The proposed model can be used to produce a large variety of novel semantic masks that can then be used in conjunction with any SIS model to generate previously unseen photo-realistic RGB images. This is achieved by designing an architecture composed by an encoder that embeds each of the semantic mask parts separately, a recurrent module composed by a series of bi-directional LSTMs [11] that learns the relationships between the shape of different mask parts and, finally, a decoder that maps the latent representation back into a realistic semantic mask. The model is trained as a Variational Autoencoder (VAE), so combining a reconstruction loss with a KL divergence in order to induce a specific distribution in the latent space. This enables the generation, interpolation or perturbation of semantic classes; these specific features, to the best of our knowledge, are still unexplored in the literature. Overall, the main contributions of this paper are the following:

- we explore the novel problem of automatic generation and editing of local semantic classes in segmentation masks, independently from the others;
- we propose a novel architecture combining a VAE and a recurrent module that learns spatial relationships among semantic classes by treating them as elements of a sequence, under the observation that the shape of each part has an influence on the surrounding ones. More in detail, each part embedding is subsequently fed into the LSTM block so to account for shape dependencies, and then employed by the decoder to generate the final mask. The proposed architecture can finally be used in combination with any SIS architecture to boost the shape diversity of the generated samples;
- we quantitatively and qualitatively validate our proposal in the task of face parts editing, and report and extensive analysis of the advantages, limitations and challenges.

2 Related Works

Given that no prior works addressed the problem presented in this paper, in the following we summarize some recent literature works on semantic image synthesis and variational autoencoders.

Semantic Image Synthesis. Semantic Image Synthesis approaches can be divided into two main categories: diversity-driven and quality-driven. Both of them take inspiration and improve upon the seminal work of Park *et al.*, named SPADE [9], where semantic image synthesis is achieved by means of custom, spatially-adaptive normalization layers. Methods in the former category focus on the task of generating samples having the shape conditioned over semantic masks, but the style is generated randomly in order to achieve an high degree of multi-modality. Some examples of these approaches are [10,16]. The trend here points towards increasing the granularity of the generated texture; for example, in CLADE [13] styles are generated at the class-level, while INADE [12] is able to generate instance-specific style by sampling from a class-wise estimated distribution. On the other side, quality-driven methods try to extract a specific style from a target image and to apply it over the generated results, in the attempt of both maintaining the shape defined by the mask and the texture defined by a reference image. An example of paper falling in this category is MaskGAN [7], in which a style mapping between an input mask and a target image is achieved using instance normalization. Also in this case, efforts are put into finding solutions to increase the precision and granularity of the style control. To this aim, Zhu *et al.* developed SEAN [18], a method that is able to extract the style class-wise from each of the different semantic part of an image and map it locally over the corresponding area of the input mask. Another work following the same trend is SC-GAN [16]. Overall, it turns out clearly that none of the recent literature works deals with the problem of locally manipulating the face shape by acting on segmentation masks.

Variational Autoencoders. Autoencoders introduced in [8] were proposed as a way to achieve a compressed latent representation of a set of data, but they lack generation capabilities. On the contrary, Variational Autoencoders (VAE) [6] described data generation through a probabilistic distribution. Indeed, once trained using a combination of reconstruction loss and Kullback-Leibler divergence, they can generate new data by simply sampling a random latent distribution and feeding it to the decoder. There exist several variations of VAE such as Info-VAE [17], β-VAE [5] and many more [1,14].

3 Network Architecture

The main objective that guided the design of the model architecture is that of performing automatic manipulation and generation of semantic masks, independently for each class. A semantic segmentation mask can be represented as C-channel image, where each channel is a binary image containing the shape of a specific object class *i.e.* $M \in [0,1]^{C \times H \times W}$. So, each pixel belongs to a unique class *i.e.* has value 1 only in a single channel, and each class shape is complementary to all the others, *i.e.* there is no intersection between the semantic classes.

The challenge behind manipulating or generating a specific semantic class in a segmentation mask is that its shape, and the shape of all its surrounding classes, need to be adapted so that the above properties are maintained. At the

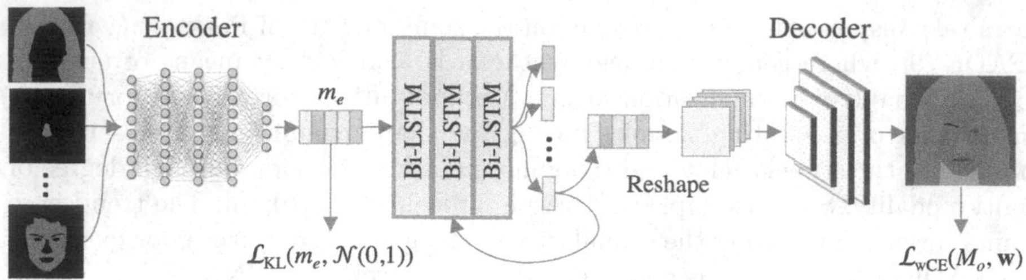

Fig. 1. Proposed architecture: the segmentation mask $M \in [0,1]^{C\times 256\times 256}$ is processed so that each channel (class) $c \in C$ is flattened and passes through a MLP encoder to obtain class-wise embeddings m_e^c for each semantic class. The embeddings $m_e = [m_e^1, \cdots, m_e^C]$ then pass through a set of three bi-directional LSTM layers followed by a feed-forward block that learn relationships across the classes. The processed embeddings are finally reshaped to form a set of feature maps $m_d \in \mathbb{R}^{C\times 16\times 16}$, and then fed to a convolutional decoder which outputs a new mask $M_o \in [0,1]^{C\times 256\times 256}$. The model is trained with (1) a pixel-wise weighted cross-entropy loss ($\mathcal{L}_{wCE}$), and (2) a KL-divergence loss $\mathcal{L}_{KL}$ applied to the embeddings m_e so to push them towards following a $\mathcal{N}(0,1)$ distribution, enabling their generation from noise or manipulation.

same time, the spatial arrangement of each class have also to be realistic, since it is a scenario-dependent property. In the case of facial features, the spatial relations of the different face parts need to be preserved; as example, the nose should be mostly centered between eyes.

3.1 Architecture

To account for the above challenges, we designed our proposed architecture (Fig. 1) to have 4 main components: (1) an MLP $\mathcal{M}$ to independently encode the mask channels into a latent representation m_e. This allows us to operate on the mask channels directly in the compressed space; (2) an LSTM-Feed Forward block $\mathcal{L}$ composed of three bi-directional LSTM layers to process the encoded mask channels m_e^j and account for possible misalignments resulting from manipulating a semantic class, and a feed-forward block $\mathcal{F}$ to further re-arrange the processed mask encodings; (3) finally, a convolutional decoder $\mathcal{D}$ to reconstruct the complete semantic mask M_o.

MLP Encoder. The encoder $\mathcal{M}$ is a simple MLP made up of three linear layers, each followed by a ReLU activation function. Each mask channel is first flattened so that the input mask has size $M \in \mathbb{R}^{C\times H^2}$ where $H = W = 256$ is the spatial size of the mask; each linear layer of the encoder has a hidden size of 256, so that $m_e = [m_e^1, \cdots, m_e^C] = \mathcal{M}(M) \in \mathbb{R}^{C\times 256}$.

Bi-directional LSTM Block. The bi-directional LSTM [11] block was designed to process the encoded mask channels m_e one after another, as if they were frames of a temporal sequence. The goal is that of correcting possible inconsistencies resulting from manipulating or generating a class embedding m_e^c, based on the information of the other classes. Intuitively, if we change the shape of a facial part in the mask *e.g.* nose, the surrounding parts need to be adjusted so that the combined result looks realistic and artifact-free. One problem arising from using a recurrent module is that of choosing the order in which the channels are processed. Temporal sequences have a unique ordering implicitly defined by the time flow, whereas in our scenario there is no clear nor unique way of choosing the order by which processing the face parts, being them simply parts of a spatial layout. This motivated us to opt for the bi-directional variant of the LSTM; indeed, the latter processes the sequence in both directions (first to last, and last to first), so that each class embedding is influenced by all other classes, not only by the previously processed ones. Each class embedding m_e^c is thus processed, and provided as hidden state both for the subsequent m_e^{c+1} and previous m_e^{c-1} classes. In addition, differently from the standard use of LSTMs where only the last processed embedding keeps flowing through the network, we also store the embeddings at intermediate steps m_e^c. In doing so, once all the C embeddings have been processed, we end up with the same number of C embeddings, one for each class. Finally, following the same principle of [15], a feed-forward block composed of two linear layers equipped with GeLu [4] activation function is stacked after the LSTMs so to make the embeddings better fit the input to the decoder.

Decoder. Finally, the convolutional decoder $\mathcal{D}$ is responsible for learning to reconstruct the segmentation mask from the C embeddings resulting from the previous steps. In particular, the C embeddings are reshaped into a set of C feature maps $m_d \in \mathbb{R}^{C\times16\times16}$. These are processed by 4 residual blocks, equipped with SiLU [2] activation function and group normalization. The decoder outputs the reconstructed segmentation mask $M_o \in \mathbb{R}^{C\times256\times256}$.

3.2 Loss Functions

The model is trained to self-reconstruct the input segmentation mask, without any other specific strategy to guide the manipulation process. The output mask is generated by minimizing a pixel-wise class prediction, using a cross-entropy loss. In particular, we used a weighted variant of the standard cross entropy $\mathcal{L}_{CE}$. More in detail, we observed that the problem resembles a highly imbalanced classification problem; indeed, smaller parts such as, for face masks, the eyes or the nose, are significantly under-represented in the data *i.e.* occupy a smaller number of pixels, with respect to larger parts such as skin or hair, ultimately weighing less in the overall loss computation. So, the weights are set considering this imbalance; smaller weights will be assigned to bigger parts, and bigger weights will be assigned to smaller parts. We calculate the weights $\mathbf{w} = [w_0, \cdots, w_C]$

based on the overall training set statistics, in the following way:

$$\mathbf{w} = 1 - \frac{1}{NHW} \sum_{N}^{i} \sum_{H}^{j} \sum_{W}^{k} x_{c,i,j,k} \; \forall \, c \in C \tag{1}$$

where N is the number of samples in the training set, and H and W are the height and width of the semantic mask, respectively. Given that each of the mask channels can contain only one or zero values, this equation provides a series of C weights that rank each of the semantic parts by their average size. The equation of the final weighted cross entropy $\mathcal{L}_{wCE}$ therefore becomes:

$$\mathcal{L}_{wCE} = -\sum_{x} \mathbf{w}(y(x)) y(x) log(\hat{y}(x)) \tag{2}$$

where $y(x)$, $\hat{y}(x)$, and $\mathbf{w}(y(x))$ are the ground-truth class labels, the predicted labels, and the weight for the ground-truth class at pixel x, respectively.

In addition to the weighted cross entropy, a KL-Loss $\mathcal{L}_{KL}$ is used to push the latent codes of each of the parts to have zero mean and unit variance and allow the generation process where a random latent code is sampled from $\mathcal{N}(0, 1)$. Ultimately, the full loss utilized to train the model is:

$$\mathcal{L} = \mathcal{L}_{wCE} + \lambda \mathcal{L}_{KL} \tag{3}$$

where λ is the KL weight and is set to 0.0005 in all the experiments.

4 Experimental Results

In this section, we report the results of an experimental validation. We show both quantitative and qualitative results, in terms of reconstruction accuracy and different generation or manipulation tasks. In fact, despite our goal being that of performing editing of semantic masks at the class level, we also need to make sure the reconstruction process does not degrade the segmentation accuracy of the input masks and in turn compromise the subsequent image synthesis.

As dataset to train and test our model, we used the CelebAMask-HQ [7], which is composed by 30K high resolution face images (1024×1024) along with the corresponding segmentation masks. Out of the 30K samples, 28K were used for training and 2K for testing.

4.1 Reconstruction, Generation and Perturbation

Given that no prior works addressed this particular problem, before analyzing the ability of the model to manipulate the mask parts, we compare our solution with some baseline architectural designs in terms of reconstruction accuracy, in a sort of an extended ablation study. Reconstruction results are reported in Table 1 in terms of pixel-wise classification accuracy (Acc) and Mean Intersection over Union (mIoU). In particular, the following configurations were explored: a simple

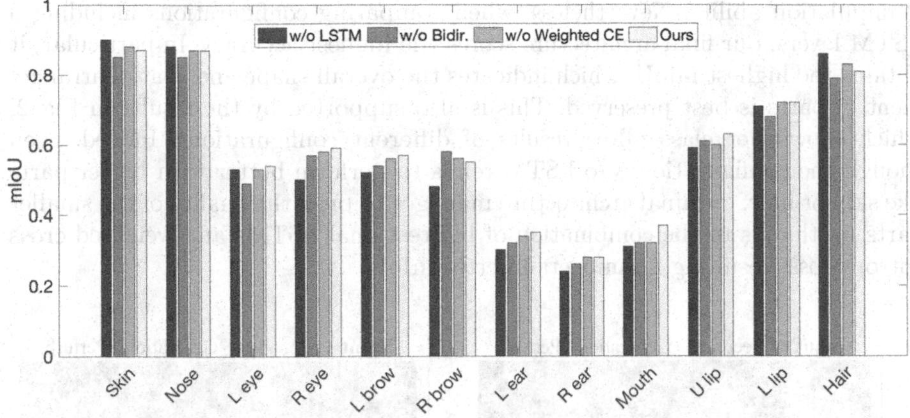

Fig. 2. mIoU results per class.

encoder-decoder trained with the standard cross-entropy (row 1), the model with 1 or 3 standard LSTMs trained with standard cross entropy (rows 2 and 3), our final model with 3 standard LSTMs trained with the weighted cross entropy (row 4), our final model with 3 bidirectional LSTMs trained with regular cross entropy (row 5), and the final architecture (bottom row).

Quantitatively, we observe a generally-high reconstruction accuracy in all the cases. The simplest architecture (w/o LSTM and weighted CE) achieves the highest accuracy but lower mIoU. A visual inspection of the results suggests that the the additional processing due to the LSTM block induces a slight smoothing of high-frequency details such as the hair contour. This is caused by the compression of each semantic part in the encoding phase, and also by the Bi-directional LSTM block pass which makes more difficult for the decoder to exactly reproduce the corresponding input. This hypothesis is supported if looking at the results obtained with either 1 or 3 LSTM layers; indeed the two measures decrease when stacking more LSTM layers. On the other hand though, we will show (Fig. 4) that removing such layers severely compromises the

Table 1. Reconstruction results comparing our solution with different baselines.

Method	mIoU ↑	Acc ↑
W/o LSTM block	68.49	**94.24**
1 LSTM w/o Bidir. w/o weigthed CE	68.15	<u>93.85</u>
3 LSTMs w/o Bidir. w/o weigthed CE	67.35	90.91
3 LSTMs w/o Bidir	68.34	90.94
3 LSTMs w/o weigthed CE	<u>69.56</u>	92.12
Ours	**70.31**	92.39

manipulation ability. Nevertheless, when comparing configurations including 3 LSTM layers, our final architecture scores the highest accuracy. In particular, it obtains the highest mIoU, which indicates the overall shape and spatial arrangement of parts is best preserved. This is also supported by the results in Fig. 2, which shows per class mIoU results of different configurations. Indeed, even though the configuration w/o LSTM tends to perform better with bigger parts like skin or hair, our final architecture manages to push the quality of the smaller parts up thanks to the combination of bidirectional LSTMs and weighted cross entropy loss, resulting in an overall better mIoU.

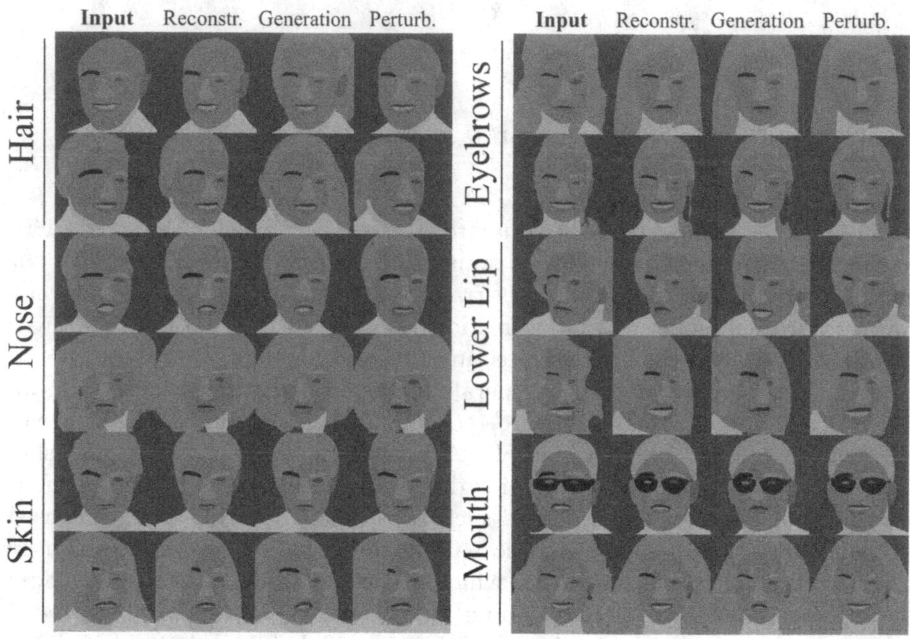

Fig. 3. Results for reconstruction, generation and perturbation of different mask parts.

In Fig. 3 some results for both reconstruction, generation and perturbation of different parts in the semantic masks are presented. More in detail, we refer to *generation* when a novel latent code drawn from the normal distribution $\hat{m}_e^j \sim \mathcal{N}(0,1)$ is substituted to its encoded counterpart m_e^j and passed to the bi-directional LSTM block in order to generate a particular part c. On the other side, we refer to *perturbation* when a random noise vector drawn from the normal distribution $z \sim \mathcal{N}(0,1)$ is added to an existing latent code *i.e.* $\hat{m}_e^j = m_e^j + z$. Indeed, in the latter, usually the shape of the generated parts is more similar to the original input, while in the first case the generated shape can be (and usually is) completely different.

Regarding reconstruction, we can see how the proposed method manages to maintain the overall shape of the semantic mask parts, supporting the results in

Table 1. Nevertheless, as discussed above, a certain degree of smoothing in the results can be noted. This represents a minor limitation of the current proposal. On the other side, results when generating parts from scratch, or by perturbing an existing latent code, are impressive. Our method is not only able to generate realistic parts independently from one another, but also, thanks to the recurrent part of the model, is able to adapt the shape of the parts surrounding the one that is being generated in order to produce a realistic final result. This can be particularly appreciated for example when perturbating the nose latent code in Fig. 3 in the third row: indeed, the nose is made longer by the perturbation and as a consequence the mouth is deformed accordingly.

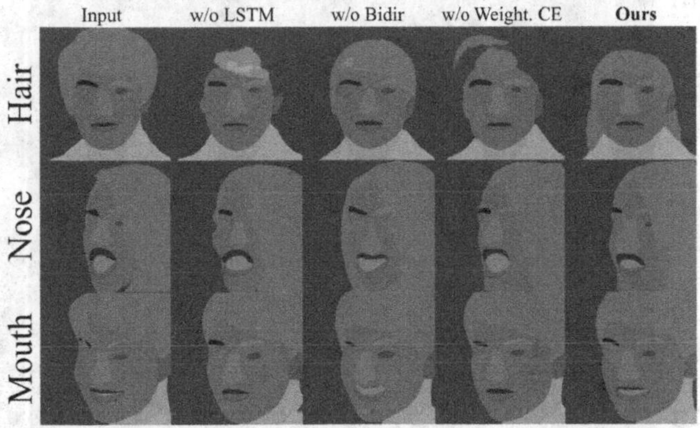

Fig. 4. Qualitative results of different ablations experiments.

Finally, in Fig. 4 we show some qualitative results to prove that the final architecture is indeed better in the generative task which is the main purpose of this paper. Starting from the top, it is clear how when generating hair the proposed model is much more capable of producing a realistic results without generating undesired classes (like the pink part in the model without LSTM). Then, in the second row, is proved how our model is much better at rearranging all the semantic parts in order to create a realistic mask with a newly generated part. Finally, in the last row, we can see how the mouth part is generated correctly by almost every configuration, but, at the same time, our model is able to generate much more varied and diverse results.

4.2 Interpolation

In Fig. 5 interpolation results are presented. Interpolation is done by choosing a part c from a source and a target mask and merging together the corresponding latent vectors using an interpolation factor α. More in detail, the interpolation equation is the following:

$$m_c^{int} = \alpha \cdot m_c^t + (1 - \alpha) \cdot m_c^s \quad (4)$$

where m_c^t and m_c^s are the latent codes of the part c of the target and source images, respectively. In addition, $\alpha = 0$ is equal to reconstructing the source image, while $\alpha = 1$ represents a sort of "face part swapping", that is a specific face part is swapped from a target face to a source one.

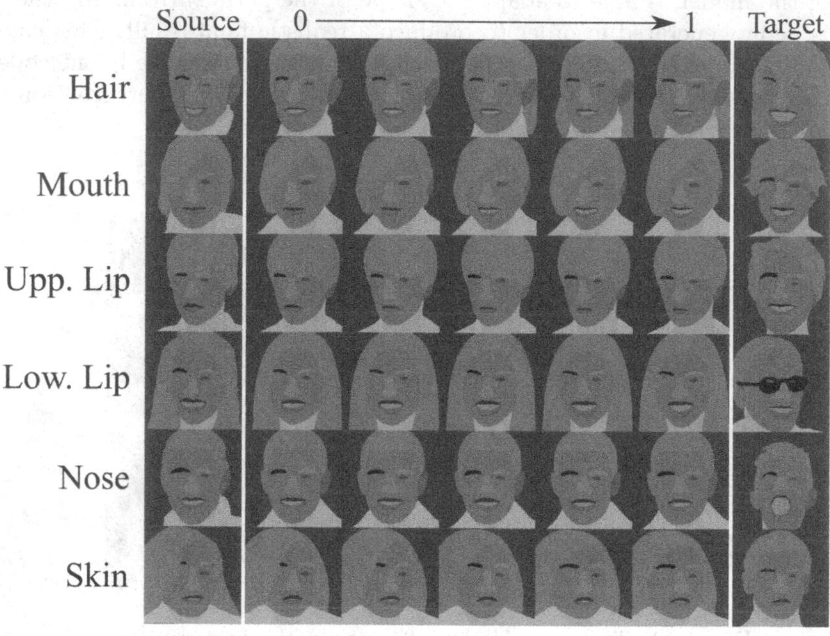

Fig. 5. Interpolation results of different parts taken from a source and target mask (first and last columns, respectively). Values of the interpolation factor α go from 0 to 1, where 0 means no interpolation.

Indeed, it is evident how the KL loss, that pushes the latent codes to have almost zero mean and unit variance, allows to easily interpolate every mask part. In particular, while increasing the interpolation factor α, the shape changes continuously. The only previous method that we are aware of capable of performing a similar task is MaskGAN [7]; however, MaskGAN can only perform global mask interpolations, and can not independently manipulate individual parts.

4.3 Semantic Image Synthesis with Shape Control

In this section we qualitatively show results for the main purpose of our model, that is equipping SIS generators with a module to enable automatic shape control. In Fig. 6, several mask with automatically generated parts are fed to a state-of-the-art SIS model in order to produce new and diverse face images. We chose to use the SEAN [9] generator to this aim because SEAN can very precisely control the image generation thanks to its semantic region-adaptive normalization layers. Previous to our proposal, the editing of masks could only be done manually. Results in Fig. 6 clearly show that, provided a generator that is accurate enough to handle local shape changes, the shape of the generated faces can be automatically edited by means of our solution. This paves the way to a very efficient way of employing SIS models, for example, for data augmentation which can be very helpful for task like re-identification, classification or detection.

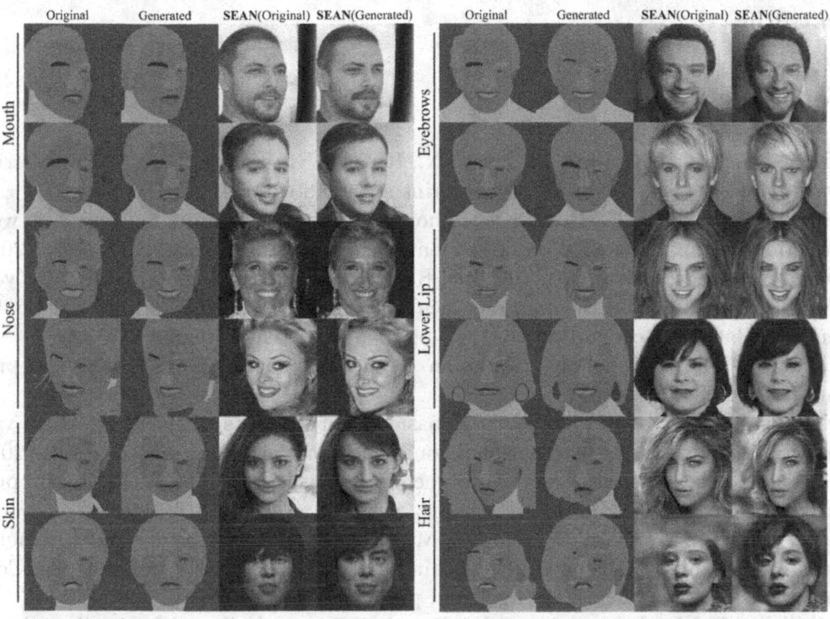

Fig. 6. RGB results obtained when feeding the original mask and a mask automatically edited with our method to a SIS model.

5 Conclusion

In this paper, we introduced the problem of automatic manipulation of semantic segmentation masks, and presented a preliminary novel architecture to achieve this goal, with a specific application to face part editing. The proposed system is able do generate or manipulate any semantic part by just feeding random noise to the LSTM block in the place of the latent representation of the corresponding part. We show the efficacy of our architecture through a series of quantitative and qualitative evaluations. Even if we observed the tendency of smoothing the shapes of the generated results, still our method is able to generate realistic semantic parts, and can be readily used in combination with potentially any SIS models so to generate a virtually infinite number of RGB results.

Finally, we believe there is still large room for improvements. For example, extending the proposal to different scenarios with less constrained objects layout or more classes would represent a valuable feature for a SIS model. Also, currently, the shape manipulation is not controlled, meaning that it is not yet possible to generate parts with a specific shape or attributes, *e.g.* long nose or curly hair. All the above are features that we plan to investigate in future works.

Acknowledgments. This work was supported by PRIN 2020 "LEGO.AI: LEarning the Geometry of knOwledge in AI systems", grant no. 2020TA3K9N funded by the Italian MIUR.

References

1. Davidson, T.R., Falorsi, L., De Cao, N., Kipf, T., Tomczak, J.M.: Hyperspherical variational auto-encoders. arXiv preprint arXiv:1804.00891 (2018)
2. Elfwing, S., Uchibe, E., Doya, K.: Sigmoid-weighted linear units for neural network function approximation in reinforcement learning. Neural Netw. **107**, 3–11 (2018)
3. Ferrari, C., Serpentoni, M., Berretti, S., Del Bimbo, A.: What makes you, you? Analyzing recognition by swapping face parts. In: 2022 26th International Conference on Pattern Recognition (ICPR), pp. 945–951. IEEE (2022)
4. Hendrycks, D., Gimpel, K.: Gaussian error linear units (gelus). arXiv preprint arXiv:1606.08415 (2016)
5. Higgins, I., et al.: beta-VAE: learning basic visual concepts with a constrained variational framework. In: International conference on learning representations (2017)
6. Kingma, D.P., Welling, M.: Auto-encoding variational bayes. arXiv preprint arXiv:1312.6114 (2013)
7. Lee, C.H., Liu, Z., Wu, L., Luo, P.: MaskGAN: towards diverse and interactive facial image manipulation. In: Proceedings of the IEEE/CVF Conference on Computer Vision and Pattern Recognition, pp. 5549–5558 (2020)
8. McClelland, J.L., Rumelhart, D.E., Group, P.R., et al.: Parallel Distributed Processing, Volume 2: Explorations in the Microstructure of Cognition: Psychological and Biological Models. vol. 2. MIT press (1987)
9. Park, T., Liu, M.Y., Wang, T.C., Zhu, J.Y.: Semantic image synthesis with spatially-adaptive normalization. In: Proceedings of the IEEE/CVF Conference on Computer Vision and Pattern Recognition, pp. 2337–2346 (2019)
10. Richardson, E., et al.: Encoding in style: a styleGAN encoder for image-to-image translation. In: Proceedings of the IEEE/CVF Conference on Computer Vision and Pattern Recognition, pp. 2287–2296 (2021)
11. Schuster, M., Paliwal, K.K.: Bidirectional recurrent neural networks. IEEE Trans. Sig. Process. **45**(11), 2673–2681 (1997)
12. Tan, Z., et al.: Diverse semantic image synthesis via probability distribution modeling. In: IEEE/CVF Conference on Computer Vision and Pattern Recognition, pp. 7962–7971 (2021)
13. Tan, Z., et al.: Efficient semantic image synthesis via class-adaptive normalization. IEEE Trans. Pattern Anal. Mach. Intell. **44**(9), 4852–4866 (2021)
14. Van Den Oord, A., Vinyals, O., et al.: Neural discrete representation learning. In: Advances in Neural Information Processing Systems. vol. 30 (2017)
15. Vaswani, A., et al.: Attention is all you need. In: Advances in Neural Information Processing Systems. vol. 30 (2017)

16. Wang, Y., Qi, L., Chen, Y.C., Zhang, X., Jia, J.: Image synthesis via semantic composition. In: Proceedings of the IEEE/CVF International Conference on Computer Vision, pp. 13749–13758 (2021)
17. Zhao, S., Song, J., Ermon, S.: Infovae: Information maximizing variational autoencoders. arXiv preprint arXiv:1706.02262 (2017)
18. Zhu, P., Abdal, R., Qin, Y., Wonka, P.: SEAN: image synthesis with semantic region-adaptive normalization. In: IEEE/CVF Conference on Computer Vision and Pattern Recognition (CVPR) (June 2020)

Improved Bilinear Pooling for Real-Time Pose Event Camera Relocalisation

Ahmed Tabia, Fabien Bonardi(✉), and Samia Bouchafa(✉)

IBISC, University Evry, Universite Paris-Saclay, 91025 Evry, France
ahmed.tabia@universite-paris-saclay.fr,
{fabien.bonardi,samia.bouchafabruneau}@univ-evry.fr

Abstract. Traditional methods for estimating camera pose have been replaced by more advanced camera relocalization methods that utilize both CNNs and LSTMs in the field of simultaneous localization and mapping. However, the reliance on LSTM layers in these methods can lead to overfitting and slow convergence. In this paper, a novel approach for estimating the six degree of freedom (6DOF) pose of an event camera using deep learning is presented. Our method begins by preprocessing the events captured by the event camera to generate a set of images. These images are then passed through two CNNs to extract relevant features. These features are multiplied using an outer product and aggregated across different regions of the image after adding L2 normalization to normalize the combining vector. The final step of the model is a regression layer that predicts the position and orientation of the event camera. The effectiveness of this approach has been tested on various datasets, and the results demonstrate its superiority compared to existing state-of-the-art methods.

Keywords: 6-DOF · Deep Learning · Event-based Camera · Pose Estimation

1 Introduction

Determining the camera pose, which involves estimating the position and orientation of the camera based on an observed scene [17], is a crucial challenge in numerous computer vision applications. These applications include autonomous vehicle navigation, robotics, augmented reality, and pedestrian visual positioning systems. Traditional computer vision relocalization techniques can be divided into two main categories: (1) geometric-based and (2) learning-based approaches. Geometric-based methods [14] primarily focus on local feature matching. The typical procedure involves extracting local features from an input image, performing a 2D-3D matching with corresponding 3D points, and subsequently calculating the camera pose with six degrees of freedom, typically using Perspective-n-Point algorithms [9]. The success of these approaches heavily depends on the accuracy of feature extraction and matching processes, which may not always be

G. L. Foresti et al. (Eds.): ICIAP 2023, LNCS 14233, pp. 222–231, 2023.
https://doi.org/10.1007/978-3-031-43148-7_19

satisfactory, particularly in situations with varying illumination conditions [10]. In recent years, the resurgence of deep learning, particularly Convolutional Neural Networks (CNN), has led to a re-examination of many computer vision applications using data-driven approaches. These novel methods have demonstrated high performance across various tasks, including object recognition [4], image classification [12], and segmentation [1]. Although deep learning-based approaches excel at extracting robust features [7], they necessitate a significant amount of training data (typically thousands of images) and substantial computational resources (high-powered and costly GPUs). Consequently, alternative strategies that do not require constant re-computation on such extensive datasets may be more appealing. Techniques such as transfer learning and integration could offer viable solutions to mitigate these challenges. Additionally, both categories of conventional camera relocalization methods continue to struggle with issues such as illumination changes, blur, and featureless images, which make feature extraction difficult. These challenges primarily stem from the characteristics of the input images captured by traditional cameras and subsequently utilized by these methods. Also referred to as neuromorphic cameras, event cameras are imaging sensors designed to detect local changes in brightness. Unlike conventional cameras, their raw output consists of a sequence of asynchronous events (distinct pixel-wise brightness alterations) that correspond to fluctuations in scene illumination. Event cameras offer several advantages over traditional cameras, including high temporal resolution, a broad dynamic range, and the absence of motion blur. These benefits make event cameras particularly well-suited for robotic applications, especially in the realm of pose estimation. Leveraging these benefits, Rebecq et al. [18] introduced a method that combines IMU and event data to estimate the 6DOF camera pose. More recently, a technique called SP-LSTM has been proposed for estimating the 6DOF of an event camera [15]. This approach employs a VGG16 architecture [20] trained from scratch using a stochastic gradient descent algorithm, along with two stacked spatial LSTM layers. While the method demonstrates promising results in camera pose estimation, it requires extensive training time due to the need for retraining the entire network model with the LSTM layer. In this paper, we introduce a novel method that capitalizes on the advantages of both CNNs and bilinear pooling while incorporating normalization layers after the combined vector to enhance the performance of deep learning models when handling event camera data.

Bilinear pooling is a technique that computes the outer product of two feature maps, generating a matrix of features. This process captures the interactions between the features in both feature maps and can be employed to merge information from different temporal scales [11]. Additionally, we apply L2 normalization to normalize the combined vector, which involves dividing each element of the vector by its Euclidean norm, yielding a vector with unit length. Furthermore, we exploit recent advancements in deep learning and utilize the ADAM optimizer [8] alongside the ELU activation function [2]. Our experiments on various datasets demonstrate superior results compared to state-of-the-art methods.

The remainder of this paper is organized as follows: Sect. 2 presents and details our proposed method. Extensive experimental results are provided in Sect. 3. Finally, we conclude the paper and discuss future work.

2 Proposed Method

We propose a novel attention-based enhanced bilinear pooling method for event camera pose relocalization. The suggested attention mechanism allows the model to concentrate on motion-relevant regions within event images. Given a sequence of events captured by an event camera, we first process the raw events and create a set of event images following [15]. The image preprocessing is discussed in Sect. 2. Once the events are converted into images, as depicted in Fig. 2, they are fed into a convolutional neural network. The extracted features are then aggregated using a bilinear pooling vector [11] for pose estimation, followed by the addition of an L2 normalization layer to refine the event image representation by ensuring a consistent scale across different layers of the model. The model architecture details are provided in Sect. 2.1.

Image Preprocessing. In contrast to traditional cameras that capture images at fixed time intervals, event cameras only record a single event whenever there is a change in brightness at a particular pixel. This work aims to address the pose relocalization problem [6,7,15] by transforming the event stream into an event image $I \in \mathbb{R}^{h*w}$, where h and w. The event e is represented as a tuple that captures the change in brightness at a specific location.

$$e =< e_t, (e_x, e_y), e_p >$$

where e_t is the timestamp of the event, (e_x, e_y) is the pixel coordinate and $e_p = \pm 1$ is the polarity that denotes the brightness change at the current pixel. The event image is computed from the event stream as follows:

$$I(e_x, e_y) = \begin{cases} 0 & if\ e_p = -1 \\ 1 & if\ e_p = 1 \end{cases} \tag{1}$$

The second step is to enlarge the image to have 224×224 pixel size, in accordance with the original image's aspect ratio and give it as input to the CNNs. Figure 2 shows an example of event images obtained after the preprocessing from event stream [5]. The preprocessing step plays an important role since it affects the quality of the event images, which are used to train the CNN and estimate the camera relocalisation (Fig. 1).

2.1 The Network Model

In our method, we propose to extract different sets of features from the event image. We employ two CNNs denoted respectively A and B (see Fig. 2). Two

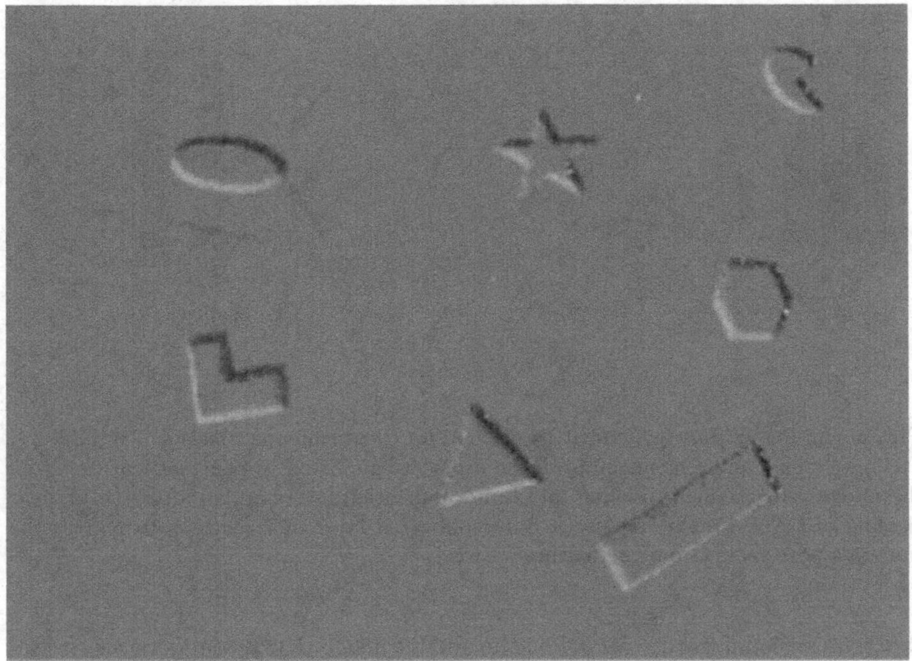

Fig. 1. The results of our event-camera image after preprosessing from point cloud events.

feature maps are extracted from the networks A and B which apply several pooling and non-linear transformations to the original event image. The intuition, behind using two CNNs, is that A and B learn different features from the input image. Then the output of both A and B are combined by a bilinear pooling layer. This layer provides a powerful representation which fuses the two sets of features by leveraging the higher-order information captured in the form of pairwise correlations between the extracted features. In our experiments, we use the pretrained MobileNetV2 [19] model as a first feature extractor A and the pretrained VGG16 [20] as a second feature extractor B. The used MobileNetV2 and VGG16 have already been trained on a very large collection of images from ImageNet [3].

Let us denote the event camera pose by $y = [p, q]$, where $p \in \mathbb{R}^3$ represents the three dimensional camera position and the quaternion $q \in \mathbb{R}^4$ codes the camera orientation. In our experiments, the used CNNs have been pretrained on the ImageNet dataset [3] with input dimensions of $224 \times 224 \times 3$. Deep features are learned from the input event images obtained from the preprocessing step. Both network A and B outputs feature maps represented respectively by the matrix V of dimensionality $n \times d$, and the matrix U of size $m \times d$. Here, n and m are the number of kernels in the output layers of the networks A and B, respectively. The dimensionality of each filter is d; it is obtained by flattening

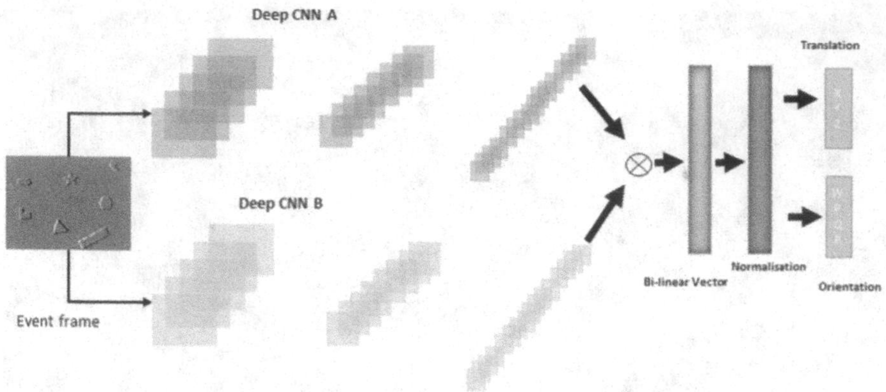

Fig. 2. Our 6DOF pose relocalization method for event cameras consists of three steps. First, we create an event image from the event stream. Next, we extract features from the image using Bilinear pooling and l2 normalization layers added after the bilinear pooling of CNN. Finally, we use a fully connected layer of 7 neurons to regress the camera's pose vector from the feature vector.

the 2-dimensional feature map, i.e., the output image that has undergone several kernel convolutions and pooling transformations. The bilinear pooling operation is then defined as:

$$X = UV^T, U \in \mathbb{R}^{m*d}, V \in \mathbb{R}^{n*d}, X \in \mathbb{R}^{m*n} \tag{2}$$

The connection between the CNN outputs and the bilinear pooling is preceded by an ELU [2] activation function $F(x)$. It is defined as:

$$F(x) = \begin{cases} x & x > 0 \\ \alpha(e^x - 1) & x <= 0 \end{cases} \tag{3}$$

In order to regress the seven-dimensional pose vector, a linear regression layer is added at the end of the model (see Fig. 2 the output layer).

Training settings and loss function
In our method, we train our model using Adam [8] optimizer (with parameters $\beta_1 = 0.9$, and $\beta_2 = 0.999$) to obtain high performance by calculating the adaptive learning rate of each hyper parameter, and to prevent redundancy and get faster gradient update.

We choose the smooth $l1$ loss instead of the mean square loss function in our implementation. The smooth $l1$ loss over n samples is defined as:

$$smoothl1(x, y) = \frac{1}{n} \sum_{i=1}^{n} z_i \tag{4}$$

where z_i is given by :

$$z_i = \begin{cases} 0.5(x_i - y_i)^2/\beta, & \text{if } |x_i - y_i| < \beta \\ |x_i - y_i| - 0.5 * \beta, & \text{otherwise} \end{cases} \tag{5}$$

where x and y are the ground truth and the target camera pose vectors, respectively. β is an optional parameter which specifies the threshold at which to change between l1 and l2 loss. As β varies, the l1 segment of the loss has a constant slope of 1. In our implementation we set β equals to 1.

Following Kendall et al. [7] work, at the test phase we normalize the quaternion to unit length, and utilize Euclidean distance to assess the difference between two quaternions. The distance should be measured in spherical space in theory, but in reality, the deep network produces a predicted quaternion $\widehat{q}$ that is close enough to the groundtruth quaternion q. This makes the difference between the spherical and Euclidean distance insignificant.

3 Experimental Results

Our evaluation of the proposed method used a collection of six real event datasets. In this section, we'll outline the datasets, training setup, and present the results of our experiments.

3.1 Dataset

We conducted experiments on the event camera dataset that was collected by [13]. The dataset includes a collection of scenes captured by a DAVIS240C from minilabs. They contain the cloud of events, images, IMU measurements, and camera calibration from the DAVIS. The groundtruth camera poses are collected from a motion-capture system with sub-millimeter precision at 200Hz. We adopt the timestamp of the motion-capture system to build event frames. All the events with the timestamps between t and $t+1$ of the motion-capture system is grouped as one event image. Without using the loss of generality, we consider the ground-truth pose of this event image as the camera pose shot by the motion capture system at instant $t+1$. This method technically limits the speed of the event camera to the speed of the motion capture system. We Follow the same evaluation protocol as in [15]. The protocol includes two type of splits:

- **The random split** we divided the event images into 70% for training and 30% for testing.
- **The novel split** we used the first 70% of each event for training and the remaining 30% for testing, resulting in two independent sequences of the same scene. The training sequence was selected from timestamp t_0 to t_{70}, and the testing sequence from t_{71} to t_{100}, following Nguyen et al. [15].

Table 1. Comparison between our method results and the results of PoseNet [7] and SP-LSTM [15]. The evaluation is performed using the random split protocol.

	PoseNet [7]		SP-LSTM [15]		Ours	
	Median Error	Average Error	Median Error	Average Error	Median Error	Average Error
shapes rotation	0.109 m, 7.388°	0.137 m, 8.812°	0.025 m, 2.256°	0.028 m, 2.946°	0.014 m, 1.952°	0.017 m, 2.144°
shapes translation	0.238 m, 6.001°	0.252 m, 7.519°	0.035 m, 2.117°	0.039 m, 2.809°	0.020 m, 1.396°	0.027 m, 1.835°
box translation	0.193 m, 6.977°	0.212 m, 8.184°	0.036 m, 2.195°	0.042 m, 2.486°	0.024 m, 1.132°	0.026 m, 1,219°
dynamic 6dof	0.297 m, 9.332°	0.298 m, 11.242°	0.031 m, 2.047°	0.036 m, 2.576°	0.025 m, 1.730°	0.028 m, 2.267°
hdr poster	0.282 m, 8.513°	0.296 m, 10.919°	0.051 m, 3.354°	0.060 m, 4.220°	0.038 m, 2.154°	0.047 m, 2,686°
poster translation	0.266 m, 6.516°	0.282 m, 8.066°	0.036 m, 2.074°	0.041 m, 2.564°	0.024 m, 1.557°	0.032 m, 1.959°
Average	0.231 m, 7.455°	0.246 m, 9.124°	0.036 m, 2.341°	0.041 m, 2.934°	**0.024 m, 1.653°**	**0.029 m, 1.951°**

Table 2. Comparison between our method results and the results of PoseNet [7] and SP-LSTM [15]. The evaluation is performed using the novel split protocol.

	PoseNet [7]		SP-LSTM [15]		Ours	
	Median Error	Average Error	Median Error	Average Error	Median Error	Average Error
shapes rotation	0.201 m, 12.499°	0.214 m, 13.993°	0.045 m, 5.017°	0.049 m, 11.414°	0.029 m, 3.281°	0.038 m, 5.713°
shapes translation	0.198 m, 6.969°	0.222 m, 8.866°	0.072 m, 4.496°	0.081 m, 5.336°	0,060 m, 4,033°	0,067 m,4,943°
shapes 6dof	0.320 m, 13.733°	0.330 m, 18.801°	0.078 m, 5.524°	0.095 m, 9.532°	0.068 m, 4.886°	0.081 m, 6.653°
Average	0.240 m, 11.067°	0.255 m, 13.887°	0.065 m, 5.012°	0.075 m, 8.761°	**0.052 m, 4.066°**	**0.062 m, 5,769°**

3.2 Training Environment

In order to evaluate the performance of our proposed method, we conduct several experiments and compare the results with deep learning architectures utilizing state-of-the-art LSTM models and CNN . Once the preprocessing stage is complete, patches of 224*224 pixels are taken from each frame and fed into the CNNs in a patchlevel dataset. Our experiments were implemented using Pytorch [16] and conducted on a platform with an Intel(R) Xeon(R) CPU @ 2.00GHz processor, 24GB of CPU memory, and a single Tesla T4 GPU. The networks were trained over 350 epochs with a learning rate of $2\exp-3$ with momentum-decay equals to $4\exp-3$ and a weight decay set to 0.

3.3 Results

We use the same protocol of comparison reported in [15] and used in PoseNet [7]. As quantitative evaluation, we choose to calculate the median and average error of the predicted pose in position and orientation. The Euclidean distance is used to compare the predicted position to the groundtruth, and the anticipated orientation is normalized to unit length before being compared to the groundtruth. For location and orientation, the median and average error are recorded in m and deg(°), respectively.

Comparison with State-of-the-art Methods. We report the comparison results between our method explained on the Sect. 2.1 and the state of the art models namely PoseNet [7] and SP-LSTM [15] using CNN and LSTM.

Random Split. In the experiment described in Table 1, the results were obtained through the use of a random split strategy. The study involved 6 sequences, including shapes rotation, box translation, shapes translation, dynamic 6DOF, HDR poster, and poster translation. Our proposed model showed the best results in all sequences with the lowest mean and average errors. Our model achieved an average median error of 0.024 m and 1.653° in the real dataset, compared to the SP-LSTM with an error of 0.036 m and 2.341° and PoseNet with an error of 0.231 m and 7.455°.

Novel Split. Table 2 presents comparison results from the novel split presented in Sect. 3.1. One can notice from this table that the novel split is more difficult to handle than the random split the errors from all methods are bigger than errors reported with the random split. We use three sequences from the shapes scene (**shapes rotation, shapes translation, shapes 6dof**) in this novel split. The results of our method are superior to the results obtained with state of the art methods. It achieves 0.061 m and 4, 066° in average median error of all sequence of the real dataset while the most recently method SP-LSTM result 0.065 m, 5.012°, and 0.240 m, 11.067° from PoseNet.

We recall that in the novel split, the testing set is selected from the last 30% of the event images. This means we do not have the "neighborhood"' relationship between the testing and training images. In the random split, the testing images can be very close to the training images since we select the images randomly from the whole sequence for training/testing.

To conclude, the extensive experimental results from both the random split and novel split setup show that our method successfully relocalizes the event camera pose using only the event image coming from the cloud of polarity. The critical reason for the improvement is using bilinear pooling to learn the spatial relationship features in the event image . The experiments using the novel split setup also confirm that our approach successfully encodes the scene's geometry during the training and generalizes well during the testing. Furthermore, our network also has a speedy inference time and requires only the event image as the input to relocalize the camera pose.

4 Conclusion

This paper introduces a deep learning-based method for estimating the 6DOF pose of an event camera. The process begins with preprocessing the events to generate event images, which are then processed using a deep convolutional neural network with Bilinear pooling to extract features. The features are aggregated and undergo L2 normalization before being fed into a fully connected layer for pose regression. Our method uses the Adam optimizer and ELU activation functions during training, leading to fast inference and accurate pose relocalization using only the event image. Our approach outperforms recent works, including LSTM-based architectures, and demonstrates good generalization performance on publicly available datasets.

References

1. Badrinarayanan, Vijay, Kendall, Alex, Cipolla, Roberto: SegNet: a deep convolutional encoder-decoder architecture for image segmentation. IEEE Trans. Pattern Anal. Mach. Intell. **39**(12), 2481–2495 (2017)
2. Clevert, D.-A., Unterthiner, T., Hochreiter, S.: Fast and accurate deep network learning by exponential linear units (ELUs). arXiv preprint arXiv:1511.07289 (2015)
3. Deng, J., Dong, W., Socher, R., Li, L.-J., Li, K., Fei-Fei, L.: ImageNet: a large-scale hierarchical image database. In: 2009 IEEE Conference on Computer Vision and Pattern Recognition, pp. 248–255. IEEE (2009)
4. Eitel, A., Springenberg, J.T., Spinello, L., Riedmiller, M., Burgard, W.: Multimodal deep learning for robust RGB-D object recognition. In: 2015 IEEE/RSJ International Conference on Intelligent Robots and Systems (IROS), pp. 681–687. IEEE (2015)
5. Gallego, Guillermo, Scaramuzza, Davide: Accurate angular velocity estimation with an event camera. IEEE Robot. Autom. Lett. **2**(2), 632–639 (2017)
6. Kendall, A., Cipolla, R.: Modelling uncertainty in deep learning for camera relocalization. In: 2016 IEEE International Conference on Robotics and Automation (ICRA), pp. 4762–4769. IEEE (2016)
7. Kendall, A., Grimes, M., Cipolla, R.: PoseNet: a convolutional network for real-time 6-DOF camera relocalization. In: Proceedings of the IEEE International Conference on Computer Vision, pp. 2938–2946 (2015)
8. Kingma, D.P., Ba, J.: Adam: A method for stochastic optimization. arXiv preprint arXiv:1412.6980 (2014)
9. Lepetit, Vincent, Moreno-Noguer, Francesc, Fua, Pascal: EPnP: an accurate o(n) solution to the PnP problem. Int. J. Comput. Vis. **81**(2), 155–166 (2009)
10. Li, Ming, Chen, Ruizhi, Liao, Xuan, Guo, Bingxuan, Zhang, Weilong, Guo, Ge.: A precise indoor visual positioning approach using a built image feature database and single user image from smartphone cameras. Remote Sens. **12**(5), 869 (2020)
11. Lin, T.-Y., RoyChowdhury, A., Maji, S.: Bilinear CNN models for fine-grained visual recognition. In: Proceedings of the IEEE International Conference on Computer Vision, pp. 1449–1457 (2015)
12. Mahajan, D., et al.: Exploring the limits of weakly supervised pretraining. In: Proceedings of the European Conference on Computer Vision (ECCV), pp. 181–196 (2018)
13. Mueggler, Elias, Rebecq, Henri, Gallego, Guillermo, Delbruck, Tobi, Scaramuzza, Davide: The event-camera dataset and simulator: event-based data for pose estimation, visual odometry, and slam. The Int. J. Robot. Res. **36**(2), 142–149 (2017)
14. Mur-Artal, R., Tardós, J.D.: ORB-SLAM2: an open-source slam system for monocular, stereo, and RGB-D cameras. IEEE Trans. Robot. **33**(5), 1255–1262 (2017)

15. Nguyen, A., Do, T.-T., Caldwell, D.G., Tsagarakis, N.G.: Real-time 6DOF pose relocalization for event cameras with stacked spatial LSTM networks. In: Proceedings of the IEEE/CVF Conference on Computer Vision and Pattern Recognition Workshops, pp. 0–0 (2019)
16. Paszke, A., et al.: PyTorch: an imperative style, high-performance deep learning library. Adv. Neural Inf. Process. Syst. **32**, 8024–8035 (2019)
17. Qu, C., Shivakumar, S.S., Miller, I.D., Taylor, C.J.: DSOL: A fast direct sparse odometry scheme. arXiv preprint arXiv:2203.08182 (2022)
18. Rebecq, H., Horstschaefer, T., Scaramuzza, D.: Real-time visual-inertial odometry for event cameras using keyframe-based nonlinear optimization (2017)
19. Sandler, M., Howard, A., Zhu, M., Zhmoginov, A., Chen, L.-C.: MobileNetv2: inverted residuals and linear bottlenecks. In: Proceedings of the IEEE Conference on Computer Vision and Pattern Recognition, pp. 4510–4520 (2018)
20. Simonyan, K., Zisserman, A.: Very deep convolutional networks for large-scale image recognition. arXiv preprint arXiv:1409.1556 (2014)

End-to-End Asbestos Roof Detection on Orthophotos Using Transformer-Based YOLO Deep Neural Network

Cesare Davide Pace[1](✉), Alessandro Bria[1], Mariano Focareta[2], Gabriele Lozupone[1], Claudio Marrocco[1], Giuseppe Meoli[2], and Mario Molinara[1]

[1] Department of Electrical and Information Engineering, University of Cassino and Southern Latium, Via G. Di Biasio 43, Cassino 03043, Italy
{cesaredavide.pace,a.bria,gabriele.lozupone,c.marrocco, m.molinara}@unicas.it

[2] MAPSAT srl, c.da Piano Cappelle 129, 82100 Benevento, BN, Italy
{g.meoli,m.focareta}@mapsat.it

Abstract. Asbestos, a hazardous material associated with severe health issues, requires accurate identification for safe management and removal. This study presents a novel end-to-end deep learning approach using a transformer-based YOLOv5 network for detecting asbestos roofs in high-resolution orthophotos, filling a gap in the scientific literature where end-to-end solutions are lacking. The model is trained on a dataset containing orthophotos with various roof types and conditions around Pisa in Italy. The transformer-based YOLO architecture enhances the detection capabilities compared to traditional CNNs. The proposed method demonstrates high accuracy in asbestos roof detection, outperforming traditional remote sensing techniques, and offers an effective, automated solution for targeting removal efforts and mitigating associated health risks. This end-to-end approach fills a gap in the existing literature and presents a promising direction for future research in asbestos roof detection.

Keywords: Asbestos · Deep Learning · Transformer · Yolo · Object Detection

1 Introduction

Asbestos is a hazardous material that poses significant health risks when its fibers become airborne. Numerous studies have demonstrated the dangers of asbestos for humans. According to the National Toxicology Program, studies in humans have shown that exposure to asbestos causes respiratory-tract cancer, mesothelioma of the lung and abdominal cavity (pleural and peritoneal mesothelioma), and cancer at other tissue sites [1]. The World Health Organization also states that all forms of asbestos are carcinogenic to humans, and exposure to

G. L. Foresti et al. (Eds.): ICIAP 2023, LNCS 14233, pp. 232–244, 2023.
https://doi.org/10.1007/978-3-031-43148-7_20

asbestos causes cancer of the lung, larynx, and ovaries, as well as mesothelioma [2]. Detecting and removing asbestos roofs is crucial for public health and safety. This article presents an end-to-end system able to detect asbestos roofs, assuming as input the entire orthophotos and generating as output a report with bounding boxes that delimitates it. When conducting studies to map roofs that contain asbestos materials, it is common practice to pre-identify the roofs within the image before verification. The solutions typically employed involve using classifiers capable of distinguishing between the various classes of roofs. The proposed approach work as an object detector based on the YOLO [3] (You Only Look Once) network, in particular with YOLOv5 [4] combined with Transformer [5] and Swin Transformer [6]. The recent success of YOLO architectures in speed and accuracy makes them an ideal choice for this task. In addition to these advanced machine-learning techniques, our approach also benefits from using a previously unused dataset of optical orthophotos. This dataset provides high-resolution aerial imagery (HRAI), whose band is only RGB, that allows for the accurate detection of asbestos roofs by combining this data with the powerful object detection capabilities of YOLO architectures. The presented results demonstrate the effectiveness of our approach in detecting asbestos roofs in orthophotos with only RGB data.

2 Related Works

Several studies have used deep learning techniques to detect asbestos roofs. Among the methods employed for detecting asbestos-containing roofing materials using remote sensing imagery and machine learning-based image analysis, three approaches have been adopted for mapping asbestos roofs: (1) pixel-based image analysis (PBIA), (2) object-based image analysis (OBIA), and (3) a Deep-Learning-based approach (DL). In PBIA methods, individual pixels are the analysis units classified based on spectral values without considering the spatial and contextual information. Finally, DL models have shown promising results in detecting asbestos-containing roofs, as cited in [7]. Other approaches for asbestos roofing recognition involve Convolutional Neural Networks (CNNs) and HRAI. This approach aims to show the feasibility of identifying asbestos cement roofs on HRAI with CNNs [8]. In [9] CNNs are used for feature extraction, applying Support Vector Machines for classification on hyperspectral images. Another study proposes using Drone-Based aerial imagery and a faster region-based CNN (Faster R-CNN) [10]. Combination of Mask R-CNN with HRAI also can provide efficient methods for the large-scale detection of asbestos-containing material as roofing [11]. YOLOv5 has shown promising results in detecting objects of interest in satellite imagery. For instance, YOLOv5 has been used to detect aircraft in satellite images with a Precision of 0.930 and Recall of 0.788, demonstrating its high accuracy and robustness [12]. Moreover, YOLOv5 has been combined with self-attention modules to improve its performance in detecting fire and smoke, achieving a Precision of 0.797, Recall of 0.672, mAP50 of 0.734, and mAP50-95 of 0.427 [13]. Similarly, YOLOv5 has been used for landslide detection, achieving a Precision of 0.784, Recall of 0.762, mAP50 of 0.740, and mAP50-95 of 0.308 [14].

3 Proposal

Object detection has become increasingly important in computer vision in recent years. YOLO stands out for its accuracy and inference speed among the various algorithms developed. This paper proposes using YOLO with some variations for large-scale object detection tasks where inference speed is critical. In most studies conducted for mapping roofs that contain asbestos materials, the roofs to be verified are previously identified within the image. The considered solutions are simply classifiers able to distinguish between the different classes of roofs. Our paper proposes a network capable of detecting and classifying roofs without any prior information about their positions, finding the roofs within the image autonomously [8,11]. During the study it was chosen to use YOLOv5 instead of YOLOv7 [15] and YOLOv8 [16]. This decision was primarily motivated by the fact that YOLOv7 was associated with numerous issues during the time of our experiments. Additionally, YOLOv8 had not yet been released at the time of our study. The following describes the dataset and the architectures used.

3.1 Dataset

Using DL for object detection in remote images presents several challenges. These include the limited availability of labeled data, the limited spatial resolution of some images, the presence of noise and distortions in the images, and the diversity of the objects to detect. These challenges can affect the performance of Machine Learning (ML) models. To identify and map asbestos roofs, a dataset supplied by MAPSAT s.r.l. located in Benevento (Italy), was utilized. The MAPSAT dataset used in this study consists of 81 HRAIs from the city of Pisa (Italy) that come from the RealVista service provided by e-Geos and have a size of 5094×3549 pixels and a spatial resolution of 0.5 m per pixel. These images contain a total of 719 asbestos roofs. An example is shown in Fig. 1

The image annotations have been provided from MAPSAT adopting the PASCAL VOC format [17]. The annotations consist of a series of XML files, one for each image in the dataset. Each XML file contains a list of object instances in the corresponding image, their class labels, and bounding box coordinates. The x and y coordinates of the top-left and bottom-right corners of the box relative to the image dimensions define the bounding boxes.

One aspect that can be analyzed using HRAI is the size of the roofs, as small objects are often present in these images. An analysis of the size of the asbestos roofs respect to COCO standard (small, medium, large), in Table 1, showed that medium-sized objects are prevalent. This information can help us understand the distribution and characteristics of asbestos roofs in the studied area.

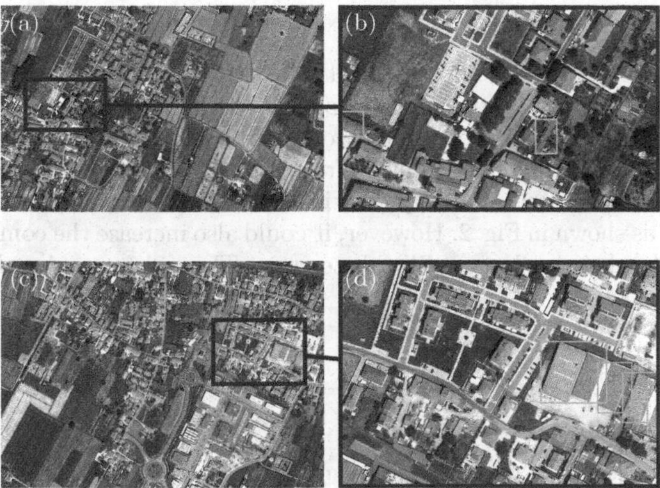

Fig. 1. Example of images of the dataset (green boxes mark the ground truth). (Color figure online)

Table 1. Table showing the size of objects in the MAPSAT dataset.

	Min scale area	Max scale area	Instances	%
Small	0×0	32×32	103	14.3 %
Medium	32×32	96×96	437	60.9 %
Large	96×96	$\infty \times \infty$	176	24.5 %

3.2 YOLOv5

YOLOv5 is an object detection architecture developed by Ultralytics and released in June 2020. It is part of a general computer vision architecture commonly used to detect objects. It uses a single stage to detect objects in an image, rather than a two-stage approach like some other object detection models, making it very fast at the inference stage. YOLOv5 consists of three main parts: a backbone, a neck, and a head. The backbone is responsible for feature extraction and is based on a Cross Stage Partial (CSP) [20] and a Spatial Pyramid Pooling Fast (SPPF), a version of Spatial Pyramid Pooling(SPP) [21] proposed by YOLOv5 that is faster than SPP. As the Neck, Yolov5 implements the Path Aggregation Network (PANet) [19] for aggregating extracted features. The head for predicting classes and bounding boxes is the same one used in YOLOv3 [18] and YOLOv4 [22] heads. YOLOv5 has five derived architectures: YOLOv5n, YOLOv5s, YOLOv5m, YOLOv5l, and YOLOv5x that share the same architecture but vary in their widths and depths. The architecture used in this study is the *s* version that has {width_multiple:0.33} and {depth_multiple:0.33}.

3.3 YOLOv5-P2

In object detection, the prediction head is a part of the model that takes the feature maps generated by the convolutional layers and produces predictions for bounding boxes and class probabilities for each object in the image. The stride of YOLO's standard architecture goes from 16 to 64. In trying to improve the detection of small objects, a new prediction head with stride 4 (termed P2) has been added, as shown in Fig. 2, However, it could also increase the computational cost and make the model more difficult to train. The width remains the same as version *s*.

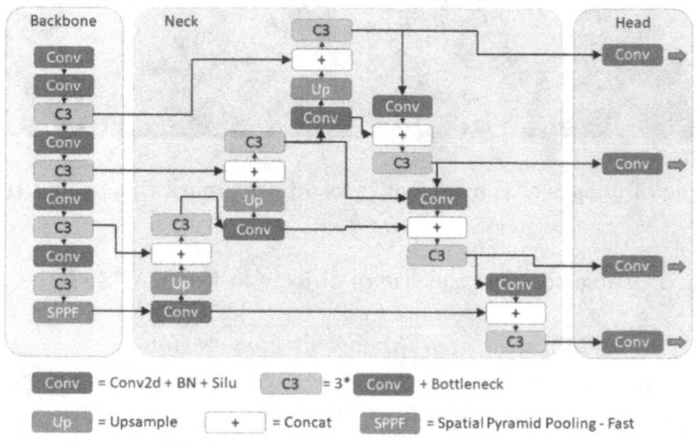

Fig. 2. YOLOv5-P2 architecture.

3.4 YOLOv5 and YOLOv5-P2 with Transformer

Recently, transformer-based architectures have achieved impressive results in various tasks, including natural language processing [5] and computer vision [26]. Transformer models use self-attention mechanisms to capture long-range dependencies in the input data, which allows them to outperform traditional CNNs on specific tasks. In object detection, transformer-based models have been shown to improve performance and reduce the number of parameters compared to CNNs [23]. This study proposes to integrate a transformer module(C3TR) into the backbone of YOLOv5 to improve the performance of the object detection system, substituting it for the last C3 module at the end of the backbone. This change was used in both YOLOv5 and YOLOv5-P2 architectures.

C3TR Module. The C3TR module utilizes the Patch Embedding operation to divide the input image into blocks of a specified size, which are then combined into a sequence and passed through the Transformer Encoder for feature extraction as shown in Fig. 3a. The transformer encoder blocks, shown in Fig. 3b, are composed of two sub-layers: a multi-head attention layer and a fully-connected layer, connected with residual connections. The vision transformer inspired this modification. It was intended to improve the model's ability to capture global and contextual information, as well as its ability to handle occluded objects.

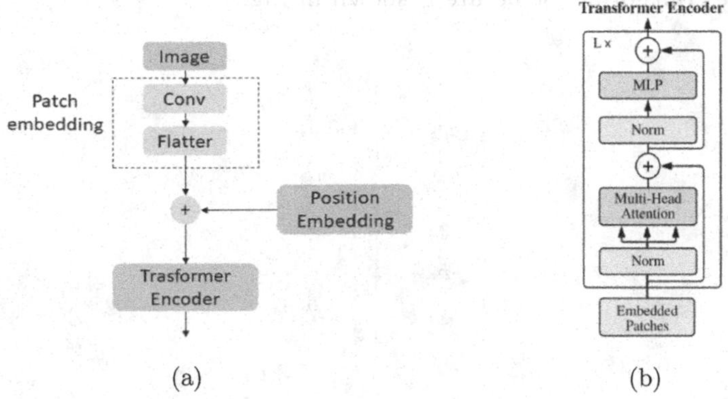

Fig. 3. (a) C3TR structure diagram. (b) Transformer encoder block

3.5 YOLOV5 with Swin Transformer

Our experiments with the Transformer Encoder Block indicated that it consumes significant computational resources when processing high-resolution images, such as satellite images because the computational complexity of Multi-head Attention in a transformer is proportional to the quadratic image size. To improve this complexity, the Transformer Encoder Block has been substituted the Swin Transformer Encoder Block [6] as an alternative to the Transformer Encoder Block, which has a reduced computational complexity since the window size is much smaller than the image size.

C3STR Module. The C3STR module has the same structure as the C3TR but with the Swin Transformer Encoder block.

The Swin Transformer encoder block consists of self-attention layers interleaved with shift-and-widen operations. The shift-and-widen operation works by shifting the channels of the input feature map along the spatial dimensions and then concatenating the shifted versions with the original input to create a more comprehensive feature map. This operation allows the Swin Transformer to efficiently model long-range dependencies in the input data while also increasing the model's capacity.

In each self-attention layer, the Swin Transformer encoder block applies multi-head self-attention to the input feature map, dividing the input feature map into multiple "heads" and applying self-attention separately to each head. The output of the self-attention layers is then combined using a linear projection, followed by a layer normalization and a ReLU activation function.

The Swin Transformer encoder block also includes skip connections, allowing it to incorporate low-level and high-level features from the input data. These skip connections are implemented as residual connections, where the output of the self-attention layers is added to the input feature map. The schematic diagram of the Swin Transformer structure is shown in Fig. 4.

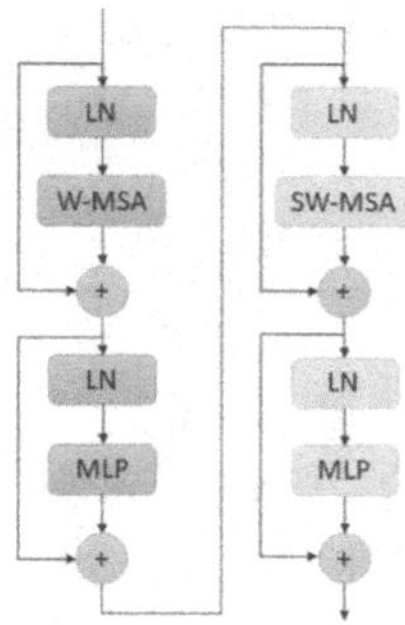

Fig. 4. Swin Transformer structure.

3.6 YOLOv5 with CBAM Module

In this new architecture, we propose to improve the YOLOv5 model by integrating the Convolutional Block Attention Module (CBAM) [24] into its architecture. Numerous studies have demonstrated that incorporating CBAM modules as attention blocks can enhance the performance of satellite imagery applications [25,27]. So compared to previous architectures, the current study has incorporated the CBAM module, C3TR, and C3STR blocks in the neck, as shown in Fig. 5, that for the version with C3STR just change the C3TR module to C3STR.

4 Methods and Material

4.1 Splitting

Given the high dimension of the images in the dataset, it was necessary to use the splitting technique. *Splitting* is a technique that involves dividing a large satellite image into smaller overlapping regions or overlays. This can be useful for deep learning object detection, as it allows the model to be trained and tested

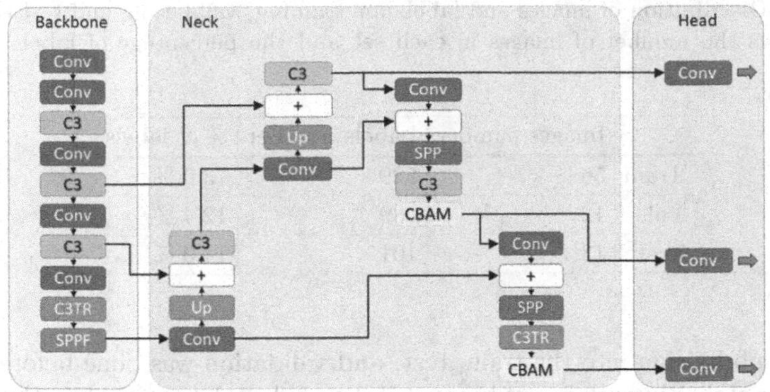

Fig. 5. YOLOv5 architecture with CBAM and Transformer block.

on smaller, more manageable pieces of data. There are several benefits to using splitting for object detection on large images. First, it allows the model to be trained and tested on smaller datasets, which can be more efficient and require fewer computational resources. This can be particularly useful when working with limited hardware or when the data size needs to be smaller to be processed reasonably. Second, splitting allows the model to learn local features and patterns in the data, which can help detect objects in different contexts. By dividing the image into smaller overlays, the model can learn to detect objects in different parts rather than relying on global features that may not be relevant in all parts of the image. Finally, splitting can also improve the performance of the object detection model by reducing the impact of noise and distortions in the data. By dividing the image into smaller overlays, the model can be trained on more homogeneous data, which can help to reduce the impact of noise and distortions on the model's performance.

This study has tested using different sizes for the subdivision of large orthophoto images for deep-learning object detection. The goal was to find a balance between two competing problems: using a patch size that is too large, which can make the images too large to be processed by the network, and using a patch size that is too small, which can lead to a fragmentary division of the ground truths and potentially miss some objects.

4.2 Train, Validation, and Test

The data were split into training, validation, and testing. The distribution chosen is 70% for training, 15% for validation, and 15% for testing. Table 2 illustrates the distribution of annotations among the different sets.

Table 2. Distribution of images and labels for training, validation, and test sets. The table shows the number of images in each set and the percentage of labels for each image.

	Images number	Labels number	% of labels
Train	56	529	73.6 %
Val	12	89	12.4 %
Test	13	101	14.0 %

This subdivision into the train, test, and validation was done before identifying the patches in each image because if it was done later, avoiding that, given the overlap, a label present in an image in the training set could also be found in the validation set or test set.

4.3 Material

Models were trained and tested on a workstation with Ubuntu Linux 20.04, 256 GB of RAM, an Intel(R) Xeon(R) Silver 4110 CPU @ 2.10 GHz, and NVIDIA Tesla V100-PCIE GPU with 16 GB memory. As for the hyperparameters, they were selected with a set of preliminary experiments and kept constant across all the experiments in this paper except the batch size. This was modified using the auto-batch function to automatically select the largest possible batch size to fully utilize the available RAM memory on the GPU, given the varying image sizes.

5 Experiments and Results

5.1 Splitting into Patches and YOLOv5 Results

As previously discussed in Sect. 4.1, several challenges are associated with choosing the size of images during the splitting. To address these issues, various sizes were tested, and a subdivision with a 50% overlap was used to avoid losing any objects.

The experiments for determining the optimal image dimension for the dataset were conducted using the standard architecture of YOLOv5 by varying input images. The results are shown in Table 3 and show how the choice of subdivision in 1280×1280 is the most effective. The tests are displayed in Precision, Recall, Average Precision at Intersect over Union (IoU) 0.5, and Average Precision over different IoU thresholds, from 0.5 to 0.95.

An additional experiment was conducted to improve the detection of small labels by maintaining the image size at 1280×1280 but increasing the size of the first layer to 1920×1920. As shown in Table 4, this modification resulted in an improvement in the detection of objects.

Analyzing the results of varying patch sizes in this section, it was used a resolution of 1280×1280 and the size of the first layer of 1920×1920 for the

Table 3. Results by varying the size of the patches.

Model	Image size	P	R	AP^{50}	AP^{50} - TTA
yolov5s	640×640	0.327	0.383	0.304	0.314
yolov5s	1280×1280	0.392	0.400	0.340	0.345
yolov5s	1920×1920	0.330	0.525	0.333	0.338

Table 4. Results maintaining the image size at 1280×1280 but increasing the size of the first layer to 1920×1920.

Model	Image size	P	R	AP^{50}	AP^{50} - TTA
yolov5s	1280×1280 -> 1920×1920	0.392	0.453	0.377	0.352

network. Using these parameters, we evaluated the performance of the different architectures.

5.2 YOLOv5 Results

In Table 5, the reported results were performed in the normal mode and the Test-Time Augmentation(TTA), which involves creating multiple augmented copies of each image in the test set, having the model predict each, then returning ensemble of those predictions. In this case, TTA flipped the images left and right and processed them at three different resolutions.

Table 5. YOLOv5 results.

Model	Normal				TTA			
	P	R	AP^{50}	AP^{50-95}	P	R	AP^{50}	AP^{50-95}
yolov5s	0.392	0.453	0.377	0.166	0.396	0.432	0.365	0.128
yolov5s-p2	0.380	0.336	0.289	0.126	0.461	0.323	0.300	0.137
yolov5s + Transf	0.397	**0.505**	**0.413**	0.201	0.363	0.436	0.380	0.206
yolov5s-p2 + Transf	0.401	0.436	0.363	0.159	**0.472**	0.327	0.371	0.179
yolov5s + Swin	**0.454**	0.362	0.354	0.137	0.465	0.389	0.337	0.128
yolov5s + Transf + CBAM	0.338	0.500	0.374	0.199	0.454	0.411	0.382	**0.218**
yolov5s + Swin + CBAM	0.388	0.468	0.395	**0.204**	0.439	**0.515**	**0.414**	0.214

6 Conclusions and Future Works

This paper presents an end-to-end system for asbestos roof detection from orthophotos based on YOLO networks. The object detection system can work directly on large-scale images without any prior indication, producing bounding boxes containing the candidate asbestos roofs as output. One of the system's goals was to have a low computational burden, capable of processing very large areas in a reasonable time. The images in the test set are 13 with a size of 5049×3549 and a resolution of 0.5 m per pixel covering a surface of about 58 km^2. Considering that this test set is processed in 13s, we estimated that an area as large as an Italian region like Tuscany (22 987,04 km^2) could be processed in less than two hours. An important element that emerged from the experiments is that, upon further verification, about 3% of the roofs identified as False Positives in this initial experiment were, in fact, asbestos roofs. This was confirmed through additional checks carried out by MAPSAT experts and, in some cases, through an actual site visit. To improve the performances, different directions will be explored. First, the collaboration between the University of Cassino and MAPSAT will aim to increase the number of images available in the dataset. A new version of the ground truth will also be generated based on the new True Positives from the experiments and verified by MAPSAT experts. Other network architectures will be used for comparisons, such as Fast R-CNN, Region-based CNN, Single Shot Detector (SSD), and those based on Focal Loss like RetinaNet. Different network initialization methodologies will also be tested, including training from scratch. Finally, a FP reduction module based on cadastral maps will be adopted to help reduce the FPs generated by gray surfaces (asphalt, concrete, etc.).

References

1. National Toxicology Program. RoC Profile: Asbestos; 15th RoC (2021). https://ntp.niehs.nih.gov/ntp/roc/content/profiles/asbestos.pdf
2. World Health Organization. Asbestos: elimination of asbestos-related diseases, 15 February 2018. https://www.who.int/news-room/fact-sheets/detail/asbestos-elimination-of-asbestos-related-diseases
3. Redmon, J., Divvala, S., Girshick, R., Farhadi, A.: You only look once: unified, real-time object detection. arXiv preprint arXiv:1506.02640 (2015)
4. Jocher, G.: ultralytics/yolov5: v3.1 - bug fixes and performance improvements. Zenodo, October 2020. https://doi.org/10.5281/zenodo.4154370
5. Vaswani, A.: Attention is all you need. arXiv preprint arXiv:1706.03762 (2017)
6. Liu, Z.: Hierarchical vision transformer using shifted windows. arXiv:2103.14030 (2021)
7. Abbasi, M., Mostafa, S., Vieira, A.S., Patorniti, N., Stewart, R.A.: Mapping roofing with asbestos-containing material by using remote sensing imagery and machine learning-based image classification: a state-of-the-art review. Sustainability **14**, 8068 (2022). https://www.mdpi.com/2071-1050/14/13/8068
8. Raczko, E., Krówczyńska, M., Wilk, E.: Asbestos roofing recognition by use of convolutional neural networks and high-resolution aerial imagery. Testing different scenarios. Comput. Educ. (2022). https://doi.org/10.1016/j.buildenv.2022.109092
9. Teng-To, Yu., Lin, Y.-C., Lan, S.-C., Yang, Y.-E., Pei-Yun, W., Lin, J.-C.: Mapping asbestos-cement corrugated roofing tiles with imagery cube via machine learning in Taiwan. Remote Sens. **14**(14), 3418 (2022). https://doi.org/10.3390/rs14143418

10. Seo, D.-M., Woo, H.-J., Kim, M.-S., Hong, W.-H., Kim, I.-H., Baek, S.-C.: Identification of asbestos slates in buildings based on faster region-based convolutional neural network (faster R-CNN) and drone-based aerial imagery. Drones **6**, 194 (2022). https://doi.org/10.3390/drones6080194
11. Hikuwai, M.V., Patorniti, N., Vieira, A.S., Frangioudakis Khatib, G., Stewart, R.A.: Artificial intelligence for the detection of asbestos cement roofing: an investigation of multi-spectral satellite imagery and high-resolution aerial imagery. Sustainability **15**, 4276 (2023). https://www.mdpi.com/2071-1050/15/5/4276
12. Jindal, M., Raj, N., Saranya, P., Sundarabalan, V.: Aircraft detection from remote sensing images using YOLOV5 architecture. In: 2022 6th International Conference on Devices, Circuits and Systems (ICDCS), pp. 332–336 (2022). https://doi.org/10.1109/ICDCS54290.2022.9780777
13. Zhang, S., Zhang, F., Ding, Y., Li, Y.: Swin-YOLOv5: research and application of fire and smoke detection algorithm based on YOLOv5. Comput. Intell. Neurosci. **2022**, 6081680 (2022). https://doi.org/10.1155/2022/6081680
14. Wang, T., Liu, M., Zhang, H., Jiang, X., Huang, Y., Jiang, X.: Landslide detection based on improved YOLOv5 and satellite images. In: 2021 4th International Conference on Pattern Recognition and Artificial Intelligence (PRAI), pp. 367–371 (2021). https://doi.org/10.1109/PRAI53619.2021.9551067
15. Wang, C., Bochkovskiy, A., Liao, H.: YOLOv7: trainable bag-of-freebies sets new state-of-the-art for real-time object detectors (2022). https://doi.org/10.48550/arXiv.2207.02696
16. Jocher, G., Chaurasia, A., Qiu, J.: YOLO by Ultralytics (2023). https://github.com/ultralytics/ultralytics
17. Everingham, M., Van Gool, L., Williams, C., Winn, J., Zisserman, A.: The PASCAL Visual Object Classes (VOC) challenge. Int. J. Comput. Vis. **88**, 303–338 (2010)
18. Redmon, J., Farhadi, A.: YOLOv3: an incremental improvement. CoRR. abs/1804.02767 (2018). http://arxiv.org/abs/1804.02767
19. Liu, S., Qi, L., Qin, H., Shi, J., Jia, J.: Path aggregation network for instance segmentation (2018). https://doi.org/10.48550/arXiv.1803.01534
20. Wang, C., Liao, H., Yeh, I., Wu, Y., Chen, P., Hsieh, J.: CSPNet: a new backbone that can enhance learning capability of CNN (2019). https://arxiv.org/abs/1911.11929
21. He, K., Zhang, X., Ren, S., Sun, J.: Spatial pyramid pooling in deep convolutional networks for visual recognition. In: Fleet, D., Pajdla, T., Schiele, B., Tuytelaars, T. (eds.) ECCV 2014. LNCS, vol. 8691, pp. 346–361. Springer, Cham (2014). https://doi.org/10.1007/978-3-319-10578-9_23
22. Bochkovskiy, A., Wang, C., Liao, H.: YOLOv4: optimal speed and accuracy of object detection. CoRR. abs/2004.10934 (2020). https://arxiv.org/abs/2004.10934
23. Carion, N., Massa, F., Synnaeve, G., Usunier, N., Kirillov, A., Zagoruyko, S.: End-to-end object detection with transformers. arXiv (2020). https://arxiv.org/abs/2005.12872
24. Woo, S., Park, J., Lee, J., Kweon, I.: CBAM: convolutional block attention Module. CoRR. abs/1807.06521 (2018). http://arxiv.org/abs/1807.06521
25. Zhu, X., Lyu, S., Wang, X., Zhao, Q.: TPH-YOLOv5: improved YOLOv5 based on transformer prediction head for object detection on drone-captured scenarios. CoRR. abs/2108.11539 (2021). https://arxiv.org/abs/2108.11539

26. Betancourt Tarifa, A.S., Marrocco, C., Molinara, M., et al.: Transformer-based mass detection in digital mammograms. J. Ambient Intell. Hum. Comput. **14**, 2723–2737 (2023). https://doi.org/10.1007/s12652-023-04517-9
27. Gong, H., et al.: Swin-transformer-enabled YOLOv5 with attention mechanism for small object detection on satellite images. Remote Sens. **14** (2022). https://www.mdpi.com/2072-4292/14/12/2861

OpenFashionCLIP: Vision-and-Language Contrastive Learning with Open-Source Fashion Data

Giuseppe Cartella[1], Alberto Baldrati[2,3], Davide Morelli[1,3], Marcella Cornia[1(✉)], Marco Bertini[2], and Rita Cucchiara[1]

[1] University of Modena and Reggio Emilia, Modena, Italy
{giuseppe.cartella,davide.morelli,marcella.cornia, rita.cucchiara}@unimore.it

[2] University of Florence, Florence, Italy
{alberto.baldrati,marco.bertini}@unifi.it

[3] University of Pisa, Pisa, Italy

Abstract. The inexorable growth of online shopping and e-commerce demands scalable and robust machine learning-based solutions to accommodate customer requirements. In the context of automatic tagging classification and multimodal retrieval, prior works either defined a low generalizable supervised learning approach or more reusable CLIP-based techniques while, however, training on closed source data. In this work, we propose OpenFashionCLIP, a vision-and-language contrastive learning method that only adopts open-source fashion data stemming from diverse domains, and characterized by varying degrees of specificity. Our approach is extensively validated across several tasks and benchmarks, and experimental results highlight a significant out-of-domain generalization capability and consistent improvements over state-of-the-art methods both in terms of accuracy and recall. Source code and trained models are publicly available at: https://github.com/aimagelab/open-fashion-clip.

Keywords: Fashion Domain · Vision-and-Language Pre-Training · Open-Source Datasets

1 Introduction

In the era of digital transformation, online shopping, and e-commerce have experienced an unprecedented surge in popularity. The convenience, accessibility, and variety offered by these platforms have revolutionized the way consumers engage with retail. Such digital shift creates an immense volume of data, therefore, the need for scalable and robust machine learning-based solutions to accommodate customer requirements becomes increasingly vital [42,48,49]. In the fashion domain, this includes tasks such as cross-modal retrieval [18,27], recommendation [10,21,39], and visual product search [2–4,32,44], which play a crucial

G. L. Foresti et al. (Eds.): ICIAP 2023, LNCS 14233, pp. 245–256, 2023.
https://doi.org/10.1007/978-3-031-43148-7_21

role in enhancing user experience, optimizing search functionality, and enabling efficient product recommendation systems.

To address these challenges, innovative solutions that combine vision-and-language understanding have been proposed [19,30,50]. Although prior works have made noteworthy contributions in the fashion domain, they still suffer from some deficiencies. Approaches like [4] are able to well fit a specific task but struggle to adapt to unseen datasets and exhibit sub-optimal performance when faced with domain shifts. This results in poor zero-shot capability.

On the contrary, other techniques have employed CLIP-based methods [26, 47], which offer better generalization capabilities thanks to the pre-training on large-scale datasets. Some works as [8], have often relied on closed-source data, limiting their applicability and hindering the ability to reproduce and extend results. Therefore, there remains a need for a scalable and reusable method that can leverage open-source fashion data with varying levels of detail while demonstrating improved generalization and performance. In response to the aforementioned challenges, in this paper, we propose OpenFashionCLIP, a vision-and-language contrastive learning method that stands out from previous approaches in several ways. We adopt open-source fashion data from multiple sources encompassing diverse styles and levels of detail. Specifically, we adopt four publicly available datasets for the training phase, namely FashionIQ [44], Fashion-Gen [36], Fashion200K [20], and iMaterialist [17]. We believe this approach not only enhances transparency and reproducibility but also broadens the accessibility and applicability of our technique to a wider range of users and domains.

The contrastive learning framework employed in OpenFashionCLIP enables robust generalization capabilities, ensuring consistent performance even in the presence of domain shifts and previously unseen data. Our method adopts a fashion-specific prompt engineering technique [6,16,35] and is able to effectively learn joint representations from multiple domains. OpenFashionCLIP overcomes the limitations of supervised learning approaches and closed-source data training, facilitating seamless integration between visual and textual modalities.

Extensive experiments have been conducted to evaluate the effectiveness of OpenFashionCLIP across diverse tasks and benchmarks. We provide a comparison against CLIP [35], OpenCLIP [43] and a recent CLIP-based method fine-tuned on closed-source fashion data, namely FashionCLIP [8]. The experimental results highlight the significant out-of-domain generalization capability of our method. Notably, our fine-tuning strategy on open-source fashion data yields superior performance compared to competitors in several metrics, thus underscoring the benefits of leveraging open-source datasets for training.

2 Related Work

The ever-growing interest of customers in e-commerce has made the introduction of innovative solutions essential to enhance the online experience. On this basis, recommendation systems play a crucial role and numerous works have been introduced [10,11,21,39]. An illustrative example is the automatic creation of capsule wardrobes proposed in [21], where given an ensemble of garments and accessories the proposed method provided some possible visually compatible outfits. One of the most significant challenges of this task is the understanding of what visual compatibility means. To this aim, Cucurull *et al.* [10] addressed the compatibility prediction problem by exploiting the context information of fashion items, whereas Sarkar *et al.* [39] exploited a Transformer-based architecture to learn an outfit-level token embedding which is then fed through an MLP network to predict the compatibility score. In addition, De Divitiis *et al.* [11] introduced a more fine-grained control over the recommendations based on shape and color.

Generally, users desire to seek a specific article in the catalog with relative ease, therefore, designing efficient multimodal systems represents another important key to success for the fashion industry. A considerable portion of user online interactions fall into the area of multimodal retrieval, the task of retrieving an image corresponding to a given textual query, and vice versa. Prior works range from more controlled environments [23,27] to in-the-wild settings [18], where the domain shift between query and database images is a challenging problem.

Beyond recommendations and retrieval, another research line that is currently attracting attention is the one of virtual try-on, both in 3D [29,37,38] and 2D [13–15,24,31,33,46]. Virtual try-on aims to transfer a given in-shop garment onto a reference person while preserving the pose and the identity of the model. A related area is the one marked by fashion image editing [5,12,34]. While Dong *et al.* [12] conditioned the fashion image manipulation process on sketches and color strokes, other approaches [5,34] introduced for the first time a multimodal fashion image editing conditioned on text.

Specifically, Pernuš *et al.* [34] devised a GAN-based iterative solution to change specific characteristics of the given image based on a textual query. Baldrati *et al.* [5], instead, focused on the creation of new garments exploiting latent diffusion models and conditioning the generation process on text, sketch, and model's pose. Solving the aforementioned downstream tasks has been made possible due to large-scale architectures explicitly trained on fashion data which effectively combine vision-and-language modalities to learn more powerful representations [19,50]. Recent approaches exploit CLIP embeddings [35] to obtain more scalable and robust solutions able to generalize to different domains without supervision [8], but the closed source data training represents the main flaw.

3 On the Adaptation of CLIP to the Fashion Domain

3.1 Fashion-Oriented Contrastive Learning

Despite the significant scaling capability of large vision-and-language models such as CLIP, such a property comes at a cost. The pre-training of these models is usually conducted on datasets that contain million [35], or even billion [40] image-text pairs that, however, are gathered from the web and thus very noisy. Unfortunately, such coarse-grained annotations have been shown to lead to sub-optimal performance for vision-and-language learning [9,25]. Moreover, the adaptation of CLIP to the specific domain of fashion is far from trivial. Indeed, a significant

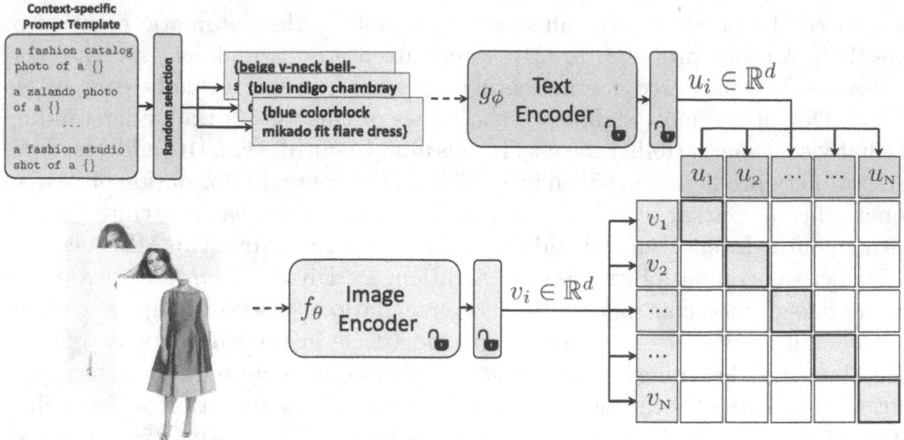

Fig. 1. Overview of our proposed method. We fine-tune both encoders and the linear projection layers toward the embedding space.

part of the images contained in these datasets is associated with incomplete captions or even worse, with simple and basic tags collected exploiting posts uploaded on the web by general and non-fashion-expert users. Considering these flaws, an adaptation of CLIP to a specific domain, uniquely relying on a vanilla pre-trained version, would not enable the attainment of optimal results. In our context, training on fashion-specific datasets containing fine-grained descriptions of garments and fashion accessories becomes crucial to obtain powerful representations while guaranteeing generalization and robustness to solve the tasks demanded by the fashion industries.

3.2 CLIP Preliminaries

Contrastive learning is a self-supervised machine learning technique that aims to learn data representations by constructing a powerful embedding space where semantically related concepts are close while dissimilar samples are pushed apart. On this line, the vision-and-language domain has already capitalized on such a learning technique. The CLIP model [35] represents the most common and illustrative method for connecting images and text in a shared multimodal space. The CLIP architecture consists of a text encoder g_ϕ and an image encoder f_θ, trained on image-caption pairs $\mathcal{S} = \{(x_i, t_i)\}_{i=1}^N$.

The image encoder f_θ embeds an image $x \in \mathcal{X}$ obtaining a visual representation $\boldsymbol{v} = f_\theta(x)$. In the same manner, the text encoder g_ϕ takes as input a tokenized string $\tilde{t}$ and returns a textual embedding $\boldsymbol{u} = g_\phi(\tilde{t})$. For each batch $\mathcal{B}$ of image-caption pairs $\mathcal{B} = \{(x_i, t_i)\}_{i=1}^L$, where L is the batch size, the objective is to maximize the cosine similarity between $\boldsymbol{v}_i$ and $\boldsymbol{u}_i$ while minimizing the cosine similarity between $\boldsymbol{v}_i$ and $\boldsymbol{u}_j$, $\forall j \neq i$. The CLIP loss can be formally expressed as the sum of two symmetric terms:

$$\mathcal{L}_{contrastive} = \mathcal{L}_{T2I} + \mathcal{L}_{I2T}, \tag{1}$$

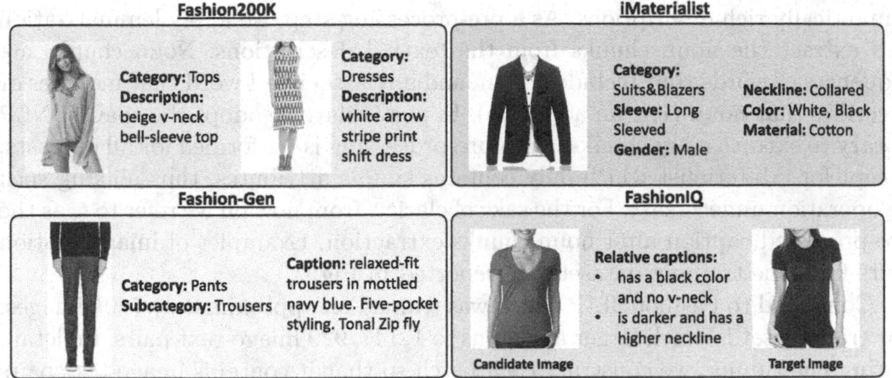

Fig. 2. Qualitative samples from the training datasets.

$$\mathcal{L}_{T2I} = -\frac{1}{L}\sum_{i=1}^{L}\log\frac{\exp(\tau \boldsymbol{u}_i^T \boldsymbol{v}_i)}{\sum_{j=1}^{L}\exp(\tau \boldsymbol{u}_i^T \boldsymbol{v}_j)}, \tag{2}$$

$$\mathcal{L}_{I2T} = -\frac{1}{L}\sum_{i=1}^{L}\log\frac{\exp(\tau \boldsymbol{v}_i^T \boldsymbol{u}_i)}{\sum_{j=1}^{L}\exp(\tau \boldsymbol{v}_i^T \boldsymbol{u}_j)}, \tag{3}$$

where τ represents a temperature parameter.

3.3 Open Source Training

In the fashion domain, several datasets, characterized by multimodal annotations from human experts, have been introduced. Differently from prior work [8] that fine-tuned CLIP on a private dataset, we devise a contrastive learning strategy entirely based on open-source data. An overview of the proposed CLIP-based fine-tuning is shown in Fig. 1. In detail, we adopt four publicly available datasets:
Fashion-Gen [36]. The dataset contains a total of 325,536 high resolution images (1360 × 1360) with 260,480 samples for the training set and 32,528 images both for validation and test set. In addition, 48 main categories and 121 fine-grained categories (*i.e. subcategory*) are defined.
Fashion IQ [44]. There are 77,684 images, divided into three main categories (dresses, shirts, and tops&tees), with product descriptions and attribute labels.
Fashion200K [20]. It contains 209,544 clothing images from five categories (dresses, tops, pants, skirts, and jackets) and an associated textual description.
iMaterialist [17]. It is a multi-label dataset containing over one million images and 8 groups of 228 fine-grained attributes.

These datasets are characterized by different levels of detail of the image annotations. FashionIQ has been proposed to accomplish the task of interactive image retrieval, therefore, the captions are relative to what should be modified in the source image to retrieve the target image. On the contrary, iMaterialist only contains attributes while Fashion-Gen and Fashion200K present more semantically rich descriptions. As a pre-processing step, we apply lemmatization and extract the noun chunks from the textual descriptions. Noun chunks are sequences of words that include a noun and any associated word that modifies or describes that noun (*e.g.* an adjective). In particular, we adopt the spaCy[1] NLP library to extract noun chunks. Data pre-processing is performed for all datasets, except for iMaterialist which only contains simple attributes, thus making such an operation unnecessary. For the sake of clarity, from now on we refer to t_i as the pre-processed caption after noun chunks extraction. Examples of image-caption pairs from the training datasets are reported in Fig. 2.

Compared to FashionCLIP which was trained on approximately $700k$ images, our training set is much larger and sums to $1,147,929$ image-text pairs. In detail, during fine-tuning, we construct each batch so that it contains image-text pairs from all the different data sources. Considering the great number of pairs of the complete training dataset, we fine-tune all the pre-trained weights of the CLIP model. Indeed, only training the projections toward the embedding space would not allow to fully effectively capture the properties of the data distribution.

3.4 Prompt Engineering

Prompt engineering is the technique related to the customization of the prompt text for each task. Providing context to the model has been shown to work well in a wide range of settings. Following prior works [6,16,35], we provide our model with a fashion-specific context defining a template of prompts related to our application domain. Specifically, given a template of prompts $\mathcal{P} = \{(p_i)\}_{i=1}^{|\mathcal{P}|}$, at each training step, we select a random $p_i \in \mathcal{P}$ for each image-caption pair $(x_i, t_i) \in \mathcal{B}$. The caption t_i is concatenated to p_i obtaining the final CLIP input.

The complete fashion-specific template includes the following prompts: `"a photo of a"`, `"a photo of a nice"`, `"a photo of a cool"`, `"a photo of an expensive"`, `"a good photo of a"`, `"a bright photo of a"`, `"a fashion studio shot of a"`, `"a fashion magazine photo of a"`, `"a fashion brochure photo of a"`, `"a fashion catalog photo of a"`, `"a fashion press photo of a"`, `"a zalando photo of a"`, `"a yoox photo of a"`, `"a yoox web image of a"`, `"an asos photo of a"`, `"a high resolution photo of a"`, `"a cropped photo of a"`, `"a close-up photo of a"`, `"a photo of one"`.

4 Experimental Evaluation

In this section, we describe the open-source datasets used as benchmarks together with the tasks performed to assess the scalability and robustness of our approach.

4.1 Benchmark Datasets

We validate our approach across three different datasets:

[1] https://github.com/explosion/spaCy.

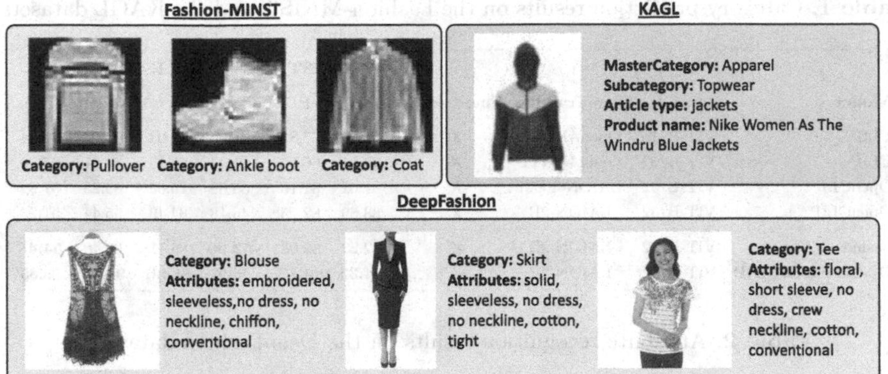

Fig. 3. Samples from the benchmark datasets.

DeepFashion [27] contains over 800, 000 images and is divided into several benchmarks. In our experiments, we employ the attribute prediction subset which contains 40, 000 images and 1, 000 different attributes.

Fashion-MNIST [45] is based on the Zalando catalog and consists of 60, 000 training images, a test set of 10, 000 examples, and 10 categories. All images are in grayscale and have a 28×28 resolution. Following [8], we apply image inversion, thus working on images with a white background.

KAGL is a subset of [1] and contains 44, 441 images equipped with textual annotations including the master category, the sub-category, the article type, and the product description. In detail, we filter out all those images not belonging to the *'apparel'* master category and kept the images depicting humans, resulting in a total of 21, 397 samples, 8 sub-categories, and 58 article types. Qualitative examples of the adopted benchmarks are reported in Fig. 3.

4.2 Implementation Details

We train the final model for 60 epochs using a batch size of 2048. To save memory, we adopt the gradient checkpointing technique [7]. AdamW [28] is employed as optimizer, with β_1 set to 0.9 and β_2 equal to 0.98, epsilon of $1e-6$, and weight decay equal to 0.2. A learning rate of $5e-7$ and automatic mixed precision are applied. For a fair comparison with competitors, we select the ViT-B/32 backbone as the image encoder. During training, we apply the prompt engineering strategy described in Sect. 3.4.

As a pre-trained CLIP model, we refer to the OpenCLIP implementation [22] trained on LAION-2B [40] composed of 2 billion image-text pairs. In the evaluation phase, following the pre-processing procedure of [35], we resize the image along the shortest edge and apply center crop.

4.3 Zero-Shot Classification

In our context, zero-shot classification refers to the task of classification on unseen datasets characterized by different data distributions compared to the training datasets. The task is crucial to assess the transfer capability of the model to adapt to new and

Table 1. Category prediction results on the Fashion-MNIST and the KAGL datasets.

				F-MNIST		KAGL			
Model	Backbone	Pre-Training	Fine-tuned	Acc@1	F1	Acc@1	Acc@5	Acc@10	F1
CLIP	ViT-B/16	OpenAI WIT	✗	69.29	67.88	31.54	70.08	90.09	36.04
CLIP	ViT-B/32	OpenAI WIT	✗	69.51	66.56	21.44	66.13	84.97	27.70
OpenCLIP	ViT-B/32	LAION-400M	✗	81.62	81.16	33.69	76.60	89.23	37.89
OpenCLIP	ViT-B/32	LAION-2B	✗	83.69	82.75	46.18	84.49	95.44	51.23
FashionCLIP	ViT-B/32	LAION-2B	✓	82.23	82.03	**52.90**	85.41	93.40	**54.48**
OpenFashionCLIP	ViT-B/32	LAION-2B	✓	**84.33**	**84.19**	45.97	**88.30**	**96.46**	53.85

Table 2. Attribute recognition results on the DeepFashion dataset.

				Overall Recall			Per-Class Recall		
Model	Backbone	Pre-Training	Fine-tuned	R@3	R@5	R@10	R@3	R@5	R@10
CLIP	ViT-B/16	OpenAI WIT	✗	8.00	11.40	17.54	13.31	17.42	24.54
CLIP	ViT-B/32	OpenAI WIT	✗	7.35	10.30	16.60	11.39	15.13	21.67
OpenCLIP	ViT-B/32	LAION-400M	✗	12.58	17.22	25.64	17.9	22.81	30.71
OpenCLIP	ViT-B/32	LAION-2B	✗	13.07	17.70	26.13	19.35	24.31	32.51
FashionCLIP	ViT-B/32	LAION-2B	✓	15.19	20.83	32.37	17.30	22.27	30.56
OpenFashionCLIP	ViT-B/32	LAION-2B	✓	**24.47**	**32.97**	**45.77**	**28.67**	**36.07**	**47.28**

unseen domains. Following the standard CLIP evaluation setup [35], we perform classification by embedding the image and all k categories. Regarding prompt engineering, we always append every category `{label}` to the same generic prompt `"a photo of a"`. We feed each category prompt through the CLIP textual encoder g_ϕ obtaining a set of feature vectors $\mathcal{U} = \{(\boldsymbol{u}_i)\}_{i=1}^N$. In the same manner, we feed the image x_i through the CLIP image encoder to get the embedded representation $\boldsymbol{v} = f_\phi(x_i)$. To classify the image we compute the cosine similarity between $\boldsymbol{v}$ and each text representation $\boldsymbol{u}_i$. The predicted category is the one with the highest similarity. Experiments have been conducted on the test splits of Fashion-MNIST, KAGL, and on the attribute prediction benchmark of DeepFashion. Our model is compared against the original CLIP model [35], which was trained on the private WIT dataset, OpenCLIP [43] pre-trained on LAION-400M [41] and LAION-2B [40], and FashionCLIP [8] that was fine-tuned on closed source data from *Farfetch*. Note that we have reproduced the results of FashionCLIP by exploiting the source code released by the authors and adapting it to our tasks and settings. The task is evaluated considering three well-known metrics, namely accuracy@k, recall@k, and weighted $F1$ score. The accuracy@k computes the number of times the correct label is among the top k labels predicted by the model. The recall@k, instead, measures the number of relevant retrieved items with respect to the total number of relevant items for a given query. The weighted F1 score accounts for the class distribution in the dataset by calculating the F1 score for each class individually and then averaging based on the class frequencies.

Quantitative results on Fashion-MNIST and KAGL are summarized in Table 1. The first aspect to mention is the improvement against CLIP and OpenCLIP on all metrics and both datasets, indicating the effectiveness of our fine-tuning strategy enabling a strong generalization and adaptation of our model to the specific fashion domain.

Table 3. Cross-modal retrieval results on the KAGL dataset.

				Image-to-Text			Text-to-Image		
Model	**Backbone**	**Pre-Training**	**Fine-tuned**	**R@1**	**R@5**	**R@10**	**R@1**	**R@5**	**R@10**
FashionCLIP	ViT-B/32	LAION-2B	✓	6.61	19.23	28.66	6.97	19.14	27.49
OpenFashionCLIP	ViT-B/32	LAION-2B	✓	**7.57**	**20.72**	**30.38**	**7.73**	**20.58**	**28.56**

Table 4. Ablation study to assess the validity of the prompt engineering technique.

	F-MNIST		**KAGL**		**DeepFashion**		**KAGL**	
Model	Acc@1	F1	Acc@1	F1	R@3	R@3 (cls)	R@1 (I2T)	R@1 (T2I)
w/o prompt engineering	83.21	82.99	**47.51**	47.3	20.34	25.21	7.47	**7.73**
OpenFashionCLIP	**84.33**	**84.19**	45.97	**53.85**	**24.47**	**28.67**	**7.57**	**7.73**

Compared to FashionCLIP, our model shows better performance on Fashion-MNIST, while when tested on the 58 article types of KAGL, the results are comparable. OpenFashionCLIP performs better with the increase of the number of considered categories.

Table 2 shows the results on the attribute prediction benchmark of the DeepFashion dataset. Categories of this dataset are attributes describing different garment characteristics (*e.g.* v-neck, sleeveless, etc.), therefore we leverage the recall metric in this setting to account for the multi-label nature of the dataset. In particular, we evaluate both the per-class recall@k and the overall recall@k among all attributes. In this case, our solution outperforms FashionCLIP by a consistent margin, highlighting the effectiveness of our training strategy with data of different annotation detail granularity.

4.4 Cross-Modal Retrieval

Cross-modal retrieval refers to the task of retrieving relevant contents from a multimodal dataset using multiple modalities such as text and images. Different modalities should be integrated to enable an effective search based on the user's input query. Cross-modal retrieval can be divided into two sub-tasks: image-to-text and text-to-image retrieval. In the first setting, given a query image x, we ask the model to retrieve the first k product descriptions that better match the image. On the opposite, in text-to-image retrieval, given a text query, the first k images that better correlate with the input query are returned. In Table 3, we evaluate our fine-tuning method on the KAGL dataset in terms of recall@k with $k = 1, 5, 10$. OpenFashionCLIP performs better compared to FashionCLIP on both settings and according to all recall metrics, thus further confirming the effectiveness of our proposal.

4.5 Effectiveness of Prompt Engineering

Finally, in Table 4, we evaluate the individual contribution of prompt engineering in our fine-tuning method. We present the ablation study on all considered benchmarks. The first line of the table (*i.e.* *w/o* prompt engineering) refers to the case where we perform fine-tuning without using the fashion-specific template described in Sect. 3.4 but employing a fixed prompt (*i.e.* `"a photo of a"`). Notably, Fashion-MNIST is used for the classification task, DeepFashion for retrieval, and KAGL for both. As the results demonstrate, the idea to construct a fashion-specific set of prompts clearly performs well across all cases except for the KAGL classification benchmark. We argue that in general, domain-specific prompt engineering represents a key factor to obtain greater domain adaptation of the CLIP model.

5 Conclusion

In this paper, we introduced OpenFashionCLIP, a vision-and-language contrastive learning method designed to address the scalability and robustness challenges posed by the fashion industry for online shopping and e-commerce. By leveraging open-source fashion data from diverse sources, OpenFashionCLIP overcomes limitations associated with closed-source datasets and enhances transparency, reproducibility, and accessibility. Our strategy, characterized by the fine-tuning of all pre-trained weights across all CLIP layers together with the adoption of a context-specific prompt engineering technique, effectively enables better adaption to our specific domain. We evaluated our strategy on three benchmarks and demonstrated that the proposed solution led to superior performance over the baselines and competitors achieving better accuracy and recall in almost all settings.

Acknowledgements. This work has partially been supported by the European Commission under the PNRR-M4C2 (PE00000013) project "FAIR - Future Artificial Intelligence Research" and the European Horizon 2020 Programme (grant number 101004545 - ReInHerit), and by the PRIN project "CREATIVE: CRoss-modal understanding and gEnerATIon of Visual and tExtual content" (CUP B87G22000460001), co-funded by the Italian Ministry of University.

References

1. Aggarwal, P.: Fashion Product Images (Small). https://www.kaggle.com/datasets/paramaggarwal/fashion-product-images-small
2. Baldrati, A., Agnolucci, L., Bertini, M., Del Bimbo, A.: Zero-shot composed image retrieval with textual inversion. arXiv preprint arXiv:2303.15247 (2023)
3. Baldrati, A., Bertini, M., Uricchio, T., Del Bimbo, A.: Conditioned image retrieval for fashion using contrastive learning and CLIP-based features. In: ACM Multimedia Asia (2021)
4. Baldrati, A., Bertini, M., Uricchio, T., Del Bimbo, A.: Conditioned and composed image retrieval combining and partially fine-tuning CLIP-based features. In: CVPR Workshops (2022)
5. Baldrati, A., Morelli, D., Cartella, G., Cornia, M., Bertini, M., Cucchiara, R.: Multimodal garment designer: human-centric latent diffusion models for fashion image editing. arXiv preprint arXiv:2304.02051 (2023)
6. Brown, T., et al.: Language models are few-shot learners. In: NeurIPS (2020)
7. Chen, T., Xu, B., Zhang, C., Guestrin, C.: Training deep nets with sublinear memory cost. arXiv preprint arXiv:1604.06174 (2016)
8. Chia, P.J., et al.: Contrastive language and vision learning of general fashion concepts. Sci. Rep. **12**(1), 18958 (2022)

9. Cornia, M., Baraldi, L., Fiameni, G., Cucchiara, R.: Universal captioner: inducing content-style separation in vision-and-language model training. arXiv preprint arXiv:2111.12727 (2022)
10. Cucurull, G., Taslakian, P., Vazquez, D.: Context-aware visual compatibility prediction. In: CVPR (2019)
11. De Divitiis, L., Becattini, F., Baecchi, C., Del Bimbo, A.: Disentangling features for fashion recommendation. ACM TOMM **19**(1s), 1–21 (2023)
12. Dong, H., et al.: Fashion editing with adversarial parsing learning. In: CVPR (2020)
13. Fenocchi, E., Morelli, D., Cornia, M., Baraldi, L., Cesari, F., Cucchiara, R.: Dual-branch collaborative transformer for virtual try-on. In: CVPR Workshops (2022)
14. Fincato, M., Cornia, M., Landi, F., Cesari, F., Cucchiara, R.: Transform, warp, and dress: a new transformation-guided model for virtual try-on. ACM TOMM **18**(2), 1–24 (2022)
15. Fincato, M., Landi, F., Cornia, M., Cesari, F., Cucchiara, R.: VITON-GT: an image-based virtual try-on model with geometric transformations. In: ICPR (2021)
16. Gao, T., Fisch, A., Chen, D.: Making pre-trained language models better few-shot learners. In: ACL (2021)
17. Guo, S., et al.: The iMaterialist fashion attribute dataset. In: ICCV Workshops (2019)
18. Hadi Kiapour, M., Han, X., Lazebnik, S., Berg, A.C., Berg, T.L.: Where to buy it: matching street clothing photos in online shops. In: ICCV (2015)
19. Han, X., Yu, L., Zhu, X., Zhang, L., Song, Y.Z., Xiang, T.: FashionViL: fashion-focused vision-and-language representation learning. In: Avidan, S., Brostow, G., Cissé, M., Farinella, G.M., Hassner, T. (eds.) ECCV. LNCS, vol. 13695, pp. 634–651. Springer, Cham (2022). https://doi.org/10.1007/978-3-031-19833-5_37
20. Han, X., et al.: Automatic spatially-aware fashion concept discovery. In: ICCV (2017)
21. Hsiao, W.L., Grauman, K.: Creating capsule wardrobes from fashion images. In: CVPR (2018)
22. Ilharco, G., et al.: OpenCLIP (2021). https://doi.org/10.5281/zenodo.5143773
23. Kuang, Z., et al.: Fashion retrieval via graph reasoning networks on a similarity pyramid. In: ICCV (2019)
24. Lee, S., Gu, G., Park, S., Choi, S., Choo, J.: High-resolution virtual try-on with misalignment and occlusion-handled conditions. In: Avidan, S., Brostow, G., Cissé, M., Farinella, G.M., Hassner, T. (eds.) ECCV. LNCS, vol. 13677, pp. 204–219. Springer, Cham (2022). https://doi.org/10.1007/978-3-031-19790-1_13
25. Li, J., Li, D., Xiong, C., Hoi, S.: BLIP: bootstrapping language-image pre-training for unified vision-language understanding and generation. In: ICML (2022)
26. Li, Y., et al.: Supervision exists everywhere: a data efficient contrastive language-image pre-training paradigm. In: ICLR (2022)
27. Liu, Z., Luo, P., Qiu, S., Wang, X., Tang, X.: DeepFashion: powering robust clothes recognition and retrieval with rich annotations. In: CVPR (2016)
28. Loshchilov, I., Hutter, F.: Decoupled weight decay regularization. In: ICLR (2019)
29. Majithia, S., Parameswaran, S.N., Babar, S., Garg, V., Srivastava, A., Sharma, A.: Robust 3D garment digitization from monocular 2D images for 3D virtual try-on systems. In: WACV (2022)
30. Moratelli, N., Barraco, M., Morelli, D., Cornia, M., Baraldi, L., Cucchiara, R.: Fashion-oriented image captioning with external knowledge retrieval and fully attentive gates. Sensors **23**(3), 1286 (2023)

31. Morelli, D., Baldrati, A., Cartella, G., Cornia, M., Bertini, M., Cucchiara, R.: LaDI-VTON: latent diffusion textual-inversion enhanced virtual try-on. arXiv preprint arXiv:2305.13501 (2023)
32. Morelli, D., Cornia, M., Cucchiara, R.: FashionSearch++: improving consumer-to-shop clothes retrieval with hard negatives. In: CEUR Workshop Proceedings (2021)
33. Morelli, D., Fincato, M., Cornia, M., Landi, F., Cesari, F., Cucchiara, R.: Dress code: high-resolution multi-category virtual try-on. In: Avidan, S., Brostow, G., Cissé, M., Farinella, G.M., Hassner, T. (eds.) ECCV. LNCS, vol. 13668, pp. 345–362. Springer, Cham (2022). https://doi.org/10.1007/978-3-031-20074-8_20
34. Pernuš, M., Fookes, C., Štruc, V., Dobrišek, S.: FICE: text-conditioned fashion image editing with guided GAN inversion. arXiv preprint arXiv:2301.02110 (2023)
35. Radford, A., et al.: Learning transferable visual models from natural language supervision. In: ICML (2021)
36. Rostamzadeh, N., et al.: Fashion-gen: the generative fashion dataset and challenge. arXiv preprint arXiv:1806.08317 (2018)
37. Santesteban, I., Otaduy, M., Thuerey, N., Casas, D.: ULNeF: untangled layered neural fields for mix-and-match virtual try-on. In: NeurIPS (2022)
38. Santesteban, I., Thuerey, N., Otaduy, M.A., Casas, D.: Self-supervised collision handling via generative 3D garment models for virtual try-on. In: CVPR (2021)
39. Sarkar, R., et al.: OutfitTransformer: learning outfit representations for fashion recommendation. In: WACV (2023)
40. Schuhmann, C., et al.: LAION-5B: an open large-scale dataset for training next generation image-text models. In: NeurIPS (2022)
41. Schuhmann, C., et al.: LAION-400M: open dataset of CLIP-filtered 400 million image-text pairs. In: NeurIPS Workshops (2021)
42. Shiau, R., et al.: Shop the look: building a large scale visual shopping system at Pinterest. In: KDD (2020)
43. Wortsman, M., et al.: Robust fine-tuning of zero-shot models. In: CVPR (2022)
44. Wu, H., et al.: Fashion IQ: a new dataset towards retrieving images by natural language feedback. In: CVPR (2021)
45. Xiao, H., Rasul, K., Vollgraf, R.: Fashion-MNIST: a novel image dataset for benchmarking machine learning algorithms. arXiv preprint arXiv:1708.07747 (2017)
46. Xie, Z., et al.: GP-VTON: towards general purpose virtual try-on via collaborative local-flow global-parsing learning. In: CVPR (2023)
47. Yao, L., et al.: FILIP: fine-grained interactive language-image pre-training. In: ICLR (2022)
48. Zhai, A., Wu, H.Y., Tzeng, E., Park, D.H., Rosenberg, C.: Learning a unified embedding for visual search at Pinterest. In: KDD (2019)
49. Zhang, Y., et al.: Visual search at Alibaba. In: KDD (2018)
50. Zhuge, M., et al.: Kaleido-BERT: vision-language pre-training on fashion domain. In: CVPR (2021)

UAV Multi-object Tracking by Combining Two Deep Neural Architectures

Pier Luigi Mazzeo[1(✉)], Alessandro Manica[2], and Cosimo Distante[1]

[1] ISASI - CNR, Via Monteroni sn, 73100 Lecce, Italy
pierluigi.mazzeo@cnr.it
[2] Università del Salento, Via Monteroni sn, 73100 Lecce, Italy

Abstract. Detecting and tracking multiple objects from unmanned aerial vehicle (UAV) videos is an high challenging task in a wide range of practical applications. Almost all traditional trackers meet some issues on UAV images due to camera movements causing view change in a 3D directions. In this work, we propose a Convolutional Neural Network specialized in multi-object tracking (MOT) for images captured from UAV. The architecture we introduced is composed by two main blocks: i) an object detection block based on YOLOv8 architecture; ii) an association block based on strongSORT architecture. We investigated different versions of YOLOv8 architectures with the strongSORT as association trackers. Experimental results on the VisDrone2019 dataset show that the proposed solution outperforms the up to date state-of-the-art tracking algorithms performance on UAV videos reaching the 42.03% in Multi-Object Tracking Accuracy (MOTA).

Keywords: Multi-object tracking · UAV · Convolutional Neural Network

1 Introduction

Many applications such as autonomous driving, intelligent transportation system and advanced video analysis, need multi-object tracking (MOT), which is a fundamental and challenging task in computer vision [22,26]. Most of the MOT methodologies are based on the tracking by detection model that consists typically of two main phases: detection and data association [3,35]. In the detection phase are generated bounding-box predictions candidate in each frame processed whilst in the data association phase predicted bounding boxes are matched across different frames by using appearance and motion features [8]. In the last years MOT through UAV views has became a warm topic in the researchers community thanking to the employing and spreading UAVs in a huge number of applications [1,24,29,38]. Unlike the advances made in traditional multiobject tracking (generally evaluated on video captured from static views) muti object tracking performed on moving UAV views is still a very open challenge. Two crucial

G. L. Foresti et al. (Eds.): ICIAP 2023, LNCS 14233, pp. 257–268, 2023.
https://doi.org/10.1007/978-3-031-43148-7_22

problems have to be solved as sooner as possible in detection and data association phases. The detection is the preliminary step, in which multiple categories of object in a moving UAV view should be detected and classified. Often the object numbers in each category are strongly imbalanced, making the training of the detection model extremely difficult. Considering that most of the object in UAV image are very small due to the high altitude of UAV flight aggravating the difficulties in the detection task. The data association task is still challenging due to the varying appearance and the consistent changing of motion parameters of the tracked objects. These continuous variations are caused by the not predictable and fast motion of the camera located on UAV that procure huge ID switches. The motion of the target objects, the occlusions and the superposition of their trajectories in addition the movement of the UAV camera, all these issues together make the multi object tracking hard to be modelled by the traditional methodologies. In this paper we propose a new architecture which births from a combination of two well-known methodologies that have been designed for a different domain from UAV multi object tracking. In order to enhance the tiny moving target object detection we optimize the employment of the YOLOv8 architecture [7]. For enhancing the ID embedding features for objects we use an extremely advanced StrongSORT architecture [6]. This way detection step is executed by the YOLOv8 architecture that is capable of handling object detection, image classification, and instance segmentation tasks with exceptional accuracy, all while maintaining a compact model size. The association step for composing the trajectory of each detected object has been executed by a lighter version of StrongSORT [6], using plug-and-play algorithms that helping missing association and missing detection achieving a good balance between speed and accuracy. We conduct different experiments on one of the most challenging public available dataset known as Visdrone 2019 [38] in order to evaluate the proposed architecture. Obtained experimental results demonstrate that the proposed method outperforms the state of art in terms of accuracy for the MOT task from UAV views. The key success factor of this work is the idea of concatenating two deep neural architectures that have been designed for different contexts and training them specifically for UAVs videos. The whole architecture taking into account the characteristics of the UAV videos, that comprise the small object dimensions, occlusions and appearance changing, reaches very good performance in multi-object tracking task. The contributions of this work lie three-fold:

- We introduce a Multi-object tracking architecture for video images captured from UAV perspectives, constituted by a concatenation of two deep architectures that have been designed for different contexts.
- We investigate different configurations of the proposed framework training it on a well-known public dataset, for finding the set up that reach the best performance in terms of tracking accuracy.
- We conduct all the experiments on one of the most challenging dataset demonstrating that the best configuration of the proposed framework outperform the up to date state of the art in term of multi-object tracking accuracy on UAV videos.

2 Related Work

In this section we discuss recent outcomes in multi-object tracking methods focusing on UAV videos. Early MOT algorithms follow Tracking-by-detection approach due to excellent performance, which formulating tracking task as a data association problem. Associations are made using optimization algorithms such as the Hungarian algorithm [30] or max-flow min-cut [18] based on appearance or motion information. GOG approach [18] uses the min-cost flow algorithm to associate detected objects with cost functions based on appearance and motion information to determine the number of trajectories and their birth and death states. SORT [30] leverages high-quality detection and combines position, size, and appearance information to track objects through long-term occlusions with fewer identity switches. IOU [2] uses the intersection-over-union (IOU) measure to compute the similarities of detection in consecutive frames. MOTDT [5] addresses unreliable detection by collecting candidates from both detection and tracking and designing a scoring function based on CNN to select optimal candidates in real-time. TrackletNet [25] constructs a graph model with tracklets as vertices to exploit temporal information and reduce computational complexity, followed by a clustering operation to generate object trajectories. Integration of object detection and tracking into a joint framework, known as joint tracking and detection approach, has emerged as a recent trend in multiple object tracking (MOT).

With the disrupting starting of the deep neural networks era many approaches has been developed for learning discriminating appearance and motion features of targets in multiple object tracking (MOT). Various methods for appearance modeling have been proposed for VisDrone-VDT2018, VisDrone-MOT2019, and VisDrone-MOT2020 datasets. In [21] is proposed a method that extracts multi-scale features to describe objects and infers the identities of detection at different frames by analyzing exhaustive permutations of extracted features. Omni-scale representation [37] (OSNet) method extracts omni-scale feature representation using the residual block with multiple convolutional streams. A unified aggregation gate is designed to dynamically fuse multi-scale features. ReID representation [15] introduces various training tricks, such as warmup learning, random erasing augmentation, label smoothing, smaller last stride, BNNeck, and center loss, to improve performance without excessive computational consumption.

Multiple granularity network [27] is a multi-branch deep network architecture that integrates discriminating information with various granularity. This method consists of one branch for global feature representation and two branches for local feature representations, allowing for the integration of discriminating information at multiple levels. Appearance modeling techniques show promising outcomes in learning discriminating features for object tracking, and their effectiveness has been demonstrated on benchmark datasets. Motion features also play a crucial role in MOT in fact, low-level motion cues such as forward-backward flow [10] and optical flow [9] can provide important information for MOT. Motion networks have been developed to learn the complex, long-term temporal dependencies of targets. These networks are often more effective than predefined motion patterns. In [17] has been proposed the first end-to-end learning approach for online MOT that does not require prior knowledge about target dynamics or clutter distributions. In [19] has been employed LSTM networks to track the motion and interactions of targets over longer periods, making it suitable for situations with long-term occlusions [38]. When video data are acquired by UAV during flight for Multiple Object Tracking (MOT), the detection results may be compromised by high levels of noise, false alarms, and missed detection. These issues can be attributed to changes in the UAV's motion, ambient light, and the unavoidable jitter.

To improve the performance of MOT, it is feasible to design a network structure that can memorize historical trajectory information and learn matching similarity measurements based on this information. Long Short-Term Memory (LSTM) [12] networks have demonstrated reliable performance in many sequence problems and can overcome the gradient disappearance and explosion problems of standard Recurrent Neural Networks (RNNs) [32].

3 Materials and Method

In Fig. 1 is schematized the proposed architecture. It can be noticed that is composed by two network architectures one for the object detection task and the second one for the association task.

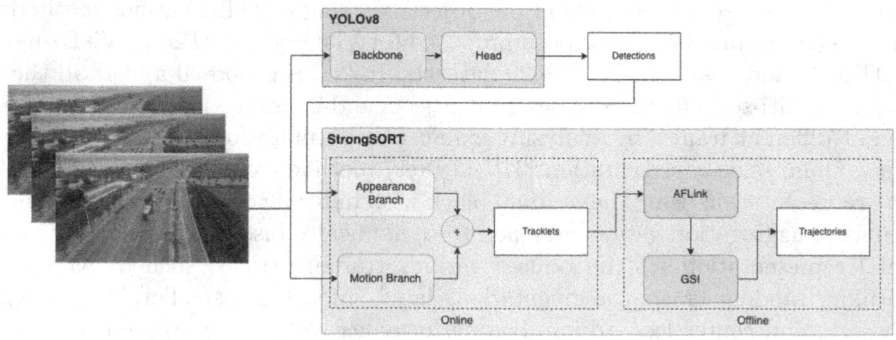

Fig. 1. Proposed Solution scheme.

3.1 YOLOv8 Architecture

Ultralytics [23] has recently released YOLOv8, the latest addition to their YOLO (You Only Look Once) model series, which is highly regarded in the computer vision field. YOLOv8 is capable of handling object detection, image classification, and instance segmentation tasks with exceptional accuracy, all while maintaining a compact model size. Ultralytics, who also developed the well-known YOLOv5 model, created YOLOv8 with various architectural and developer experience enhancements. With YOLO models, even a single GPU can be used for training, making it an affordable option for both edge hardware and cloud deployment, which has made it popular among machine learning practitioners [20]. The latest iteration of the YOLO object detection and image segmentation model is Ultralytics YOLOv8. This cutting-edge, state-of-the-art model builds upon the successes of its predecessors and introduces new features and improvements to enhance its performance, flexibility, and efficiency.

One of the key strengths of YOLOv8 is its strong emphasis on speed, size, and accuracy, making it a highly compelling option for a variety of vision AI tasks. Its performance surpasses that of earlier versions thanks to innovative enhancements such as a new backbone network, a new anchor-free split head, and new loss functions. These upgrades enable YOLOv8 to achieve superior outcomes while maintaining a small size and exceptional speed [23]. The model of YOLOv8 offers a range of size options using N/S/M/L/X scales, which are similar to those seen in YOLOv5, to accommodate different scenarios. Several modifications have been made to the backbone network and neck module.

In terms of the N/S/M/L/X models, the scaling factors for the N/S and L/X models have changed, but the number of channels in the S/M/L backbone network varies and does not follow the same scaling factor rule. This was done because using the same channel settings under a set of scaling factors was not optimal.

3.2 StrongSORT Architecture

StrongSORT is an improved version of the DeepSORT tracker that enhances its detection, embedding, and association capabilities. It has achieved impressive results on the MOT17 and MOT20 datasets by setting new records for the HOTA and IDF1 metrics. DeepSORT [30] is a two-branch framework consisting of an appearance branch and a motion branch. The *appearance branch* employs a simple CNN, which has been pre-trained on the person re-identification dataset MARS [36], to extract appearance features from the detections in each frame.

OSNet. In the appearance branch, StrongSORT uses a robust appearance feature extractor to extract highly discriminative features. This feature extractor is particularly well-suited for re-identification of individuals in a given scene, and for this purpose, the OSNet (Omni-Scale Network) represents the best solution available. Person re-identification is a complex task that faces two major challenges. The first challenge arises from the large intra-class variations in appearance that occur due to changes in camera viewing conditions. For example, two individuals carrying backpacks may look very different when viewed from different camera angles, making it challenging to match them accurately. The second challenge is related to the small inter-class variations that occur because people in public spaces often wear similar clothing, making it difficult to distinguish between them even at a distance, as is typical in surveillance videos. These challenges can be overcome by learning discriminative features that cover different scales [37]. OSNet is a novel convolutional neural network architecture that has been specifically designed to learn omni-scale feature representations. OSNet achieves this by stacking multiple convolutional streams with different receptive field sizes.

4 Experiments

For our purpose, we used two datasets: VisDrone-DET and VisDrone-MOT. We used the former to train YOLOv8 and the latter to train StrongSORT. Below, we will analyze the two datasets to understand the main differences between them.

VisDrone-DET. The DET dataset comprises 10,209 images of complex and uncontrolled scenes, divided into subsets for training (6,471 images), validation (548 images), test-challenge (1,580 images), and test-dev (1,610 images). However, the detection performance is significantly affected by the class imbalance issue. For instance, there are over 40 fewer instances of awning-tricycles than cars, which could impact detection accuracy. This research track focuses on ten object categories that are commonly found in daily life, including pedestrian, person (classified as pedestrian if walking or standing, otherwise as a person), car, van, bus, truck, motor, bicycle, awning-tricycle, and tricycle. Uncommon vehicles such as forklift trucks and tanker trucks are ignored. The dataset also includes two attributes for each annotated bounding box, namely occlusion and truncation ratios, which are used to analyze algorithms thoroughly.

VisDrone-MOT. The MOT dataset is a challenging dataset for multiple object tracking (MOT) with 96 video clips. This dataset includes 56 training clips (24,198 frames), 7 validation clips (2,846 frames), 16 challenge testing clips (6,322 frames), and 17 dev testing clips (6,635 frames). The distribution of object categories in the dataset shows a severe class imbalance issue that poses a challenge for algorithm performance. Specifically, the number of car trajectories in the training set is more than 50% of the number of bus trajectories, making it difficult for tracking algorithms to achieve optimal performance. Moreover, the object trajectories vary significantly in length, ranging from 1 to 1,255 frames, necessitating tracking algorithms to perform well in both short-term and long-term tracking scenarios. Similar to the DET track, annotations are provided for occlusion and truncation ratios of each object, as well as ignored regions in each video frame [38].

Training Configuration. For the detection step we use three models: YOLOv8n, YOLOv8x, and YOLOv8x6. The first two models, YOLOv8n and YOLOv8x, had pre-trained on the COCO dataset, while YOLOv8x6 was trained from scratch. For the first two models, an alternative training was carried out

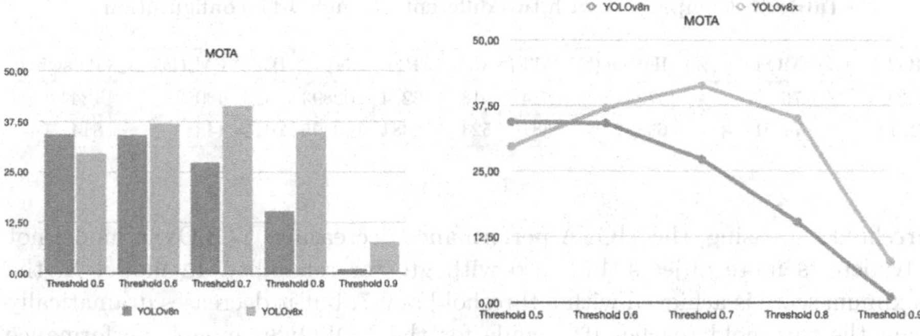

Fig. 2. MOTA score trend

by selective freezing backbone levels. A parameter was included in the training phase which allowed us to freeze specific layers. For all the experiments we carried out we use softmax function as loss function, AMSGrad algorithm as optimizer which is an extension of Adam with an extra step, was used with a learning rate of 0.0015. We also use cosine learning rate scheduler, which adjusts the learning rate according to a predefined schedule during the training process.

4.1 Results

Table 1 shows how the proposed architecture performance improve varying the confidence threshold parameter in the object detection step. It can be noticed that MOTA and score varies by changing the threshold during inference between 0.5 and 0.9.

Table 1. Tracking accuracy performance with varying threshold values

Threshold	Method	MOTA↑ (%)	MOTP↑ (%)	IDF1↑ (%)	MT↑	ML↓	FP↓	FN↓	IDs↓	FM (Frag)↓	ID switch
0.5	YOLOv8n	34.35	75.86	52.41	568	581	35001	116000	1493	4984	1535
	YOLOv8x	29.67	75.60	57.42	843	314	82899	78784	2058	5775	1735
0.6	YOLOv8n	34.13	77.18	49.00	432	743	17607	134304	1196	4251	1141
	YOLOv8x	37.01	76.07	58.50	777	391	57284	87608	1814	5304	1465
0.7	YOLOv8n	27.25	79.28	38.97	283	947	7756	160519	879	3366	775
	YOLOv8x	**41.31**	76.97	57.65	684	518	32244	102893	1527	4805	1244
0.8	YOLOv8n	15.28	82.33	22.51	117	1248	2361	194123	526	1804	377
	YOLOv8x	35.01	79.19	48.01	445	796	11607	138692	1078	3351	715
0.9	YOLOv8n	1.12	89.26	1.90	0	1672	103	229652	98	107	17
	YOLOv8x	7.78	87.09	12.65	58	1459	1142	213020	359	616	117

Table 1 data has been also drawn for helping the reader to better understand as the MOTA score changing in according to the confidence threshold (Fig. 2).

It has been observed that StrongSORT with model N and a threshold of 0.5 performs better than StrongSORT with model X. It can be caused by the model X that detects a greater number of objects. However, as the confidence

Table 2. Comparison with two different StrongSORT configuration

MOTA↑ (%)	MOTP↑ (%)	IDF1↑ (%)	MT↑	ML↓	FP↓	FN↓	IDs↓	FM (Frag)↓	ID switch
41.31	76.97	57.65	684	518	32244	102893	1527	4805	1244
42.03	77.01	**61.27**	681	523	30381	103495	2015	4458	814

threshold increasing, the MOTA performance decreasing. YOLOv8x model not only detects more objects, but also with greater reliability. In model X, the maximum score is achieved with a threshold of 0.7, but it decreases dramatically when the threshold reaches 0.9, while for the YOLOv8n model, performance continues to decrease as the threshold increases. The YOLOv8x model with a threshold of 0.7 reach the best performance in terms of MOTA.

During the inference process, a non-maximum suppression (NMS) threshold of 0.7 and a detection confidence threshold of 0.5 are applied. The latest configuration of StrongSORT involves setting a matching threshold of 0.20, a gating threshold of 0.50, a maximum limit of 30 missed detection before a track is deleted, and a limit of 0 unmatched predictions. The number of frames that a track remains in initialization phase is 3, and the maximum size of the appearance descriptors gallery is 100. Additionally, the warp mode for ECC is set to MOTION EUCLIDEAN, the momentum term α in EMA is 0.9, and the weight factor for appearance cost λ is 0.98. Experiment with this different StrongSORT configuration reached a new goal improving of 0.72 % in terms of MOTA and an improvement of 3.62 % in terms of IDF1 score as shown in Table 2.

Comparison with the State of the Art. In this section, we compare the best results obtained with the proposed solution against the up to date state-of-art MOT algorithms one the Visdrone dataset. Table 3 summarized results obtained in the object detection step in comparison with 5 different models that have been trained on Visdrone-DET dataset. It can be noticed that YOLOv8x and YOLOv8x6 outperformed all the compared methodologies. Table 3 shows that YOLOv8x is the most performing candidate to combine with StrongSORT archi-

Table 3. Comparison with state-of-the-art methods on the VisDrone-DET dataset

Method	AP (%)	AP50 (%)
Cascade R-CNN [4]	21.80	37.84
DetNet [11]	20.07	37.54
RefineDet [34]	19.89	37.27
RetinaNet [13]	18.94	31.67
Yolov8n	15.20	27.10
Yolov8n-backbone	11.40	20.60
Yolov8x	**22.10**	**37.40**
Yolov8x-backbone	18.80	32.30
Yolov8x6	**21.80**	**37.00**

Table 4. Comparison with state-of-the-art methods on the VisDrone-MOT dataset

Method	MOTA↑ (%)	MOTP↑ (%)	IDF1↑ (%)	MT↑	ML↓	FP↓	FN↓	IDs↓	FM (Frag)↓
GOG [18]	28.7	76.1	36.4	346	836	17706	144657	1387	2237
IOUT [2]	28.1	74.7	38.9	467	670	36158	126549	2393	3829
SORT [30]	14.0	73.2	38.0	506	545	80845	112954	3629	4838
MOTDT [5]	−0.8	68.5	21.6	87	1196	44548	185453	1437	3609
MOTR [33]	22.8	72.8	41.4	272	825	28407	147937	959	3980
TrackFormer [16]	25.0	73.9	30.5	385	770	25856	141526	4840	4855
DAN [21]	28.9	74.8	37.7	535	602	–	–	1952	5634
JDE [28]	26.6	74.1	34.9	516	751	–	–	3200	3176
FairMOT [35]	30.8	74.3	41.9	577	697	–	–	3007	2996
FPUAV [31]	34.3	74.2	45.0	585	688	–	–	2138	2577
UAVMOT [14]	36.1	74.2	51.0	520	574	27983	115925	2775	7396
Ours	**42.03**	**77.01**	**61.27**	**681**	**523**	**30381**	**10345**	**2015**	**4458**

Fig. 3. Tracking on VisDrone dataset

tecture for improving the state of the art performance. Finally Table 4 demonstrate how our proposed solution substantially improve the state of the art MOT algorithms on VisDrone-MOT dataset. Combining YOLOv8x with StrongSORT and after some experiments testing different configuration, we achieved a MOTA of 5.93%, higher than the best result obtained by UAVMOT [14]. Additionally, comparing our solution to the best MOTP value reached by DAN [21], we outperform that value by 2.21%, with an IDF1 of 61.27%, which is more than 10 points higher than all compared methods. Proposed solution also achieved the best results in terms of MT, ML, and FN.

We demonstrated that combining YOLOv8x + StrongSORT architectures we obtained the best solution in terms of performance for challenging dataset such as VisDrone in the task of multi-object tracking. In Fig. 3 are depicted some video frames showing the output of proposed solution demonstrating how it work in real-time tracking task.

5 Conclusions

This work proposed a novel solution for multi-object tracking (MOT) task in unmanned aerial vehicles (UAVs). We demonstrated that combining YOLOv8 architecture for the object detection step and StrongSORT architecture for the association step, we outperform the up to date state of the art, fixing a new achievement on VisDrone2019 Dataset. Our solution reaching a MOTA of 42.03 % increased referred state of the art performance of more than 5%.

Acknowledgement. This research was funded in part by Future Artificial Intelligence Research-FAIR CUP B53C220036 30006 grant number PE0000013, and in part by the Ministry of Enterprises and Made in Italy with the grant ENDOR "ENabling technologies for Defence and mOnitoring of the foRests" - PON 2014-2020 FESR - CUP B82C21001750005. The authors would like to thank Mr. Arturo Argentieri from CNR-ISASI Italy for his technical contribution on the multi-GPU computing facilities.

References

1. Azimi, S.M., Kraus, M., Bahmanyar, R., Reinartz, P.: Multiple pedestrians and vehicles tracking in aerial imagery using a convolutional neural network. Remote. Sens. **13**, 1953 (2021)
2. Bochinski, E., Eiselein, V., Sikora, T.: High-speed tracking-by-detection without using image information. In: 2017 14th IEEE International Conference on Advanced Video and Signal Based Surveillance (AVSS), pp. 1–6. IEEE (2017)
3. Braso, G., Leal-Taixe, L.: Learning a neural solver for multiple object tracking. In: 2020 IEEE/CVF Conference on Computer Vision and Pattern Recognition (CVPR), pp. 6246–6256 (2020)
4. Cai, Z., Vasconcelos, N.: Cascade R-CNN: delving into high quality object detection. In: Proceedings of the IEEE Conference on Computer Vision and Pattern Recognition, pp. 6154–6162 (2018)
5. Chen, L., Ai, H., Zhuang, Z., Shang, C.: Real-time multiple people tracking with deeply learned candidate selection and person re-identification. In: 2018 IEEE International Conference on Multimedia and Expo (ICME), pp. 1–6. IEEE (2018)
6. Du, Y., et al.: StrongSORT: make DeepSORT great again. IEEE Trans. Multimedia (2023)
7. Glenn, J., Ayush, C., Jing, Q.: YOLO by ultralytics (2023). https://github.com/ultralytics/ultralytics, software
8. Huang, C., Wu, B., Nevatia, R.: Robust object tracking by hierarchical association of detection responses. In: Forsyth, D., Torr, P., Zisserman, A. (eds.) ECCV 2008. LNCS, vol. 5303, pp. 788–801. Springer, Heidelberg (2008). https://doi.org/10.1007/978-3-540-88688-4_58

9. Ilg, E., Mayer, N., Saikia, T., Keuper, M., Dosovitskiy, A., Brox, T.: FlowNet 2.0: evolution of optical flow estimation with deep networks. In: Proceedings of the IEEE Conference on Computer Vision and Pattern Recognition, pp. 2462–2470 (2017)
10. Kalal, Z., Mikolajczyk, K., Matas, J.: Forward-backward error: automatic detection of tracking failures. In: 2010 20th International Conference on Pattern Recognition, pp. 2756–2759. IEEE (2010)
11. Li, Z., Peng, C., Yu, G., Zhang, X., Deng, Y., Sun, J.: DetNet: design backbone for object detection. In: Ferrari, V., Hebert, M., Sminchisescu, C., Weiss, Y. (eds.) ECCV 2018. LNCS, vol. 11213, pp. 339–354. Springer, Cham (2018). https://doi.org/10.1007/978-3-030-01240-3_21
12. Liang, Y., Zhou, Y.: LSTM multiple object tracker combining multiple cues. In: 2018 25th IEEE International Conference on Image Processing (ICIP), pp. 2351–2355. IEEE (2018)
13. Lin, T.Y., Goyal, P., Girshick, R., He, K., Dollár, P.: Focal loss for dense object detection. In: Proceedings of the IEEE International Conference on Computer Vision, pp. 2980–2988 (2017)
14. Liu, S., Li, X., Lu, H., He, Y.: Multi-object tracking meets moving UAV. In: Proceedings of the IEEE/CVF Conference on Computer Vision and Pattern Recognition, pp. 8876–8885 (2022)
15. Luo, H., Gu, Y., Liao, X., Lai, S., Jiang, W.: Bag of tricks and a strong baseline for deep person re-identification. In: Proceedings of the IEEE/CVF Conference on Computer Vision and Pattern Recognition Workshops (2019)
16. Meinhardt, T., Kirillov, A., Leal-Taixe, L., Feichtenhofer, C.: TrackFormer: multi-object tracking with transformers. In: Proceedings of the IEEE/CVF Conference on Computer Vision and Pattern Recognition, pp. 8844–8854 (2022)
17. Milan, A., Rezatofighi, S.H., Dick, A., Reid, I., Schindler, K.: Online multi-target tracking using recurrent neural networks. In: Proceedings of the AAAI Conference on Artificial Intelligence, vol. 31 (2017)
18. Pirsiavash, H., Ramanan, D., Fowlkes, C.C.: Globally-optimal greedy algorithms for tracking a variable number of objects. In: CVPR 2011, pp. 1201–1208. IEEE (2011)
19. Sadeghian, A., Alahi, A., Savarese, S.: Tracking the untrackable: learning to track multiple cues with long-term dependencies. In: Proceedings of the IEEE International Conference on Computer Vision, pp. 300–311 (2017)
20. Solawetz, J.: What is YOLOv8? The ultimate guide. https://blog.roboflow.com/whats-new-in-yolov8/
21. Sun, S., Akhtar, N., Song, H., Mian, A., Shah, M.: Deep affinity network for multiple object tracking. IEEE Trans. Pattern Anal. Mach. Intell. **43**(1), 104–119 (2019)
22. Tang, Z., et al.: CityFlow: a city-scale benchmark for multi-target multi-camera vehicle tracking and re-identification. 2019 IEEE/CVF Conference on Computer Vision and Pattern Recognition (CVPR), pp. 8789–8798 (2019)
23. ultralytics: Ultralytics YOLOv8: the state-of-the-art YOLO model. https://ultralytics.com/yolov8
24. Varga, L.A., Kiefer, B., Messmer, M., Zell, A.: SeaDronesSee: a maritime benchmark for detecting humans in open water. In: 2022 IEEE/CVF Winter Conference on Applications of Computer Vision (WACV), pp. 3686–3696 (2021)
25. Wang, G., Wang, Y., Zhang, H., Gu, R., Hwang, J.N.: Exploit the connectivity: multi-object tracking with TrackletNet. In: Proceedings of the 27th ACM International Conference on Multimedia, pp. 482–490 (2019)

26. Wang, G., Yuan, X., Zheng, A., Hsu, H.M., Hwang, J.N.: Anomaly candidate identification and starting time estimation of vehicles from traffic videos. In: CVPR Workshops (2019)
27. Wang, G., Yuan, Y., Chen, X., Li, J., Zhou, X.: Learning discriminative features with multiple granularities for person re-identification. In: Proceedings of the 26th ACM International Conference on Multimedia, pp. 274–282 (2018)
28. Wang, Z., Zheng, L., Liu, Y., Li, Y., Wang, S.: Towards real-time multi-object tracking. In: Vedaldi, A., Bischof, H., Brox, T., Frahm, J.-M. (eds.) ECCV 2020. LNCS, vol. 12356, pp. 107–122. Springer, Cham (2020). https://doi.org/10.1007/978-3-030-58621-8_7
29. Wen, L., et al.: Detection, tracking, and counting meets drones in crowds: a benchmark. In: 2021 IEEE/CVF Conference on Computer Vision and Pattern Recognition (CVPR), pp. 7808–7817 (2021)
30. Wojke, N., Bewley, A., Paulus, D.: Simple online and realtime tracking with a deep association metric. In: 2017 IEEE International Conference on Image Processing (ICIP), pp. 3645–3649. IEEE (2017)
31. Wu, H., Nie, J., He, Z., Zhu, Z., Gao, M.: One-shot multiple object tracking in UAV videos using task-specific fine-grained features. Remote Sens. **14**(16), 3853 (2022)
32. Wu, X., Li, W., Hong, D., Tao, R., Du, Q.: Deep learning for unmanned aerial vehicle-based object detection and tracking: a survey. IEEE Geosci. Remote Sens. Mag. **10**(1), 91–124 (2021)
33. Zeng, F., Dong, B., Zhang, Y., Wang, T., Zhang, X., Wei, Y.: MOTR: end-to-end multiple-object tracking with transformer. In: Avidan, S., Brostow, G., Cissé, M., Farinella, G.M., Hassner, T. (eds.) Computer Vision-ECCV 2022: 17th European Conference, Tel Aviv, Israel, 23–27 October 2022, Proceedings, Part XXVII, pp. 659–675. Springer, Cham (2022). https://doi.org/10.1007/978-3-031-19812-0_38
34. Zhang, S., Wen, L., Bian, X., Lei, Z., Li, S.Z.: Single-shot refinement neural network for object detection. In: Proceedings of the IEEE Conference on Computer Vision and Pattern Recognition, pp. 4203–4212 (2018)
35. Zhang, Y., Wang, C., Wang, X., Zeng, W., Liu, W.: FairMOT: on the fairness of detection and re-identification in multiple object tracking. Int. J. Comput. Vis. **129**, 3069–3087 (2021)
36. Zheng, L., et al.: MARS: a video benchmark for large-scale person re-identification. In: Leibe, B., Matas, J., Sebe, N., Welling, M. (eds.) ECCV 2016. LNCS, vol. 9910, pp. 868–884. Springer, Cham (2016). https://doi.org/10.1007/978-3-319-46466-4_52
37. Zhou, K., Yang, Y., Cavallaro, A., Xiang, T.: Omni-scale feature learning for person re-identification. In: Proceedings of the IEEE/CVF International Conference on Computer Vision, pp. 3702–3712 (2019)
38. Zhu, P., et al.: Detection and tracking meet drones challenge. IEEE Trans. Pattern Anal. Mach. Intell. **44**(11), 7380–7399 (2021)

GLR: Gradient-Based Learning Rate Scheduler

Maria Ausilia Napoli Spatafora(✉), Alessandro Ortis, and Sebastiano Battiato

Department of Mathematics and Computer Science, University of Catania, Catania, Italy
maria.napolispatafora@phd.unict.it, {ortis,battiato}@dmi.unict.it

Abstract. Training a neural network is a complex and time-consuming process because of many combinations of hyperparameters that have to be adjusted and tested. One of the most crucial hyperparameters is the learning rate which controls the speed and direction of updates to the weights during training. We proposed an adaptive scheduler called Gradient-based Learning Rate scheduler (GLR) that significantly reduces the tuning effort thanks to a single user-defined parameter. GLR achieves competitive results in a very wide set of experiments compared to the state-of-the-art schedulers and optimizers. The computational cost of our method is trivial and can be used to train different network topologies.

Keywords: Neural network · Optimization · Hyperparameters

1 Introduction

Training a neural network can be a complex and time-consuming task because there are many hyperparameters to be properly tuned. Indeed, the design of an ANN-based approach involves the definition of an architecture of the model, a cost function and an optimizer to perform the training. Then, a number of epochs or a target accuracy is set as training stop criteria. Moreover, the training process involves other choices like input normalization, batch size, etc. Such a wide variability can lead to the need of a preliminary hyperparameter investigation, often times referred as grid-search. However, although grid-search of a few hyperparameters is a common approach for methods involving a few parameters such as classic machine learning models (e.g., SVM, K-nn, etc.), in the deep learning settings an exhaustive hyperparameter search is not feasible. Each method or function selected, has a set of hyperparameters that must be tuned. For example, Stochastic Gradient Descent (SGD) [3], that is the core of many training algorithms, is defined as:

$$\theta_{i+1} = \theta_i - \lambda_i g(\theta_i) \tag{1}$$

where θ denotes the network parameters (e.g., weights and biases), λ is the learning rate and $g(\theta)$ stands for the derivative of the cost function.

G. L. Foresti et al. (Eds.): ICIAP 2023, LNCS 14233, pp. 269–281, 2023.
https://doi.org/10.1007/978-3-031-43148-7_23

The selection of the exact value for each hyperparameter follows some best practices. Generally, most of the hyperparameters are set randomly and then fine-tuned [6,8]. Such a tuning process is time-consuming and the outcomes have to be confirmed by averaging the results of several settings.

The learning rate is one of the most crucial hyperparameter of a deep learning model that controls the speed and direction of updates to the weights during training [3]. A well-tuned learning rate plays a crucial role in achieving optimal model performance. However, selecting an appropriate learning rate for a neural network can be challenging, as it depends on various factors such as the structure of the model, the type of dataset, and the employed optimizer. Moreover, the performance of a model can plateau or deteriorate if the learning rate is too low or too high respectively. One way to tackle this problem is by employing learning rate schedulers, which adjust the learning rate during training to optimize model performance. In the most of the approaches, the learning rate is set to a value obtained by hyperparameter tuning, usually done on a subset of the data for fast design, and then decreased at specific time or kept constant during the training of the model. Motivated by the need for better optimization techniques, this paper proposes a new automatic learning rate scheduler based on the gradient to improve the model convergence and generalisation. We introduce a Gradient-based Learning Rate scheduler (GLR) that combines the learning rate with the norm of the gradient by adjusting automatically the value throughout the training process. Our scheduler follows the gradient introducing a booster or a blocker factor according to the trend of the training. GLR requires only the initial value of learning rate. Our method achieves competitive and better results compared to existing state-of-the-art schedulers with multiple user-defined parameters and Adam optimizer algorithm which takes into account the gradient during training [10]. Thus, our contributions are as follows. First, we propose a novel automatic learning rate scheduler based on the norm of the gradient with a single user-defined parameter, i.e. initial learning rate value, and low tuning complexity; the algorithm achieves good results compared to hand-tuned schedulers, and has low computational cost. Second, we extend our experiments to the optimizer Adam whose performances are worse than our algorithm. Finally, we will release publicly our code to encourage future comparisons of our work.

The remainder of the paper is organised as follows. Section 2 contains an overview about the complexity of learning rate tuning, common techniques to adjust it and optimizers. Section 3 explains the proposed method. Section 4 provides the experiments results and Sect. 5 provides conclusions and future works.

2 Learning Design

Among the high number of choices to be determined in the design of a model training, in this section we focus on the complexity of the learning rate tuning, learning rate scheduling and the optimizer contribute to the outcomes of the training.

2.1 Complexity of the Learning Rate Tuning

Learning rate is one of the essential parameters employed in deep learning algorithms, which determines how much contribution each training example has on the overall model updates. In Eq. 1 SGD iteratively updates the network parameters θ by multiplying the learning rate λ by the derivative of the cost function $g(\theta)$ and subtracting it from the parameters. The learning rate allows to control the magnitude of the change in network parameters. A small learning rate causes smaller changes at each step of the training, while a large learning rate results in more significant changes to the parameters. For this reason, the learning rate can be considered as a hyperparameter that controls the speed at which a model is trained. It is a crucial factor that can significantly influence the performance of a neural network [14,15]. Moreover, it is not possible to know the best learning rate a priori [18]. Since real-world network have tens of millions of parameters [5,7,21,22], guessing a good learning rate is challenging and expensive task [16].

The error surface for a linear neuron lies in a space with an horizontal axis for each weight and one vertical axis for the model error. For a linear neuron with a squared error, it is a quadratic bowl, vertical cross-sections are parabolas (see Fig. 1). For multi-layer non-linear neural networks the error surface is much more complicated, but locally, a piece of quadratic bowl is usually a good approximation. If the learning rate is big, the weights slosh to and from across the ravine (see Fig. 1). If the learning rate is too big this oscillation diverges. For this reason, a common safe strategy consists on keeping the learning rate low and decreasing it after a number of epochs chosen a priori (i.e., time-based) or keeping it constant. Turning down the learning rate reduces the random fluctuations in the error due to the different gradients on different mini-batches. As consequence, we reach a local minimum but we also obtain a slower learning. For this reason, some learning rate decay scheduling strategies have been proposed so far. The reason is that the hyperparameters controlling the learning rate scheduling depends on the specific task, the data and the model. Each of these factors contributes on the definition of an error function corresponding to a complex high-dimensionality and not predictable manifold. Most of the

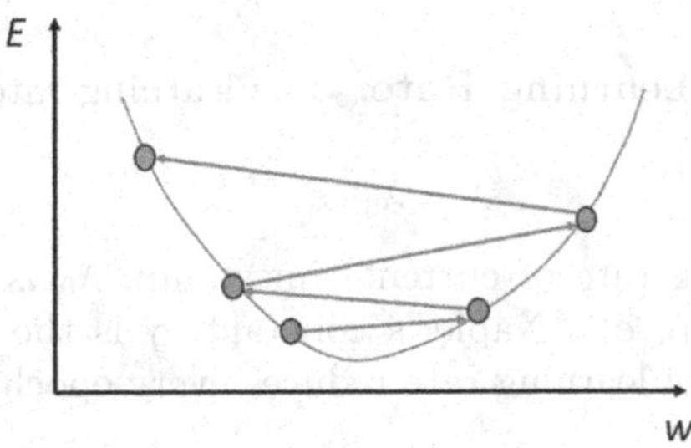

Fig. 1. Example of when the learning goes wrong due to a large learning rate.

efforts have been instead devoted to the definition of adaptive gradient descent approaches to obtain a better optimization with simple learning rate definition.

2.2 Learning Rate Schedulers

Learning rate schedulers play a crucial role in deep learning. Originally, Bottou et al. [2] proposed them to ensure the convergence of SGD in convex optimization reducing noise.

As previous said, the concept behind learning rate scheduling is to adjust the learning rate at specific intervals during the training process to balance the resolution and training speed of the model. The main motivation of employing schedulers of learning rate is to enhance the overall accuracy of a deep learning algorithm by adjusting the learning rate as the training process progresses [13]. The learning rate scheduling strategy can be implemented in various ways. Each scheduler has its own advantages and disadvantages and they all require hyper-parameter tuning. In the following, we provide a brief review of the most common approaches for learning rate tuning.

Constant Learning Rate. If we do not employ any scheduler, the learning rate is a fixed number. The right choice of the value is a difficult task that can be determined through experiments [4]. It is a time-consuming and labor-intensive task.

Step Decay Learning Rate. The learning rate gradually decreases after a predefined number of epochs:

$$\lambda = \lambda_0 \cdot \gamma^{\lfloor \frac{epoch}{step} \rfloor} \tag{2}$$

where λ is the learning rate of current timestamp, λ_0 is the learning rate of the previous timestamp, γ is the constant factor by which learning rate drops each time, *step* is the number of epochs after which learning rate will drop, *epoch* is the number of epoch. Generally, we choose a relatively high value that reduces during the training as shown in Fig. 2a.

Exponential Decay Learning Rate. The learning rate decreases exponentially following the rule

$$\lambda = \lambda_0 \cdot e^{-\gamma t} \tag{3}$$

where λ is the learning rate of current timestamp, λ_0 is the learning rate of the previous timestamp, e is Napier's constant, γ is the constant factor, t is timestamp. The value of learning rate reduces every epoch as shown in Fig. 2b.

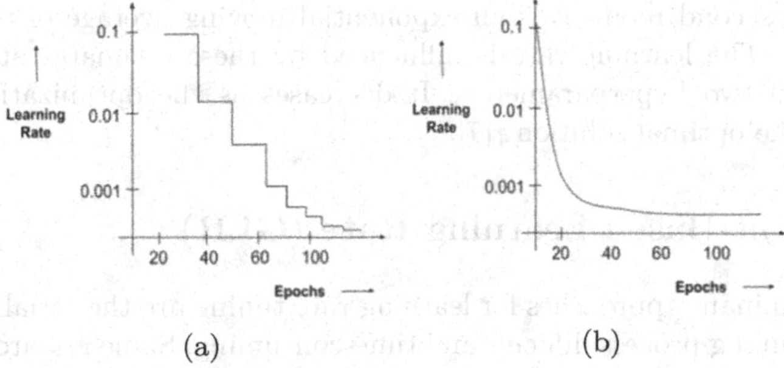

(a) (b)

Fig. 2. Most common learning rate schedulers available in all deep learning libraries: **(a)** step decay; **(b)** exponential decay. Images from work of Konar et al. [11].

ASLR. Khodamoradi et al. [9] proposed an Adaptive Scheduler for Learning Rate (ASLR) that adjusts the learning rate during the training process. If the validation error plateaus, the learning rate is adjusted through a search algorithm. The value of the learning rate can be increased or decreased according to the direction of a search algorithm as shown in Fig. 3. Its value is multiplied by a factor $s = 10\lfloor log_{10}\lambda \rfloor$ after every next epoch and stops when there is an improvement in the validation error.

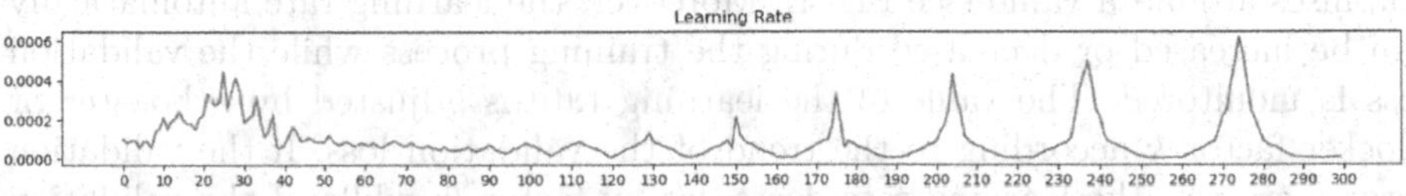

Fig. 3. ASLR scheduler with initial learning rate = 0.0001.

2.3 Optimizer

Choosing the optimizer is considered to be among the most crucial design decisions in deep learning. This task is not easy as that of learning rate scheduler [20]. The growing literature now lists hundreds of optimization methods. SGD and Adaptive Moment Estimation (Adam) are commonly employed to minimize the cost function during training.

SGD works by computing the gradient of the loss function for the model parameters, and then adjusting the parameters in the direction of the negative gradient as explained in Sect. 2. However, SGD has some limitations that often lead to slow convergence and poor generalization. Adam addresses these issues of SGD converging quickly towards the optimal solution [19]. The Adam optimizer estimates the first and the second moments of the gradients to update the network parameters: the first moment is an exponential moving average of the gradient; the second moment is an exponential moving average of the squared gradient [10]. The learning rate is influenced by these estimates at each iteration through two hyperparameters. It decreases as the optimization process approaches the optimal solution [17].

3 Gradient-Based Learning Rate (GLR)

The main dominant approaches for learning rate tuning are the "trial and error" making the tuning process difficult and time-consuming. Some researchers investigated self-adaptive learning rate algorithms to adjust the learning rate according to certain rules. These rules denote the so-called learning rate schedulers. These are effective but most of them are designed manually or kept static during the training process. Other researchers dealt with self-adaptive optimizers that influence the learning rate. Nevertheless, both approaches introduce further hyperparameters to optimize the training process and still hardly provide good generalisation performance for all datasets [1].

Our proposed scheduler addresses these limitations. The Algorithm 1 is the pseudo-code of GLR; while Fig. 4 illustrates an example of learning rate adjustment with GLR. The user chooses the starting value for the learning rate. During the training, the learning rate is multiplied by the value of the average of the norm of gradient extracted for each layer of the neural network. We observed that the norm of the gradient rises when the error is falling steeply and then stabilises around a value (see Fig. 4). Moreover, the learning rate automatically can be increased or decreased during the training process while the validation loss is monitored. The value of the learning rate is adjusted by a booster or blocker factor k according to the trend of the validation loss. If the validation error increases, the learning rate decreases by factor k; while, if the validation error decreases, the learning rate increases by factor k. The intuition behind this adjustment is to accelerate if the loss is approaching a minimum (i.e. the booster factor when the validation error decreases) and to decelerate if it is moving away (i.e. the blocker factor when the validation error increases). This adjustment is made after every next epoch unless the validation loss value remains unchanged (this case is very rare). In addition, there is a check on the value of the current learning rate lr that must not exceed 1: if it happens, the starting value is restored.

Heuristic Observations. During the assessment of our method, we observed some useful behaviours in its application. We suggested $k = 10$ in order to have feasible changes in the current learning rate lr. Generally, state-of-the-art schedulers usually start from a big value of learning rate because it always decreases during the training. In our case, this is not guaranteed. Moreover, we observed that a good starting value of the learning rate is 10^{-3} which is a

Algorithm 1. GLR algorithm

Require: initial learning rate lr_0
  $lr \leftarrow lr_0$ ▷ The starting value of learning rate is saved in lr_0 while lr will be updated
  minimum_error ← 1
  while training **do**
    process one epoch
    $grd \leftarrow$ average of $\|gradient\|$ of each layer
    $lr \leftarrow lr \times grd$
    error ← validation error
    if error > minimum_error **then**
      $lr = lr \times \frac{1}{k}$
    else if error < minimum_error **then**
      $lr = lr \times k$
    end if
    if $lr >= 1$ **then**
      $lr \leftarrow lr_0$
    end if
    minimum_error ← error
  end while

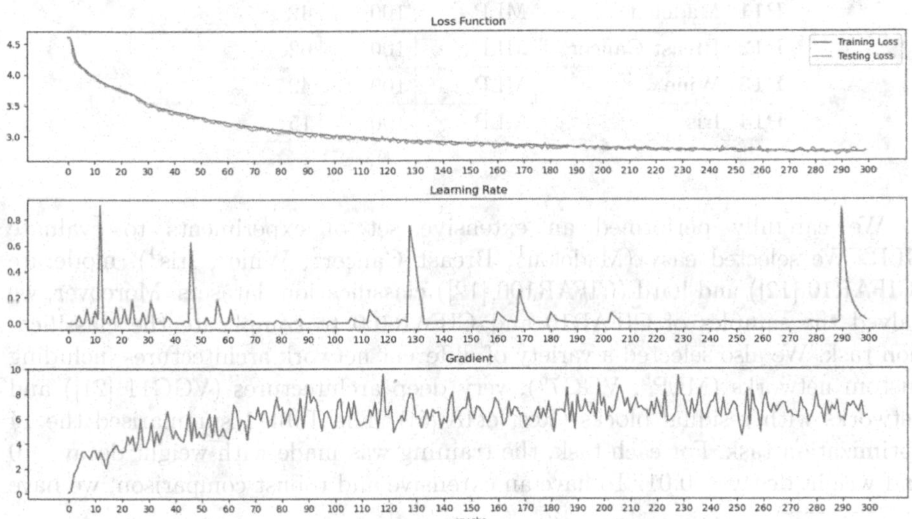

Fig. 4. GLR algorithm. The learning rate is changed according to the validation loss; if it exceeds 1, the starting value is restored. The norm of the gradient rises when the loss is falling steeply and then stabilises around a value.

relatively small value; if the value is greater than 10^{-3}, the adjustments lead to a learning rate too big to skip the minimum; while, if the value is lower than 10^{-3}, the target accuracy is achieved slowly.

4 Results and Experiments

Table 1. Summary of problems employed in our experiments.

	Dataset	Model	Epochs	No. parameters
P1	100% CIFAR10	VGG11	300	128807306
P2	100% CIFAR100	VGG11	300	128807306
P3	100% CIFAR10	VGG7	300	107085028
P4	100% CIFAR100	VGG7	300	107085028
P5	50% CIFAR10	VGG11	300	128807306
P6	50% CIFAR100	VGG11	300	128807306
P7	50% CIFAR10	VGG7	300	107085028
P8	50% CIFAR100	VGG7	300	107085028
P9	100% CIFAR100	ResNet18	300	11227812
P10	50% CIFAR100	ResNet18	300	11227812
P11	Madelon	MLP	100	42
P12	Breast Cancer	MLP	100	62
P13	Wine	MLP	100	42
P14	Iris	MLP	100	15

We carefully performed an extensive set of experiments to evaluate GLR. We selected easy (Madelon[1], Breast Cancer[2], Wine[3], Iris[4]), moderate (CIFAR10 [12]) and hard (CIFAR100 [12]) classification datasets. Moreover, we halved the samples of CIFAR10 and CIFAR100 to complicate the classification task. We also selected a variety of different network architectures including custom networks (MLP[5], VGG7[6]), very deep architectures (VGG11 [21]) and networks with residual blocks (ResNet18 [5]). The Table 1 summarised the 14 optimization task. For each task, the training was made with weight decay = 0 and weight decay = 0.01. To have an extensive and robust comparison, we have

[1] https://scikit-learn.org/stable/modules/generated/sklearn.datasets.make_classification.html.

[2] https://scikit-learn.org/stable/modules/generated/sklearn.datasets.load_breast_cancer.html.

[3] https://scikit-learn.org/stable/modules/generated/sklearn.datasets.load_wine.html.

[4] https://scikit-learn.org/stable/modules/generated/sklearn.datasets.load_iris.html.

[5] It consists of only one linear layer.

[6] It is generated from VGG11 removing 4 convolutional layers.

the following configurations for each problem: constant learning rate with SGD, constant learning rate with Adam, step learning rate with SGD ($\gamma = 0.1$; step = 75 from P1 to P10 and step = 30 from $P11$ to $P14$), GLR with SGD ($k = 10$). In this way, our setting covers the following combinatorial space resulting in 112 experiments:

$$\begin{Bmatrix} \text{P1} \\ \text{P2} \\ \ldots \\ \text{P14} \end{Bmatrix}_{14} \times \begin{Bmatrix} wd = 0.0 \\ wd = 0.1 \end{Bmatrix}_2 \times \begin{Bmatrix} \text{constant with SGD} \\ \text{constant with Adam} \\ \text{step with SGD} \\ \text{GLR with SGD} \end{Bmatrix}_4 \tag{4}$$

In all of our experiments, we employed the Cross Entropy Loss Function as loss function and the accuracy as metric. Additionally, we set learning rate = 0.0001 and batch size = 32. The Table 2 shows the results of our set of experiments.

We observed that our method does not degrade the results for all network topologies and task complexity. Actually, it achieves results comparable to the other configurations and in some cases improves on them. The Table 2 depicts that in 13 configurations the best algorithm is GLR with SGD, in 7 it is step with SGD, in 5 it is constant with SGD and finally in 3 it is constant with Adam. Our method brings the following benefits that combine in some cases (for simplicity, subscript 0 indicates wd = 0 and subscript 1 wd = 1.):

- increased accuracy: see $P1_0$, $P2_1$, $P5_0$, $P7_0$, $P8_0$, $P9_0$, $P11_1$, $P13_1$, $P14_0$ and $P14_1$;
- faster attainment of the same accuracy: see $P1_0$, $P5_0$, $P7_0$, $P12_1$, $P13_0$ and $P13_1$;
- removal or delay of overfitting: see $P1_0$, $P5_0$, $P7_0$, $P11_1$ $P9_0$, $P8_0$, $P14_0$ and $P14_1$.

The Fig. 5 is a comparison of all four configurations for the $P7_0$ task in which we can observe the combination of all improvements delivered by our scheduler. While the Fig. 6 shows the same comparison for the toy $P13_1$ in which there are increased accuracy and faster attainment of it. Finally, we observed that our method does not affect the computational cost.

Table 2. Results of all set of experiments. The best results are highlighted in bold. The phenomenon cell denotes the behaviour that affects the training: O stands for overfitting from a certain epoch; GE stands for gradient's explosion; no corresponds to a good training.

	wd	Configuration	Best Accuracy	Best Epoch	Phenomenon		wd	Configuration	Best Accuracy	Best Epoch	Phenomenon
P1	0.0	constant SGD	0.8046	295	O ep. 90	P2	0.01	**constant SGD**	**0.1769**	**300**	**no**
		constant Adam	0.8009	298	O ep. 60			constant Adam	0.4713	133	O ep. 40
		step SGD	0.7939	278	O. ep. 60			step SGD	0.5157	129	O ep. 60
		GLR SGD	**0.8064**	**288**	**O ep. 90**			GLR SGD	0.5051	129	O ep. 70
	0.01	constant SGD	0.7140	263	O ep. 170		0.01	constant SGD	0.0145	33	GE
		constant Adam	0.5966	297	no			constant Adam	0.1197	300	no
		step SGD	**0.7338**	**236**	**no**			step SGD	0.0538	263	no
		GLR SGD	0.6514	221	no			**GLR SGD**	**0.1455**	**297**	**no**
P3	0.0	**constant SGD**	**0.8085**	**292**	**O ep. 60**	P4	0.0	constant SGD	0.5419	285	O epo. 45
		constant Adam	0.7945	231	O ep. 30			constant Adam	0.5251	130	O ep. 25
		step SGD	0.8081	251	O ep. 60			**step SGD**	**0.5465**	**287**	**O ep. 40**
		GLR SGD	0.8035	298	O ep. 60			GLR SGD	0.5406	300	O ep. 45
	0.1	constant SGD	0.6344	213	O ep. 60		0.1	constant SGD	0.3186	283	no
		constant Adam	0.6304	215	no			constant Adam	0.2317	259	no
		step SGD	**0.6898**	**242**	**no**			**step SGD**	**0.3373**	**261**	**no**
		GLR SGD	0.6344	214	no			GLR SGD	0.3171	289	no
P5	0.0	constant SGD	0.7522	261	O ep. 70	P6	0.0	constant SGD	0.4214	160	O ep. 60
		constant Adam	0.7456	247	O ep. 40			constant Adam	0.3850	136	O ep. 40
		step SGD	0.7426	296	O ep. 60			**step SGD**	**0.4220**	**179**	**O ep. 55**
		GLR SGD	**0.7554**	**203**	**O ep. 70**			GLR SGD	0.4200	157	O 60
	0.1	**constant SGD**	**0.6550**	**220**	**no**		0.1	constant SGD	0.0440	210	no
		constant Adam	0.5354	278	no			**constant Adam**	**0.1272**	**284**	**no**
		step SGD	0.6854	300	O ep. 150			step SGD	0.0464	167	no
		GLR SGD	0.6520	287	no			GLR SGD	0.0920	291	no
P7	0.0	constant SGD	0.7522	264	O ep. 70	P8	0.0	constant SGD	0.0606	299	no
		constant Adam	0.7372	192	O ep. 20			constant Adam	0.4352	210	O ep. 30
		step SGD	0.7518	229	O ep. 60			step SGD	0.4328	263	O ep. 50
		GLR with SGD	**0.7540**	**186**	**O ep. 70**			**GLR SGD**	**0.4446**	**277**	**O ep. 50**
	0.1	**constant SGD**	**0.6416**	**277**	**no**		0.1	constant SGD	0.0220	265	no
		constant Adam	0.6292	281	no			constant Adam	0.2022	273	no
		step SGD	0.6230	248	no			step SGD	0.2510	224	no
		GLR SGD	0.6340	244	no			**GLR SGD**	**0.2956**	**293**	**no**
P9	0.0	constant SGD	0.4248	267	O ep. 40	P10	0.0	constant SGD	0.3460	291	O epoch 35
		constant Adam	0.4360	248	O ep. 30			constant Adam	0.3546	115	O ep. 30
		step SGD	0.4283	163	O ep. 40			**step SGD**	**0.3540**	**145**	**O ep. 35**
		GLR SGD	**0.4258**	**285**	**O ep. 40**			GLR SGD	0.3472	247	O ep. 35
	0.1	constant SGD	0.3628	245	no		0.1	constant SGD	0.3636	284	no
		constant Adam	**0.4227**	**285**	**no**			**constant Adam**	**0.4046**	**278**	**no**
		step SGD	0.5376	277	O ep. 150			step SGD	0.4674	274	O ep. 120
		GLR SGD	0.3629	294	no			GLR SGD	0.3606	294	no
P11	0.0	**constant SGD**	**0.8800**	**99**	**no**	P12	0.0	constant SGD	0.9371	51	no
		constant Adam	0.6800	89	no			constant Adam	0.3357	1	no
		step SGD	0.7600	35	no			**step SGD**	**0.9371**	**24**	**no**
		GLR SGD	0.8560	60	no			GLR SGD	0.9371	53	no
	0.1	constant SGD	0.8960	79	no		0.1	constant SGD	0.9371	61	no
		constant Adam	0.5600	61	no			constant Adam	0.3357	1	no
		step SGD	0.7200	58	O			step SGD	0.9371	24	no
		GLR SGD	**0.9280**	**90**	**no**			**GLR SGD**	**0.9371**	**2**	**no**
P13	0.0	constant SGD	0.9333	100	no	P14	0.0	constant SGD	0.6842	50	no
		constant Adam	0.4222	1	no			constant Adam	0.3421	1	no
		step SGD	0.8889	34	no			step SGD	0.6579	30	O
		GLR SGD	**0.9333**	**64**	**no**			**GLR SGD**	**0.9474**	**71**	**no**
	0.1	constant SGD	0.9333	95	no		0.1	constant SGD	0.8421	83	no
		constant Adam	0.4222	1	no			constant Adam	0.2632	1	O
		step SGD	0.7778	16	no			step SGD	0.5263	14	no
		GLR SGD	**0.9333**	**64**	**no**			**GLR SGD**	**0.9474**	**97**	**no**

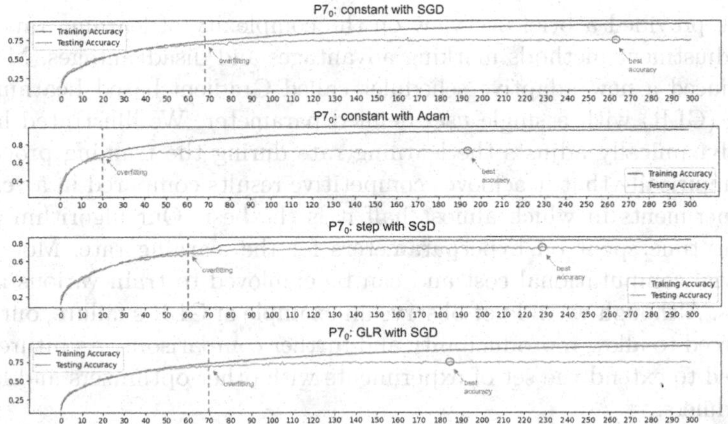

Fig. 5. Accuracy curves of the four configuration accuracy for $P7_0$ task (i.e. VGG7 on 50% CIFAR 10) of high complexity. The red dotted line marks the onset of overfitting, while the red ellipse highlights the best accuracy. $P7_0$ has all three benefits of GLR. Actually, we can observe that GLR delays the occurrence of overfitting significantly in constant with Adam configuration; moreover, GLR achieves higher accuracy with fewer epochs than other configurations. (Color figure online)

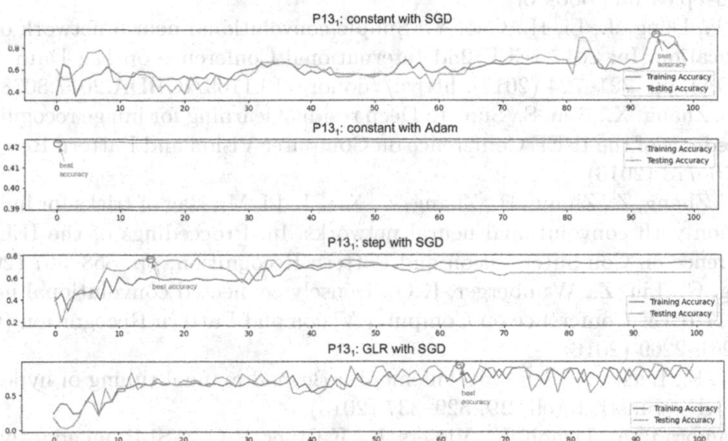

Fig. 6. Accuracy curves of the four configuration accuracy for the toy $P13_1$ task (i.e. MLP on Wine dataset). The red ellipse highlights the best accuracy. Here, we can observe that GLR achieves higher accuracy with fewer epochs than other configurations. (Color figure online)

5 Conclusions

This work provided a brief overview on the complexity of learning rate tuning and its adjustment methods marking advantages and disadvantages. Moreover, we introduced a new adaptive scheduler called Gradient-based Learning Rate scheduler (GLR) with a single user-defined parameter. We illustrated how our method dynamically adjusts the learning rate during the training process and showed empirically that it achieves competitive results compared in a very wide set of experiments in which almost half it is the best. Our algorithm reduces the tuning time spent on hyperparameters for the learning rate. Moreover, it has a trivial computational cost and can be employed to train various network topologies. Although we did not observe an example of GLR's failure, our code is open-sourced to allow reproducibility and further comparisons. As future works, we planned to extend the set of experiments with other optimizers and learning rate schedulers.

Acknowledge financial support from. PNRR MUR project PE0000013-FAIR

References

1. Andrychowicz, M., et al.: Learning to learn by gradient descent by gradient descent. In: Advances in Neural Information Processing Systems, vol. 29 (2016)
2. Bottou, L.: Online learning and stochastic approximations. Online Learn. Neural Netw. **17**, 142 (1998)
3. Goodfellow, I., Bengio, Y., Courville, A.: Deep Learning. MIT Press (2016). http://www.deeplearningbook.org
4. Guo, T., Dong, J., Li, H., Gao, Y.: Simple convolutional neural network on image classification. In: 2017 IEEE 2nd International Conference on Big Data Analysis (ICBDA), pp. 721–724 (2017). https://doi.org/10.1109/ICBDA.2017.8078730
5. He, K., Zhang, X., Ren, S., Sun, J.: Deep residual learning for image recognition. In: Proceedings of the IEEE Conference on Computer Vision and Pattern Recognition, pp. 770–778 (2016)
6. He, T., Zhang, Z., Zhang, H., Zhang, Z., Xie, J., Li, M.: Bag of tricks for image classification with convolutional neural networks. In: Proceedings of the IEEE/CVF Conference on Computer Vision and Pattern Recognition, pp. 558–567 (2019)
7. Huang, G., Liu, Z., Weinberger, K.Q.: Densely connected convolutional networks. In: 2017 IEEE Conference on Computer Vision and Pattern Recognition (CVPR), pp. 2261–2269 (2016)
8. Hutter, F., Lücke, J., Schmidt-Thieme, L.: Beyond manual tuning of hyperparameters. KI - Künstl. Intell. **29**, 329–337 (2015)
9. Khodamoradi, A., Denolf, K., Vissers, K., Kastner, R.C.: ASLR: an adaptive scheduler for learning rate. In: 2021 International Joint Conference on Neural Networks (IJCNN), pp. 1–8 (2021). https://doi.org/10.1109/IJCNN52387.2021.9534014
10. Kingma, D.P., Ba, J.: Adam: a method for stochastic optimization. CoRR (2015)
11. Konar, J., Khandelwal, P., Tripathi, R.: Comparison of various learning rate scheduling techniques on convolutional neural network. In: 2020 IEEE International Students' Conference on Electrical, Electronics and Computer Science (SCEECS) (2020). https://doi.org/10.1109/SCEECS48394.2020.94

12. Krizhevsky, A.: Learning multiple layers of features from tiny images. Toronto University, ON, Canada - Master's thesis (2009)
13. Lewkowycz, A.: How to decay your learning rate. ArXiv abs/2103.12682 (2021)
14. Martens, J.: Deep learning via hessian-free optimization. In: International Conference on Machine Learning (2010)
15. Martens, J., Grosse, R.: Optimizing neural networks with Kronecker-factored approximate curvature. In: International Conference on Machine Learning (2015)
16. Nocedal, J., Wright, S.J.: Numerical Optimization. Springer, New York (1999). https://doi.org/10.1007/978-0-387-40065-5
17. Reddi, S.J., Kale, S., Kumar, S.: On the convergence of ADAM and beyond. ArXiv abs/1904.09237 (2018)
18. Reed, R., MarksII, R.J.: Neural Smithing: Supervised Learning in Feedforward Artificial Neural Networks. MIT Press (1999)
19. Ruder, S.: An overview of gradient descent optimization algorithms. ArXiv abs/1609.04747 (2016)
20. Schmidt, R.M., Schneider, F., Hennig, P.: Descending through a crowded valley-benchmarking deep learning optimizers. In: International Conference on Machine Learning (2021)
21. Simonyan, K., Zisserman, A.: Very deep convolutional networks for large-scale image recognition. In: Proceedings of the 3rd International Conference on Learning Representations (ICLR 2015) (2015)
22. Zagoruyko, S., Komodakis, N.: Wide residual networks. In: Proceedings of the British Machine Vision Conference (BMVC) (2016). https://doi.org/10.5244/C.30.87

A Large-scale Analysis of Athletes' Cumulative Race Time in Running Events

David Freire-Obregón(✉), Javier Lorenzo-Navarro, Oliverio J. Santana, Daniel Hernández-Sosa, and Modesto Castrillón-Santana

SIANI, Universidad de Las Palmas de Gran Canaria, Las Palmas de Gran Canaria, Spain

david.freire@ulpgc.es

Abstract. Action recognition models and cumulative race time (CRT) are practical tools in sports analytics, providing insights into athlete performance, training, and strategy. Measuring CRT allows for identifying areas for improvement, such as specific sections of a racecourse or the effectiveness of different strategies. Human action recognition (HAR) algorithms can help to optimize performance, with machine learning and artificial intelligence providing real-time feedback to athletes. This paper presents a comparative study of HAR algorithms for CRT regression, examining two important factors: the frame rate and the regressor selection. Our results indicate that our proposal exhibits outstanding performance for short input footage, achieving a mean absolute error of 11 min when estimating CRT for runners that have been on the course for durations ranging from 8 to 20 h.

Keywords: Sports Analytics · Ultra-distance competition · Human Action Recognition

1 Introduction

Action recognition models have become increasingly important in recent years due to their ability to identify and analyze specific movements or actions executed by athletes during competition or training. These models can provide valuable insights into technique, form, and performance, making it possible for coaches and trainers to develop more effective training programs and improve overall performance.

Machine learning and artificial intelligence techniques have revolutionized how data is collected and utilized for sports analytics. These technologies have enabled coaches, trainers, and analysts to gain deeper insights into athlete performance, identify areas for improvement, and develop more effective strategies.

This work is partially funded by the the Spanish Ministry of Science and Innovation under project PID2021-122402OB-C22, and by the ACIISI-Gobierno de Canarias and European FEDER funds under project, ProID2021010012, ULPGC Facilities Net, and Grant EIS 2021 04.

G. L. Foresti et al. (Eds.): ICIAP 2023, LNCS 14233, pp. 282–292, 2023.
https://doi.org/10.1007/978-3-031-43148-7_24

Fig. 1. Samples of a runner's footage at each recording point. The runner in focus is enclosed within a green container. We utilized various HAR backbones on each footage for feature extraction to obtain the runner's embeddings. Subsequently, these extracted embeddings were utilized as inputs for a model to estimate CRT at a particular recording point.

Combining action recognition algorithms and classical regression techniques can provide a powerful toolset for analyzing and improving athlete performance. From the runner's perspective, coaches and trainers can develop more targeted training programs, optimize race strategies, and improve overall performance by accurately measuring the cumulative race time (CRT) and identifying key actions and movements.

Recently, ultra-distance competitions have emerged as a challenging scenario to evaluate runner CRT [5,6]. In contrast to previous research on action quality assessment (AQA) in sports, the focus is not only on measuring the performance of methods based on ground truth and predicted score series but instead on CRT. At a specific recording point RP_i, CRT can be defined as the split time that shows the cumulative race time up to that recording point in the race, see Fig. 1. Accurately measuring CRT is critical in many sports, such as motorsports, cycling, and skiing. This measure can help to select areas for improvement in athlete performance, such as identifying the fastest sections of a racecourse or comparing lap times to determine the effectiveness of different strategies.

An algorithm that can estimate a runner's race time by observing them has several potential advantages, including (i) real-time feedback, (ii) reduced cost and time, and (iii) enhanced safety. With an algorithm that can estimate a runner's race time, coaches or race officials can provide real-time feedback to the runner during a race. This can help the runners adjust their pace and strategy to optimize their performance. Traditional methods of estimating a runner's race time often involve collecting and analyzing data from multiple sources, which can be time-consuming and costly. An algorithm that can estimate a runner's race time by observing them could reduce the need for additional data collection and analysis, saving time and resources. Real-time monitoring of a runner's race time could also enhance safety by identifying when a runner is at risk of overexertion or other health issues. This information could be used to intervene and prevent injury or other adverse outcomes.

Consequently, integrating action recognition models with sports analytics can revolutionize athlete training, enhancing their technique, efficiency, and performance. However, implementing this approach in ultra-distance races poses significant challenges due to race duration, dynamic scenes, runners' appearance variability, occlusions, and multiple scenarios. To overcome these obstacles, our study proposes a framework that utilizes seven action recognition backbones to outperform existing state-of-the-art proposals. The backbones were trained on different frame rate inputs to assess their performance under various temporal resolutions. Additionally, traditional machine learning regressors were combined with the backbones. The effectiveness of the proposed framework was evaluated on the TGC20ReID dataset, an ultra-distance runner's benchmark dataset. The study's results revealed that the proposed method achieved a mean absolute error of roughly 11 min when estimating CRT for runners who had been active for 8 to 20 h, surpassing state-of-the-art approaches.

2 Related Work

In recent years, deep learning has gained significant attention for AQA in sports due to its potential to provide more accurate and efficient analysis than traditional methods. AQA in sports is a challenging task that requires accurate recognition of athletes' movements and poses. Recently, action recognition networks have emerged as a promising solution to this problem. These networks can automatically recognize and classify the movements performed by athletes and can also be used to assess the quality of their movements [4]. However, one of the key challenges in AQA is the variability and complexity of the movements involved. Traditional methods for analyzing movements in sports often rely on manual annotations or feature engineering, which can be time-consuming and prone to errors. Deep learning approaches offer a promising alternative: they can automatically learn relevant features from raw data and model the temporal dependencies in the movements [3]. Although deep learning AQA methods have been proposed for various sports, including basketball, soccer, and swimming, our work focuses on a different aspect of sports analytics. We propose a CRT regression model that predicts the time a runner has spent in a race rather than assessing the quality of the running technique. This approach can provide insights into the athlete's performance and workload during a competition, which can be valuable for training and injury prevention. Both AQA and CRT regression share common characteristics, such as belonging to the domain of sports analytics and utilizing a similar pipeline scheme.

Most traditional AQA approaches analyze the quality of actions in videos by examining their frames, but they often overlook the unique characteristics of pose dynamics. To address this limitation, some researchers have proposed novel techniques incorporating pose information to improve AQA performance. For example, the Pose + DCT method utilizes independent subspace analysis and DCT/DFT features from poses to evaluate action quality [15]. Pan et al. developed the AQA-7 system, which extracts motion features from patch videos of

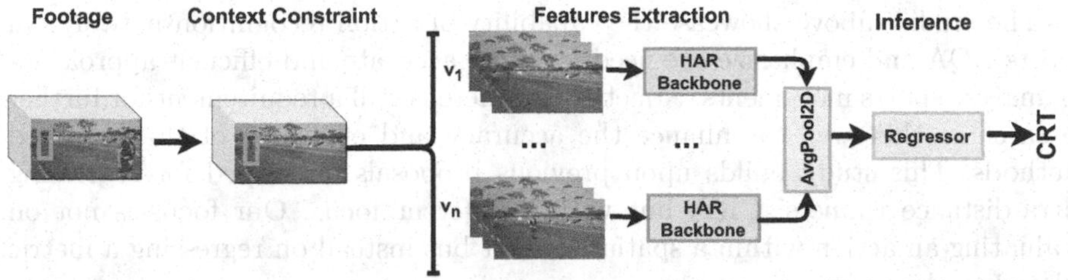

Fig. 2. The proposed pipeline for regressing runner's CRT. The process we have developed comprises three key components: the footage pre-processing block, the feature extraction block and the regression block. The first component involves using a tracker to remove the background activity and isolate the runner. The second component involves breaking the footage into n small clips through down-sampling, which are then sent to a HAR backbone to extract features. The features are further synthesized using an average pooling technique to obtain the final feature tensor. This tensor serves as the input to the regressor for predicting CRT.

major body joints, using joint difference and commonality modules to capture better joint dependency [12]. However, this method is time-intensive, requiring cropping local patches around joints. In contrast, the FALCONS approach includes a difficulty estimator that utilizes a look-up table to evaluate the task's difficulty level during diving [10]. The EAGLE-Eye method employs a joint coordination estimator on the pose heatmaps to learn temporal dependencies using multiscale convolution [11].

Skeleton pose estimation is a challenging task in sports due to various factors such as occlusions, unusual poses of humans, missing key points due to blurry footage, and changes in lighting or clothing. Appearance-based approaches have emerged as an alternative to tackle this problem. Parmar and Morris proposed the C3D-AVG-MTL multitask model, which learns spatiotemporal feature vectors to identify actions, provide quality scores, and generate captions for videos [13]. They introduced the MTL-AQA benchmark dataset, which is now the standard dataset for AQA in diving videos. C3D-AVG-MTL employs the 3D ConvNet architecture (C3D) to extract features from videos, which is successful in representing spatiotemporal data hierarchically [18]. The extracted features are fed into the regression modeling architecture to obtain scores. The two-stream inflated 3D ConvNets (I3D) build upon the C3D architecture by incorporating an optical-flow stream component, providing better representations. This I3D architecture is also utilized by Freire et al., who address CRT regression in ultra-distance runners. Their approach uses the I3D architecture to extract video features; several regressors are tested, mapping the features to performance [5] and to CRT [6]. In a related study, Yu et al. reframe the AQA problem as regressing the relative scores by referencing another video with shared characteristics (such as category and difficulty) to obtain more precise AQA during training and inference [20].

The studies above showcase the capability of action recognition networks in sports AQA and emphasize the need for more accurate and efficient approaches to analyze sports movements. Nonetheless, there is still a requirement for further research in this area to enhance the accuracy and efficiency of the proposed methods. This study builds upon previous proposals developed for regressing ultra-distance runners' CRT, but with a different focus. Our focus is not on evaluating an action within a spatial context but instead on regressing a metric related to the temporal context.

3 Methodology

The proposed process consists of three main components, as shown in Fig. 2: the footage pre-processing block, the feature extraction block, and the regression block. Firstly, the tracker eliminates extraneous background activity to focus on the runner of interest. This pre-processing block prepares the raw input data for further analysis. Next, the footage is down-sampled and segmented into n smaller clips, which are then processed by HAR backbones to extract features. Finally, the resulting features are synthesized through an average pooling process to create the embedding, which is fed into the regressor.

Context Removal. To ensure accurate inference, input footage for HAR networks must be free of unwanted objects, such as other athletes, race personnel, and cars. Therefore, the initial block of our proposed network removes these elements by pre-processing the raw data and focusing only on the runner of interest. We used ByteTrack [21], a multi-object tracking network, to track the runner in each frame. We then performed a context-constrained pre-processing step to generate the footage used in our experiments. This step involves taking the bounding box area $BB_i(t, RP)$ of the runner i at a given time $t \in [0, T]$ in a recording point $RP \in [0, P]$, along with the average number of frames needed to create clean footage with a still background in which the runner appears. This can be represented formally as:

$$F_i'[RP] = BB_i(t, RP) \cup \tau(RP) \tag{1}$$

The notation $\tau(RP)$ denotes the average frame needed to generate clear footage where the runner is present in a still background.

Feature Extraction and Regression. After down-sampling and dividing the input footage into n video clips (v_1, ..., v_n), each clip containing q consecutive frames representing an activity snapshot (see Fig. 2), the next step is to extract features. For this purpose, each video clip v_i is passed through a pre-trained HAR network, resulting in a feature vector with 192 dimensions. The HAR models have been pre-trained on the Kinetics dataset [9], which contains 400 different action categories. After obtaining all the feature vectors, an average pooling layer is applied to ensure that each clip's information is considered equally. Ultimately, the extracted features are utilized to train a CRT regressor; multiple regression models are experimented with.

3.1 HAR Backbones

The selection of an appropriate model is crucial for accurately classifying actions videos of sporting events, as it must be capable of efficiently extracting spatial and temporal features from video sequences. This subsection provides an overview of the human action recognition models considered for this work.

The **C2D** (Convolutional 2D) model for action recognition is a deep learning architecture designed to classify actions in videos [16]. This model leverages the effectiveness of 2D CNNs in extracting spatial features from video frames, which is a key factor in recognizing actions. The architecture of the C2D model usually includes convolutional layers, pooling layers, and fully connected layers. The convolutional layers extract spatial features from the input frames, while the pooling layers reduce the dimensionality of the features to prevent overfitting. The C2D model operates on each frame of a video sequence independently by employing a CNN to extract spatial features. Subsequently, these features are merged across frames to capture the temporal dynamics of the action.

Unlike the C2D model, the **SlowFast** model is designed based on the concept that different video parts have varying temporal resolutions, which contain essential information for recognizing human actions [2]. For instance, some actions happen rapidly and require high temporal resolution for detection, while others occur slowly and can be recognized using a lower temporal resolution. To overcome this, the SlowFast model employs two separate pathways, the fast and slow pathways, which process video data at different temporal resolutions.

Consequently, **Slow**, adopts a two-stream architecture to capture both short-term and long-term temporal dynamics in videos [3]. Its slow pathway operates on high-resolution frames but at a lower frame rate, similar to the C2D model. Additionally, Slow includes a temporal-downsampling layer to capture longer-term temporal dynamics.

Fig. 3. TGC20ReID Dataset Collage.

The Inflated 3D ConvNet (**I3D**) model processes short video clips as 3D spatiotemporal volumes to capture both appearance and motion cues through a two-stream approach [1]. The first stream processes RGB images, initialized with weights pre-trained on large-scale image classification datasets like ImageNet, while the second stream processes optical flow images, fine-tuned along with the RGB stream.

A modified version of I3D, known as **I3D NLN**, integrates non-local operations to improve video spatiotemporal dependency modeling [19]. I3D NLN also operates on 3D spatiotemporal volumes using a two-stream architecture with RGB and optical flow streams. However, instead of the Inception module, it uses non-local blocks that can learn long-range dependencies between any two positions in the input feature maps. By computing a weighted sum of input features from all positions based on the similarity between every other position in the feature maps, I3D NLN captures global context information and leads to improved modeling of temporal dynamics.

4 Dataset and Experiments

While many works in this sporting domain rely on statistical data, manually gathering sufficient multimedia data is expensive. This is a key challenge in performing AQA, as our pipeline needs athlete's data to regress CRT properly. To address this issue, we employ a collection derived from the TGC20ReID dataset [14] provided by the authors that contains seven-second 25 fps clips at each recording point for each participant (see Fig. 3). The dataset includes annotations for nearly 600 participants across six recording points. However, given the varying performances of the runners, the gap between the leaders and the last runners increases along the course as the number of active participants decreases. As a result, only a subset of 214 runners is eligible for estimating CRT - those runners that have covered the last three recording points during the dataset recording time.

Metric. The observation $o_i[RP]$ of an athlete i at a recording point $RP \in [0, P]$ in the AQA pipeline includes a pre-processed footage $F'_i[RP]$ and a CRT value $\phi_i[RP]$, which is normalized between [0,1] using Eq. 2.

$$\phi'_i[RP] = \frac{\phi_i[RP] - min(\phi_i[0])}{max(\phi_i[RP])} \tag{2}$$

The objective we aim to achieve is to minimize the following end-to-end regression technique:

$$min\ L(\phi'_i[RP], \psi_i[RP]) = \frac{1}{N}\sum_{j=0}^{N} |\phi'_i[RP]_j - \psi_i[RP]_j| \tag{3}$$

Here, $\psi_i[RP]$ represents the predicted value at recording point RP for runner i based on seven seconds of movement observation, and N represents the batch

Table 1. Mean average error (MAE) achieved by each configuration. The first column displays the considered HAR model. The second column shows the number of frames per video clip, the rest of the columns show the achieved MAE on the considered classifiers. Lower is better.

HAR model	#Frames	LR ↓	k-NN ↓	GB ↓	SVM ↓	MLP ↓
C2D [16]	8	0.020	0.011	0.018	0.019	0.019
I3D [1]	8	0.018	0.010	0.016	0.014	0.019
I3D NLN [19]	8	0.019	0.011	0.014	0.014	0.020
Slow4x16 [3]	4	0.021	0.010	0.020	0.022	0.019
Slow8x8 [3]	8	0.021	**0.009**	0.016	0.016	0.020
SlowFast4x16 [2]	32	0.020	0.011	0.015	0.014	0.020
SlowFast8x8 [2]	32	0.020	0.011	0.015	0.016	0.021

size. For performance evaluation purposes, we compute the average Mean Absolute Error (MAE) over 20 repetitions of 10-fold cross-validation. On average, the training dataset consists of 410 samples, while the testing dataset comprises 46 samples.

4.1 Experimental Analysis

We evaluate our pipeline's performance using different regression methods. In addition, we present rates for the following regression techniques: Linear Regression (LR), k-NN, Gradient Boosting (GB), Support Vector Machines (SVM), and Multi-Layer Perceptron (MLP). The results of this evaluation provide insights into the effectiveness of the proposed approach in estimating CRT, which can be helpful for coaches, athletes, and race organizers in the ultra-distance running community.

Table 1 presents the MAE obtained by different regression models on several video architectures. Lower MAE values indicate better model performance. The architectures tested in the table include C2D, I3D, I3D NLN, Slow4x16, Slow8x8, SlowFast4x16, and SlowFast8x8. Each architecture uses a different number of frames per prediction, with some using 4, 8, or 32 frames. Overall, the results show that the k-NN model provides the best performance, achieving the lowest MAE values in the seven architectures. The SVM and GB models also perform well, achieving interesting MAE values in some of the seven architectures.

The Slow and SlowFast architectures achieve low MAE values with the k-NN model, indicating that these architectures may be particularly well-suited for this type of regression task. The Slow8x8 architecture achieves the lowest MAE value of 0.009 with the k-NN model, which is a strong result. On the other hand, the C2D architecture consistently performs worse than the other architectures, with the highest MAE values in most cases. This suggests that the C2D architecture may not be as effective as the other architectures for this type of task. These results provide insight into which regression models perform well for video-based

Table 2. Comparison of different architectures on the dataset used in the present work. The first column shows the considered pre-trained architectures, whereas the second shows the MAE, respectively. Lower is better.

Architecture	MAE ↓
C3D [17]	0.038
3D ResNets-D30 [8]	0.036
3D ResNets-D50 [8]	0.033
3D ResNets-D101 [8]	0.032
3D ResNets-D200 [8]	0.031
I3D-800SB [6]	0.019
I3D-2048SB [6]	0.015
X3D-XS [7]	0.010
Slow8x8 (Ours)	**0.009**

regression tasks and which architectures may be well-suited for these tasks. The k-NN model is a strong choice across most architectures, but other models, such as SVM and GB, may also perform well in some instances.

Table 2 presents the performance comparison of different state-of-the-art architectures for HAR in terms of their MAE values. The architectures are evaluated on the TGC20ReID dataset, and the MAE values are reported in the table. The architectures include C3D, four 3D ResNet architectures (3D ResNets-D30, 3D ResNets-D50, 3D ResNets-D101, and 3D ResNets-D200), I3D-800SB, I3D-2048SB, and our Slow8x8b architecture. The MAE values range from 0.038 for C3D to 0.009 for Slow8x8.

The results indicate that the proposed Slow8x8 architecture outperforms all the state-of-the-art architectures on the benchmark dataset. It achieves the lowest MAE value of 0.009, a 40% better than the best-performing architecture I3D-2048SB with a MAE value of 0.015. Moreover, Slow8x8 also outperforms the popular 3D ResNet architectures, which are known to achieve high accuracy on various computer vision tasks for CRT regression.

5 Conclusions

This paper presents an ultra-distance runner CRT regressor developed using seven different HAR pre-trained backbones. Our large-scale analysis has shown that k-NN is the best regressor in this scenario, achieving an impressive MAE of 0.009. This means that our model can predict CRT with an error of roughly 11 min for a race that can take between 8 to 20 h, a remarkable achievement.

Our findings suggest that using HAR architectures for CRT regression is highly effective. Furthermore, our comparative analysis has shown that different backbones significantly impact the performance of the CRT regressor. The findings of this study can help researchers and practitioners in the field of ultra-distance running to develop more accurate models for predicting CRT, which can be used to inform training and race strategies.

References

1. Carreira, J., Zisserman, A.: Quo Vadis, action recognition? A new model and the kinetics dataset. In: 2017 IEEE Conference on Computer Vision and Pattern Recognition (CVPR), pp. 4724–4733 (2017)
2. Feichtenhofer, C., Fan, H., Malik, J., He, K.: Slowfast networks for video recognition. In: 2019 IEEE/CVF International Conference on Computer Vision (ICCV), pp. 6201–6210 (2018)
3. Feichtenhofer, C., Fan, H., Xiong, B., Girshick, R.B., He, K.: A large-scale study on unsupervised spatiotemporal representation learning. In: 2021 IEEE/CVF Conference on Computer Vision and Pattern Recognition (CVPR), pp. 3298–3308 (2021)
4. Freire-Obregón, D., Barra, P., Castrillón-Santana, M., de Marsico, M.: Inflated 3D ConvNet context analysis for violence detection. Mach. Vis. Appl. **33**(15) (2022). https://doi.org/10.1007/s00138-021-01264-9
5. Freire-Obregón, D., Lorenzo-Navarro, J., Castrillón-Santana, M.: Decontextualized I3D ConvNet for ultra-distance runners performance analysis at a glance. In: International Conference on Image Analysis and Processing (ICIAP), pp. 242–253 (2022)
6. Freire-Obregón, D., Lorenzo-Navarro, J., Santana, O.J., Hernández-Sosa, D., Castrillón-Santana, M.: Towards cumulative race time regression in sports: I3D ConvNet transfer learning in ultra-distance running events. In: International Conference on Pattern Recognition (ICPR), pp. 805–811 (2022)
7. Freire-Obregón, D., Lorenzo-Navarro, J., Santana, O.J., Hernández-Sosa, D., Castrillón-Santana, M.: An X3D neural network analysis for runner's performance assessment in a wild sporting environment. In: International Conference on Machine Vision Applications (MVA) (2023)
8. Hara, K., Kataoka, H., Satoh, Y.: Can spatiotemporal 3D CNNs retrace the history of 2D CNNs and ImageNet. In: Proceedings of the IEEE Conference on Computer Vision and Pattern Recognition (CVPR), pp. 6546–6555 (2018)
9. Kay, W., et al.: The kinetics human action video dataset. CoRR (2017)
10. Nekoui, M., Cruz, F., Cheng, L.: Falcons: fast learner-grader for contorted poses in sports. In: 2020 IEEE/CVF Conference on Computer Vision and Pattern Recognition Workshops (CVPRW), pp. 3941–3949 (2020)
11. Nekoui, M., Cruz, F., Cheng, L.: EAGLE-Eye: extreme-pose action grader using detaiL bird's-Eye view. In: 2021 IEEE Winter Conference on Applications of Computer Vision (WACV), pp. 394–402 (2021)
12. Pan, J., Gao, J., Zheng, W.: Action assessment by joint relation graphs. In: 2019 IEEE/CVF International Conference on Computer Vision (ICCV), pp. 6330–6339 (2019)
13. Parmar, P., Morris, B.T.: What and How Well You Performed? A Multitask Learning Approach to Action Quality Assessment. In: 2019 IEEE/CVF Conference on Computer Vision and Pattern Recognition (CVPR), pp. 304–313 (2019),
14. Penate-Sanchez, A., Freire-Obregón, D., Lorenzo-Melián, A., Lorenzo-Navarro, J., Castrillón-Santana, M.: TGC20ReId: a dataset for sport event re-identification in the wild. Pattern Recogn. Lett. **138**, 355–361 (2020). https://doi.org/10.1016/j.patrec.2020.08.003

15. Pirsiavash, H., Vondrick, C., Torralba, A.: Assessing the quality of actions. In: European Conference on Computer Vision (2014)
16. Simonyan, K., Zisserman, A.: Two-stream convolutional networks for action recognition in videos. ArXiv abs/1406.2199 (2014)
17. Tran, D., Bourdev, L.D., Fergus, R., Torresani, L., Paluri, M.: C3D: generic features for video analysis. CoRR abs/1412.0767 (2014)
18. Tran, D., Bourdev, L.D., Fergus, R., Torresani, L., Paluri, M.: Learning spatiotemporal features with 3D convolutional networks. In: 2015 IEEE International Conference on Computer Vision (ICCV), pp. 4489–4497 (2014)
19. Wang, X., Girshick, R.B., Gupta, A.K., He, K.: Non-local neural networks. In: 2018 IEEE/CVF Conference on Computer Vision and Pattern Recognition, pp. 7794–7803 (2017)
20. Yu, X., Rao, Y., Zhao, W., Lu, J., Zhou, J.: Group-aware contrastive regression for action quality assessment. In: 2021 IEEE/CVF International Conference on Computer Vision (ICCV), pp. 7899–7908 (2021)
21. Zhang, Y., et al.: Bytetrack: multi-object tracking by associating every detection box. In: European Conference on Computer Vision (2021)

Uncovering Lies: Deception Detection in a Rolling-Dice Experiment

Laslo Dinges[1(✉)], Marc-André Fiedler[1], Ayoub Al-Hamadi[1], Ahmed Abdelrahman[1], Joachim Weimann[2], and Dmitri Bershadskyy[2]

[1] Neuro-Information Technology Group, Otto-von-Guericke University Magdeburg, 39106 Magdeburg, Germany
{laslo.dinges,marc-andre.fiedler,ayoub.al-hamadi, ahmed.abdelrahman}@ovgu.de

[2] Faculty of Economics and Management, Otto-von-Guericke University Magdeburg, 39106 Magdeburg, Germany
{joachim.weimann,dmitri.bershadskyy}@ovgu.de
https://www.nit.ovgu.de

Abstract. Deception detection is a challenging and interdisciplinary field that has garnered the attention of researchers in psychology, criminology, computer science, and even economics. While automated deception detection presents more obstacles than traditional polygraph tests, it also offers opportunities for novel economic applications. In this study, we propose a novel multimodal approach that combines deep learning with discriminative models to automate deception detection. We tested our approach on two datasets: the Rolling-Dice Experiment, an economically motivated experiment, and a real-life trial dataset for comparison. We utilized video and audio modalities, with video modalities generated through end-to-end learning (CNN). However, for actual deception detection, we employed discriminative approaches due to limited training data in this field. Our results show that the use of multiple modalities and feature selection improves detection results, particularly in the Rolling-Dice Experiment. Furthermore, we observed that due to minimized reactions, deception detection is much more difficult in the Rolling-Dice Experiment than in the high-stake dataset, quantified with an AUC of 0.65 compared to 0.86. Our study highlights the challenges and opportunities of automated deception detection for economic experiments, and our novel multimodal approach shows promise for future research in this field.

Keywords: Deception detection · Rolling-Dice Experiment · Multimodal approach

1 Introduction

Deception detection is a complex and interdisciplinary field with various techniques, including traditional methods like polygraph tests and modern

---

This research was funded by the German Research Foundation (DFG) project AL 638/13-1 and AL 638/14-1.

G. L. Foresti et al. (Eds.): ICIAP 2023, LNCS 14233, pp. 293–303, 2023.
https://doi.org/10.1007/978-3-031-43148-7_25

approaches like computer vision, natural language processing, and machine learning. While historically used in criminalistics, recent developments have shown the need for automated and contact-free deception detection, particularly in border security. However, the use of such techniques for conviction or restriction of freedom should be approached with caution due to the potential for unreliable or discriminatory decisions [16].

Nowadays, machine learning-based deception detection is also of interest in experimental economicss [19]. This offers potentially less invasive application scenarios. Salespeople misrepresenting their products can cause significant economic harm, and an automated deception detection system could assist customers during virtual sales meetings. In this regard, automized deception detection might help customers and honest salespeople, while false-positives would have less drastic consequences than a conviction or stricter inspection.

This paper proposes a multi-modal approach to evaluate contact-free deception detection on a given high-stake dataset as well as on new sample data of a Rolling-dice Experiment (RDE).

1.1 Related Works

Polygraph testing is widely used to detect deception by measuring physiological responses, but its accuracy is debated in the scientific community. Control Question Technique (CQT) and Concealed Information Test (CIT) are the two main types of polygraph tests, but both have been criticized for their questionable assumptions and unethical testing conditions [11]. The scientific community generally views polygraph tests as less suitable for lie detection compared to organizations such as the American Polygraph Association. Deceptive answers may not produce unique cues and may be influenced by cultural or contextual factors. The illusion of transparency further complicates polygraph testing, as truth-tellers may react aggressively if they feel they are not believed [22]. While traditional polygraph tests and automized deception detection are better than chance, they only indicate the possibility of deception and do not provide definitive identification. Purely manual attempts, however, are generally not effective.

Machine-Learning Based Deception Detection. When it comes to Machine-Learning based deception detection, a proper baseline in the form of control questions is typically not used. This is because the system needs to be able to automatically work on unseen data, which makes the task of lie detection even more challenging. However, this approach significantly reduces the expected costs. Additionally, since only software and readily available devices, such as a webcam, are required, it is now possible to apply lie detection technology in situations where hiring a professional polygraph expert is not feasible.

Typical contact-free modalities which are used for automized deception detection are micro-expressions, macro-expressions, thermal images, gaze, gestures, voice features (tone and pitch), or the transcription of what was said [15].

Body poses and gestures may not always be effective features for detecting lies, especially in seated subjects. However, head pose and gaze direction can serve as potential indicators of the user's state in most situations. Two types of methods are commonly used for head pose estimation: landmark-based and landmark-free. Landmark-free methods perform better on challenging datasets [5–7,14]. Gaze estimation can be achieved by conventional regression-based methods, but approaches based on CNNs are favored [21]. Several studies have utilized gaze features for deception detection, including statistical features and pupil dilation. Combining features from multiple modalities can improve accuracy.

Gaze has also be used as main or only feature of deception detection in some works. Kumar et al. uses statistical gaze features of a group that plays the game Resistance [8]. They discovered that deceivers demonstrate a reduced frequency of focus shifts compared to non-deceivers, which suggests lower levels of engagement in the game. Pasquali et al. report that deception and truthfulness among children who play a game can be classified by the mean pupil dilation. They employed only this single feature and achieved a promising F1-Score of 56.5% for Random-Forest (RF), respectively 67.7% for Support-Vector-Machine (SVM) [12].

In the case of facial expressions, micro-expressions (ME) theoretically have the potential to reveal underlying, hidden emotions. However, due to the limited availability of databases, training deep learning models on ME is challenging, and cross-database experiments to detect deceptions are even more difficult [18]. Furthermore, ME occur infrequently, making them an unreliable feature [22].

On the other hand, macro-expressions can be trained on comprehensive in-the-wild databases, making them more robust. Similar to head pose detection, landmark-free detection of action units (AUs) is becoming increasingly common in facial expression analysis [4,20]. Additionally, end-to-end learning has proven effective in predicting basic emotions on challenging yet comprehensive datasets [10]. Chang et al. [1] for example used a Convolutional Neural Network (CNN) to estimate AUs in a first step, followed by regression of intensities for valence and arousal values. These values cover – in contrast to basic emotion classes – subtle intensities of the shown emotions, which is vital for many real world problems.

Gupta et al. achieved 66.17% Accuracy on a subset of Bag-of-lies (BgL) using EEG, gaze, video, and audio features, and 60.09% (without EEG) on the full dataset (where random guess would be 50%). The best accuracy using single modalities was achieved with gaze features, at 61.7%/57.11% (subset/full set) while using video features only achieved 56.2%/55.26% [3]. Perez et al. used manually extracted features of the Real-life trial (RL) dataset to train a multimodal deception detection system that achieved accuracies in the range of 60–75% using RF and regular decision trees. They report that facial features performed best.

Existing Datasets for Deception Detection. To the best of our knowledge, there is no deception dataset with economical background. However, there are four available datasets with different contexts, that we will outline in the following.

BgL is a deception detection dataset that includes video, audio, and EEG data (for a subset). It contains 35 subjects and 325 samples with an even distribution of truth and lies [3]. The dataset aims to facilitate the development of better deception detection algorithms for real-world scenarios. Samples are collected by displaying a photo and asking the participant to describe it truthfully or deceptively without any motivation for deception.

Box-of-lies (BxL) is a dataset based on a late night TV show game where the guest and host take turns describing an object truthfully or deceptively. It includes linguistic, dialog, and nonverbal features, manually ground-truthed, including features from gaze, eye, mouth, eyebrows, face, and head [17].

Miami University Deception Detection Database (MU3D) is a free resource with 320 videos of individuals telling truths and lies, featuring 80 different targets of different ethnic backgrounds [9]. Each target provided four videos containing both truthful and deceptive statements, both positive (related to someone they liked) and negative. This resulted in a fully crossed dataset for research on deception. The videos were transcribed and evaluated by naive raters, with descriptive analyses of the video characteristics and subjective ratings provided. The MU3D offers standardized stimuli for replication among labs and facilitates research on the interactive effects of race and gender in deception detection. The dataset includes veracity ground truth, allowing analysis of truth proportion in relation to features such as anxiety, word count, and attractiveness.

The RL dataset created by Perez et al. consists of videos from public court trials [13]. This dataset is important for identifying deception in court trial data, given the high stakes involved. The dataset includes both verbal and non-verbal modalities to discriminate between truthful and deceptive statements made by defendants and witnesses.

2 Methodology

Deception detection is a complex and highly context-sensitive task, which makes it challenging to solve using end-to-end learning or transfer learning. Available datasets for deception detection are limited, and signs of deception are often ambiguous, individual, and weakly expressed, especially in low-stake contexts. Instead of using end-to-end learning, a more flexible strategy involves using separate CNNs for multiple modalities to generate meaningful features, which are then classified by discriminative machine learning approaches as SVM or RF, which perform better on limited datasets.

We use head pose, gaze estimation, and action unit regression based on ResNet50 backbones. Additionally, we train a multitask network based on EfficientNet (small) that simultaneously classifies basic emotion and regresses continuous valence arousal (VA) values as basic features for deception detection. From each feature, as for example the valence values, we derive several statistical features such as the min, max, mean, or skewness. Additional to the visual features, we also use several audio features such as speech intensity, pitch, speech rate, spectrum, and mel cepstrum coefficients.

2.1 Feature Selection

In the next step, we reduce the number of (derived) features used in our classification model to address the curse of the many dimensions. Therefore, we first rank all features according to their relevance and then use different fraction α of the ranked features using Repeated Random Train-Test Splits ($n = 50$). We employed the permutation importance technique to assess the relative importance of features for all our models. We have observed that in general, retaining approximately 30%- 50% of the most relevant features yields good results and reduces computational costs. However, the optimal number of features depends on the dataset and the type of classifier used.

2.2 The Rolling-Dice Experiment

Fischbacher et al. conducted a Rolling-dice Experiment (RDE) to investigate the economic factors that motivate individuals to lie. According to their findings, only 20% of participants roll the dice and lie to the fullest extent possible. Conversely, 39% of individuals are completely truthful. It is important to note that the participants were aware that they were not being monitored or recorded, making it impossible to detect lies [2].

Our study, on the other hand, records both the participants and the results of the dice rolls. Prior to the experiment, participants are informed that their data will be labeled for analysis and research purposes. However, the experimenter and their group will not have access to the records to ensure the integrity of the data. Only authorized individuals outside the research team will have access to the records.

The experimental setup involves the use of a standard six-sided dice with possible outcomes of 1, 2, 3, 4, 5, or 6. Participants roll the die once, and the reward is 1 euro times the outcome, except for a result of 6, which yields no reward. We have 100 participants, and we analyze the results to determine the frequency of each outcome.

2.3 Deception Context

The significance of all features may depend on the data and as a consequence, on the context of the deception. Table 1 shows some statistics of the four available deception databases and our RDE. We decided to also evaluate our approach on the RL dataset, which has a comparable size and average sample length but a different context, consisting of real video samples from court.

Hypothesis. Although participants were monitored, we expect only moderate signs of guilt in response to fraudulent claims in the case of the RDE compared to RL, which has a high-stake context.

Table 1. Statistics for the four availible datasets and our Rolling-dice Experiment (RDE) with ×: No, ✓ : Yes, Unb: unbiased, IW: in-the-wild.

Database	Samples	Subj	Average length	Unb	IW	Context
BgL	325	35	9.2 ± 4.8s	✓	×	Describing photo
BxL	25	25	154.9 ± 45.2s	×	✓	Game show
MU3D	320	82	35.7 ± 3.7s	✓	×	Social
RL	118	56	24.9 ± 14.2s	✓	✓	Court
RDE	101	101	28.31 ± 2.78	×	×	Economy

3 Experimental Results

In the experimental section, we first compare the results of our modified RDE with those of Fischbacher et al [2]. Thereafter, we evaluate our approach of automatic deception detection on the RDE samples and for comparison also on the RL dataset.

3.1 Statistical Analysis of the RDE

The results presented in Fig. 1 reveal that in our RDE only 58.42% of the participants were truthful, in contrast to the 39% rate of honesty in the experiment conducted by [2]. This suggests that the presence of a record to indicate deceptive behavior had only a moderate influence on the participants.

Moreover, the results suggest that the actual value of the dice roll strongly affected the participants' behavior. Specifically, participants who rolled a 1 or 6 (the two worst possible rolls) were more likely to falsely claim to have rolled a 4 or 5 (the two best possible rolls). On the other hand, most participants who actually rolled a 2, 3, 4, or 5 were honest about their roll. Additionally, a few participants claimed to have rolled a lower value than they actually did, or they claimed a value without rolling at all (as indicated by a tick label of 0 in Fig. 1).

These findings imply that the accuracy of self-reported data in experiments can be influenced by multiple factors, such as the presence of a record to reveal deception and the actual value of the outcome. Thus, researchers should take these factors into account when interpreting self-reported data in experiments.

3.2 Evaluation of the Proposed Deception Detection Method

For the following experiments, we perform Repeated Random Train-Test Splits ($n_r = 50$) to increase the reliability. Therefore, we use 70% of the samples to train a SVM with linear kernel and the remaining for testing.

Feature Selection. To evaluate the influence of selecting the best features, we computed the Receiver Operating Characteristic (RoC) curves for both datasets

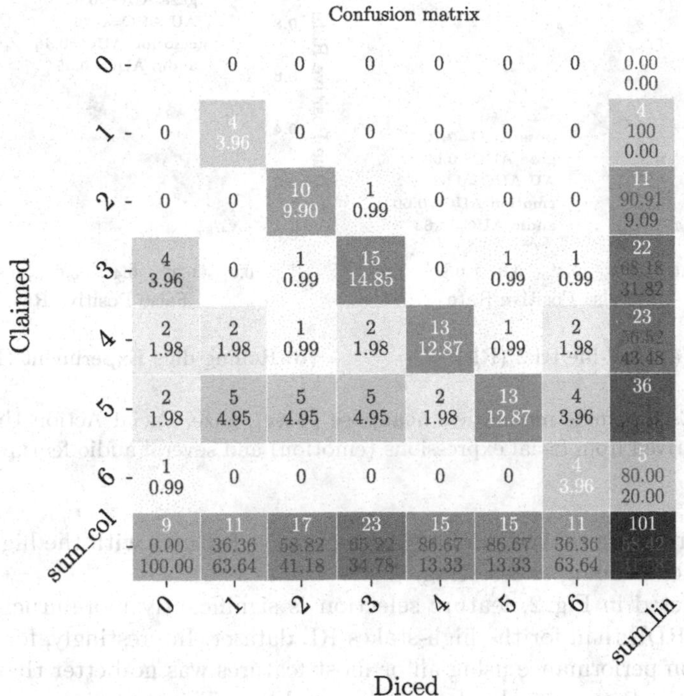

Fig. 1. Confusion Matrix of our RDE showing the claimed rolls over the actual rolls

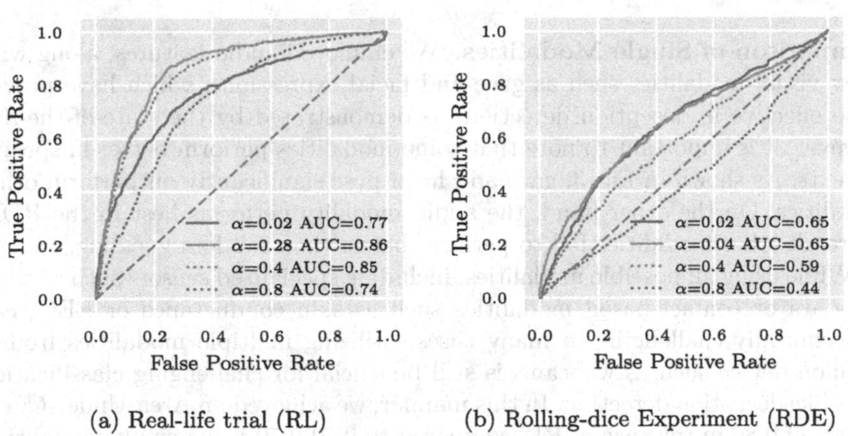

(a) Real-life trial (RL) (b) Rolling-dice Experiment (RDE)

Fig. 2. RoC curves for different fractions of the most relevant features, where α represents the fraction of the used features.

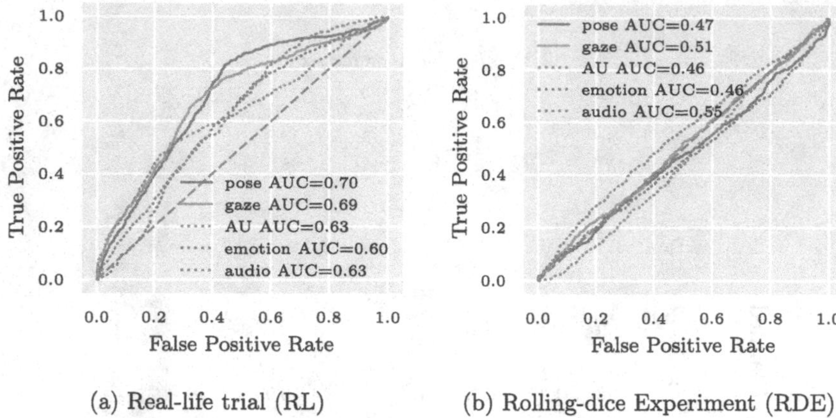

Fig. 3. RoCs for single modalities: headpose (pose), gaze, Facial Action Units (AU), emotions derived from facial expressions (emotion) and several audio features (audio).

All n features were ranked, and a subset of $n \times \alpha$ features with the highest rank were used for training and testing.

As depicted in Fig. 2, feature selection is significantly more crucial for the low-stakes RDE than for the high-stakes RL dataset. Interestingly, for RDE the classification performance using all or most features was no better than random guessing, as indicated by the dashed diagonal line. This may be due to the fact that individuals in low-stakes situations may not experience a sense of guilt or have minimal reactions when deceiving. Consequently, it is reasonable to anticipate that numerous features may not aid in distinguishing between deception and truth.

Comparison of Single Modalities. We employed audio features, along with a few visual modalities such as gaze and facial expressions, which have proven to be effective in deception detection, as demonstrated by the state-of-the-art. However, it is important to note that some modalities perform better in specific contexts. As shown in Fig. 3, gaze and head pose significantly outperform other modalities. On the other hand, the audio modality performs best in the RDE, where all other modalities fail to perform better than random guessing.

While using all possible modalities, including specialized sensors such as thermal data or contact-based modalities such as skin conductance or EEG, can be technically challenging in many cases, utilizing multiple modalities from a common device such as webcams is still beneficial for challenging classification tasks like deception detection. In this manner, we achieved an Area-Under-Curve (AUC) of 0.86 in the case of RL, compared to 0.60-0.70 using single modalities. In the case of RDE, we achieved an AUC of 0.65 instead of 0.46-0.55.

Comparison of the Classifier. In our final experiment, we investigated the influence of the choice of classifier on the performance. To this end, we trained two different classifiers on our datasets: a SVM with linear kernel, which we had used in previous experiments, and RF.

As shown in Fig. 4, the results indicate that SVM with a linear kernel outperforms RF on RL and is equally well on RDE, while SVM with an RBF kernel performs worse. This outcome is in line with expectations since RF tends to perform better on datasets with high-dimensional feature spaces and complex feature interactions, while SVMs may be more suitable for datasets with a smaller number of features and a linearly separable boundary. Moreover, the advantage of the linear kernel over RBF can be attributed to the linear separability of the data, which the linear kernel exploits while being less sensitive to noise and overfitting.

To optimize the SVM's performance, we employed grid search to determine the optimal values of the hyperparameters C and γ. Additionally, we experimented with different values for the maximal number of bags of RF. However, due to the lower performance, we suspect that the number of useful features in the deception context may be too small to train a proper ensemble of decision trees. Furthermore, Pasquali et al. [12] have also observed that SVM outperformed RF for a deception detection problem.

In terms of accuracy, we achieved $68.90 \pm 1.19\%$ for RDE, where $\pm$ indicates the standard error. Regarding RL, a purely manual approach resulted in an accuracy of 62%. The semi-automated approach reported in [13], which requires carefully annotated manual features, achieved an accuracy range of $60-75\%$. In comparison, our fully automated approach yielded an accuracy of $81.39 \pm 0.89\%$.

In Fig. 5 we illustrate some of the used features for a sample of RL (video data of the RDE must not be published due to data protection regulations).

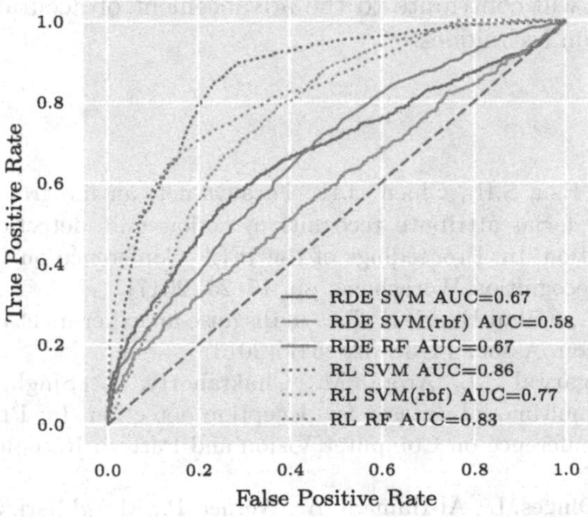

Fig. 4. Receiver Operating Characteristic (RoC) comparing SVM with linear and RBF kernel and Random-Forest (RF) classifier on the RDE and RL dataset.

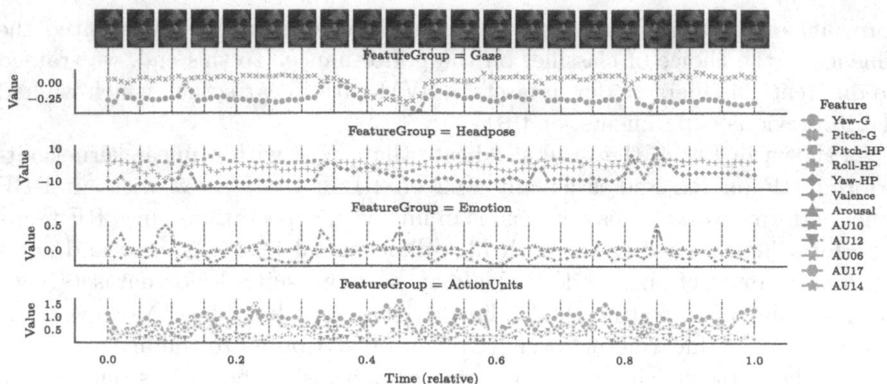

Fig. 5. Qualitative Result: Plots of the most important features over time for a sample of the Real-life trial (RL) dataset. Legend notes: G: Gaze, HP: Headpose.

4 Conclusion

In summary, this paper introduced the Rolling-Dice experiment as an innovative method for automated deception detection. By utilizing both visual and audio data, we developed a robust multimodal approach that demonstrated promising outcomes, particularly in low-risk situations. We anticipate that this experiment can provide a solid basis for future research, specifically in the field of deception detection during sales meetings.

Currently, we are in the process of acquiring a more extensive database that will surpass the existing four deception datasets, encompassing roughly 1000 samples. Our upcoming research will concentrate on optimizing our proposed approach for this specific context and integrating deception detection directly into the experiment. Through continued research and development, we hope that our method will contribute to the advancement of accurate and effective deception detection techniques.

References

1. Chang, W.Y., Hsu, S.H., Chien, J.H.: Fatauva-net: an integrated deep learning framework for facial attribute recognition, action unit detection, and valence-arousal estimation. In: Proceedings of the IEEE Conference on Computer Vision and Pattern Recognition Workshops, pp. 17–25 (2017)
2. Fischbacher, U., Föllmi-Heusi, F.: Lies in disguise-an experimental study on cheating. J. Eur. Econ. Assoc. **11**(3), 525–547 (2013)
3. Gupta, V., Agarwal, M., Arora, M., Chakraborty, T., Singh, R., Vatsa, M.: Bag-of-lies: a multimodal dataset for deception detection. In: Proceedings of the IEEE/CVF Conference on Computer Vision and Pattern Recognition Workshops (2019)
4. Handrich, S., Dinges, L., Al-Hamadi, A., Werner, P., Al Aghbari, Z.: Simultaneous prediction of valence/arousal and emotions on AffectNet, Aff-Wild and AFEW-VA. Procedia Comput. Sci. **170**, 634–641 (2020)

5. Hempel, T., Abdelrahman, A.A., Al-Hamadi, A.: 6d rotation representation for unconstrained head pose estimation. In: 2022 IEEE International Conference on Image Processing (ICIP), pp. 2496–2500. IEEE (2022)
6. Hsu, H.W., Wu, T.Y., Wan, S., Wong, W.H., Lee, C.Y.: QuatNet: quaternion-based head pose estimation with multiregression loss. IEEE Trans. Multimedia **21**(4), 1035–1046 (2019). https://doi.org/10.1109/TMM.2018.2866770
7. Huang, B., Chen, R., Xu, W., Zhou, Q.: Improving head pose estimation using two-stage ensembles with top-k regression. Image Vis. Comput. **93**, 103827 (2020)
8. Kumar, S., Bai, C., Subrahmanian, V., Leskovec, J.: Deception detection in group video conversations using dynamic interaction networks. In: Proceedings of the International AAAI Conference on Web and Social Media, vol. 15, pp. 339–350 (2021)
9. Lloyd, E.P., Deska, J.C., Hugenberg, K., McConnell, A.R., Humphrey, B.T., Kunstman, J.W.: Miami university deception detection database. Behav. Res. Methods **51**, 429–439 (2019)
10. Mollahosseini, A., Hasani, B., Mahoor, M.H.: AffectNet: a database for facial expression, valence, and arousal computing in the wild. IEEE Trans. Affect. Comput. **10**(1), 18–31 (2017)
11. Nortje, A., Tredoux, C.: How good are we at detecting deception? a review of current techniques and theories. S. Afr. J. Psychol. **49**(4), 491–504 (2019)
12. Pasquali, D., Gonzalez-Billandon, J., Aroyo, A.M., Sandini, G., Sciutti, A., Rea, F.: Detecting lies is a child (robot)'s play: gaze-based lie detection in HRI. Int. J. Soc. Robot. **15**(4), 583–598 (2021)
13. Pérez-Rosas, V., Abouelenien, M., Mihalcea, R., Burzo, M.: Deception detection using real-life trial data. In: Proceedings of the 2015 ACM on International Conference on Multimodal Interaction, pp. 59–66 (2015)
14. Ruiz, N., Chong, E., Rehg, J.M.: Fine-grained head pose estimation without keypoints. In: 2018 IEEE/CVF Conference on Computer Vision and Pattern Recognition Workshops (CVPRW), pp. 2155–215509 (2018)
15. Saini, R., Rani, P.: LDM: a systematic review on lie detection methodologies (2022)
16. Sánchez-Monedero, J., Dencik, L.: The politics of deceptive borders:'biomarkers of deceit'and the case of iborderctrl. Inf. Commun. Soc. **25**(3), 413–430 (2022)
17. Soldner, F., Pérez-Rosas, V., Mihalcea, R.: Box of lies: multimodal deception detection in dialogues. In: Proceedings of the 2019 Conference of the North American Chapter of the Association for Computational Linguistics: Human Language Technologies, Volume 1 (Long and Short Papers), pp. 1768–1777 (2019)
18. Talluri, K.K., Fiedler, M.A., Al-Hamadi, A.: Deep 3d convolutional neural network for facial micro-expression analysis from video images. Appl. Sci. **12**(21), 11078 (2022)
19. Weimann, J., Brosig-Koch, J.: Methods in Experimental Economics. STBE, Springer, Cham (2019). https://doi.org/10.1007/978-3-319-93363-4
20. Werner, P., Saxen, F., Al-Hamadi, A.: Facial action unit recognition in the wild with multi-task CNN self-training for the emotionet challenge. In: Proceedings of the IEEE/CVF Conference on Computer Vision and Pattern Recognition Workshops, pp. 410–411 (2020)
21. Zhang, X., Sugano, Y., Fritz, M., Bulling, A.: Mpiigaze: real-world dataset and deep appearance-based gaze estimation. IEEE Trans. Pattern Anal. Mach. Intell. **41**(1), 162–175 (2017)
22. Zloteanu, M.: Reconsidering facial expressions and deception detection. Handbook Facial Expr. Emot. **3**, 238–284 (2020)

Active Class Selection for Dataset Acquisition in Sign Language Recognition

Manuele Bicego[1(✉)], Manuel Vázquez-Enríquez[2], and José L. Alba-Castro[2]

[1] Computer Science Department, University of Verona, Verona, Italy
manuele.bicego@univr.it

[2] atlanTTic research center, University of Vigo, Vigo, Spain
{mvazquez,jalba}@gts.uvigo.es

Abstract. Dataset collection for Sign Language Recognition (SLR) represents a challenging and crucial step in the development of modern automatic SLR systems. Typical acquisition protocols do not follow specific strategies, simply trying to gather equally represented classes. In this paper we provide some empirical evidences that alternative, more clever, strategies can be really beneficial, leading to a better performance of classification systems. In particular, we investigate the exploitation of ideas and tools of Active Class Selection (ACS), a peculiar Active Learning (AL) context specifically devoted to scenarios in which new data is labelled *at the same time* it is generated. In particular, differently from standard AL where a strategy asks for a specific label from an available set of unlabelled data, ACS strategies define *from which class* it is more convenient to acquire a new sample. In this paper, we show the beneficial effect of these methods in the SLR scenario, where these concepts have never been investigated. We studied both standard and novel ACS approaches, with experiments based on a challenging dataset recently collected for an ECCV challenge. We also preliminary investigate other possible exploitations of ACS ideas, for example to select which would be, for the classification system, the most beneficial *signer*.

Keywords: Active Class Selection · Sign Language Recognition · Active Learning

1 Introduction

Automatic Sign Language Recognition (SLR) represents a classic Pattern Recognition problem, which interest has increased in recent years due to the latest advances in deep learning models with flexible spatial-temporal representation capacities [18]. SLR can be broadly categorized into Isolated (ISLR) and Continuous (CSLR) sign language recognition. ISLR is the most extensively researched scenario, and more annotated datasets are available for it due to the relatively simple, but tedious, annotation protocols of a discrete and predefined set of signs. On the other hand, CSLR is much more complex, and much fewer annotated datasets exist. This is due to different challenges related to annotation:

G. L. Foresti et al. (Eds.): ICIAP 2023, LNCS 14233, pp. 304–315, 2023.
https://doi.org/10.1007/978-3-031-43148-7_26

i) co-articulation between signs significantly increases the variability of the sign realization with respect to isolated signs; ii) the speed is much higher, and iii) there is an interplay of non-manual components such as torso, head, eyebrows, eyes, mouth, lips, and even tongue [8].

Still, it is widely acknowledged that proper SLR systems require massive amounts of data [18]. In typical scenarios, datasets are collected so that all possible class instances are equally represented, and performed by all the available experts (the signers); however, in some contexts, this is not a doable strategy, sometimes because of restrictions in the availability of signers, sometimes because signs are simply gathered from real-world videos [2,22]. The definition of a proper strategy to acquire data is therefore crucial, and deserves more attention than that which is typically devoted to it: actually, the standard solution is to simply try to gather equally represented classes, labelled by all signers. However, it is highly possible that having a non-homogeneous distribution of classes is more useful for the classifier, since it is quite normal that some signs have more intrinsic or per-signer variability, and they are more complicated for the classification. Moreover, also the number of samples which are assessed by each signer should be carefully decided: for example, in the distributed Isolated Sign Language Recognition (ISLR) acquisition system described in [23], the collaboration of deaf individuals is voluntary, so it is crucial to minimize the number of instances required per sign to prevent signers from becoming disinterested in contributing their time.

Given all these observations, it seems evident that it would be very beneficial to define a proper dataset construction strategy: and this aspect, never considered in the SLR scenario, represents the main goal of this paper. In particular, here we investigate the usefulness of a particular class of Active Learning strategies [20], called Active Class Selection (ACS – [15]), for the construction of an SLR dataset. ACS represents a very specific Active Learning paradigm, introduced for a very peculiar scenario, which goes beyond the classic Active Learning assumption that getting objects is cheap, whereas getting labels is not. Actually, in ACS the assumption is that labels are directly obtained when getting data, since the experiments are designed *given the labels.* The example, reported in [15], is odour classification with an electronic nose. In a typical setup, the odorant is let to flow over the array of sensors, from which the signal is recorded. In this scenario, the label is decided in advance (the odorant), and the recording is acquired together with the label. Therefore, differently from standard AL, where we have at disposal a set of unlabelled data, and the AL strategy asks for a specific label, ACS strategies aim at selecting the best "class" from which we should get a sample, without having unlabelled data. ACS strategies have not been largely studied, due to their very specific applicability: examples can be found in the already cited odour classification [15,19], Brain-Computer Interfaces [17,24,25], geology [14] and robotics [6]. Also, from a theoretical and methodological point of view, Active Class Selection has not been largely investigated, with an increasing interest only in recent years, with the work of Kottke and colleagues [13], and, especially, with the very recent theoretical works of Bunse and colleagues [4,5].

In this paper, we investigate for the first time the usefulness of ACS techniques in the SLR context: in particular, we show that ACS strategies can be very beneficial in this context, and can improve the classic option of simply getting equally numbered classes. We tested different ACS options, also proposing a novel computationally cheaper version based on Random Forests. We designed a pipeline to test all strategies, evaluating them with a real-world dataset [22], in which classification is based on a pipeline defined with Graph Convolutional Networks [21], showing that ACS strategies can be very useful in the SLR context. We also preliminary study the possible exploitation of the ACS ideas and strategies to face another aspect: the selection of the signers. Signers are very different, can be true deaf people or sign-language interpreters. Differences are large, and selecting the "best" signer for a given sample can be possibly very important. Results, in this case, are mixed, and further investigations are needed.

2 Sign Language Recognition

2.1 Dataset Creation for SLR

The acquisition of 3D spatial-temporal information from sign-language gestures has traditionally involved a multitude of devices such as motion-capturing systems [12], depth-based sensors [1], ultrasound-based sensors [7], and even wearable sensors [26]. Thus, the collection of samples required lab facilities and a slow and costly process of recording deaf signers one by one [10]. However, recent research has shown a trend towards the utilization of only RGB inputs, without any supplementary capturing device. With the emergence of Deep Learning methods, accurate depth estimation can be achieved through learned body models [11], so data acquisition with cumbersome devices can be avoided. Raw RGB-based approaches typically employ Convolutional Neural Network (CNN)-based backbones for spatial feature extraction and Recurrent Neural Network (RNN) for temporal encoding or a spatio-temporal feature extractor (3DCNN) commonly used in Human Action Recognition (HAR) tasks. Current research suggests that spatio-temporal graph convolutional networks (ST-GCN) [27] is the state-of-the-art approach for this purpose.

The class samples used for this research on ACS are extracted from the LSE_eSaude_UVIGO dataset used for the 2022 Sign Spotting Challenge at ECCV'22 [22]. Each sign instance annotated in the sign language dataset contains a variable number of frames. The duration of the co-articulated signs starts from 120 ms (3 frames) up to 2 sec (50 frames), with a mean duration of 520 ms (13 frames). We will provide more specific information in the experimental part.

2.2 A DL Pipeline for SLR

The ACS scheme can be seen as a procedure that defines from which class the next object has to come. In our study, we aim at assessing the potential classification improvements due to such careful selection of classes: in the following, we describe the Deep Learning pipeline which we used as basis for computing classification accuracy. This pipeline, summarized in Fig. 1, represents the first part of the inference scheme provided as the baseline of the cited Challenge, and it is explained in more details in Sect. 3.3 of [22]. The pipeline takes a sliding temporal window of 400 ms containing the whole sign instance (or a part of it) and outputs an embedding vector with spatial-temporal information. The embedding vector is collected at the output of an ST-GCN network and just before the softmax layer; each annotated sign instance is then represented with a variable number of embedding vectors (of dimension 60) that span from 400 ms to 2 sec of spatial-temporal sign information. More formally, each sign x in the dataset X is a collection of 60-dimensional vectors (the number of vectors

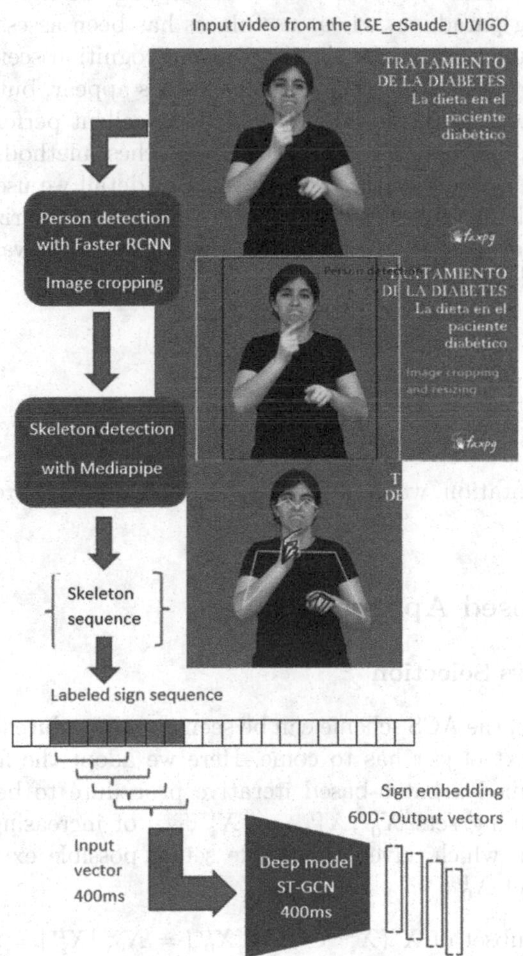

Fig. 1. Pipeline to process an RGB sign instance and convert it to a sequence of embedding spatial-temporal vectors.

in x depends on the length of the sign x), i.e.

$$x_k = [x_k^1, x_k^2, \cdots, x_k^l], \qquad x_k^i \in \Re^{60}$$

We call each vector x_k^j as *frame*, even if it derives from the analysis of a subsequence of 400 msec of the original video, as explained above. To perform classification, instead of using the procedure described for the Challenge baseline [22] (which involves a second computationally demanding ST-GCN), we investigate a simpler approach based on a Multiple Instance Learning scheme, a recent learning paradigm [9] which extends classical supervised learning. In this paradigm, each object is represented by an unordered set of feature vectors, called instances. This set of instances, called a bag, has a unique label. By considering the frames inside a sign as the instances, and the sign as a bag, we can apply this learning paradigm, whose usefulness has been assessed in many different fields, but not (yet) in the sign language recognition scenario. Of course, we are losing the order in which the different frames appear, but, in many applications, this restriction still permits us to have excellent performances [16]. In particular, here we used simple bag-based approaches, methods which summarize each bag with a single feature vector: more in detail we used the Max pool strategy: in this scheme, the sign $x = [x^1, x^2, \cdots, x^l]$ is summarized with a single vector z, in which each entry $z(h)$ is just the max over all the values of the entry h of the vectors $x^1, x^2, \cdots, x^l$:

$$z = [z(1), z(2), \cdots z(60)]$$

where

$$z(h) = \max_{d=1,\ldots,l} x^d(h)$$

Given this representation, we used as classifier the Random Forest Classifier [3], with 100 trees.

3 The Proposed Approach

3.1 Active Class Selection

As described above, the ACS scheme can be seen as a procedure that defines from which class the next object has to come. Here we adopt the formulation used in [15], which defined a batch-based iterative procedure to be used to define a sequence of training sets $X_0^{tr}, X_1^{tr}, \cdots, X_i^{tr}, \cdots$ of increasing size $N_0, N_0 + N, N_0 + 2N \cdots$, in which, given the whole set of possible examples X and a starting training set X_0^{tr}:

- each X_i^{tr} is a subset of X ($X_i^{tr} \subset X$), $|X_0^{tr}| = N_0$, $|X_i^{tr}| = N_0 + iN$, where $|\cdot|$ denotes the cardinality of a set.
- X_i^{tr} includes X_{i-1}^{tr} ($X_{i-1}^{tr} \subset X_i^{tr}$). In other words, each dataset X_i^{tr} is obtained by adding to X_{i-1}^{tr} N elements sampled from $X \setminus X_{i-1}^{tr}$.

The **Active Class Selection** (ACS) strategy defines how to get X_i^{tr} from X_{i-1}^{tr}, i.e. defines how to sample N new objects. In particular, every ACS strategy defines the proportion of classes at time i, i.e. which classes deserve more objects and which less. This is typically formalized as a multinomial $\mathbf{p}^i = [p_1^i, p_2^i, \cdots p_C^i]$, where C is the number of classes and $\sum_c p_c^i = 1$. To define this multinomial $\mathbf{p}^i$ we can use all the samples in X_{i-1}^{tr}; the new objects are sampled from $X \setminus X_{i-1}^{tr}$ according to $\mathbf{p}^i$: in particular, if we have to sample N objects, for each of them we select a class according to $\mathbf{p}^i$, taking one random object from that class.

3.2 The ACS Strategies

In the paper we implemented different strategies, some already presented in [15] and some others adapted for the specific context.

Baseline-Random: this represents the baseline and the usual way of creating the dataset, where all classes are equiprobable (i.e. we try to have all classes with the same number of signs). In this case $\mathbf{p}^i = [1/C, 1/C, \cdots 1/C]$.

Baseline-Natural: in this case, $\mathbf{p}^i$ reflects the *true* distribution of the classes of the problem, which is typically unknown. In our application, however, the signs are extracted from real life video sequences, thus we can estimate it by checking the composition of the whole dataset X.

ACS-Inverse: in this case, each p_c^i is defined as inversely proportional to the accuracy of the classifier on the class c. The idea is that if a class is poorly classified, then we need more samples for that class. We compute the accuracy of the classifier described in previous section on each of the classes of the set X_{i-1}^{tr}, using a 5-fold cross validation protocol.

ACS-AccImprovement: this is the Accuracy improvement method proposed in [15], in which the idea is to give less importance to classes for which there was not a significant improvement in accuracy in the previous iteration. The idea is that we do not need objects for classes which did not benefit from additional samples in the previous iteration.

ACS-Redistricting: this is the Redistricting method proposed in [15], in which the idea is to ask more sample for classes which are more "unstable", i.e. for which there has been a large change in label assignments in the previous iteration.

ACS-RF-impurity. This represents a novel ACS strategy which we propose here, in order to have a computationally simpler approach with respect to ACS-Inverse. The main idea is that we can learn the proportions $\mathbf{p}^i$ by observing the posterior probabilities of a Random Forest, exploiting the usual out-of-bag mechanism. In this way we are not required to perform the whole classification task (as in the ACS-Inverse), but we can extract $\mathbf{p}^i$ from a single trained RF. In particular:

- we train a single RF on the whole dataset;
- each object in the out-of-bag sample (i.e. the objects not used for training a tree) is let fall down each tree and classified by assigning it to the class most probable in the leaf where it arrives. The probability of each class, at

leaf level, is determined during training, and is basically proportional to the number of training objects of each class falling in such leaf. The posterior of the object x, with respect to class k, is denoted as $\text{Prob}(k|x)$.
- the object of class c would contribute to the uncertainty of class c with the factor
$$U(x) = 1 - \sum_{k \neq c} \text{Prob}(k|x) \qquad (1)$$
- the total uncertainty of class c is obtained by averaging $U(x)$ over the objects of class c;
- $\mathbf{p}^i$ is proportional to uncertainty: the more uncertain a class, the larger the number of objects which are required.

4 Experimental Evaluation

In our experiments, we used part of the dataset LSE_eSaude_UVIGO from the 2022 Sign Spotting Challenge at ECCV'22 [22]. This dataset has 100 signs located within 10 h of Spanish sign language videos. The challenge had two different Tracks, and we focus here on the Track1. The number of instances per sign is largely not uniform, so we selected a subset of signs with at least 30 samples. We create two different versions, each one including 28 different classes: in the first (**ACSDataset1**), for each class we extract 30 random samples, resulting in 840 samples. This represents a dataset with equally represented classes. In the second, (**ACSDataset2**), for each class we kept all samples, resulting in a total of 1639 samples. The smallest class ("EPOCA") contains 30 samples, the largest ("ENFERMO") 176.

To assess the usefulness of the ACS methods we used the protocol proposed in [15], which, starting from a given dataset X, simulates an ACS scenario. In particular, we split the dataset X in two random parts: X^{tr}, used for training and X^{val}, used for validation. The goal is to compute the classification accuracy on X^{val} of the classifier described in the previous section, trained using training sets of increasing size $X_0^{tr}, X_1^{tr}, \cdots, X_i^{tr}, \cdots$ defined in Sect. 3. By plotting the accuracy obtained for the different training sets, we have the ACS curve. More precisely, in our experiments we perform a random 50%-50% splitting of the dataset X – the random split is so that the class proportions in X are maintained in X^{tr} and X^{val}. The initial set X_0^{tr} contains 100 elements uniformly sampled from X^{tr}, whereas at every iteration we added 50 samples. For each of them, i) we select a class, according to the $\mathbf{p}^i$ given by the ACS strategy, and ii) we take one random object from that class. If in the dataset the selected class does not have any further object, we select another random class from those classes which still have objects (this is the same scheme adopted in [15] and subsequent papers). Due to this, the last part of the ACS curve is the less representative. In order to avoid random fluctuations, the whole procedure is repeated 100 times, and results are averaged.

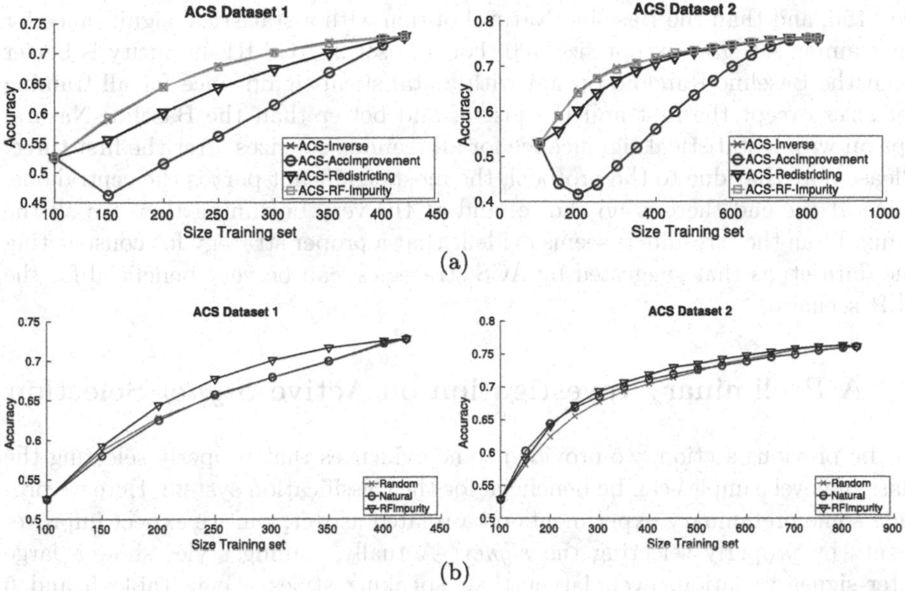

Fig. 2. Active Class Selection results on the two datasets: (a) comparison between different ACS strategies; (b) comparison between the RF-impurity ACS strategy and conventional approaches.

4.1 Results

The results are shown in Fig. 2. In particular, we performed two analyses: in the former, reported in part (a) of the figure, we compare the 4 different ACS strategies, for the two datasets. From the plot it seems evident that the ACS-Inverse and the ACS-RF-impurity represent the best option, for both datasets[1]. It is important to note that ACS-RF-impurity represents a computationally cheaper version, with respect to the ACS-Inverse one, reaching a comparable level of accuracies. In order to get an idea of this equivalence, we performed a statistical analysis: in particular, for each training set size, we made a t-test on the 100 repetitions of the two methods, checking if the two populations are different in a statistically significant way (threshold 0.05): for all sizes of training set the two methods are equivalent, with one exception: training set size equal to 450 for Dataset 2. This confirms that the lighter ACS-RF-impurity represents a viable alternative to the ACS-Inverse scheme. In the second part of the plot (part (b)), we compare our proposed ACS strategy (ACS-RF-impurity) with the two conventional alternatives of "Baseline-Random" and "Baseline-Natural". Also in this case we performed a statistical analysis: after a Bonferroni correction for multiple tests, we observed that ACS-RF-impurity is better than the Baseline-Random option with a statistical significance for all training set sizes, except

[1] We are still trying to understand the strange behaviour of the ACS-AccImprovement strategy, which at the very beginning decreases the performances.

size 150, and than the Baseline-Natural option with a statistical significance for all training set sizes except size 400. For dataset 2, ACS-RF-impurity is better than the Baseline-Random option with a statistical significance for all training set sizes except the first and the last 3, and better than the Baseline-Natural option with a statistical significance for all training set sizes after the first three. Please note that due to the protocol, the most significant part is the central one, since at the end there is no choice, and at the very beginning they are all the same. From these results it seems evident that a proper strategy for constructing the dataset, as that suggested by ACS strategies, can be very beneficial for the SLR scenario.

5 A Preliminary Investigation on Active *Signer* Selection

In the previous section, we provided some evidences that properly selecting the class of novel samples can be beneficial for the classification system. Here we provide some preliminary experiments on a related aspect: can we expect improvements by properly selecting the *signer*? Actually, signing styles show a large inter-signer variation, even larger than speaking styles – e.g. Tables 5 and 6 of [22] show this effect on the F1-score of Track1 for all the classification systems compared. To investigate this aspect we devise a set of strategies, which we call Active Signer Selection (ASS) strategies, which are very similar to ACS ones, but do not select a class, but a signer. Even if we can also think to perform selection of the signer *and* the class, here we investigate a simpler option: once the signer is selected the class is then selected random from signs of that signer. The strategies we investigated are:

Baseline-RandomSigner, Baseline-NaturalSigner: the baselines, in which we select the signer randomly (Baseline-RandomSigner) or according to the natural proportions of signers (Baseline-NaturalSigner).

ASS-Inverse: inspired from the ACS-Inverse strategy, we give a higher probability to be extracted to the signer who is performing worst. To compute the performance, we select in X_{i-1}^{tr} all the signs scored by a given signer, computing how good is the classifier on such dataset. We also implemented another version, which we called **ASS-Inverse(AllSigners)**, in which we created a single classifier, dividing then the errors signer by signer. This seems to be more adequate when X_{i-1}^{tr} is small.

ASS-RF-Impurity: similar to ACS-RF-Impurtity, but now Random Forest uncertainties are computed and aggregated signer by signer. Also in this case we tested another version, called **ASS-RF-Impurity(AllSigners)**, which implements the same modification discussed for ASS-Inverse(AllSigners).

5.1 ASS Results

To test the strategies, we extracted again signs from the dataset LSE_eSaude_UVIGO from the 2022 Sign Spotting Challenge at ECCV'22 [22]. In particular, to get reasonable results, we extracted two datasets with two and three signers, respectively, who covered a reasonable number of signs for each class: **ASS-Dataset1**, with three signers, 10 classes, and at least 9 samples per signer for each class (652 signs in total), and **ASSDataset2**, with two signers, 16 classes, and at least 8 samples per signer for each class (652 signs in total). The protocol is identical to that employed for ACS experiments: results are shown in Fig. 3, following the same formatting of Fig. 2: in part (a) we reported the comparison between the different ASS methods, for the two datasets. It seems that the four strategies are almost equivalent: this is confirmed by the statistical tests, from which it holds that, on dataset 1, the different variants are all equivalent according to a t-test followed by a Bonferroni correction; the same holds for dataset 2, except for ASS-RF-Impurity(Allsigners) which is better than alternatives in the central range. In part (b) we reported the comparison between ASS-RF-Impurity(Allsigners) and the standard baselines Baseline-RandomSigner and Baseline-NaturalSigner. In this case the advantage is less evident, with the ASS strategy not permitting to improve conventional strategies: the only statistically significant improvement is in the central part of the plot (250 and 200 objects in the training set for dataset 1 and 2, respectively). Probably, a strategy based on class information is not enough, and we should derive methods which more explicitly exploit signer information – this being the object of current research.

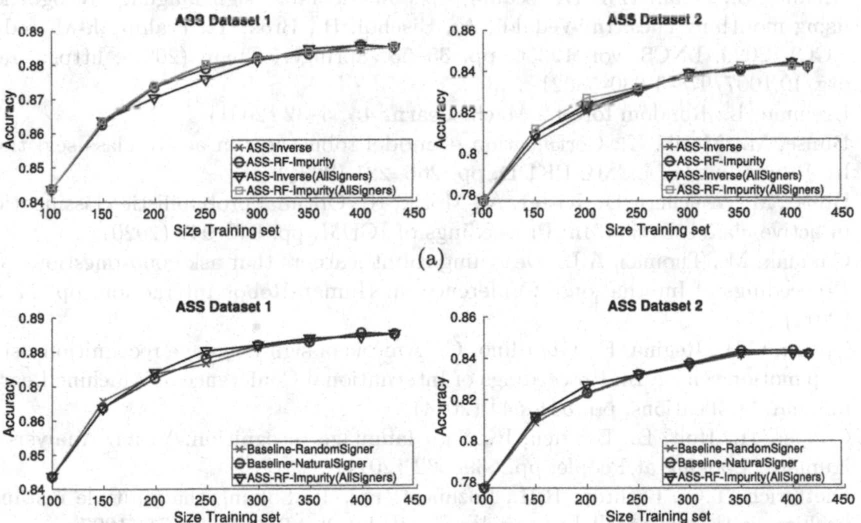

Fig. 3. Active Signer Selection results on the two datasets: (a) comparison between the different ACS strategies; (b) comparison between the best ASS strategy and conventional methods.

6 Conclusions

In this paper we investigated the usefulness, in the Sign Language Recognition scenario, of ideas and tools of Active Class Selection for the construction of a dataset. We investigated both standard and novel ACS approaches, with experiments on a recent, challenging dataset. Results are promising, showing that a proper dataset construction strategy can be very beneficial for the accuracy of automatic SLR systems. Moreover, we also preliminary investigate other possible exploitations of ACS ideas in this scenario, for example to select which would be, for the classification system, the most beneficial *signer*.

Acknowledgements. This research has been funded by the Spanish Ministry of Science and Innovation through the project PID2021-123988OB-C32 and by the Galician Regional Government under project ED431B 2021/24GPC. Manuel Vazquez is also funded by the Spanish Ministry of Science and Innovation through the predoc grant PRE2019-088146. This work has been started during the visit of Manuele Bicego to the University of Vigo in 2022, partially supported by Fondazione Cariverona under the NEUROCONNECT grant.

References

1. Agarwal, A., Thakur, M.K.: Sign language recognition using Microsoft Kinect. In: Proceedings of International Conference on Contemporary Computing, pp. 181–185 (2013)
2. Albanie, S., et al.: BSL-1K: scaling up co-articulated sign language recognition using mouthing cues. In: Vedaldi, A., Bischof, H., Brox, T., Frahm, J.-M. (eds.) ECCV 2020. LNCS, vol. 12356, pp. 35–53. Springer, Cham (2020). https://doi.org/10.1007/978-3-030-58621-8_3
3. Breiman, L.: Random forests. Mach. Learn. **45**, 5–32 (2001)
4. Bunse, M., Morik, K.: Certification of model robustness in active class selection. In: Proceedings of ECML-PKDD, pp. 266–281 (2021)
5. Bunse, M., Weichert, D., Kister, A., Morik, K.: Optimal probabilistic classification in active class selection. In: Proceedings of ICDM, pp. 942–947 (2020)
6. Cakmak, M., Thomaz, A.L.: Designing robot learners that ask good questions. In: Proceedings of International Conference on Human-Robot Interaction, pp. 17–24 (2012)
7. Chuan, C.H., Regina, E., Guardino, C.: American sign language recognition using leap motion sensor. In: Proceedings of International Conference on Machine Learning and Applications, pp. 541–544 (2014)
8. Cooper, H., Holt, B., Bowden, R.: Sign language recognition. Visual Analysis of Humans: Looking at People, pp. 539–562 (2011)
9. Dietterich, T.G., Lathrop, R.H., Lozano-Pérez, T.: Solving the multiple instance problem with axis-parallel rectangles. Artif. Intell. **89**(1–2), 31–71 (1997)
10. Docío-Fernández, L., et al.: LSE_UVIGO: a multi-source database for Spanish Sign Language recognition. In: Proceedings of International Workshop on the Representation and Processing of Sign Languages, pp. 45–52 (2020)
11. Grishchenko, I., et al.: Blazepose GHUM holistic: real-time 3d human landmarks and pose estimation. CoRR abs/2206.11678 (2022)

12. Jedlička, P., Krňoul, Z., Kanis, J., Železný, M.: Sign language motion capture dataset for data-driven synthesis. In: Proceedings International Conference on Language Resources and Evaluation (2020)
13. Kottke, D., et al.: Probabilistic active learning for active class selection. In: NeurIPS Workshop on the Future of Interactive Learning Machine (2016)
14. Liu, S., Ding, W., Gao, F., Stepinski, T.F.: Adaptive selective learning for automatic identification of sub-kilometer craters. Neurocomputing **92**, 78–87 (2012)
15. Lomasky, R., Brodley, C.E., Aernecke, M., Walt, D., Friedl, M.: Active class selection. In: Proceedings of ECML, pp. 640–647 (2007)
16. Naik, N., et al.: Deep learning-enabled breast cancer hormonal receptor status determination from base-level H&E stains. Nat. Commun. **11**(1), 1–8 (2020)
17. Parsons, T.D., Reinebold, J.L.: Adaptive virtual environments for neuropsychological assessment in serious games. IEEE Trans. Consum. Electron. **58**(2), 197–204 (2012)
18. Rastgoo, R., Kiani, K., Escalera, S.: Sign language recognition: a deep survey. Expert Syst. Appl. **164**, 113794 (2021)
19. Rodriguez-Lujan, I., Fonollosa, J., Vergara, A., Homer, M., Huerta, R.: On the calibration of sensor arrays for pattern recognition using the minimal number of experiments. Chemom. Intell. Lab. Syst. **130**, 123–134 (2014)
20. Settles, B.: Active learning literature survey. University of Wisconsin-Madison Department of Computer Sciences, Technical report (2009)
21. Vazquez-Enriquez, M., Alba-Castro, J.L., Docío-Fernández, L., Rodriguez-Banga, E.: Isolated sign language recognition with multi-scale spatial-temporal graph convolutional networks. In: CVPR Workshops, pp. 3462–3471 (2021)
22. Vázquez Enríquez, M., Castro, J.L.A., Fernandez, L.D., Jacques Junior, J.C.S., Escalera, S.: ECCV 2022 sign spotting challenge: Dataset, design and results. In: Karlinsky, L., Michaeli, T., Nishino, K. (eds.) Computer Vision–ECCV 2022 Workshops. ECCV 2022. LNCS, vol. 13808, pp. 225–242. Springer, Cham (2023). https://doi.org/10.1007/978-3-031-25085-9_13
23. Vazquez-Enriquez, M., Losada-Rodríguez, P., González-Cid, M., Alba-Castro, J.L.: Deep learning and collaborative training for reducing communication barriers with deaf people. In: Proceedings of Conference on Information Technology for Social Good (2021)
24. Wu, D., Lance, B.J., Parsons, T.D.: Collaborative filtering for brain-computer interaction using transfer learning and active class selection. PLoS ONE **8**(2), e56624 (2013)
25. Wu, D., Parsons, T.D.: Active class selection for arousal classification. In: Proceedings International Conference on Affective Computing and Intelligent Interaction, pp. 132–141 (2011)
26. Wu, J., Tian, Z., Sun, L., Estevez, L., Jafari, R.: Real-time American sign language recognition using wrist-worn motion and surface EMG sensors. In: Proceedings of International Conference on Wearable and Implantable Body Sensor Networks, pp. 1–6 (2015)
27. Yan, S., Xiong, Y., Lin, D.: Spatial temporal graph convolutional networks for skeleton-based action recognition. In: Proceedings of AAAI (2018)

MC-GTA: A Synthetic Benchmark for Multi-Camera Vehicle Tracking

Luca Ciampi[1(✉)], Nicola Messina[1], Gaetano Emanuele Valenti[2], Giuseppe Amato[1], Fabrizio Falchi[1], and Claudio Gennaro[1]

[1] Institute of Information Science and Technologies, ISTI-CNR, Via G. Moruzzi 1, 56124 Pisa, Italy
{luca.ciampi,nicola.messina,giuseppe.amato,fabrizio.falchi, claudio.gennaro}@isti.cnr.it

[2] Department of Information Engineering, University of Pisa, Via G. Caruso 16, 56122 Pisa, Italy

Abstract. Multi-camera vehicle tracking (MCVT) aims to trace multiple vehicles among videos gathered from overlapping and non-overlapping city cameras. It is beneficial for city-scale traffic analysis and management as well as for security. However, developing MCVT systems is tricky, and their real-world applicability is dampened by the lack of data for training and testing computer vision deep learning-based solutions. Indeed, creating new annotated datasets is cumbersome as it requires great human effort and often has to face privacy concerns. To alleviate this problem, we introduce MC-GTA - Multi Camera Grand Tracking Auto, a synthetic collection of images gathered from the virtual world provided by the highly-realistic Grand Theft Auto 5 (GTA) video game. Our dataset has been recorded from several cameras recording urban scenes at various crossroads. The annotations, consisting of bounding boxes localizing the vehicles with associated unique IDs consistent across the video sources, have been automatically generated by interacting with the game engine. To assess this simulated scenario, we conduct a performance evaluation using an MCVT SOTA approach, showing that it can be a valuable benchmark that mitigates the need for real-world data. The MC-GTA dataset and the code for creating new ad-hoc custom scenarios are available at https://github.com/GaetanoV10/GT5-Vehicle-BB.

Keywords: Multi-Camera Vehicle Tracking · Multi-Target Multi-Camera Tracking · Synthetic Data · Deep Learning · Computer Vision

1 Introduction

Intelligent transportation systems (ITS) constitute an essential pillar of modern smart cities, playing a crucial role in traffic management, urban areas planning, pollution reduction, and, in general, improving urban mobility and sustainability. In particular, automated video analysis is emerging as one of the more attractive ITS smart applications, aided by the ubiquity of city-camera networks and the recently astonishing progress in computer vision.

G. L. Foresti et al. (Eds.): ICIAP 2023, LNCS 14233, pp. 316–327, 2023.
https://doi.org/10.1007/978-3-031-43148-7_27

In this context, multi-target multi-camera tracking (MTMCT) is essential for such applications since it provides crucial information for scene understanding. Specifically, it aims at tracking objects over large areas in multiple, possibly non-overlapping, surveillance camera networks. Among its branches, multi-camera vehicle tracking (MCVT) is beneficial for city-scale traffic analysis and management and for tasks in modern security, e.g., tracing a felonious vehicle between different cameras in a city. However, designing and developing MCVT techniques is tricky since several challenging computer vision tasks are involved – object detection, single-camera multiple object tracking, and object re-identification. A more critical challenge is the lack of suitable datasets for training and testing the deep learning models on which SOTA computer vision solutions rely. Indeed, to enable MCVT development and assessment, data needs to include a comprehensive ground truth covering heterogeneous scenarios, different illumination and weather conditions, large variations in camera distance, resolution, and view angle, as well as provide consistent vehicle IDs across all the cameras. Such datasets require a tremendous human effort for data acquisition and curation, and often it is not even possible to collect them due to violations of data protection rights.

In this paper, we tackle the data scarcity problem affecting the MCVT task by introducing and making freely available MC-GTA - Multi Camera Grand Tracking Auto, a collection of synthetic images gathered from the virtual world provided by the highly-realistic Grand Theft Auto 5 (GTA) video game. Our dataset has been collected by several overlapping and non-overlapping cameras recording urban scenes located at various crossroads, as shown in Fig. 1. Annotations are automatically generated by interacting with the game engine and consist of bounding boxes localizing the vehicles with associated unique IDs that remain consistent across all the cameras, thus making our dataset suitable for training/testing deep learning-based MCVT models. Furthermore, we provide the code[1] needed for easily creating and recording new ad-hoc scenarios resembling real-world scenes that practitioners wish to simulate, where it is possible to vary not only camera locations but also other factors of interest, such as weather conditions and time of the day. This simulated environment has been exploited as a benchmark where we conduct a baseline performance evaluation using a SOTA deep learning-based approach for MCVT. The results show that our simulator can be a helpful and versatile tool for assessing MCVT techniques.

By summarizing, we list below the main contributions of this paper:

- we propose MC-GTA - Multi Camera Grand Tracking Auto, a new synthetic benchmark suitable for training/testing deep learning-based multi-camera vehicle tracking techniques, collected by exploiting the highly-realistic Grand Theft Auto 5 (GTA) video game;
- we release the code for designing and implementing new ad-hoc scenarios where practitioners have control of several factors such as camera locations and weather conditions;
- we conduct a performance evaluation using a SOTA MCVT model and our dataset as a testing ground;

[1] https://github.com/GaetanoV10/GT5-Vehicle-BB.

- results show that our simulated scenarios can be a valuable tool for measuring the performances of MCVT techniques in a controlled environment, mitigating the need for new real-world annotated data.

2 Related Work

In this section, we report some works present in the literature that are relevant to our. Specifically, we focus on methods suitable for the Multi-Camera Vehicle Tracking task. Then we describe some of the most influential synthetic datasets exploited for tackling the data scarcity problem.

2.1 Multi-Camera Vehicle Tracking

Multi-camera vehicle tracking (MCVT) [24,28] is usually tackled using a combination of different computer vision techniques ranging from object detection, single-camera multi-object tracking, and object re-identification.

Object Detection. Object detection is one of the fundamental tasks of computer vision aiming at localizing instances of semantic objects belonging to several classes, such as people, bikes, or vehicles, in digital images and videos. Current approaches rely on deep learning, leveraging different approaches that can be classified as (i) anchor-based that rely on anchors, i.e., prior bounding boxes with various scales and aspect ratios, either directly regressing from pixels to bounding boxes (such as YOLO family [16,25,29] and RetinaNet [19] algorithm) or by refining a bunch of region of interest computed in a preliminary step (such as Faster R-CNN [26] and Mask R-CNN [15]); (ii) anchor-free that rely on predicting key points, such as corners or center points, instead of using anchor boxes and their inherent limitations (such as CenterNet [32] and YOLOX [14]); (iii) transformer-based that rely on the recently introduced attention modules in processing image feature maps (such as DEtection TRansformer (DETR) [4] and one of its evolution Deformable DETR [33]).

Single-camera Multi-Object Tracking. Multi-object tracking (MOT) aims to trace multiple targets in video frames. Popular implementations of MOT algorithms are SORT [3], which uses object detection and Kalman filtering, and DeepSORT [30], which improves SORT by adding a feature similarity matching strategy. Another notable architecture is Towards-Realtime-MOT [18], which unifies detection and feature into a single model. More recently, architectures relying on transformers, such as TrackFormer [23], are also becoming available.

Object Re-identification. Object re-identification (ReID) is usually considered a retrieval task that aims at matching targets in different scenes. Previously, most works addressed person ReID; on the other hand, vehicle ReID is even more challenging since the same vehicle models have the same appearance. Therefore, to enhance accuracy, some methods resorted to multiple information formats like license plates, vehicle color information, time and space metadata, etc [20–22,31].

2.2 Synthetic Datasets

Synthetic datasets have recently received considerable interest since they represent an appealing solution to mitigate the data scarcity problem. Usually, data is gathered by creating ad-hoc scenarios using simulators based on the Unreal or Unity graphical engines; some examples of these collections of images are [9,10,12,13,27], suitable for autonomous driving and pedestrian and tracking. Furthermore, another option is to use the Grand Theft Auto 5 (GTA) video game, which exhibits a higher level of realism and variability among scenarios; some notable existing works present in the literature are Joint Track Auto (JTA) [11] for pedestrian pose estimation and tracking, CrowdVisorPPE [2] for personal protective equipment detection, Virtual World Fallen People (VWFP) [5] for fallen people detection, Grand Traffic Auto (GTA) [8] for vehicle segmentation and counting, and Virtual Pedestrian Dataset (ViPeD) [1,6] for pedestrian detection and tracking. In this work, we fill the gap determined by the lack of a synthetic dataset collected from the GTA5 video game suitable for the multi-camera vehicle tracking task.

3 The MC-GTA - Multi Camera Grand Tracking Auto Dataset

In this section, we deeply describe our Multi Camera Grand Tracking Auto (MC-GTA) dataset, providing details about the procedure employed for acquiring and processing data and illustrating some statistics.

3.1 Overview

Our proposed dataset, which we called MC-GTA - Multi Camera Grand Tracking Auto, includes high-quality imagery collected from the virtual world provided by the highly-realistic Grand Theft Auto 5 (GTA) video game. Specifically, it has been gathered from two different simulated urban scenarios, where several overlapping and non-overlapping cameras have been placed at various crossroads: (i) the first scenario consists of six cameras divided into three overlapping camera-pairs located in three consecutive crossroads (Fig. 1a); (ii) the second scenario includes two non-overlapping cameras placed in two consecutive crossroads (Fig. 1b).

Annotations are automatically gathered by interacting with the game engine and rely on bounding boxes localizing the vehicles together with associated IDs that remain unique and consistent among the cameras belonging to a specific scenario. In Table 1, we report some statistics from each acquired scenario. Specifically, each scenario was recorded with a final frame rate of around 12 FPS. The first scenario lasts 79 s with 70 unique vehicles, while the second scenario lasts 253 s and contains 109 different vehicles. The detailed statistics on the number of cameras traversed by each vehicle are reported in Fig. 2b. Specifically, most of the vehicles are captured by two cameras in both environments. This is most

(a) Scenario #1

(b) Scenario #2

Fig. 1. The considered scenarios of our MC-GTA dataset: (a) the first one includes three pairs of overlapping cameras located at three crossroads; (b) the second one includes two non-overlapping cameras placed at two crossroads.

likely in scenario #1, where every crossing is captured by two cameras. In this case, the few vehicles captured by only one camera are the ones that are leaving the intersection when the video is started. In Fig. 2a, we can better appreciate the distribution of tracks' lifespan, either within single cameras (for standard single-camera vehicle tracking) or across cameras for MCVT.

The core reason that motivated the creation of this dataset is the need to have a well-labeled benchmark resembling with high fidelity the real world, helpful for assessing the performance of multi-camera multi-vehicle tracking models in a simulated and controlled environment. Since labels are generated through an automated procedure, MC-GTA helps in mitigating and contrasting the data scarcity problem, also considering that practitioners can create new custom ad-hoc scenarios by using the provided freely available code. Furthermore, since data come from a virtual environment, our dataset can alleviate possible privacy concerns with the depicted subjects. Secondly, MC-GTA can also be exploited as training data for the supervised learning of deep learning models, given that the limitation represented by the domain gap between synthetic training data and real-world test data is less pronounced in MTMCT [17], or can eventually be contrasted with domain adaptation techniques [6–8].

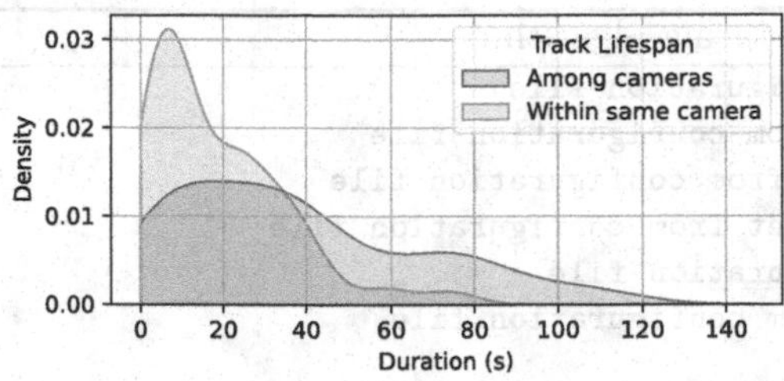

(a) Distribution of track lifespan for both single-camera tracks and tracks transiting within multiple cameras.

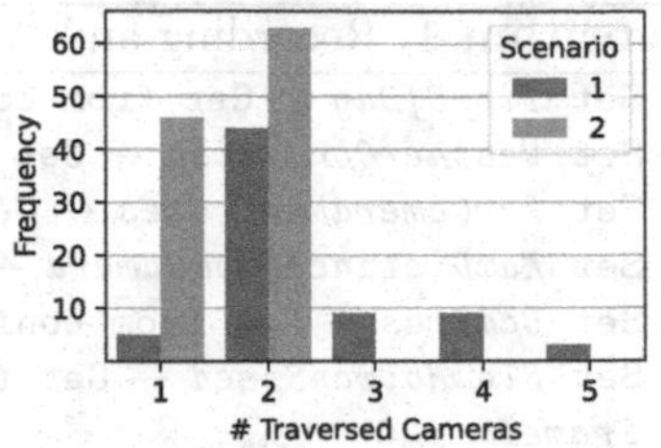

(b) Distribution of the number of cameras traversed by each vehicle for both scenarios.

Fig. 2. MC-GTA dataset statistics.

Table 1. MC-GTA dataset statistics for each scenario.

Feature	Scenario 1	Scenario 2
Num Cameras	6	2
Video duration	79 s	253 s
FPS	12.6	11.8
Unique vehicles	70	109
Average Vehicles per Camera	5.3 ± 2.5	7.7 ± 4.0
Track lengths	28.0 ± 21.6 s	43.9 ± 32.1 s

3.2 Dataset Creation

For the creation of the MC-GTA dataset, we created a GTA plugin using the Script Hook V library[2] to interface with the GTAV environment, similarly to [11,17]. The same plugin can be used by practitioners to create new ad-hoc scenarios. Specifically, the plugin implements an interface with a set of functionalities for creating, deleting, and navigating through the network of cameras in the desired scenarios. Compared to [11,17], we added four more parameters to have finer control over the virtual environments: (i) *TimeOfDay*, which allows practitioners to select the time of the day; (ii) *MaxDistanceFromCamera*, determining the maximum distance at which framed vehicles should be annotated; (iii) *PerCameraNumFrames*, which sets the number of frames to be recorded per camera; (iv) *WeatherCondition*, to set the weather conditions.

The framed vehicle collection and annotation algorithm, shown in detail in Algorithm 1, consists of an iterative procedure that queries the video game's engine to obtain the vehicles present at a given time in the views of a network of cameras. Since GTA is a single-player video game, it does not support the synchronization of multiple cameras; therefore, we exploited a workaround similar to the one used in [17], recording camera frames one after the other by changing

[2] http://www.dev-c.com/gtav/scripthookv/.

Algorithm 1. Recording and Annotating algorithm

```
Set TimeOfDay ← Get from configuration file
Set WeatherCondition ← Get from configuration file
Set PerCameraNumFrames ← Get from configuration file
Set MaxDistanceFromCamera ← Get from configuration file
Set Cameras ← Get from configuration file
Set SlowMotionSpeed ← Get from configuration file
FrameID = 0
while FrameID < PerCameraNumFrames do
   for each Camera in Cameras do
      CameraID ← Get ID of the current camera
      Set Camera as the main camera
      Teleport player to Camera coordinates
      Vehicles ← Get vehicles in the scene
      for each Vehicle in Vehicles do
         VehicleDistance ← Distance between camera and vehicle
         if VehicleDistance < MaxDistanceFromCamera then
            if Vehicle not occluded and Vehicle engine is on then
               LicensePlate ← Get vehicle license plate
               BoundingBox3D ← Get vehicle 3D bounding box coordinates
               BoundingBox2D ← Project 3D bounding box to 2D screen
               Save [FrameID, CameraID, LicensePlate,
               BoundingBox2D, BoundingBox3D] to file
```

camera positions and angles between shots. The drawback of this strategy is that a slight offset-time occurs each time the camera position is changed. To reduce this offset to only a few milliseconds, we activated the slow-motion mode by setting a playback speed, arguing that a few milliseconds delay is negligible in this applicative scenario.

Once all vehicles are extracted from a single camera view, they are filtered based on several criteria. These include vehicle-to-camera distance, which cannot be greater than the *MaxDistance*, and vehicles not present in the camera's field of view or occluded by buildings. Vehicles with their engine off are also excluded, given that fixed vehicles can never travel between different cameras.

In contrast to the ray-tracing methodology used by [17] to obtain vehicle bounding boxes, we simply projected the 3D bounding box coordinates of the vehicles into the camera frame. While this technique creates not tight-fitting 2D bounding boxes around vehicles, it (i) reduces computational complexity and enables scaling data acquisition to many cameras to possibly handle many recording hours, and (ii) mitigates the variability of the bounding-box shape, which in the case of ray-tracing tends to depend on the positioning of the rigging bones that changes across different vehicle classes. Furthermore, we consider as the unique ID that must remain consistent among the cameras the vehicle license plate. We provide the final format of the saved annotations concerning each collected frame in Table 2.

Table 2. Final MC-GTA dataset annotation format.

Metadata	Description
Timestamp	Time information about current frame
FrameID	Index frame to establish temporal order
LicensePlate	Vehicle unique license plate
$\text{Center}_{x,y}$	Coordinates of the *i-th* vehicle's 2D bounding box center
Width	Width of the *i-th* vehicle's 2D bounding box
Height	Height of the *i-th* vehicle's 2D bounding box
$\text{Vehicle}_{x,y,z}$	Coordinates of the *i-th* vehicle's 3D bounding box

4 Experimental Evaluation

In this section, we perform an assessment of our MC-GTA dataset, exploiting a SOTA MCVT approach[3] and our synthetic data as a testing benchmark. The adopted methodology was one of the top solutions proposed in the City-Scale Multi-Camera Vehicle Tracking track of the AI City 2022 Challenge[4], a popular competition about applying AI to several ITS tasks. It uses the YOLOv5 object detector [16], three ReID models for vehicle detection and feature extraction, the ByteTrack deep learning-based algorithm [31] for MOT, and a post-processing procedure that exploits geometrical and temporal information.

More in detail, we performed the experimental evaluation only over the second scenario of our MC-GTA dataset, since it resembles the real-world scenario addressed in the AI City 2022 Challenge. In order to exploit the above-mentioned MCVT approach and to be compliant with it, we manually defined a set of Regions Of Exclusion (ROE), i.e., we filtered out vehicle tracklets ending in some pre-defined regions, and we exploited temporal filters (TFs), i.e., we filtered out tracklets vehicles that are traced from temporal intervals not coherent with pre-defined temporal intervals computed on the basis of camera distances.

We measured the performance by using two golden standard MTMCT evaluators, i.e., the MOTA - Multi-Object Tracking Accuracy, the MOTP - Multi-Object Tracking Precision, and the IDF1, defined as:

$$MOTA = 1 - \frac{|FN| + |FP| + |IDSW|}{|GT_{Dets}|} \tag{1}$$

$$MOTP = \frac{1}{|TP|}\sum_{TP} S \tag{2}$$

$$IDF1 = \frac{|IDTP|}{|IDTP| + 0.5 \times |IDFN| + 0.5 \times |IDFP|} \tag{3}$$

[3] https://github.com/royukira/AIC22_Track1_MTMC_ID10.
[4] https://www.aicitychallenge.org/.

Table 3. The obtained results with and without the Region Of Exclusions (ROEs) and Temporal Filters (TFs).

	MOTA ↑	MOTP ↑	IDF1 ↑	IDP ↑	IDR ↑	FN ↓	IDSW ↓
No ROEs, No TFs	71.11	24.58	76.33	84.58	69.54	11,267	27
2nd Cam ROEs, TFs	73.14	24.58	77.51	84.58	71.52	9,925	27
ROEs, TFs	73.50	24.67	78.55	84.65	73.27	9,107	26

where FN are False Negatives, FP are False Positives, TP are True Positives, IDSW are ID Switches, GT_{Dets} are the ground truth detections, IDTP is Identity True Positives, IDFP is Identity False Positives, IDFN is Identity False Negatives, and S is a similarity function (in this case the IoU) used to consider matches with a value greater than a threshold. For a finer analysis we also computed the IDP and the IDR defined as $IDP = \frac{|IDTP|}{|IDTP|+|IDFP|}$ and $IDR = \frac{|IDTP|}{|IDTP|+|IDFN|}$. We report the obtained results in Table 3, where we also show an ablation study of the considered methodology with and without the ROEs and the temporal filters used for the post-processing procedure.

As we can see, with the best setting we obtained an IDF1 score of 78.55, a value comparable with the ones obtained in the AI City 2022 Challenge, demonstrating that our synthetic benchmark can be a valuable tool for testing MCVT models. In particular, the same methodology applied to the scenarios of the AI City 2022 Challenge obtained an IDF1 score of 81.7. Furthermore, it is worth noting that the inclusion of the ROEs and the temporal filters implies an improvement in performance, as expected. Specifically, we obtained a reduction of 19% concerning the number of FNs and, consequently, a boost in the IDR of 4%, a slight increment in the IDF1 and MOTA metrics, and, finally, a small improvement in the number of IDSW.

5 Conclusion

In this paper, we introduced MC-GTA, a synthetic, freely available dataset gathered from the popular GTA5 video game. Gathering data from virtual worlds is an appealing solution contrasting the data scarcity problem since annotations are automatically generated by interacting with the graphical engine, thus considerably reducing human effort. Specifically, we collected images and labels from several overlapping and non-overlapping cameras located at various crossroads of simulated urban scenarios; labels correspond to bounding boxes localizing the vehicles present in the scenes, together with associated unique IDs persistent among the different video sources. Thus, MC-GTA is suitable for training/testing deep learning models performing multi-camera vehicle tracking, a challenging task extremely helpful for city-scale traffic analysis and security but whose real-world applicability is often constrained by the lack of data. Furthermore, future practitioners can use the code we released to create new ad-hoc scenarios resembling specific custom scenes they want to simulate. To assess our new dataset, we performed a performance evaluation exploiting a SOTA MCVT deep learning methodology using MC-GTA as a testing ground; the obtained results showed that it can be a valuable benchmark that alleviates the need for real-world data. We hope the data and reference results will spark further activities in the field.

Acknowledgements. Supported by: MOST - Sustainable Mobility National Research Center, funded by the European Union Next-GenerationEU (Piano Nazionale di Ripresa E Resilienza (PNRR) - Missione 4 Componente 2, Investimento 1.4 - D.D. 1033 17/06/2022, CN00000023); AI4Media – A European Excellence Centre for Media, Society, and Democracy (EC, H2020 No. 951911); SUN – Social and hUman ceNtered XR (EC, Horizon Europe No. 101092612).

References

1. Amato, G., Ciampi, L., Falchi, F., Gennaro, C., Messina, N.: Learning pedestrian detection from virtual worlds. In: Ricci, E., Rota Bulò, S., Snoek, C., Lanz, O., Messelodi, S., Sebe, N. (eds.) ICIAP 2019. LNCS, vol. 11751, pp. 302–312. Springer, Cham (2019). https://doi.org/10.1007/978-3-030-30642-7_27
2. Benedetto, M.D., Carrara, F., Ciampi, L., Falchi, F., Gennaro, C., Amato, G.: An embedded toolset for human activity monitoring in critical environments. Expert Syst. Appl. **199**, 117125 (2022). https://doi.org/10.1016/j.eswa.2022.117125
3. Bewley, A., Ge, Z., Ott, L., Ramos, F., Upcroft, B.: Simple online and realtime tracking. In: 2016 IEEE International Conference on Image Processing (ICIP). IEEE, September 2016. https://doi.org/10.1109/icip.2016.7533003
4. Carion, N., Massa, F., Synnaeve, G., Usunier, N., Kirillov, A., Zagoruyko, S.: End-to-end object detection with transformers. In: Vedaldi, A., Bischof, H., Brox, T., Frahm, J.-M. (eds.) ECCV 2020. LNCS, vol. 12346, pp. 213–229. Springer, Cham (2020). https://doi.org/10.1007/978-3-030-58452-8_13
5. Carrara, F., Pasco, L., Gennaro, C., Falchi, F.: Learning to detect fallen people in virtual worlds. In: International Conference on Content-based Multimedia Indexing. ACM, September 2022. https://doi.org/10.1145/3549555.3549573
6. Ciampi, L., Messina, N., Falchi, F., Gennaro, C., Amato, G.: Virtual to real adaptation of pedestrian detectors. Sensors **20**(18), 5250 (2020). https://doi.org/10.3390/s20185250
7. Ciampi., L., Santiago., C., Costeira., J., Falchi., F., Gennaro., C., Amato., G.: Unsupervised domain adaptation for video violence detection in the wild. In: Proceedings of the 3rd International Conference on Image Processing and Vision Engineering - IMPROVE, pp. 37–46. INSTICC, SciTePress (2023). https://doi.org/10.5220/0011965300003497
8. Ciampi, L., Santiago, C., Costeira, J., Gennaro, C., Amato, G.: Domain adaptation for traffic density estimation. In: Proceedings of the 16th International Joint Conference on Computer Vision, Imaging and Computer Graphics Theory and Applications. SCITEPRESS - Science and Technology Publications (2021). https://doi.org/10.5220/0010303401850195
9. Deschaud, J.: KITTI-CARLA: a kitti-like dataset generated by CARLA simulator. CoRR abs/2109.00892 (2021)

10. Dosovitskiy, A., Ros, G., Codevilla, F., López, A.M., Koltun, V.: CARLA: an open urban driving simulator. In: 1st Annual Conference on Robot Learning, CoRL 2017, Mountain View, California, USA, November 13–15, 2017, Proceedings. Proceedings of Machine Learning Research, vol. 78, pp. 1–16. PMLR (2017)
11. Fabbri, M., Lanzi, F., Calderara, S., Palazzi, A., Vezzani, R., Cucchiara, R.: Learning to detect and track visible and occluded body joints in a virtual world. In: Ferrari, V., Hebert, M., Sminchisescu, C., Weiss, Y. (eds.) ECCV 2018. LNCS, vol. 11208, pp. 450–466. Springer, Cham (2018). https://doi.org/10.1007/978-3-030-01225-0_27
12. Foszner, P., et al.: CrowdSim2: an open synthetic benchmark for object detectors. In: Proceedings of the 18th International Joint Conference on Computer Vision, Imaging and Computer Gràphics Theory and Applications. SCITEPRESS - Science and Technology Publications (2023). https://doi.org/10.5220/0011692500003417
13. Foszner, P., et al.: Development of a realistic crowd simulation environment for fine-grained validation of people tracking methods. In: Proceedings of the 18th International Joint Conference on Computer Vision, Imaging and Computer Graphics Theory and Applications. SCITEPRESS - Science and Technology Publications (2023). https://doi.org/10.5220/0011691500003417
14. Ge, Z., Liu, S., Wang, F., Li, Z., Sun, J.: YOLOX: exceeding YOLO series in 2021. arXiv preprint arXiv:2107.08430 (2021)
15. He, K., Gkioxari, G., Dollár, P., Girshick, R.B.: Mask R-CNN. In: IEEE International Conference on Computer Vision, ICCV 2017, pp. 2980–2988. IEEE Computer Society (2017). https://doi.org/10.1109/ICCV.2017.322
16. Jocher, G., et al.: ultralytics/yolov5: v7.0 - YOLOv5 SOTA Realtime Instance Segmentation, November 2022. https://doi.org/10.5281/zenodo.7347926
17. Kohl, P., Specker, A., Schumann, A., Beyerer, J.: The MTA dataset for multi target multi camera pedestrian tracking by weighted distance aggregation. In: 2020 IEEE/CVF Conference on Computer Vision and Pattern Recognition Workshops (CVPRW). IEEE, June 2020. https://doi.org/10.1109/cvprw50498.2020.00529
18. Li, Y., Hilton, A., Illingworth, J.: Towards reliable real-time multiview tracking. In: Proceedings 2001 IEEE Workshop on Multi-Object Tracking. IEEE Computer Society. https://doi.org/10.1109/mot.2001.937980
19. Lin, T., Goyal, P., Girshick, R.B., He, K., Dollár, P.: Focal loss for dense object detection. IEEE Trans. Pattern Anal. Mach. Intell. **42**(2), 318–327 (2020). https://doi.org/10.1109/TPAMI.2018.2858826
20. Liu, C., et al.: City-scale multi-camera vehicle tracking guided by crossroad zones. In: 2021 IEEE/CVF Conference on Computer Vision and Pattern Recognition Workshops (CVPRW). IEEE, June 2021. https://doi.org/10.1109/cvprw53098.2021.00466
21. Liu, H., Tian, Y., Wang, Y., Pang, L., Huang, T.: Deep relative distance learning: tell the difference between similar vehicles. In: 2016 IEEE Conference on Computer Vision and Pattern Recognition (CVPR). IEEE, June 2016. https://doi.org/10.1109/cvpr.2016.238
22. Liu, X., Liu, W., Mei, T., Ma, H.: PROVID: progressive and multimodal vehicle reidentification for large-scale urban surveillance. IEEE Trans. Multimed. **20**(3), 645–658 (2018). https://doi.org/10.1109/tmm.2017.2751966
23. Meinhardt, T., Kirillov, A., Leal-Taixe, L., Feichtenhofer, C.: TrackFormer: multi-object tracking with transformers. In: 2022 IEEE/CVF Conference on Computer Vision and Pattern Recognition (CVPR). IEEE, June 2022. https://doi.org/10.1109/cvpr52688.2022.00864

24. Qian, Y., Yu, L., Liu, W., Hauptmann, A.G.: Electricity: an efficient multi-camera vehicle tracking system for intelligent city. In: Proceedings of the IEEE/CVF Conference on Computer Vision and Pattern Recognition Workshops, pp. 588–589 (2020)
25. Redmon, J., Farhadi, A.: Yolov3: an incremental improvement. arXiv preprint arXiv:1804.02767 (2018)
26. Ren, S., He, K., Girshick, R., Sun, J.: Faster r-CNN: towards real-time object detection with region proposal networks. IEEE Trans. Pattern Anal. Mach. Intell. **39**(6), 1137–1149 (2017). https://doi.org/10.1109/tpami.2016.2577031
27. Staniszewski, M., et al.: Application of crowd simulations in the evaluation of tracking algorithms. Sensors. **20**(17), 4960 (2020). https://doi.org/10.3390/s20174960
28. Tan, X., et al.: Multi-camera vehicle tracking and re-identification based on visual and spatial-temporal features. In: CVPR Workshops, pp. 275–284 (2019)
29. Wang, C., Bochkovskiy, A., Liao, H.M.: Yolov7: trainable bag-of-freebies sets new state-of-the-art for real-time object detectors. CoRR abs/2207.02696 (2022). arXiv:2207.02696
30. Wojke, N., Bewley, A., Paulus, D.: Simple online and realtime tracking with a deep association metric. In: 2017 IEEE International Conference on Image Processing (ICIP). IEEE, September 2017. https://doi.org/10.1109/icip.2017.8296962
31. Zhang, Y., et al.: ByteTrack: multi-object tracking by associating every detection box. In: Avidan, S., Brostow, G., Cissé, M., Farinella, G.M., Hassner, T. (eds) Computer Vision – ECCV 2022. LNCS, vol. 13682, pp. 1–21. Springer, Cham (2022). https://doi.org/10.1007/978-3-031-20047-2_1
32. Zhou, X., Wang, D., Krähenbühl, P.: Objects as points. arXiv preprint arXiv:1904.07850 (2019)
33. Zhu, X., Su, W., Lu, L., Li, B., Wang, X., Dai, J.: Deformable DETR: deformable transformers for end-to-end object detection. In: 9th International Conference on Learning Representations, ICLR 2021. OpenReview.net (2021)

A Differentiable Entropy Model for Learned Image Compression

Alberto Presta[1(✉)], Attilio Fiandrotti[1], Enzo Tartaglione[2], and Marco Grangetto[1]

[1] Computer Science Department, University of Turin, Turin, Italy
{alberto.presta,attilio.fiandrotti,marco.grangetto}@unito.it
[2] LTCI, Telecom Paris, Institut Polytechnique de Paris, Palaiseau, France
enzo.tartaglione@telecom-paris.fr

Abstract. In an end-to-end learned image compression framework, an encoder projects the image on a low-dimensional, quantized, latent space while a decoder recovers the original image. The encoder and decoder are jointly trained with standard gradient backpropagation to minimize a rate-distortion (RD) cost function accounting for both distortions between the original and reconstructed image and the quantized latent space rate. State-of-the-art methods rely on an auxiliary neural network to estimate the rate R of the latent space. We propose a non-parametric entropy model that estimates the statistical frequencies of the quantized latent space during training. The proposed model is differentiable, so it can be plugged into the cost function to be minimized as a rate proxy and can be adapted to a given context without retraining. Our experiments show comparable performance with a learned rate estimator and better performance when is adapted over a temporal context.

Keywords: Learned image coding · entropy estimation · differentiable entropy · autoencoder · image compression

1 Introduction and Related Works

End-to-end image compression is gaining momentum as it enables learning the encoder and decoder functions jointly, instead of relying on handcrafted transformations that standard codecs are based on [1]. Most designs depend on an autoencoder architecture, leveraging recent advances in deep artificial neural networks. At the transmitter side, a convolutional *encoder* extracts, from the image, a vector of features, known as *latent space*. This vector has lower dimensionality than the image, achieving preliminary compression. The vector is further quantized, yielding a compressed representation of the image in the form of a *bitstream*. At the receiver side, this representation is projected back to the original dimension by a *decoder* network, recovering the original image. Encoder and decoder are jointly trained via gradient backpropagation, minimizing some RD

Supplementary Information The online version contains supplementary material available at https://doi.org/10.1007/978-3-031-43148-7_28.

G. L. Foresti et al. (Eds.): ICIAP 2023, LNCS 14233, pp. 328–339, 2023.
https://doi.org/10.1007/978-3-031-43148-7_28

cost function in the form $\lambda D + R$, where D is the reconstruction error and R is the rate of the quantized latent space. Therefore, one key problem is modeling the rate of the quantized latent space as a differentiable function that can be optimized at training time through backpropagation. Several approaches have been proposed to estimate the rate of the latent space. In [2], they built a simple autoencoder with one single latent space, estimating the rate using a parametric function, while in [3] an ad-hoc neural network has been trained within the overall framework. Upon this seminal idea, more advanced architectures have been built to improve quantitative results in terms of both rate and distortion. In [3], a scale hyperprior latent space was introduced to capture spatial dependencies within an image, while in [4–6] they exploited an autoregressive context model, inspired by their success in probabilistic generative models. Other proposed solutions were for example to add graph-based modules in the autoencoder [7] to capture non-spatial correlations, to add attention modules [8], or to exploit Swin-transformers [9]. All the above approaches rely on a neural network to estimate the latent space rate, however not without drawbacks. First, this network requires training itself, adding to the encoder complexity and learning time. Second, assuming that a temporal context is available, this network must be refined, i.e. retrained, on the context. In this work, we propose a non-parametric model of the latent space entropy distribution as a proxy of the encoding rate. In a nutshell, our method estimates the latent space statistical frequencies in a differentiable way, so it can be plugged into the RD cost function as a proxy of the rate at training time. Our model not only fulfills the requirements for learning with standard gradient backpropagation, but is agnostic of the overall encoder-decoder architecture, and can be adapted without lengthy refinement procedures. As our entropy model is non-parametric, it reduces the overall complexity and accelerates convergence at training time. Moreover, whenever a temporal context is available, no additional training is required to update the entropy model. We experiment with a variety of learned image compression architectures and we achieve similar performance for a static entropy model, with a slight improvement when the model is updated over a temporal context.

2 Background

This section provides the relevant background on learnable image compression and introduces the problem of estimating the entropy of the latent space.

In most end-to-end learnable image compression schemes, an encoder-decoder pipeline is implemented as a neural network-based autoencoder, following the so-called transform coding approach [10]. The *encoder* f_a projects the image $\mathbf{x}$ into a *latent space* $\mathbf{y} = f_a(\mathbf{x}, \theta_f) \in \mathbb{R}^{N_c \times N_d}$, where N_c and N_d represent the number and the dimension of the flattened latent space *channels* respectively, while θ_f represents the learnable parameters of the encoder. This latent space has a lower dimension than the image, achieving preliminary compression. Then, the latent space is quantized using a function Q, obtaining a discretized latent space $\hat{\mathbf{y}} = Q(\mathbf{y})$. This quantized latent space is entropy coded, e.g. by arithmetic

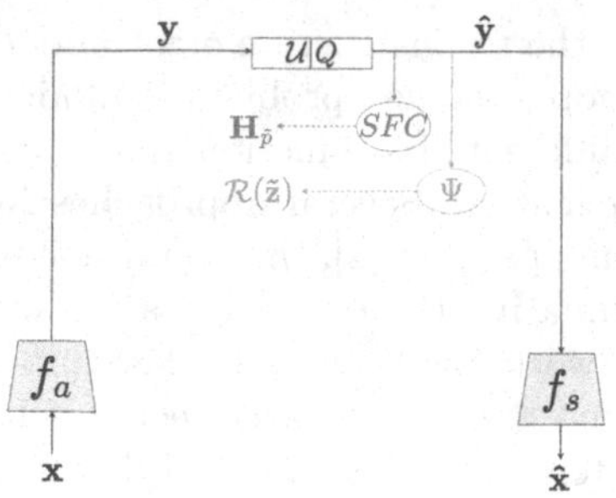

(a) *Ballé2017*[2] architecture.

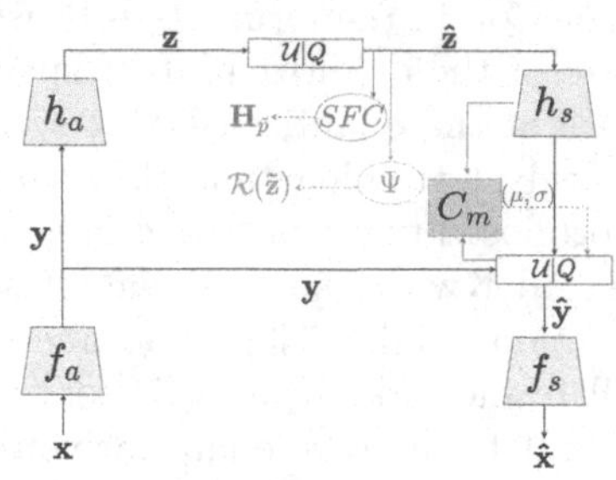

(b) Hyperprior-based architecture.

Fig. 1. Architectures considered in this work. (a)*Ballé2017* [2] with a single latent representation modeled as a fully factorized distribution. (b) Hyperprior-based architecture [3,4,8] with two latent representations: in purple, we have the context model introduced in [4,8]. Ψ (grey) represents the neural network used by baselines to estimate the rate, while SFC (red) represents our model plugged in the cost function.

coding, producing the actual bitstream. At the receiver side, the bitstream is entropy-decoded and then is fed as input to a decoder network f_s that projects it to the original image dimension and recovers the reconstruction $\hat{\mathbf{x}} = f_s(\hat{\mathbf{y}}, \theta_g)$, where θ_g represents the learnable parameters of the decoder. A blueprint of this architecture is depicted in Fig. 1(a). In the context of learned image compression framework, the entropy model used to encode the latent space is represented by a probability distribution $p_{\hat{\mathbf{y}}}$, and it has the role to approximate the real marginal distribution, which is unknown a priori.

The autoencoder described above should be trained end-to-end via standard gradient descent of the backpropagated error gradient. Namely, training the autoencoder resolves to finding the (θ_f, θ_g) that minimize the cost function

$$\mathcal{L} = \lambda \cdot d(\mathbf{x}, \hat{\mathbf{x}}) + \mathcal{R}(\hat{\mathbf{y}}) \tag{1}$$

where d is some distortion metric, $\mathcal{R}(\hat{\mathbf{y}}) = \mathbb{E}\left[-\log_2 p_{\hat{y}}(\hat{\mathbf{y}})\right]$ is an estimation of the rate of the latent space, and λ controls the RD tradeoff between these two competing terms. However, training the above autoencoder via backpropagation requires all the cost function terms to be differentiable, hence the problem of defining a differentiable proxy of the quantization function. In [2], rounding quantization is replaced by adding uniform noise to the latent space, obtaining $\tilde{\mathbf{y}} = \mathbf{y} + \Delta$, where $\Delta \sim \mathcal{U}(-0.5, 0.5)$. This approach has two advantages: first, the density function $p_{\tilde{\mathbf{y}}}$ is a continuous relaxation of the discrete density mass $p_{\hat{\mathbf{y}}}$ [2] ; second, the moments of the quantized random variable are the same as the moments of the original signal plus an additive signal-independent uniform noise [11]. Another crucial step is to define an effective proxy of the rate $\mathcal{R}$. To tackle this problem, works like [2,3,5,8] introduce a parametric function to estimate $p_{\tilde{\mathbf{y}}}$ during training, modeling it as a fully factorized model defined as follows:

$$p_{\tilde{\mathbf{y}}}(\tilde{\mathbf{y}}, \psi) = \prod_{i=1}^{c} p_{\tilde{y}_i}(\tilde{y}_i, \psi_i) = \prod_{i=1}^{c} p_{y_i}(y_i, \psi_i) * \mathcal{U}(y_i| -0.5, 0.5) \tag{2}$$

where $*$ represents the convolution operation, and ψ represents the learnable parameters related to the entropy model. In particular, [2] models each marginal of $p_{\tilde{\mathbf{y}}}$ with a piecewise linear function where the parameters represent the value of the specific sampling points. Conversely, in [3,8,12] $p_{\tilde{\mathbf{y}}}$ is modeled via its cumulative by an auxiliary neural network Ψ trained jointly with the entire image compression framework. This network must guarantee the theoretical characteristics of a cumulative function, namely the positivity and the boundedness between 0 and 1. In particular, this network is modeled as a cascade of K parametric vector functions τ_k, obtaining $\Psi = \tau_K \circ \tau_{K-1} ... \circ \tau_1$. The actual choice of τ in [3,4,8] is the following

$$\tau_k = m_k(\mathbf{T}^k\mathbf{x} + \mathbf{b}^k) \quad \text{if} \quad 1 \leq k < K \tag{3}$$

$$\tau_K = sigmoid(\mathbf{T}^k\mathbf{x} + \mathbf{b}^k) \quad \text{if} \quad k = K \tag{4}$$

where $m_k = \mathbf{x} + \mathbf{a}^k \odot tanh(\mathbf{x})$ represents the non linearity, $\odot$ the elementwise multiplication, and ($\mathbf{T}^k$, $\mathbf{b}^k$,$\mathbf{a}^k$) are vectors of trainable parameters which form ψ. To respect the cumulative conditions mentioned above, a reparametrization step is performed at each step, by applying the *softplus* and *tanh* functions to $\mathbf{T}^k$ and $\mathbf{b}^k$, respectively. All methods implemented Ψ with $K = 4$, obtaining an architecture with around 20000 parameters, depending on the dimension of the latent space. With the developments of more complex architectures also the modeling of the entropy model became more precise; in [3] they added a hyperprior representation $\mathbf{z}$ to capture spatial dependencies among $\mathbf{y}$: in particular, an auxiliary encoder h_a is applied to output $\mathbf{z} = h_a(\mathbf{y}, \theta_{h_a})$, which is then quantized obtaining $\hat{\mathbf{z}}$ and fed to an auxiliary decoder h_s that extracts the scale factor $\sigma^2 = h_s(\hat{\mathbf{z}}, \theta_{h_s})$. In this case, while the $p_{\hat{\mathbf{z}}}$ is a fully-factorized model like (2), $p(\hat{\mathbf{y}}|\hat{\mathbf{z}})$ is parameterized as a zero-mean Gaussian distribution with the scale factor equal to σ^2: a diagram of this architecture is depicted in Fig. 1(b). To train architecture [3], a further term in Eq. (1) has been introduced, representing the rate of the hyperprior space, obtaining the cost function

$$\mathcal{L} = \lambda \cdot d(\mathbf{x}, \hat{\mathbf{x}}) + \mathcal{R}(\mathbf{y}|\hat{z}) + \mathcal{R}(\hat{\mathbf{z}}) \tag{5}$$

where $\mathcal{R}(\hat{\mathbf{z}})$ is equivalent to the rate term in Eq. (1), while $\mathcal{R}(\mathbf{y}|\hat{z}) = \mathbb{E}\left[-\log_2 p_{\hat{y}}(\hat{\mathbf{y}}|\hat{\mathbf{z}})\right]$ is the rate term related to the Gaussian distributed latent space. Following that, [4,6] exploit a context model C_m, formed by mask convolution and a parameter estimation module, with a mean-scale hyperprior to extract a more accurate entropy model, while [8] modeled $p(\hat{\mathbf{y}}|\hat{\mathbf{z}})$ as a mixture of Gaussians and added attention module to the autoencoder: both of these architectures are represented in Fig. 1(b), where C_m is represented in purple. As already mentioned, the above approaches rely on a neural network for learning a parametric entropy model at training time. However, this not only impacts the training complexity but also requires a fine-tuning step if some context from where to extract updated statistics was available. We overcome such limitations by modeling the entropy as a differentiable frequency counter function as detailed below.

3 Proposed Method

In this section, we present a simple and differentiable model for the latent space entropy as a proxy of the encoding rate. We first state the problem of estimating the entropy, then we formulate our method and finally show how to plug it into the cost function minimized during training.

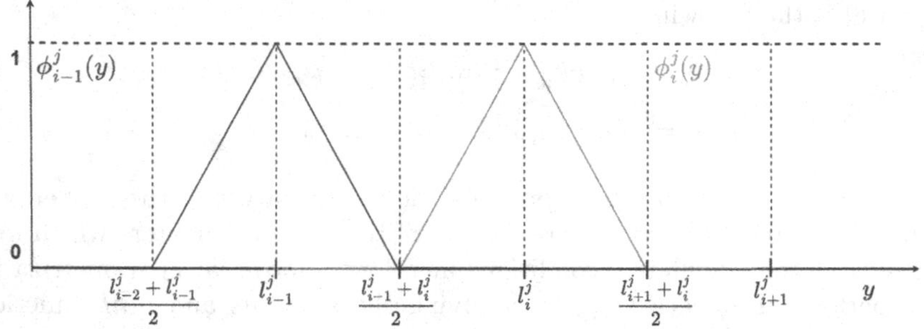

Fig. 2. The ϕ function used to associate weights to a specific quantization level, applied for l^j_i (red) and l^j_{i-1} (blue). Only values within the specific quantization level have a contribution greater than zero.

3.1 Problem Statement

At inference time we perform uniform scalar quantization over integers, considering L different quantization levels uniformly distributed in a symmetric range $\left[-\frac{L}{2}, \frac{L}{2}\right]$, obtaining a bounded latent representation, which is enforced during training. To deal with the quantization step, we add uniform noise during training as in [2] and we model the entropy model as a fully factorized prior as in Eq. (2), meaning that each channel of the latent space has its distribution over the quantization levels, and there is no correlation between different channels. Indicating with l^j_i the i-th quantization level of the j-th channel and with $\left[\frac{l^j_{i-1}+l^j_i}{2}, \frac{l^j_{i+1}+l^j_i}{2}\right]$ its associated quantization interval, the 1-st order entropy of the of $p_{\hat{\mathbf{y}}}$ is expressed as

$$\mathbf{H}_{\hat{p}} = -\frac{1}{N_c}\sum_{j=1}^{N_c} H_{\hat{p}_j} = -\frac{1}{N_c}\sum_{j=1}^{N_c}\left[\sum_{i=1}^{L}\hat{p}_j(l^j_i)\log_2 \hat{p}_j(l^j_i)\right] \tag{6}$$

Unfortunately, the fact that $p_{\hat{\mathbf{y}}}$ is a discrete distribution makes Eq. (6) non-differentiable, making it impossible to minimize it within a gradient-based optimization framework. In the next section, we introduce a differentiable approximation of (6), based on a relaxed frequency statistics counter, enabling us to minimize the entropy in the cost function and controlling thus the rate-distortion tradeoff.

3.2 A Differentiable Entropy Model

We solve the problem of non-differentiability of (6) by introducing a relaxed definition of the entropy based on a differentiable formulation of a soft frequency counter $SFC(l_i^j)$ associated to each level. Given the j-th channel and a quantization level l_i^j, the desired formulation must associate every value of the latent space y_n^j to a weight inversely proportional to the distance with l_i^j, where n varies within the same channel and ranges from 1 to N_d; in this way, by adding all the weights together we obtain a soft counter which is higher for most representative levels. To model such a mechanism, our soft counter relies on a scalar function ϕ_i^j, whose behaviour is depicted in Fig. 2; given the considered level l_i^j, any value of y that lies outside the quantization range has zero weight, thus not contributing to the final soft counter; on the contrary values within the quantization range are linearly weighted according to the distance to the center, obviously with maximum weight equal to 1 when $y = l_i^j$. Being a relaxed approximation of the frequency counter for y, we can normalize results among the different levels, obtaining thus relaxed statistical frequency for estimating probability distribution $\tilde{p}_j$ for every single channel. In particular, we have that

$$\mathbf{H}_{\tilde{p}} = -\frac{1}{N_c}\sum_{j=1}^{N_c} H_{\tilde{p}_j} = -\frac{1}{N_c}\sum_{j=1}^{N_c}\sum_{i=1}^{L} SFC(l_i^j)\log_2(SFC(l_i^j)) \tag{7}$$

where

$$SFC(l_i^j) = \frac{\sum_{n=1}^{N_d}\phi_i^j(y_n^j)}{\sum_{m=1}^{L}\sum_{n=1}^{N_d}\phi_m^j(y_n^j)} \tag{8}$$

and

$$\phi_i^j(y_n^j) = \begin{cases} 1-(2\cdot|y_n^j - l_i^j|) & \text{if } |y_n^j - l_i^j| < 0.5 \\ 0 & \text{otherwise} \end{cases} \tag{9}$$

Relaxation of the entropy represented by Eq. (7) makes it possible to directly minimize it during the training phase, so it can be inserted in RD cost functions. Since our formulation is consistent with the frequency statistics, at inference time this relaxation is replaced by the actual frequency statistics of a limited batch of images.

3.3 Rate Optimization with Entropy Model

Finally, we show how our entropy model can be plugged into the cost function minimized when training a generic learned image compression algorithm. Our model is agnostic to the underlying autoencoder architecture, so it can be in principle plugged into any learnable image compression scheme. However, here we exemplify its application to the architectures in [2–4,8], depicted in Fig. 1. Namely, we show how to replace in Eqs. (1) and (5) the terms related to $\hat{\mathbf{y}}$ and $\hat{\mathbf{z}}$ respectively, which are the latent space modeled using our relaxed frequency statistics counter. In particular, for a model based on architecture taken from [2], the cost function becomes

$$\mathcal{L} = \lambda \cdot d(\mathbf{x}, \hat{\mathbf{x}}) + \mathbf{H}_{\tilde{\mathbf{p}}_{\mathbf{y}}}. \tag{10}$$

For hyperprior-based architectures [3,4,8], the cost function turns into

$$\mathcal{L} = \lambda \cdot d(\mathbf{x}, \hat{\mathbf{x}}) + \mathcal{R}(\hat{\mathbf{y}|\hat{z}}) + \mathbf{H}_{\tilde{\mathbf{p}}_{\mathbf{z}}}. \tag{11}$$

Each architecture can then be trained via standard gradient descent as usual. To use learned image compression architectures at inference time, it is necessary to extract the entropy model for the arithmetic codec first: while standard frameworks like [3,4,8] exploits neural network Ψ trained on the whole dataset, in our model it is enough to compute the entropy model by applying Eq. (8) using as input of Eq. (9) the quantized latent space $\hat{\mathbf{y}}$ of a small subset of the training set, whose size is denoted as ω, and use them as the actual probability distribution of the latent space. In Sect. 4.4 we prove how this strategy allows for easy adaptation of the entropy model with some temporal context.

4 Experiments

In this section, we experiment with our entropy model over four state-of-the-art learnable image compression schemes [2–4,8], each of them with different characteristics and architectures. We organize the rest of the section as follows: at first, we give all the details about how we train both reference and our architectures, then we evaluate our results in terms of Rate-distortion performance on two distinct datasets, namely Kodak [13] and CLIC validation dataset [14]. We also briefly investigate how our method allows a faster convergence concerning the rate terms. In the end, we prove how it is possible to adapt in a fast way the entropy model based on simple frequency statistics, showing performance in terms of BD-Rate and BD-PSNR on the Jvet dataset [15]. As a final remark, our experiments aim at assessing the effectiveness of our entropy model over different architectures, not comparing the relative performance.

4.1 Training Details

For an unbiased comparison, each architecture used for comparison is trained from scratch as for the reference algorithm. For our models, we plugged our entropy model in Eq. (7) instead of the auxiliary neural network, and we retrained from scratch each architecture. Following standard practice, all models are trained over the Vimeo-90K dataset [16] over 256×256 patches cropped at random from the training images, and multiple RD tradeoffs are obtained by properly imposing different values of λ, which ranges between 0.0009 and 0.045. We heuristically set $L = 120$, meaning that we automatically bound the latent space in a range between -60 and 60. We set the initial learning rate to 1e–4 and we halve it whenever the cost function hits a plateau, with a patience of 20 epochs. We trained each architecture for 1–2 million steps and with batch-sizes of 32 images. At inference time, we extract the entropy model exploiting a subset of the training set, and we use a non-adaptive arithmetic coder to encode and decode the latent space, by fixing $\omega = 32$. All our experiments are performed leveraging the CompressAI library [17] codebase[1].

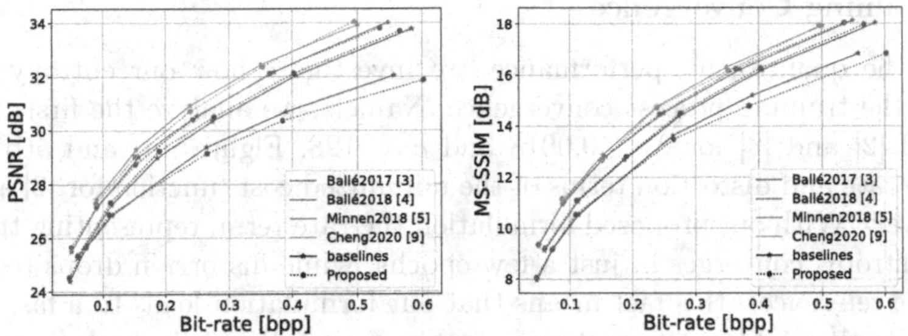

Fig. 3. R-D performance averaged on images the Kodak dataset.

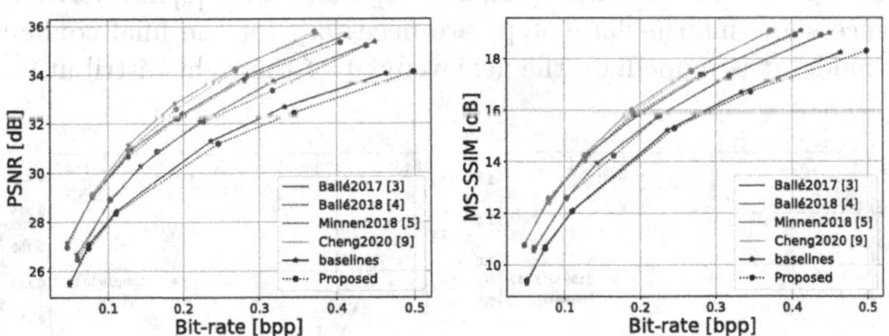

Fig. 4. R-D performance averaged on the CLIC validation dataset.

4.2 Rate-Distorsion Performance

Figure 3 and 4 show quantitative performance in terms of both peak signal-to-noise ratio (PSNR) and multiscale structural similarity (MS-SSIM) over the Kodak image dataset and the CLIC validation dataset, respectively. For clarity,

[1] The code is publicly available on https://github.com/EIDOSLAB/SFC.

we convert MS-SSIM to $-10\log_{10}(1 - \textit{MS-SSIM})$. For each considered architecture, the solid line represents the reference scheme with Ψ, whereas the dotted line represents our non-parametric entropy model. For all the considered baseline models, our method performs close if not identical to the original reference, especially at low bitrates. We point out that the performance gap is negligible for architectures where our entropy model is used to estimate the rate of a hyperprior latent space [3,4,8]. On the other hand, a little decrease in performance is visible concerning [2]: however, in Sect. 4.4 we show how adapting our entropy model with a simple statistics computation closes this gap.

4.3 Training Convergence

Besides the quantitative performance, we investigate how our entropy model impacts the training process convergence. Namely, we analyze the first 20 iterations of [2] and [8] for $\lambda = 0.0018$ and $c = 128$. Figure 5(a) and 5(b) show both the rate and distortion terms of the minimized cost function for [2] and [8], respectively. With our proposed formulation the rate term, representing the estimated entropy, converges in just a few epochs, while distortion drops regularly as for the references: this fact means that our formulation leads to a faster convergence to the stable configuration in terms of rate. Typically, such frameworks automatically allocate bits to different channels, shrinking to zero any useless ones. This also applies to our formulation, but redundant channels are discarded more quickly, as could already be imagined from Fig. 5: this fact is shown in Fig. 6, where it is shown how the probability distribution of a specific channel, in this case, the first one, changes during the epoch, taking [8] as reference. While for reference some intermediate steps are necessary for the final configuration, with our model it is immediate the achievement of the right distribution.

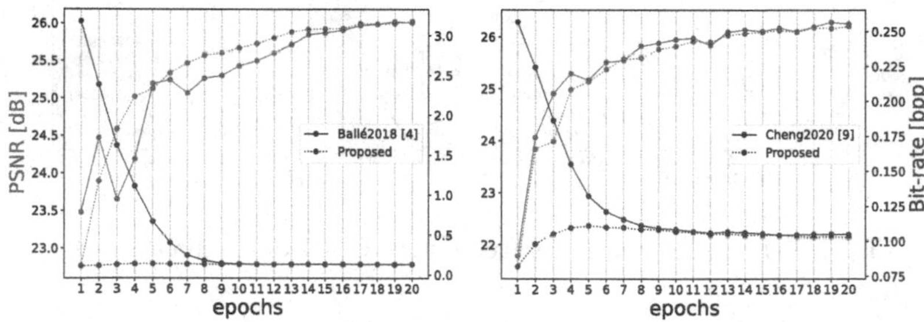

Fig. 5. Convergence of the distortion (red) and rate (blue) terms of the cost function, with $\lambda = 0.0067$ (left [2], right [8]).

4.4 Adapting the Entropy Model

While modern video codecs rely on a context-adaptive arithmetic coder (CABAC), recently learned image codecs [2–4,8] involve the use of a fixed probability distribution extracted through the auxiliary neural network, that should

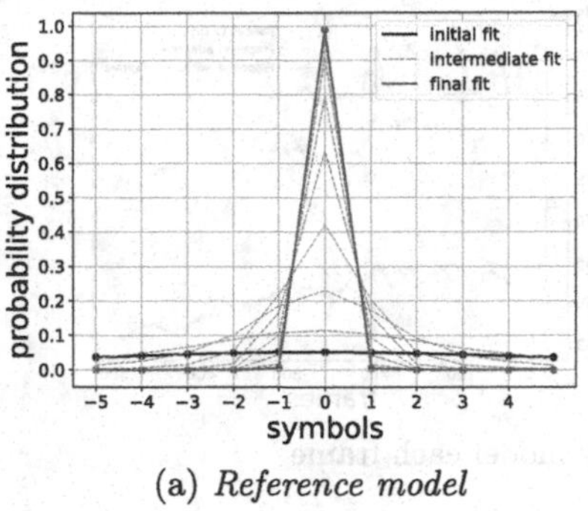

(a) *Reference model*

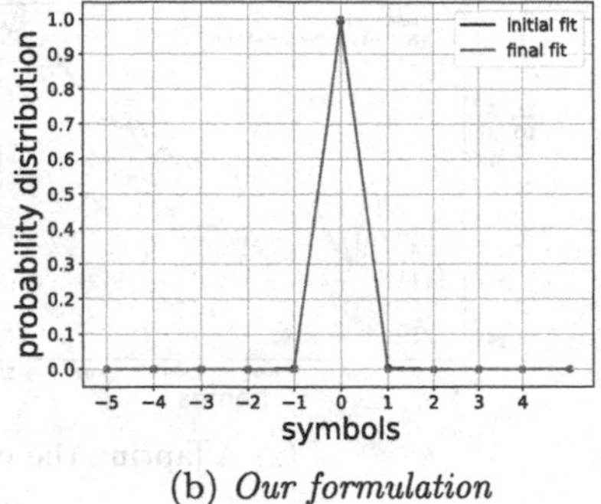

(b) *Our formulation*

Fig. 6. Probability distribution of the first hyper-channel concerning [8]. (a) Our method. (b) Reference model. Models are trained with $\lambda = 0.0067$.

Table 1. Bjontegaard metrics for the JVET dataset for Balle2017 (left) and Minnen2018 (right). We adapt the entropy model every frame, every 16th frame or we ablate it.

	Ballé17						Minnen18					
	Adapt each frame		Adapt every 16 frame		No adapt		Adapt each frame		Adapt every 16 frame		No adapt	
	BDRate	BDPSNR	BDRate	BDPSNR	BDRate	BDPSNR	BDRate	BDPSNR	BDRate	BDPSNR	BDRate	BDPSNR
A	-1.95	-0.02	-0.29	-0.13	11.28	-0.50	-1.04	0.06	-1.01	0.05	-0.44	0.02
B	-4.27	0.11	-0.09	0.06	9.04	-0.35	-0.70	0.03	-0.59	0.03	0.83	- 0.03
C	-11.18	0.41	-4.92	0.14	-1.21	0.03	- 0.20	0.01	-0.10	0.01	0.87	- 0.02
D	-11.47	0.43	-8.06	0.27	0.67	- 0.04	0.21	0.00	0.27	- 0.01	0.94	- 0.04
E	-7.23	0.30	-5.48	0.15	6.59	- 0.31	-0.8	0.05	-0.71	0.04	0.02	0.01
Avg	-7.21	0.246	-3.768	0.098	5.274	- 0.234	-0.09	0.03	-0.024	0.024	0.62	-0.012

be retrained in case of entropy model adaptation. Our entropy formulation is parametric-free since it is only based on Eq. 6, and can be updated to a given context by recomputing simple statistics. To prove our point, we encode the long JVET video sequences consisting of different contents at different resolutions (up to 4K, i.e. 2160p) using the very same architectures trained above (no retraining performed): we take [2] and [4] as baselines, to consider cases where our method is applied to latent spaces of different types. For the first frame only, we rely on the entropy model computed at training time using the Vimeo dataset, as explained in Sect. 4, since no temporal context is available. However, for the following frames, we adapt the entropy model by averaging the current entropy distribution with frequency statistics of the previous frame, calculated using Eq. (8). As two adjacent frames of the same sequence are in most cases very similar, so are expected to be the distributions of the relative latent spaces. To make a more fair comparison, we also experiment with this method with a sampling rate of 16, meaning that we adapt the entropy model every 16-th frames only instead of exploiting every single one: this configuration is more similar to the one used by classic codecs. Table 1 shows the gains in terms of BD-Rate (rate reduction for equivalent distortion, lower is better) and BD-PSNR (distortion reduction for equivalent rate, higher is better) [18], using as a reference the case

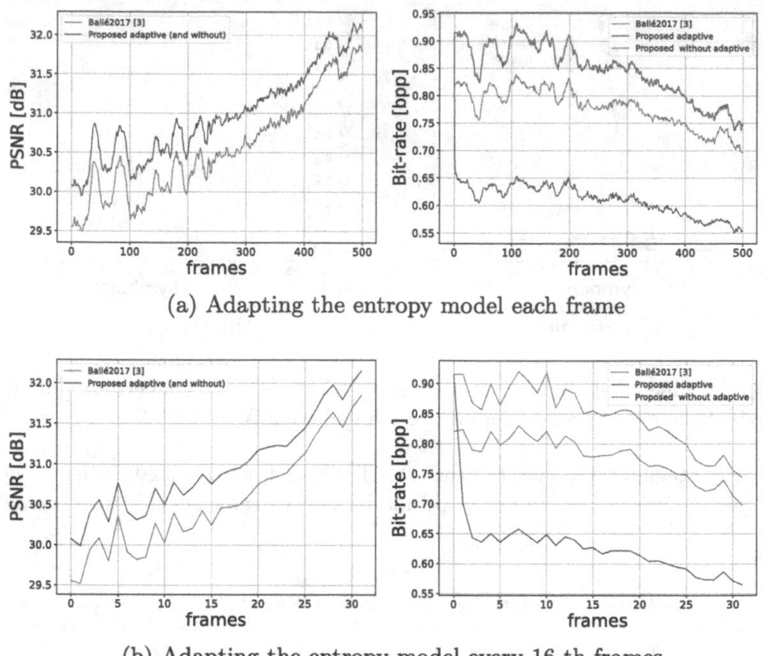

Fig. 7. Results obtained on *PartyScene* sequence from JVET, using [2] as baseline at the highest bitrates. Case (a) and (b) represent when adapting every frame or every 16-th frames, respectively.

the architecture where a neural network estimates the rate. Thanks to adaptive entropy modeling, the BD-Rate improves beyond 10% concerning Ballé2017 [2], which was previously the worst result, closing the gap with the reference model. We attribute these gains to the fact that in [2] all the information required to reconstruct the image is encoded in the latent space whose entropy our model accounts for, whereas in [4] we only model the hyperprior latent representation. Figure 7a and 7b illustrate the performance trend for *Partyscene* sequence, adapting the entropy model every frame and every 16-th frame, respectively. Blue and green lines in the right figures represent the results with and without adaptation, respectively. As it is possible to observe, in both case refining the entropy model by exploiting temporal context yields about a 10% better rate, without affecting the distortion results.

5 Conclusions

We proposed a differentiable and non-parametric model of the latent space entropy as a proxy of the rate into the RD cost function. Our model is built around a soft statistical counter that attributes to each quantization level of a value proportional to its effective frequency in a specific channel of the latent space, and which once normalization occurred could be used as a proxy of the entropy model. Experimental results with four different learned image compression architectures show performance similar to the case where a neural network estimates the latent space rate, and prove that our formulation achieves a stable solution faster to reference models. Moreover, we show how it is possible to update the entropy distribution by exploiting temporal content without any retraining, achieving overall slight improvements in the performance.

References

1. Ma, S., et al.: Image and video compression with neural networks: a review. In: IEEE TCSVT (2019)
2. Ballé, J., Laparra, V., Simoncelli, E.P.: End-to-end optimized image compression. In: ICLR, Simoncelli (2017)
3. Ballé, J., Minnen, D., Singh, S., Hwang, S.J., Johnston, N.: Variational image compression with a scale hyperprior. In: ICLR (2018)
4. Minnen, D., et al.: Joint autoregressive and hierarchical priors for learned image compression. In: Advances in Neural Information Processing Systems (2018)
5. Lee, J., et al.: Context-adaptive entropy model for end-to-end optimized image compression. In: International Conference on Learning Representations (ICLR) (2019)
6. Minnen, D., Saurabh, S.: Channel-wise autoregressive entropy models for learned image compression. In: IEEE International Conference on Image Processing (2020)
7. Yang, C., et al.: Graph-convolution network for image compression. In: IEEE International Conference on Image Processing (ICIP) (2021)
8. Cheng, Z., e al.: Learned image compression with discretized gaussian mixture likelihoods and attention modules. In: CVPR (2020)
9. Zou, R., et al.: The devil is in the details: window-based attention for image compression. In: CVPR (2022)
10. Goyal, V.K.: Theoretical foundations of transform coding. In: IEEE Signal Processing Magazine (2001)
11. Robert, M., Neuhoff, D.: Quantization. In: IEEE Transactions on Information Theory (1998)
12. Lee, J., et al.: DPICT: deep progressive image compression using trit-planes. In: IEEE/CVF CVPR (2022)
13. Eastman Kodak Company. Kodak Lossless True Color Image Suite (1999)
14. Toderici, G., et al.: Workshop and challenge on learned image compression. In: CVPR (2021)
15. Joint Video Exploration Team (JVET) of ITU-T SG16 WP3 andISO/IEC JTC1/SC29/WG11: JVET-G1010: JVET common test conditions and software reference configurations, in 7th Meeting, Torino (IT) (2017)
16. Xue, T., et al.: Video enhancement with task-oriented flow. In: International Journal of Computer Vision (IJCV) (2019)
17. Bégaint, J., et al.: CompressAI: a PyTorch library and evaluation platform for end-to-end compression research. In arXiv preprint arXiv:2011.03029 (2020)
18. Bjontegaard, G.: Calculation of average PSNR differences between RD-curves. In: VCEG-M33 (2001)

Learning Landmarks Motion from Speech for Speaker-Agnostic 3D Talking Heads Generation

Federico Nocentini[1(✉)], Claudio Ferrari[2], and Stefano Berretti[1]

[1] Media Integration and Communication Center (MICC), Università Degli Studi di Firenze, Florence, Italy
{federico.nocentini,stefano.berretti}@unifi.it

[2] Department of Engineering and Architecture, Università degli studi di Parma, Parma, Italy
claudio.ferrari2@unipr.it

Abstract. This paper presents a novel approach for generating 3D talking heads from raw audio inputs. Our method grounds on the idea that speech related movements can be comprehensively and efficiently described by the motion of a few control points located on the movable parts of the face, *i.e.*, landmarks. The underlying musculoskeletal structure then allows us to learn how their motion influences the geometrical deformations of the whole face. The proposed method employs two distinct models to this aim: the first one learns to generate the motion of a sparse set of landmarks from the given audio. The second model expands such landmarks motion to a dense motion field, which is utilized to animate a given 3D mesh in neutral state. Additionally, we introduce a novel loss function, named Cosine Loss, which minimizes the angle between the generated motion vectors and the ground truth ones. Using landmarks in 3D talking head generation offers various advantages such as consistency, reliability, and obviating the need for manual-annotation. Our approach is designed to be identity-agnostic, enabling high-quality facial animations for any users without additional data or training. Code and models are available at: S2L+S2D.

Keywords: 3D Talking Heads · Landmarks · Facial Animation · Identity-Agnostic · Landmarks Motion

1 Introduction

Speech-driven 3D talking heads generation is a rapidly growing field of research and development that has garnered significant interest in recent years. This technology involves generating realistic 3D digital avatars that can accurately replicate human speech and facial expressions. This innovation has far-reaching implications for a wide range of applications, including virtual assistants, video games, education, and entertainment. One of the most significant advantages of speech-driven 3D talking heads is the ability to create immersive and engaging user experiences. This technology can be used to enhance communication in

G. L. Foresti et al. (Eds.): ICIAP 2023, LNCS 14233, pp. 340–351, 2023.
https://doi.org/10.1007/978-3-031-43148-7_29

many different domains, from customer service to online education, and can provide a more human-like interaction than traditional text or voice-only interfaces. Furthermore, speech-driven 3D talking heads can have a significant impact on accessibility and inclusivity. By providing a visual representation of speech and language, this technology can help individuals with hearing or speech impairments to communicate more effectively. Recent advancements in speech-driven 3D facial animation have focused on two primary approaches: vertex-based animation and parameter-based animation. Vertex-based approaches utilize mappings from audio to sequences of 3D face models, with mesh vertex positions predicted to animate the model. However, the main challenge of this approach lies in the complexity of the resulting models, as they must learn to generate vertex mesh sequences containing a large number of 3D points. Parameter-based approaches generate animation curves from audio, resulting in sequences of animation parameters. However, a significant challenge of this approach is converting a sequence of 3D meshes into a sequence of parameters, typically requiring hand-annotated viseme or blendshapes as a starting point.

In this paper, we introduce a novel approach for generating 3D talking heads that decomposes the problem into two distinct sub-problems, each tackled by a separate model, as described in Fig. 1. The first model tracks the movements of scattered landmarks in response to the speech. Specifically, it takes an audio signal as input, from which it generates a frame-by-frame motion of a set of landmarks. The motion is modeled as displacement relative to a neutral configuration of 3D landmarks. The second model takes the resulting displacement of scattered landmarks and densifies them to create a dense motion field. Using the latter, the model then animates a 3D face mesh by adding the motion field to the 3D face vertices. By addressing each sub-problem independently, we aim to improve the overall performance and efficiency of our approach for generating high-quality 3D talking heads. The use of landmark displacements to model speech movements in 3D talking head generation offers several key advantages: firstly, the use of landmarks provides a consistent and reliable way to define the structure of the face, which makes it easier to generate realistic facial expressions. Secondly, landmarks displacements can be interpreted as parameters, eliminating the need for hand-annotation. Thirdly, training the model to predict landmarks displacements from audio allows for complete independence from the identity of the speaker. This enables the predicted displacements from a given audio to be used for animating multiple identities without requiring the model to be retrained. Finally, landmarks are particularly effective for representing the movement of the mouth during speech, making them an ideal choice for speech-driven 3D talking heads.

2 Related Works

In the following, we summarize the work in the literature that are closer to our proposed solution distinguishing between 2D and 3D methods. Several previous studies have focused on the generation of 2D talking head videos driven by speech. Suwajanakorn et al. [13] utilized an LSTM network trained on 19 h of video footage of former President Obama to predict his specific 2D lip landmarks from speech inputs, which was then used for image generation. Vougioukas et al. [15] proposed a method for generating facial animation from a single RGB image using a temporal generative adversarial network. Chung et al. [3] introduced a real-time approach for generating an RGB video of a talking face by directly mapping audio input to the video output space, which can be used to redub a new target identity not seen during training. Landmarks have also emerged as a powerful tool for generating 2D talking heads from speech inputs [16]. By furnishing a concise encoding of facial motion, these landmarks can be reliably estimated through computer vision methodologies. Nonetheless, their usefulness is confined to 2D rendering and fail to account for the comprehensive 3D structure of the face. In this study, we address this inadequacy of 2D landmark extraction by employing 3D landmarks obtained from meshes. Methods for 2D talking head generation focus on generating realistic lip motion, posing less attention to the geometrical face deformations which are mainly induced by texture changes. In the 3D domain instead, facial deformations are to be accounted from the geometric perspective. In earlier attempts, researchers concentrated on animating a pre-designed facial rig with the aid of procedural rules. For instance, HMM-based models generated visemes from input audio or text, and the ensuing facial animations were generated with viseme-dependent co-articulation models or through blending facial templates [5,6,8]. In particular, these methods are based on pre-trained speech models to create an abstract and generalized representation of the audio input. A CNN or autoregressive model then interprets this representation to map it either to a 3DMM space or directly to 3D meshes. For example, Karras et al. [9] learned a 3D facial animation model from 3–5 minutes of high-quality actor-specific 3D data. Similarly, VOCA [4] is trained on 3D data of multiple subjects and can animate the corresponding set of identities from input audio. Meanwhile, MeshTalk [12] learns a categorical representation for facial expressions and auto-regressively samples from this categorical space to animate a given 3D facial template mesh of a subject from audio inputs. FaceFormer [7], on the other hand, uses a transformer-based decoder to regress displacements on top of a template mesh. While both VOCA and FaceFormer require a speaker identification code for the model to choose from the training set's talking styles, our approach differs in that it is completely identity independent. In contrast to existing methods, our work aims to predict 3D facial animations from speech that can be used to animate 3D digital avatars independently of the speaker's identity.

3 Proposed Approach

We propose a novel method to generate 3D facial animations using only an audio input. Unlike existing methods, our framework is designed to be completely agnostic to the identity of the subject being animated. Our approach involves two separately trained models that work in tandem to generate realistic and expressive facial animations from audio inputs. Through the decoupling

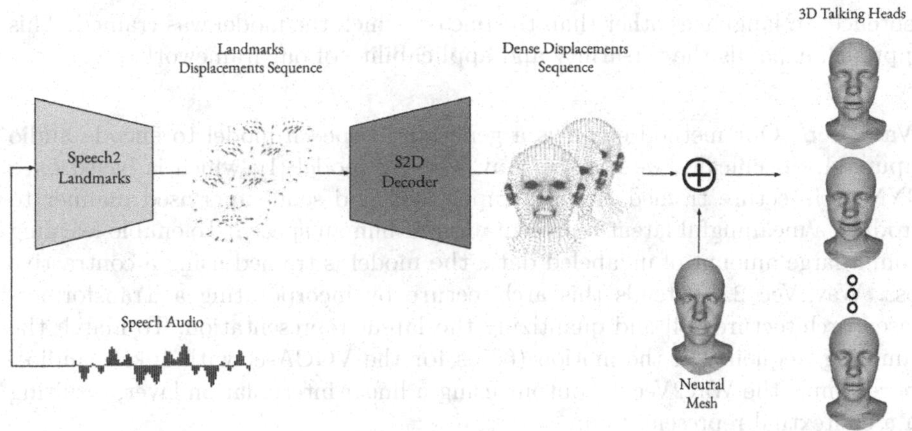

Fig. 1. Overview of our architecture: given a raw audio input and a neutral 3D face mesh as input, our proposed approach can synthesize a sequence of realistic 3D facial motions with precise lip movements.

of landmarks displacement generation and densification, our method enables the generation of high-quality facial animations for any user without the need for additional data or training. Our proposed methodology deviates from the existing literature by not directly synthesizing meshes from audio input. This novel approach offers the advantage of reducing the computational complexity of speech motion generation as the number of landmarks is significantly fewer than that of mesh vertices. Additionally, the independence of the two models allows for the generation of landmarks displacements that can animate a variety of meshes.

3.1 Speech2Landmarks (S2L)

Let $L = \{(L_i^{gt}, L_i^n, A_i)\}_{i=0}^{N}$ denote the training set comprising N samples, where A_i is an audio containing a spoken sentence, $L_i^{gt} = (l_i^0, \dots l_i^{K_i}) \in \mathbb{R}^{K_i \times 68 \times 3}$ represents the facial landmark sequence of length K_i that corresponds to the spoken sentence in audio A_i and $L_i^n \in \mathbb{R}^{68 \times 3}$ are the landmarks of the neutral face. To derive the landmarks displacement dataset with respect to the neutral configuration, we apply a transformation that results in $L_d = \{(S_i^{gt}, L_i^n, A_i)\}_{i=0}^{N}$, where $S_i^{gt} = (s_i^0, \dots s_i^{K_i}) \in \mathbb{R}^{K_i \times 68 \times 3}$ and each $s_i^k = l_i^k - L_i^n \in \mathbb{R}^{68 \times 3}$ is the landmarks displacements. Our aim is to learn a mapping function (S2L) that establishes a correspondence between the audio input A_i and the ground-truth landmark displacements S_i^{gt}, which is realized by assembling a three-part composite model comprising a Wav2Vec Encoder, a multilayer bidirectional LSTM, and a fully connected layer. The utilization of a pre-trained audio processing model enhances the generalization capabilities of our framework, thereby enabling us to animate sentences in languages other than the one on which the model was trained. This approach expands the versatility and applicability of our framework.

Wav2Vec. Our method employs a generalized speech model to encode audio inputs A. Specifically, we use the Wav2Vec 2.0 model [1], which is based on a CNN architecture trained in a self-supervised and semi-supervised manner to produce a meaningful latent representation of human speech. To enable learning from a large amount of unlabeled data, the model is trained using a contrastive loss. Wav2Vec 2.0 extends this architecture by incorporating a Transformer-based architecture [14] and quantizing the latent representation. To match the sampling frequency of the motion (60fps for the VOCAset with 16kHz audio), we resample the Wav2Vec 2.0 output using a linear interpolation layer, resulting in a contextual representation:

$$A \rightarrow \{a_i\}_{i=0}^{T}. \tag{1}$$

where T is the number of frames extracted from the audio. In this study, we utilize a pre-trained version of the Wav2Vec 2.0 encoder.

3.2 Sparse2Dense (S2D)

Otbertout et al. [10,11] presented the S2D Decoder, which is based on the spiral operator proposed in [2]. In the following, all meshes employed possess a uniform topology and are in complete point-to-point correspondence. The training set $L = \left\{(M_i^n, M_i^{gt}, L_i^n, L_i^{gt})\right\}_{i=0}^{N}$ consists of N samples, where $M_i^n \in \mathbb{R}^{M\times 3}$ represents a neutral 3D face, $M_i^{gt} \in \mathbb{R}^{M\times 3}$ represents a 3D talking head, and $L_i^n \in \mathbb{R}^{68\times 3}$ and $L_i^{gt} \in \mathbb{R}^{68\times 3}$ denote the 3D landmarks that correspond to M_i^n and M_i^{gt}, respectively. To generate the sparse-to-dense displacement dataset, we employ a transformation that yields $L_d = \{(D_i, s_i)\}_{i=0}^{N}$, where $D_i = M_i^{gt} - M_i^n$ and $s_i = L_i^{gt} - L_i^n$. The S2D Decoder takes the landmarks' displacements as input and produces the corresponding 3D mesh vertex displacements. This model transforms a sparse set of scattered displacements into a dense set of displacements by utilizing five spiral convolution layers, each of which is followed by an up-sampling layer. In order to obtain the reconstructed mesh utilizing the model prediction, we employ the following equation: $\hat{M}_i = \hat{D}_i + M_i^n$, where $\hat{D}_i$ represents the model prediction and M_i^n represents the mesh in its neutral expression.

4 Training

In order to accelerate the training process, we opted to train the two models independently. This approach resulted in improved convergence for both models. Both models were trained on the same training set of VOCAset [4], which comprises paired audio phrases and 3D talking head animations.

4.1 S2L Losses

For the training of our S2L model, we formulated a loss function comprising of four terms, which can be expressed as:

$$L_{S2L} = \lambda_1 L_{rec} + \lambda_2 L_{mouth} + \lambda_3 L_{cos} + \lambda_4 L_{vel}. \tag{2}$$

Here, L_{rec} represents the loss incurred in reconstructing the facial landmark displacements, while L_{mouth} corresponds to the loss incurred in reconstructing the mouth landmark displacements. Additionally, L_{vel} denotes the velocity loss, and L_{cos} signifies the cosine loss. The hyperparameters λ_1, λ_2, λ_3, and λ_4 control the contribution of each loss term in the overall loss function.

Reconstruction Losses: The reconstruction loss L_{rec} is defined as the L_2 norm computed between all generated landmarks displacements and their respective ground truth counterparts. Specifically, this loss function is applied uniformly across all generated landmarks displacements. In a similar vein, the reconstruction loss for the mouth region L_{mouth} is formulated as an L_2 norm, with the exception that it is only calculated for the landmarks displacements that are more important during speech, namely those of the mouth and jaw:

$$L_{rec} = \frac{1}{N}\sum_{n=1}^{N}\frac{1}{T_n}\sum_{t=1}^{T_n}\left\|d_{n,t} - \hat{d}_{n,t}\right\|_2, \tag{3}$$

$$L_{mouth} = \frac{1}{N}\sum_{n=1}^{N}\frac{1}{T_n}\sum_{t=1}^{T_n}\left\|m_{n,t} - \hat{m}_{n,t}\right\|_2. \tag{4}$$

where N refers to the number of sequences, T_n corresponds to the length of the nth sequence, while $d_{n,t}$ and $m_{n,t}$ represent the respective ground truth values for all displacements and mouth/jaw displacements. Conversely, $\hat{d}_{n,t}$ and $\hat{m}_{n,t}$ denote the model predictions for all displacements and mouth/jaw displacements, respectively. In order to enhance the convergence of our displacement prediction model, we propose the use of a cosine loss. By incorporating this loss, we aim to minimize the angle between predicted and ground truth displacements, thereby improving the overall performance of the model:

$$L_{cos} = \frac{1}{N}\sum_{n=1}^{N}\frac{1}{T_n}\sum_{t=1}^{T_n} 1 - \frac{d_{n,t}\cdot\hat{d}_{n,t}}{\left\|d_{n,t}\right\|_2\left\|\hat{d}_{n,t}\right\|_2}. \tag{5}$$

Temporal Consistency Loss: In our efforts to augment the temporal consistency of our model, we use a loss, denoted as velocity loss and inspired by [4], that aims to minimize the L_2 norm of the pairwise differences.

$$L_{vel} = \frac{1}{N}\sum_{n=1}^{N}\frac{1}{T_n}\sum_{t=2}^{T_n}\left\|(d_{n,t} - d_{n,t-1}) - (\hat{d}_{n,t} - \hat{d}_{n,t-1})\right\|_2. \tag{6}$$

4.2 S2D Losses

In order to enhance the efficacy of S2D Decoder training, we propose a three-loss framework. This framework comprises of three unique losses. The initial two losses are similar to those expounded earlier for S2L and function directly on the displacements. Meanwhile, the third loss governs the precision of the generated mesh. Again, the hyperparameters λ_5, λ_6 and λ_7 control the contribution of each loss term in the overall loss function. We define the loss as follows:

$$L_{S2D} = \lambda_5 L_{rec} + \lambda_6 L_{cos} + \lambda_7 L_{weighted}. \tag{7}$$

Reconstruction Loss: The dense displacement reconstruction is subject to a reconstruction loss, which is defined as follows:

$$L_{rec} = \frac{1}{N} \sum_{i=1}^{N} \left\| D_n - \hat{D}_n \right\|_2. \tag{8}$$

where $\hat{D}_n$ denotes the predicted value by the model, and D_n represents the corresponding ground truth.

Weighted Loss: To enhance reconstruction accuracy, we introduce an additional loss term that minimizes the discrepancy between the estimated shape $\hat{M}_i$ and the actual expressive mesh M_i. Notably, the vertices in close proximity to the landmarks are susceptible to more significant deformations, while other regions, such as the forehead, remain relatively stable. Thus, similar to [10], we propose a weighted L_2 loss, where certain regions of the mesh are assigned more weight to account for their greater importance in the reconstruction process:

$$L_{weighted} = \frac{1}{N} \sum_{i=1}^{N} w_i \left\| p_i - \hat{p}_i \right\|_2. \tag{9}$$

Following [10], we use a specific method for defining the weights, denoted as w_i, on a mesh represented by vertices p_i and landmarks l_j. Specifically, the weight of each vertex is defined as the inverse of the Euclidean distance between the vertex and its closest landmark, i.e., $w_i = \frac{1}{min\ d(p_i, l_j)}, \forall j$. This weighting scheme provides a coarse estimation of the contribution of each vertex p_i to the generation of lip movements. Given that the mesh topology is constant, we precompute the weights w_i and leverage them across all samples.

4.3 Training Details

The S2L model was trained using the Adam optimizer for 300 epochs with a learning rate of 10^{-4}. The model's bi-directional Long Short-Term Memory (Bi-LSTM) architecture comprises three layers, each with a hidden size of 64. The loss function, L_{S2L}, includes four regularization terms, namely, $\lambda_1 = 10^{-1}$,

$\lambda_2 = 1$, $\lambda_3 = 10^{-4}$, and $\lambda_4 = 10$. On the other hand, the S2D model was trained using the Adam optimizer for 300 epochs with a learning rate of 10^{-3}. The model is built by concatenating five spiral convolution layers with an upsampling layer. The loss function, L_{S2D}, includes three regularization terms, namely, $\lambda_5 = 10^{-1}$, $\lambda_6 = 10^{-4}$, and $\lambda_7 = 1$.

4.4 Inference Time

To obtain the talking heads after the training of the models, the following steps are followed:

1. **S2L** takes an audio file A_i as input and produces a sequence of landmark displacements denoted as $\hat{S}_i = S2L(A_i) = \left(\hat{s}_i^{\,0}, \ldots, \hat{s}_i^{\,K_i}\right) \in \mathbb{R}^{K_i \times 68 \times 3}$.
2. **S2D** takes the landmarks displacement generated by **S2L**, denoted as $\hat{S}_i$, and produces a sequence of vertices displacements denoted as $\hat{D}_i = S2D(\hat{S}_i) = \left(\hat{d}_i^{\,0}, \ldots, \hat{d}_i^{\,K_i}\right) \in \mathbb{R}^{K_i \times M \times 3}$.
3. The neutral 3D face M_i^n is summed with each vertex displacement $\hat{D}_i$ to generate the 3D talking heads sequence denoted as $\hat{M}_i = \hat{D}_i + M_i^n$.

5 Experiments

In this section, we present the experiments conducted to assess the efficacy of our proposed approach. Specifically, we conducted a comparative study with respect to two existing methods, namely, Faceformer [7] and VOCA [4]. Our objective is to evaluate the performance of our approach relative to the state-of-the-art.

VOCAset: Our experimental setup utilized the VOCAset, comprising of 12 actors, with an equal gender split of 6 males and 6 females. Each actor delivered 40 distinct sentences, with durations ranging from 3 to 5 s. The dataset includes high-fidelity audio recordings and 3D facial reconstructions per frame, captured at a frame rate of 60 fps. The dataset was partitioned into three distinct subsets for the purposes of training, validation, and testing. The training subset consists of 8 actors, while the validation and test subsets include 2 actors each.

5.1 Results

To evaluate the efficacy of Faceformer and VOCA, we employed the pre-trained models made available by their respective authors. As Faceformer operates at a frame rate of 30 fps, we compared its output against the ground truth at this rate. Conversely, our approach and VOCA operate at a frame rate of 60 fps, and thus we compared their outputs against the original ground truth. Landmarks play a crucial role in evaluating the animation process, and their quality directly impacts the naturalness and realism of the resulting speech. To assess the effectiveness of landmark generation, we also evaluate the landmarks obtained from meshes generated by both Faceformer and VOCA. All experiments were conducted exclusively on the test subset of the VOCAset dataset, and the presented results in Table 1 and Table 2 represent an average of all results.

Lips Error: As suggested in [7], we computed the Lips Error (LE) to assess the quality of the generated lip motion sequences. This metric is defined in [7] as the maximum of the L_2 distance between each lip vertex (or landmarks) of the generated sequence and those of the corresponding ground truth.

Displacements Errors: We compared the displacement outputs generated by Faceformer and VOCA to those produced by our approach using ground truth data, as our model operates on the displacements. To evaluate the quality of the generated results, we utilized both cosine distance and L_2 distance metrics, specifically focusing on the average L_2 distance between all displacements (DE) and the maximum angle between all displacements (DAE). We calculated the error on both the landmark displacement outputs generated by S2L and those on vertices generated by concatenating S2L and S2D. Our proposed approach surpasses both Faceformer and VOCA in generating landmarks and vertices, as demonstrated in Table 1. While the differences among models in terms of LE are insignificant, the gaps in DAE are more pronounced. This suggests that our approach produces more realistic and accurate landmark and vertex displacements, closely resembling ground truth data. This is unsurprising since our models were trained to minimize the angle between generated and ground truth displacements. Notably, the quantitative performance of landmark-based outcomes is inferior to that of vertex-based outcomes, which can be attributed to the relatively greater proportion of salient landmarks in speech compared to vertices in a mesh. Figure 2 illustrates two qualitative examples of our proposed framework. The effectiveness of our framework is evident in accurately capturing the lip closure during the pronunciation of consonants such as *"b"* , *"p"*, and *"m"*. Figure 3 shows a comparison of the meshes generated by VOCA, Faceformer, and our model, which outperforms other methods mostly when the mouth takes certain positions during speech, such as when generating displacements for phonemes like *"Sh"*, *"Wa"* or *"Gl"*. Additional quantitative results and comparisons can be found in the supplementary video.

Table 1. Comparison with the state-of-the-art. Errors are reported for Lips Error (LE) and Dense Error (DE) and displacement angle discrepancy (DAE) error.

Methods	Landmarks			Dense		
	LE (mm)	DE (mm)	DAE (Rad)	LE (mm)	DE (mm)	DAE (Rad)
VOCA [4]	8.72	7.71	0.29	7.24	6.29	0.23
Faceformer [7]	6.1	5.6	0.20	5.12	4.24	0.17
Ours	**5.01**	**4.42**	**0.13**	**4.31**	**3.42**	**0.12**

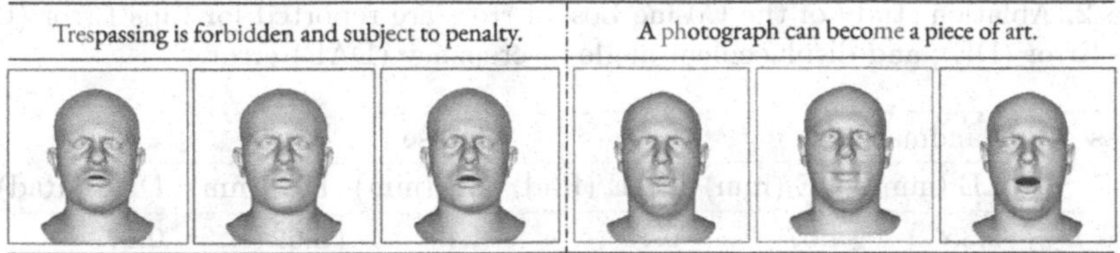
Trespassing is forbidden and subject to penalty.
A photograph can become a piece of art.

Fig. 2. Qualitative examples of the proposed framework.

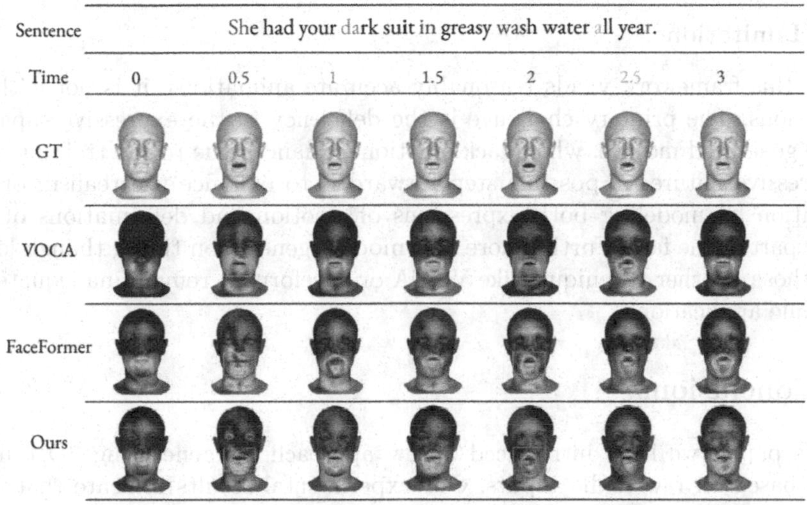

Sentence	She had your dark suit in greasy wash water all year.						
Time	0	0.5	1	1.5	2	2.5	3
GT							
VOCA							
FaceFormer							
Ours							

Fig. 3. Qualitative evaluation of our proposed framework in comparison to Faceformer and VOCA. The assessment is performed on generated meshes, where color gradation indicates the magnitude of the deviation from the groundtruth. Specifically, the color blue denotes a lower level of error, while the color red indicates a higher level of error. (Color figure online)

5.2 Ablation Study

In order to evaluate the efficacy of our proposed approach, we performed an ablation study on the selection of loss functions employed during model training. Our utilized loss functions are widely accepted in the field and have been employed in previous works, in contrast to our novel introduction of the cosine loss. Thus, to gauge the enhancement provided by the latter, we compared the errors obtained from models trained both with and without the cosine loss, utilizing the previously defined metrics for landmarks and vertices displacements. According to Table 2, incorporating the cosine loss during training of the two models enhances the fidelity of the generated displacements for both landmarks and vertices. As a result, the utilization of the cosine loss augments the potential of our framework to produce convincing talking heads. Additional quantitative results about the advantage of cosine loss usage can be found in the supplementary video.

Table 2. Ablation study of the Cosine Loss. Errors are reported for Lips Error (LE), Dense Error (DE), and displacement angle discrepancy (DAE) error.

Loss	Landmarks			Dense		
	LE (mm)	DE (mm)	DAE (Rad)	LE (mm)	DE (mm)	DAE (Rad)
w/o L_{cos}	5.35	4.67	0.19	4.48	3.62	0.17
w/ L_{cos}	**5.01**	**4.42**	**0.13**	**4.32**	**3.42**	**0.12**

5.3 Limitations

While this framework yields reasonably accurate animations, it is not without limitations. The primary challenge is the deficiency in the expressive capacity of the generated meshes, which lack emotional nuances due to the training data inexpressive nature. A possible step forward is to enhance the realism of the animation by modeling both expressions of emotion and deformations of the upper part of the face. Furthermore, our model's generation times, though lower than those of other techniques like VOCA or Faceformer, remain inadequate for real-time applications.

6 Conclusions

In this paper, we have introduced a new approach for generating 3D talking heads based on raw audio inputs. Our experimental results indicate that capturing the motion of facial landmarks is sufficient to effectively represent speech movements. Additionally, training two separate models to separate this motion from the movement of mesh vertices leads to improved realism and accuracy in lip movements. However, the generation of 3D facial animations raises ethical concerns. Creating fabricated narratives using generated 3D faces can be risky and have both intentional and unintentional consequences for individuals and society as a whole. It is important to emphasize that technology should always prioritize human-centered considerations. Therefore, it is crucial to carefully consider the social and psychological impacts of such technology.

References

1. Baevski, A., Zhou, H., Mohamed, A., Auli, M.: wav2vec 2.0: a framework for self-supervised learning of speech representations. CoRR abs/2006.11477 (2020)
2. Bouritsas, G., Bokhnyak, S., Ploumpis, S., Bronstein, M.M., Zafeiriou, S.: Neural 3d morphable models: spiral convolutional networks for 3d shape representation learning and generation. CoRR abs/1905.02876 (2019)
3. Chung, J.S., Jamaludin, A., Zisserman, A.: You said that? CoRR abs/1705.02966 (2017)
4. Cudeiro, D., Bolkart, T., Laidlaw, C., Ranjan, A., Black, M.J.: Capture, learning, and synthesis of 3d speaking styles. In: Proceedings of the IEEE/CVF Conference on Computer Vision and Pattern Recognition (CVPR), June 2019

5. De Martino, J.M., Pini Magalhães, L., Violaro, F.: Facial animation based on context-dependent visemes. Comput. Graph. **30**(6), 971–980 (2006). https://doi.org/10.1016/j.cag.2006.08.017, https://www.sciencedirect.com/science/article/pii/S0097849306001518
6. Edwards, P., Landreth, C., Fiume, E., Singh, K.: Jali: an animator-centric viseme model for expressive lip synchronization. ACM Trans. Graph. **35**(4) (2016). https://doi.org/10.1145/2897824.2925984
7. Fan, Y., Lin, Z., Saito, J., Wang, W., Komura, T.: Faceformer: speech-driven 3d facial animation with transformers. CoRR abs/2112.05329 (2021)
8. Kalberer, G., Van Gool, L.: Face animation based on observed 3d speech dynamics. In: Proceedings Computer Animation 2001. Fourteenth Conference on Computer Animation (Cat. No.01TH8596), pp. 20–251 (2001). https://doi.org/10.1109/CA.2001.982373
9. Karras, T., Aila, T., Laine, S., Herva, A., Lehtinen, J.: Audio-driven facial animation by joint end-to-end learning of pose and emotion. ACM Trans. Graph. **36**(4) (2017). https://doi.org/10.1145/3072959.3073658
10. Otberdout, N., Ferrari, C., Daoudi, M., Berretti, S., Del Bimbo, A.: Sparse to dense dynamic 3d facial expression generation. In: Proceedings of the IEEE/CVF Conference on Computer Vision and Pattern Recognition, pp. 20385–20394 (2022)
11. Otberdout, N., Ferrari, C., Daoudi, M., Berretti, S., Del Bimbo, A.: Generating multiple 4d expression transitions by learning face landmark trajectories. IEEE Trans. Affect. Comput. (2023)
12. Richard, A., Zollhöfer, M., Wen, Y., la Torre, F.D., Sheikh, Y.: Meshtalk: 3d face animation from speech using cross-modality disentanglement. CoRR abs/2104.08223 (2021)
13. Suwajanakorn, S., Seitz, S.M., Kemelmacher-Shlizerman, I.: Synthesizing obama: learning lip sync from audio. ACM Trans. Graph. **36**(4) (2017). https://doi.org/10.1145/3072959.3073640
14. Vaswani, A., et al.: Attention is all you need. In: Guyon, I., et al. (eds.) Advances in Neural Information Processing Systems, vol. 30. Curran Associates, Inc. (2017). https://proceedings.neurips.cc/paper1_files/paper/2017/file/3f5ee243547dee91fbd053c1c4a845aaPaper.pdf
15. Vougioukas, K., Petridis, S., Pantic, M.: Realistic speech-driven facial animation with gans. CoRR abs/1906.06337 (2019)
16. Zhou, Y., Li, D., Han, X., Kalogerakis, E., Shechtman, E., Echevarria, J.: Makeittalk: speaker-aware talking head animation. CoRR abs/2004.12992 (2020)

SCENE-pathy: Capturing the Visual Selective Attention of People Towards Scene Elements

Andrea Toaiari[1(✉)], Federico Cunico[1], Francesco Taioli[1], Ariel Caputo[1], Gloria Menegaz[1], Andrea Giachetti[1], Giovanni Maria Farinella[2], and Marco Cristani[1]

[1] University of Verona, Verona, Italy
{andrea.toaiari,federico.cunico,francesco.taioli,ariel.caputo, gloria.menegaz,andrea.giachetti,marco.cristani}@univr.it
[2] University of Catania, Catania, Italy
gfarinella@dmi.unict.it

Abstract. We present *SCENE-pathy*, a dataset and a set of baselines to study the visual selective attention (VSA) of people towards the 3D scene in which they are located. In practice, VSA allows to discover which parts of the scene are most attractive for an individual. Capturing VSA is of primary importance in the fields of marketing, retail management, surveillance, and many others. So far, VSA analysis focused on very simple scenarios: a mall shelf or a tiny room, usually with a *single* subject involved. Our dataset, instead, considers a *multi-person* and much more complex 3D scenario, specifically a high-tech fair showroom presenting machines of an Industry 4.0 production line, where 25 subjects have been captured for 2 min each when moving, observing the scene, and having *social interactions*. Also, the subjects filled out a questionnaire indicating which part of the scene was most interesting for them. Data acquisition was performed using Hololens 2 devices, which allowed us to get ground-truth data related to people's tracklets and gaze trajectories. Our proposed baselines capture VSA from the mere RGB video data and a 3D scene model, providing interpretable 3D heatmaps. In total, there are more than 100K RGB frames with, for each person, the annotated 3D head positions and the 3D gaze vectors. The dataset is available here: https://intelligolabs.github.io/scene-pathy.

Keywords: Visual Attention · Social Signal Processing · Gaze Estimation · Benchmark

1 Introduction

Capturing the attention of people toward specific elements in a scene is an attractive yet unsolved problem in computer vision. It is a precious skill in practical fields such as marketing and retail management [5]: knowing where the attention of most people will be directed allows one to properly set up the advertisements inside a building, as well as to charge adequately for the posting of the advertisements

G. L. Foresti et al. (Eds.): ICIAP 2023, LNCS 14233, pp. 352–363, 2023.
https://doi.org/10.1007/978-3-031-43148-7_30

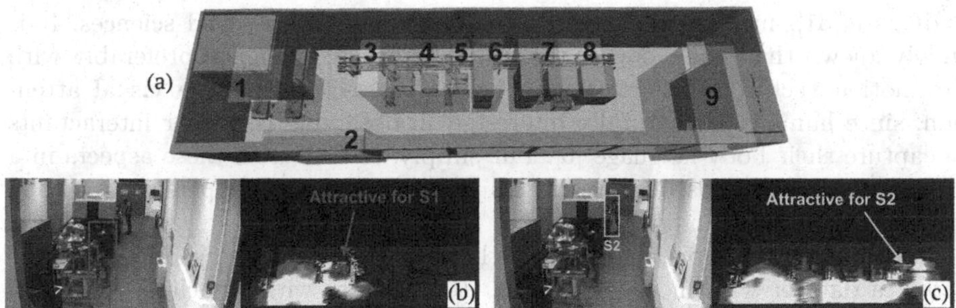

Fig. 1. (a) The 3D model of the *SCENE-pathy* showroom scenario, with the 9 possible areas of interest highlighted by colored bounding boxes. (b, c) On the left is an input video frame, in which we highlight a single subject. On the right, the corresponding VSA estimation is in the form of a 3D map, where hotter colors indicate areas that attracted more VSA. (Color figure online)

themselves. This type of attention is called *visual selective attention* (VSA), *i.e.*, the process of directing the gaze to relevant visual stimuli while ignoring the irrelevant ones in the environment [8]. VSA is generally studied by exploiting the eye gaze dynamics [7,21] in very simple 2D or 3D scenarios, where the area of interest is limited to planar scenes, such as mall shelf [4,32], or inside tiny rooms [16,29]. This is a real limitation, as visual selective attention is certainly important even in more structured and larger scenes. In this paper, we present *SCENE-pathy*, a dataset collected in a 20 by 5 m showroom of a tech fair (Fig. 1), where 9 different work areas are distinguishable, each one equipped with costly hardware to be exhibited. The showroom has been manually reconstructed beforehand as a 50K-points 3D cloud. A total of 25 unacquainted subjects have agreed to participate to build the dataset. For each participant, we have collected video sequences and sensory data, thanks to the use of Hololens 2 devices, for about 1.5 h of data footage. Hololens 2 allowed us to extract tracklets of 3D head positions and eye gaze trajectories, which we collected as annotations for the dataset.

As an additional, higher-level, annotation, the subjects filled out a simple questionnaire, communicating which parts of the scenes attracted their attention the most. A dataset with these characteristics would be a novel and useful asset to study problems related to VSA estimation and attention target detection in general, both in the 2D and 3D spaces. To demonstrate some applications on the dataset, we propose a set of baselines to capture the VSA of single and pairs of people. In general, these baselines individuate and track individuals by 3D pose estimation, using solely the RGB video stream as input. Successively, they intersect a physiologically plausible view frustum, fitted on the estimated head pose, with the 3D point cloud (+3D culling) of the scene. The resulting intersections are considered as a vote of VSA. Across time, votes are accumulated, providing a 3D weighted map that, for each person, indicates which parts of the scene have attracted their visual selective attention the most (see Fig. 1b, Fig. 1c). The use of a coloured 3D point cloud heatmap is a novel concept in this type of research. The baselines are enriched by social signal processing (SSP) findings [4,31], merging computer vision techniques with social sciences. It is widely known that while moving the visual attention is aligned preferably with the motion vector [15]. Furthermore, social interactions require visual attention, since humans are naturally interested in observing the other interactants to capture their body language [6]. Put simply, to deal with these aspects in a principled way, our baselines weigh less VSA votes when the person is moving or engaging in social interactions.

In summary, our contributions are the following: *(i)* we introduce *SCENEpathy*, a dataset with more than 100K RGB frames with associated, for each person, the 3D head positions and the 3D gaze vectors captured by Hololens 2 devices; *(ii)* a set of baselines capturing VSA from RGB frames in a complex multi-person scenario, taking into account also the social signal processing point of view; *(iii)* a new way of encoding the visual attention using a coloured 3D point cloud, manually reconstructed to allow the VSA counting even on areas occluded to the camera.

2 Related Works

Attention Target Detection. The idea of continuously estimating attention on a complex 3D heatmap from a third-person perspective is a novel task in the literature. The most similar task is attention target detection [2,9,10,32], or gaze-following [16,22,27], which aims at identifying the object looked at by a certain person in an image or video and it is useful to identify intentions and predict future actions. In [9,27,32], the task is tackled in the 2D image space, exploiting insights from the saliency maps theory [17]. The dataset proposed in [27], despite being one of the most used, presents shortcomings when considered from a 3D standpoint, such as the inability to counteract the effect of occlusions and perspective. Some recent works extend their models to handle 3D depth estimation [2,10,16,22], trying to figure out the distance from the camera of the various objects in the scene. In [2], 3D point clouds are constructed directly from 2D images and a model incorporates both the scene contextual cues and the predicted probabilities to refine the target fixation prediction. In [16], two datasets for the 3D gaze-following task are proposed, but the subject's sightlines were annotated manually, which renders them potentially inaccurate. A special case is addressed in [22], where a method is proposed to find the target of attention in 360-degree images. In [29], a single-person scenario in a lab setting is presented. A depth camera extracts the 3D human skeleton to estimate the head view vector and infer the attended object of interest. However, at least one RGB-D camera is necessary and multi-person capabilities are not tested. Our proposed dataset differs from other works by leveraging a complete 3D model of the environment, enabling us to compute attention even on parts of the scene occluded to the camera. We provide both low-level annotations, which are 3D positions and gaze vectors obtained from top-tier devices, and high-level annotations in the form of interest questionnaires, filled by the subjects after visiting the scene. Additionally, we capture VSA in a multi-person setting, through a top-down pose estimation pipeline, starting from the mere RGB video stream.

Visual Attention Mapping. One of the distinctive features of the proposed method is the use of the environment point cloud as a heatmap, with a coloring scheme that shows the areas of greatest interest. In [3], the analysis of the visual attention in a museum is carried out with a simple subject-artwork association, without considering the environment as a whole. The works of [28] and [4] exploit an external point of view of the scene and propose the concept of a colored attention heatmap applied to the environment. Differently from us, the used maps are not fine-grained, the frameworks did not consider the interplay of motor activities nor social interactions, and discriminate only a fixed discrete number of head orientations, while we have continuous orientations.

The Social Signal Processing Point of View. Humans sample the visual world by making eye movements to direct a centralized region of visual acuity towards different parts of the environment [23]. This allows to direct awareness to relevant stimuli while ignoring irrelevant ones in the environment and is called visual selective attention [14]. Eye gazing is the clearest way to capture the VSA [12]. In the absence of eye-tracking capabilities due to low-resolution images, the head pose is the second best indicator of where eyes may be fixated [24]. This happens due to the orbital reserve [13], the tendency for the eyes to be positioned centrally in the orbits, and the fact that people orient their heads towards important features in the environment [12]. Selective attention and vigilance have important interplay during activities such as walking [11] and interacting with other people [31]: it is known that without walking or the presence of social interaction, eye gazing is most probably associated with visual selective attention [14]. We, therefore, propose to incorporate these findings in our approach, demonstrating that they can help reduce the number of false positives when estimating the VSA.

3 The *SCENE-pathy* Dataset

SCENE-pathy is the first dataset for VSA estimation that features accurate 3D annotations of the position and gaze direction of multiple people moving in the scene, acquired automatically by head-mounted Hololens 2 devices. The main goal is to provide a novel benchmark to study whether VSA can be captured in a non-collaborative scenario using computer vision techniques on a monocular video stream, taking into account also the SSP point of view.

To this end, we captured two kinds of ground truth data: *low-level* and *high-level*. We define low-level data as the annotations for the 3D head position and gaze vector of the subjects, relative to a fixed starting position, captured using Hololens 2 devices (one for each person). These ground truth data were then synchronized with the RGB video frames, captured from a single wall-mounted camera, and the starting position is shifted according to the environment coordinate reference system. Since the main challenge we want to explore in this paper lies in capturing perceptual-cognitive processes, *i.e.*, how SSP findings can help reduce false positives in the VSA estimation, we also introduce high-level ground truth in the form of simple questionnaires, in which the subjects have to rank, in descending order of perceived interest, all the 9 stations in the scene. The scene is a $20l \times 5d \times 4h$ meters showroom with 9 different industrial stations, replicating a modern Industry 4.0 production line (see Fig. 1a).

We manually created an accurate 3D model of the environment using Blender, sampled it into an equally distributed point cloud, and defined 3D bounding boxes around each machine. This will be useful to measure the final VSA of each object. The biggest station measures, in meters, $2.5l \times 3d \times 3.5h$, and the smallest measures $0.3l \times 0.1d \times 0.6h$. The dataset is organized into two experimental sessions: *unsupervised* and *supervised*, depending on whether the subjects freely explored the environment or were directed to a specific industrial station. Note that the supervised experiments were performed after the unsupervised ones. Each session was further divided into *single-person* and *multi-person*. The multi-person experiments were conducted with two people at the same time, in which we require them to engage in a discussion, hence having a social interaction.

Unsupervised Experiments: In the single-person unsupervised experiments, the subjects were asked to freely explore the environment and then fill out the questionnaire. In the multi-person variant, the couple was also expected to discuss their station of interest, motivating their choices, before filling out the questionnaire. The average running time of each experiment is about 2 min.

Supervised Experiments: In the supervised experiments, for each subject, we selected *beforehand* a specific industrial station to inspect. Obviously, the subjects are now more familiar with the environment, as they have already seen it in the unsupervised session. In this case, we want to be sure which station should be the most looked at, to see if we can estimate it correctly. As in the unsupervised session, for the multi-person variant, we also asked the subjects to have a brief discussion after the exploration. Since the subjects can't undergo the unsupervised trials more than once, this second type of test is useful to further enrich the dataset and provide simpler and less randomized movement patterns.

The *SCENE-pathy* dataset contains 19 unsupervised experiments (13 single-person and 6 with pairs), and 60 supervised experiments (48 single-person and 12 with pairs). For the data collection, we hired 25 unacquainted subjects: 23 males and 2 females, average age of 27. The dataset is composed of the 3D model of the scene and more than 100K RGB frames with associated, for each person, the accurate annotation of 3D head positions, the 3D gaze vectors, and also the questionnaires as additional annotation of the actual interest of the people.

4 The Proposed Baselines

In this section, we present a modular baseline dubbed *Socio-Dynamic VSA* (*SD-VSA*) to compute the VSA on the *SCENE-pathy* dataset. Furthermore, to demonstrate the effectiveness of the involved social signal modules, we provide simpler ablative versions. The first module of the pipeline estimates the 3D pose of each subject, in the form of a 3D skeletal model. The captured videos are processed by a top-down 2D pose estimation algorithm [30], with a tracking module [33] to maintain the subject ids. The extracted 2D pose joints are then converted by a 3D pose lifter [25] to 17-joint 3D skeletal models, centered initially at the axis origin. These skeletons are then located in their correct scene locations using a homography matrix to transform points from the camera space to a top-view planimetry of the environment. The vector passing through the two 3D joints on the head, one for the nose and the other between the ears, becomes the central axis of a pyramid-shaped view frustum. This represents the direction and area of focus of the subject's gaze. As shown in Sect. 5, this estimated vector aligns sufficiently well with the ground truth sensor provided by the Hololens 2 devices, in terms of both pan and tilt angular errors.

At the start, each of the N 3D points of the point cloud representing the environment, dubbed P, is assigned a null attention value. For each frame t, and for each human subject i, the head position and the derived gaze direction are used to estimate the subset of visible points F_t^i. This is done in two steps: first, the occluded points are determined and removed using the Hidden Point Removal operator [18]. Then, a view frustum culling is performed considering a horizontal FOV of 90° and a vertical FOV of 60°, mainly based on [20]. In the last step, the attention values of the non-occluded points inside the frustum are increased according to a weighting scheme, with a maximum around the central axis and exponentially decreasing at the margins, following the idea of the attention spotlight model [26]. The view frustum is quantized into three concentric parts with different weights to speed up the computation.

Specifically, the VSA_t^i for a specific frame t and a subject i is composed by the indexes of the point cloud with an associated attention value $s_{t,k}^i = \gamma$, if $p_k \in F_t^i$, else 0, with p_k the k-th point of P. The weight γ is quadratic and it is chosen depending on which subdivision of the view frustum contains the considered point. The weights can then be further modulated according to the movement and the social activity of the subject, as described in the following subsections. Considering the interval of frames T and the subject i, the individual map VSA^i is the sum of the map scores at each frame t: $\text{VSA}^i = \sum_t^T \text{VSA}_t^i$.

The Dynamic Module. Section 2 explains that motor activity and social interaction are overriding the visual attention a subject may spend towards the environment [11,12]. In the *Dynamic* module of *SD-VSA*, the motor activity is modeled by changing the equation for $s_{t,k}^i$, substituting γ with $(\gamma \cdot w(v))^2$, where $w(v)$ is a weight inversely proportional to the velocity of the subject in the scene. In simple words, the higher the velocity, the lower the weight that a certain intersection between the view frustum and the scene does have.

The Social Interaction Module. This module copes with the fact that social interactions inhibit the visual selective attention of the individuals towards the scene elements [6,31]. Many approaches are available in the literature; the most effective ones take into account proxemics cues (the position of the interactants) and kinesics cues (body poses). In particular, [4] individuates a social interaction when people are close enough (accounting to Kendon's findings on Hall's usage of the personal space [19]). In practice, people at a distance of less than 1.2 m, that are pointing their faces at each other, individuate a social interaction.

All these cues are available: thanks to the multi-person pose estimation, this configuration can be found simply by analyzing each possible pair of stationary individuals. On each pair, if the people are close enough, two conditions are checked: if there is an intersection between the view frustums and if each one of the vision cones contains the other subject's head. When in the presence of both conditions, the counting of the scene-frustum intersections is suspended.

Ablative Baselines. Alternative baselines can be obtained by suppressing either one or both of the social signal modules. The *Vanilla* baseline indicates the scenario in which none of the modules is active, and the VSA maps are always increased by the same amount, no matter the speed of the subjects or the occurrence of social interactions. *DYN* and *SOCIAL* represent respectively the baseline in which only the *Dynamic* module is active or only the *Social* module is active. Of course, the *Social* module only makes sense in experiments where more than one subject is present.

5 Experiments and Results

In this section, we present the experiments and comment on the results obtained by applying our baselines. Firstly, we show that our 3D head pose estimation pipeline (approximating the gaze) performs sufficiently well, demonstrating quantitatively the orbital reserve effect [13] discussed in Sect. 2. Since all of the baselines of Sect. 4 depend on the gaze direction, this is a necessary step of the analysis. Then, we show the capability of *SD-VSA* and the alternative baselines in providing accurate VSA maps on the *SCENE-pathy* dataset, despite the uncertainty in the true gaze direction. Finally, we show that our baseline can provide useful VSA maps also in more crowded scenarios, using the GVEII benchmark [1].

Capturing the Human Gaze. We extracted the Hololens 2 head poses and compared them with the eye-tracking data by measuring the pan and tilt angles error, resulting in less than 1° and 3° discrepancy, respectively. Then, to assess the quality of our head pose estimation pipeline we calculated the pan and tilt error of the whole dataset in the range [0°, 360°] for the pan and [0°, 180°] for tilt. For pan, 70% of the elements are in the range of 0° and 180°, while on tilt 97% of the elements are between 60° and 135°. Globally, the average pan error is 52.36° and the average tilt error is 50.81°. We consider these results acceptable, given the size and complexity of the analyzed scenario.

Unsupervised Experiments. In this section, we show how VSA maps associated with each subject are coherent w.r.t.the high-level ground truth questionnaires discussed in Sect. 3, and the impact of the *Dynamic* and *Social* modules detailed in Sect. 4. We start by comparing the ranking from the questionnaires with the ones obtained by the scores of the computed VSA. In particular, the 9 objects are ranked taking the maximum score within the bounding boxes defined around each station. Subsequently, we compute over all the experiments the Cumulative Matching Curve (CMC), comparing the two rankings. This curve

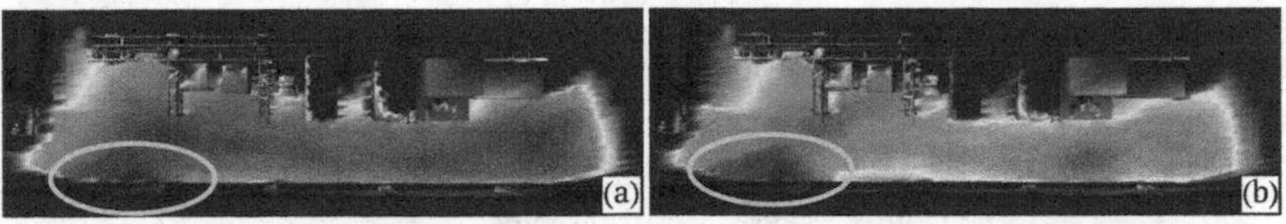

Fig. 2. Qualitative results of an unsupervised single-subject experiment. (a) *DYN* module disabled (b) *DYN* module active. Without the *DYN* module, the resulting VSA is definitely noisier. The target object was an advertising board, circled in green. (Color figure online)

Table 1. The accuracy in unsupervised experiments.

Method	CMC Single-Person				CMC Multi-Person			
	Rank-1	Rank-2	Rank-3	Rank-4	Rank-1	Rank-2	Rank-3	Rank-4
Vanilla	30.8	46.2	61.5	76.9	25.0	41.7	50.0	66.7
DYN	38.5	61.5	84.6	100.0	25.0	50.0	75.0	83.3
SOCIAL	–	–	–	–	33.3	58.3	75.0	100.0
SD-VSA	–	–	–	–	33.3	58.3	83.3	100.0

defines the probability of a specific object having the highest score, and hence being the correct object, within the first N points of the curve. We first analyze the *unsupervised single-person* sequences. Considering the presence of a single person, we evaluated the system in two configurations: one with the *DYN* module and one without (*Vanilla*). Numerically, the *DYN* module gives an accuracy boost, and Fig. 2 shows that VSA maps generated without it are definitely noisier, showing a marked trace on the floor where the subject has passed.

In the *unsupervised multi-person* sequences, together with the *Vanilla* and *DYN* module we also consider whether the accumulation on the maps is stopped by the occurrence of social interactions, dubbed *SOCIAL*. This leads to four possible combinations (see Table 1). As a metric, we report the rank 1–4 accuracies, which show the predominance of the complete *SD-VSA* model as expected.

Discussion: The single-person unsupervised experiments results are presented in Table 1. Since the subjects have never seen the environment, they usually start by navigating the whole hallway, before focusing on some specific industrial station. The accuracy of the system solely relying on people's location is low, achieving below 50% at rank-2, which is improved to over 60% introducing the *DYN* module. The significant difference in accuracy between rank-1 and rank-2 can be explained by the fact that the machines are distributed very close to each other, indicating that often the correct object is spatially close to the estimated one. Moreover, since our pipeline is modular, replacing the gaze estimation module with a more accurate one could certainly provide an additional boost in the performance described by the CMC curve.

Multi-person unsupervised experiments also present a non-trivial task to solve, as the presence of two individuals can introduce noise in the VSA estimation, due to social interactions. Incorporating an SSP module helps to improve

Table 2. The accuracy in supervised experiments. In this case, the value tells us if we were able to identify the specific station assigned to subjects.

Method	Single-Person	Multi-Person
Vanilla	56.3	50.0
DYN	62.5	54.2
SOCIAL	–	62.5
SD-VSA	–	70.8

the performance of the system. It is important to note that the results of multi-person experiments are generally lower than those of single-person ones, since the pose estimation pipeline face challenges with occlusion and identity swapping caused by multiple individuals exploring the large scene. Finally, the actual interest of the subjects, collected through the questionnaires, shows that solely relying on gaze estimation is not enough, and insights from SSP increase the chance of capturing the actual interest.

Supervised Experiments. In the supervised experiments the goal is to check whether the VSA maps can highlight the object in the scene that the subject was assigned to. In this case, the ground truth is therefore the assigned station. As an accuracy measure, we consider a success whether the peak of the VSA map is inside the bounding box surrounding the target object, 0 otherwise. The results of these experiments are presented in Table 2. As a qualitative result, Fig. 3 shows the VSA maps after a long social interaction has occurred in the hallway. Without the *SOCIAL* module, there are incorrect attention peaks: on the wall behind subject 1 (Fig. 3c) and on the machine behind subject 2 (Fig. 3a). Instead, with *SD-VSA*, we can recognize this interaction and suspend the attention weighting, focusing only on the real attention of the subjects.

Discussion: In these experiments, the subjects are now more familiar with the environment. Additionally, since we assigned beforehand the target industrial machine, subjects naturally perform less to no exploration of the room, heading straight for the given machine. This explains why the supervised experiments achieve higher results than the unsupervised ones. The results shown in Table 2 clearly reflect this difference between the two sets of experiments, providing further evidence of the usefulness of taking into account SSP insights.

5.1 Does the *SD-VSA* Scale up to Crowded Scenarios?

To prove that the SSP modules of the *SD-VSA* baseline are effective and generalize to highly crowded scenes, we use the GVEII benchmark [1], where hundreds of people walk in an outdoor mall. Positions and group formations are already available as annotations. We compute the VSA with *SD-VSA* and the *Vanilla* baseline, using a simple 3D model and projecting the frustums on the floor for visualization purposes. Note that here the view frustums are not computed through 3D pose estimation; instead, we use the oriented velocity vector derived from the ground truth trajectory data. In Fig. 4b, it is possible to note a clearly

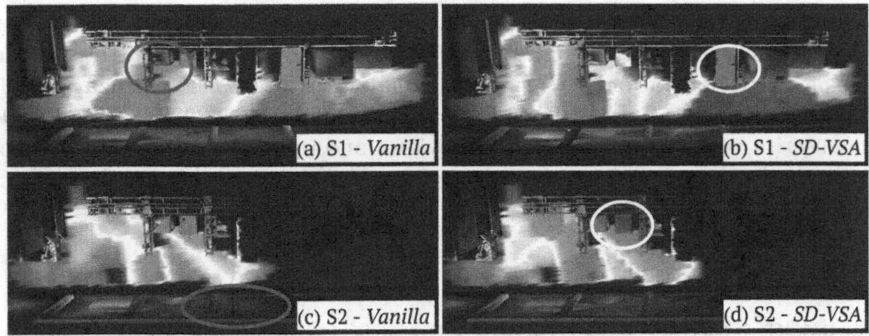

Fig. 3. Qualitative results of a supervised experiment with a pair of subjects (S1 and S2), that had a dialogue in the center of the hallway. In (b) and (d) the objects circled in green correspond to the correct most attractive points for those VSA maps. (a) and (c) show instead the false positives targets (circled in red) obtained by disabling the *DYN* and *SOCIAL* modules. (Color figure online)

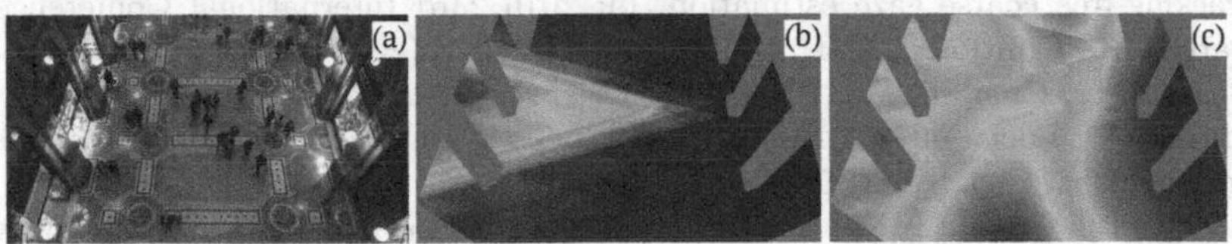

Fig. 4. (a): a frame from the GVEII dataset; (b): *SD-VSA* baseline; (c): *Vanilla* baseline: *DYN* and *SOCIAL* modules are switched off. (Color figure online)

interpretable global VSA, communicating that the people were interested in the two shop windows on the left, while the bottom-right window shop, which is closed, gather the least attention. In Fig. 4c results are incorrect, since the most interesting part of the scene appears to be the pathway, due to the vertical flow of the people.

6 Conclusions

In this paper, we propose *SCENE-pathy*, a benchmark dataset to study the visual selective attention of people towards a complex 3D scene. Along with the RGB video streams, we provide complete annotations of the 3D position and gaze direction of the involved subjects, extracted from Hololens 2 devices, and questionnaires reporting the most interesting areas visited by them. The proposed baselines, created joining pose estimation techniques and social signal processing insights, allowed us to demonstrate that more accurate VSA maps can be obtained by considering the subjects' movement speed and the possible occurrence of social interactions in multi-person scenarios.

Acknowledgments. This work was partially supported by the Italian MIUR within PRIN 2017, Project Grant 20172BH297: I-MALL - improving the customer experience in stores by intelligent computer vision and PNRR research activities of the consortium iNEST (Interconnected North-Est Innovation Ecosystem) funded by the European Union Next-GenerationEU (Piano Nazionale di Ripresa e Resilienza (PNRR) - Missione 4 Componente 2, Investimento 1.5 - D.D. 1058 23/06/2022, ECS_00000043). This manuscript reflects only the Authors' views and opinions, neither the European Union nor the European Commission can be considered responsible for them.

References

1. Bandini, S., Gorrini, A., Vizzari, G.: Towards an integrated approach to crowd analysis and crowd synthesis: a case study and first results. Pattern Recogn. Lett. **44**, 16–29 (2014)
2. Bao, J., Liu, B., Yu, J.: Escnet: gaze target detection with the understanding of 3d scenes. In: Proceedings of the IEEE/CVF Conference on Computer Vision and Pattern Recognition, pp. 14126–14135 (2022)
3. Bartoli, F., Lisanti, G., Seidenari, L., Del Bimbo, A.: User interest profiling using tracking-free coarse gaze estimation. In: 2016 23rd International Conference on Pattern Recognition (ICPR), pp. 1839–1844. IEEE (2016)
4. Bazzani, L., Cristani, M., Tosato, D., Farenzena, M., Paggetti, G., Menegaz, G., Murino, V.: Social interactions by visual focus of attention in a three-dimensional environment. Exp. Syst. **30**(2), 115–127 (2013)
5. Becattini, F., et al.: I-mall an effective framework for personalized visits. improving the customer experience in stores. In: Proceedings of the 1st Workshop on Multimedia Computing towards Fashion Recommendation, pp. 11–19 (2022)
6. Birmingham, E., Bischof, W.F., Kingstone, A.: Gaze selection in complex social scenes. Vis. Cogn. **16**(2–3), 341–355 (2008)
7. Borji, A., Itti, L.: State-of-the-art in visual attention modeling. IEEE Trans. Pattern Anal. Mach. Intell. **35**(1), 185–207 (2012)
8. Carrasco, M.: Visual attention: The past 25 years. Vis. Res. **51**(13), 1484–1525 (2011)
9. Chong, E., Wang, Y., Ruiz, N., Rehg, J.M.: Detecting attended visual targets in video. In: Proceedings of the IEEE/CVF Conference on Computer Vision and Pattern Recognition, pp. 5396–5406 (2020)
10. Fang, Y., et al.: Dual attention guided gaze target detection in the wild. In: Proceedings of the IEEE/CVF Conference on Computer Vision and Pattern Recognition, pp. 11390–11399 (2021)
11. Fotios, S., Uttley, J., Cheal, C., Hara, N.: Using eye-tracking to identify pedestrians' critical visual tasks, part 1. dual task approach. Lighting Res. Technol. **47**(2), 133–148 (2015)
12. Foulsham, T., Walker, E., Kingstone, A.: The where, what and when of gaze allocation in the lab and the natural environment. Vis. Res. **51**(17), 1920–1931 (2011)
13. Fuller, J.H.: Eye position and target amplitude effects on human visual saccadic latencies. Exp. Brain Res. **109**(3), 457–466 (1996)
14. Gordon, R.D.: Selective attention during scene perception: evidence from negative priming. Memory Cogn. **34**(7), 1484–1494 (2006)
15. Hasan, I., Setti, F., Tsesmelis, T., Del Bue, A., Galasso, F., Cristani, M.: Mx-LSTM: mixing tracklets and vislets to jointly forecast trajectories and head poses. In: Proceedings of the IEEE Conference on Computer Vision and Pattern Recognition, pp. 6067–6076 (2018)

16. Hu, Z., Yang, D., Cheng, S., Zhou, L., Wu, S., Liu, J.: We know where they are looking at from the RGB-d camera: Gaze following in 3d. IEEE Trans. Instrum. Meas. **71**, 1–14 (2022)
17. Itti, L., Koch, C.: Computational modelling of visual attention. Nat. Rev. Neurosci. **2**(3), 194–203 (2001)
18. Katz, S., Tal, A., Basri, R.: Direct visibility of point sets. In: ACM SIGGRAPH 2007 Papers, pp. 24-es. SIGGRAPH 2007, Association for Computing Machinery, New York, NY, USA (2007)
19. Kendon, A.: Conducting interaction: Patterns of behavior in focused encounters, vol. 7. CUP Archive (1990)
20. Kress, B.C.: Digital optical elements and technologies (edo19): applications to AR/VR/MR. In: Digital Optical Technologies, vol. 11062, pp. 343–355 (2019)
21. Li, Y., Liu, M., Rehg, J.: In the eye of the beholder: gaze and actions in first person video. IEEE Trans. Pattern Anal. Mach. Intell. **45**, 6731–6747 (2021)
22. Li, Y., Shen, W., Gao, Z., Zhu, Y., Zhai, G., Guo, G.: Looking here or there? gaze following in 360-degree images. In: Proceedings of the IEEE/CVF International Conference on Computer Vision, pp. 3742–3751 (2021)
23. Melcher, D.: Visual stability. Philos. Trans. R. S. B: Biol. Sci. **366**(1564), 468–475 (2011)
24. Parks, D., Borji, A., Itti, L.: Augmented saliency model using automatic 3d head pose detection and learned gaze following in natural scenes. Vis. Res. **116**, 113–126 (2015)
25. Pavllo, D., Feichtenhofer, C., Grangier, D., Auli, M.: 3D human pose estimation in video with temporal convolutions and semi-supervised training. In: Proceedings of the IEEE/CVF Conference on Computer Vision and Pattern Recognition, pp. 7753–7762 (2019)
26. Posner, M.I., Snyder, C.R., Davidson, B.J.: Attention and the detection of signals. J. Exp. Psychol. Gen. **109**(2), 160 (1980)
27. Recasens, A., Khosla, A., Vondrick, C., Torralba, A.: Where are they looking? In: Advances in Neural Information Processing Systems, vol. 28 (2015)
28. Reid, I., Benfold, B., Patron, A., Sommerlade, E.: Understanding interactions and guiding visual surveillance by tracking attention. In: Koch, R., Huang, F. (eds.) ACCV 2010. LNCS, vol. 6468, pp. 380–389. Springer, Heidelberg (2011). https://doi.org/10.1007/978-3-642-22822-3_38
29. Shi, X., Yang, Y., Liu, Q.: I understand you: Blind 3d human attention inference from the perspective of third-person. IEEE Trans. Image Process. **30**, 6212–6225 (2021)
30. Sun, K., Xiao, B., Liu, D., Wang, J.: Deep high-resolution representation learning for human pose estimation. In: Proceedings of the IEEE/CVF Conference on Computer Vision and Pattern Recognition, pp. 5693–5703 (2019)
31. Vinciarelli, A., Pantic, M., Bourlard, H.: Social signal processing: survey of an emerging domain. Image Vis. Comput. **27**(12), 1743–1759 (2009)
32. Wang, B., Hu, T., Li, B., Chen, X., Zhang, Z.: Gatector: A unified framework for gaze object prediction. In: Proceedings of the IEEE/CVF Conference on Computer Vision and Pattern Recognition, pp. 19588–19597 (2022)
33. Zhang, Y., et al.: Bytetrack: multi-object tracking by associating every detection box. In: Proceedings of the European Conference on Computer Vision (ECCV) (2022)

Not with My Name! Inferring Artists' Names of Input Strings Employed by Diffusion Models

Roberto Leotta[1], Oliver Giudice[2(✉)], Luca Guarnera[3], and Sebastiano Battiato[1,3]

[1] iCTLab Spinoff of University of Catania, Catania, Italy
roberto.leotta@ictlab.srl

[2] Applied Research Team, IT Department, Banca d'Italia, Italy
oliver.giudice@bancaditalia.it

[3] Department of Mathematics and Computer Science, University of Catania, Catania, Italy
{luca.guarnera,sebastiano.battiato}@unict.it

Abstract. Diffusion Models (DM) are highly effective at generating realistic, high-quality images. However, these models lack creativity and merely compose outputs based on their training data, guided by a textual input provided at creation time. Is it acceptable to generate images reminiscent of an artist, employing his name as input? This imply that if the DM is able to replicate an artist's work then it was trained on some or all of his artworks thus violating copyright. In this paper, a preliminary study to infer the probability of use of an artist's name in the input string of a generated image is presented. To this aim we focused only on images generated by the famous DALL-E 2 and collected images (both original and generated) of five renowned artists. Finally, a dedicated Siamese Neural Network was employed to have a first kind of probability. Experimental results demonstrate that our approach is an optimal starting point and can be employed as a prior for predicting a complete input string of an investigated image. Dataset and code are available at: https://github.com/ictlab-unict/not-with-my-name.

Keywords: Diffusion Models · Artist Recognition · Multimedia Forensics

1 Introduction

The rapid advancement of generative models, particularly Diffusion Models [7], has led to a surge in high-quality, realistic image generation. These models have demonstrated immense potential for creative applications across various domains, including art, design, and advertising. However, their ability to replicate the styles of specific artists raises concerns about Intellectual Property (IP) rights and potential copyright infringements[1] [19,20,25].

[1] https://www.theverge.com/2023/1/16/23557098/generative-ai-art-copyright-legal-lawsuit-stable-diffusion-midjourney-deviantart.

G. L. Foresti et al. (Eds.): ICIAP 2023, LNCS 14233, pp. 364–375, 2023.
https://doi.org/10.1007/978-3-031-43148-7_31

As the boundary between human creativity and machine-generated content becomes increasingly blurred, it is crucial to address the legal and ethical implications of using generative models to produce art, as well as to develop methods that evaluate the extent to which generated images are influenced by the works of real artists and ensure the protection of their intellectual property.

Several studies and articles have explored the legal ramifications of generative models and their potential to infringe on copyrights[2]. In an article dated 2021 the challenges in determining copyright ownership for AI-generated works were discussed and authors emphasized the need for legal frameworks that could adequately address the unique nature of generative models[3]. MLQ.AI also reported on a copyright infringement case involving generative AI[4], which sparked debates on the responsibilities of AI developers and users in protecting original creators' rights. In response to these concerns, legal scholars have delved into the complexities of copyright law as it pertains to AI-generated artworks. Gillotte [8] examined the challenges in assigning liability and protecting IP rights, arguing that existing copyright laws may not be sufficient to address the unique characteristics of AI-generated contents. Indeed, it is became a copyright dilemma and the need for a balance between innovation and IP protection is arising to ensure that creative works are safeguarded without stifling technological advancements [16].

It is not easy to determine how images generated by tools like DALL-E 2 or Midjourney are created by combination of images employed at training time. But, it could useful to develop tools able to deduce the textual prompts that generated an investigated image. Recently, a Kaggle competition was launched on the task[5] but it still lacks of effective methods. However, to make images that appear with an artist's style the prompt of the generating tools should necessary contain the artist's name. Indeed, the generating tool was trained with original artworks belonging to that artist, coupled (as labels) with sentences containing the artist's name. This should demonstrate at a certain level the use of artist's artworks at training time thus violating copyright. Moreover, also the use of the name of a person without his consent should arise other issues.

In order make a starting point in the state of the art for techniques able to protect artists, in this study, an introductory empirical analysis method is presented to infer if an artist's name was employed to generate an investigated image. To simplify the problem, an extremely constrained scenario was built by collecting a dataset composed of original artworks from five renowned artists and also images generated employing the OpenAI's DALL-E 2 [17] with artists' names employed as prompted textual strings. Starting from these data a Siamese Neural Network was exploited to learn a dedicated metric for the purpose. Through a series of experiments, the effectiveness of the proposed approach was demonstrated in identifying a sort of probability of the usage of an artist's

[2] https://www.theverge.com/23444685/generative-ai-copyright-infringement-legal-fair-use-training-data.

[3] https://www.oreilly.com/radar/what-does-copyright-say-about-generative-models/.

[4] https://www.mlq.ai/copyright-infringement-generative-ai-this-week-in-ai/.

[5] https://www.kaggle.com/competitions/stable-diffusion-image-to-prompts/data.

name in the generation process of an image. This study can be extremely important in forensics in order to reconstruct and analyze the history of multimedia content [3].

The remainder of this paper is organized as follows: Sect. 2 lists research papers on the topic introducing the reader to the ethical and IP problems; Sect. 3 presents the collected dataset with details on composition and how it was built; Sect. 4 describes the proposed approach starting from a discriminative solution (Sect. 4.1) to the final metric objective of the study (Sect. 4.2). Experimental results are presented and discussed in Sect. 5 and Sect. 6 concludes the paper with some hints for future development.

2 Related Works

Generative models are a type of machine learning model that use data generation techniques to create new data instances from an existing dataset. These models can be used to generate images, text, sound and other types of data, and have multiple applications in fields such as content creation, simulation and new product creation. Mainly there are two large families of modern generative techniques: the Generative Adversarial Networks (GAN) [11] and the Diffusion Models (DM) [7]. The latter are ultimately being employed by successful applications given their simplicity with which they allow the user to control how the multimedia content has to be generated.

Regarding the intellectual property of the data on which generative models have been trained, it can be an important issue as these models can be trained on large amounts of data that belong to other owners, such as images or texts on the internet. If so, the use of this data by generative models could be considered an infringement of copyright or other forms of intellectual property [1].

Furthermore, since generative models use training data to learn and create new data, they can also create potential cybersecurity and privacy issues. For example, if a generative model was trained on individuals' personal data, it could generate new personal information that could be used for improper purposes.

Given the potential risks of generated multimedia contents, several state of the art works already addressed the issue by trying to detect if a query sample is real or fake [22,24]) or even recognize the specific GAN architecture that generated it [9,13,14,23]. Contents generated by DMs are already being investigated and there are some works that try to expose them [6,18].

However, detecting a synthetic content is just the first step. When it comes to evaluate if a certain image infringes IP property different approaches have to be developed. Yet while the problem is only discussed in newspapers first research papers are being published.

The most interesting one is the work of Carlini et al [5] which analyses different DMs in order to infer if a certain image was employed in the training set given a set of constraining hypothesis. This is a great starting point giving hints on possible discriminating features, however it only addresses privacy issues.

Table 1. Number of images employed. From the second to the last column the artists' names are given. The last two rows describe the total number of synthetic and real original images collected for each artist.

	Alfred Sisley	Claude Monet	Pablo Picasso	Paul Cezanne	Pierre-Auguste Renoir
#Synthetic	1,722	1,702	1,515	1,734	1,846
#Original	470	1,044	1,019	588	1,008

A method to effectively protect the IP is yet to be proposed. This work is a first attempt at predicting if a specific text (i.e. artists's names) was employed as input prompt at generation time.

3 A Dataset of Original and Synthetic Artworks

Three different dedicated datasets were collected to evaluate the effectiveness of modern Diffusion Models in generating images of specific artists and to have a basis to train the solutions proposed in this paper.

The first two datasets, containing synthetic and original data respectively, were created as described below.

Five prominent artists were selected to generate the synthetic images: Alfred Sisley, Claude Monet, Pablo Picasso, Paul Cezanne and Pierre-Auguste Renoir. For each artist, 2,000 synthetic images with a size of 512×512 pixels, were generated using DALL-E 2 and employing 1,000 different descriptive text lines similar to the following: *A field of sunflowers with a blue sky background*; *A group of children playing at a playground in a city park*; *A group of elephants drinking at a watering hole in the savannah.*

To ensure that the generated images had the same style as the selected artist, the text "*by Artist's name*" was added to each line of input prompting text, such as follows: *A field of sunflowers with a blue sky background,* ***by Pablo Picasso***; *A group of children playing at a playground in a city park,* ***by Alfred Sisley***; *A group of elephants drinking at a watering hole in the savannah, by* ***Claude Monet***.

In this way, two images per artist were generated for each text line. In addition, different context sentences were used for each artist, thus ensuring a diverse representation of the artist's style across subjects and maximizing variability.

In order to work with synthetic data representing the style of the involved artists, validation and cleaning operations of the generated dataset were necessary because the DALL-E 2 algorithm also produces images that did not match the intended artistic style or content (e.g., photo-realistic images rather than paintings).

In addition, original paintings of the same five artists were also downloaded from WikiArt[6]. Table 1 shows the total number of original and synthetic images

[6] www.wikiart.org.

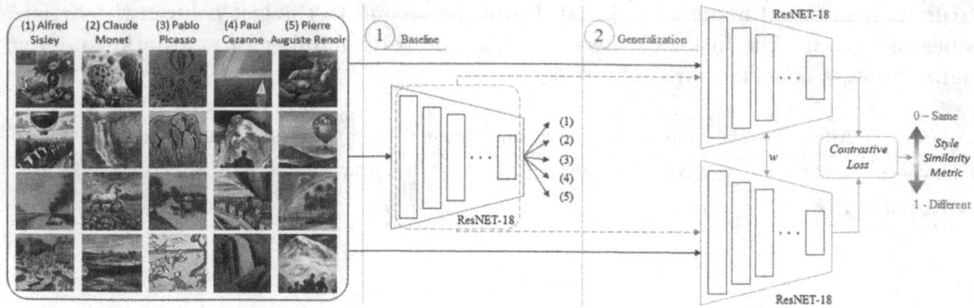

Fig. 1. Proposed approach. The dataset of artists (real and synthetic) was used in the Baseline approach (1) to discriminate against artist. To achieve more generalization, explainability and best define the similarity between different artist styles (2), a Siamese engine was trained from the pretrained model of (1). The ResNET-18 architecture was used in all experiments.

used in the experimental phase. The original images have different resolutions and, as shown in the Table 1, some artists have a limited number of works.

Finally, a third mixed dataset was created, consisting of 50% original images and 50% synthetic images. To obtain a balanced dataset, 470 images from the synthetic image dataset and 470 images from the real image dataset for each artist were randomly selected. This choice was made as Alfred Sisley has 470 images in the real image dataset.

The three datasets were split into two parts: 80% as training set and the remaining 20% of samples as validation set.

4 The Proposed Approach

The ultimate goal of this study was to develop a metric, a sort of probability that a content generated by OpenAI's DALL-E 2 was obtained prompting the tool with a sentence containing the name of one of the artists to be protected.

As a starting point, let D be the *reference dataset* to be protected containing a finite number of authors and their artworks. The presented metric will evaluate the query image by means of comparison with all the elements in D. Specifically, the objective is to estimate a set of distances $L = \widehat{d}(I, s)$ where I is the image to be investigated and $s \in D$ is an original image of the artists taken into account. The $\widehat{d}$ function is not known and has to be modelled in such a way that L has to be close to 0 if the two considered samples belong to the same author (both being original or synthetic and generated with his name). On the contrary, values of L should be 1 or bigger if the two considered samples are unluckily related.

At first, an evaluation study of a method to classify images between authors with the aim to learn the discriminative features was carried out. Further analysis was then carried out in order to model the final metric by means of a Siamese Neural Network (taking inspiration from [2,10,12] with the objective of learning a distance $\widehat{d}$. The overall study could be schematized as described in Fig. 1.

4.1 Learning to Discriminate

In order to demonstrate the possibility of discriminating the authors considered in our dataset D. Taking inspiration from [6,18], a Resnet-18 [15] was employed.

The first training experiment was to assess the Resnet-18 ability to discriminate between the different artists employing the original artworks only as training data. This set of images was split into 80% for training and 20% for testing purposes. After training the Resnet-18 model, excellent results in discriminating the artists were obtained on the original images meaning that the model successfully captured the distinctive characteristics of each artist's style, which was reflected in the high accuracy achieved during testing. However, the trained Resnet-18 model applied on the synthetic images, was not able to obtain similar good results as shown in Fig. 2(a). The model performances significantly deteriorated, with only a few artists being discriminated with reasonable accuracy. This indicated that the model struggled to generalize its understanding of the artists' styles for the synthetic images. It is clear that this *transfer learning* approach was not able to take into account features extractable from synthetic images, probably because of the generative process itself [13].

Given that the before-mentioned transfer learning attempt was not able to correctly discriminate synthetic images, a new Resnet-18 model was trained with the mixed dataset described in Sect. 3 (with the common train-test split 80%-20% as well). This time, the results obtained were way better then previous one so it can be said that this last Resnet-18 model learned not only features about authors' styles but also how DALL-E 2 alters images invisibly (i.e. by producing different distributions of frequencies of the generated images with respect to original ones). Upon analyzing the confusion matrix (Fig. 2(b)), it was possible to find an higher entropy for those artists with fewer publicly available works in the dataset. This could give an hint about DALLE-E training phase: it might have been trained on a dataset with a similar distribution and cardinality as the employed dataset D, which led to the generative model's inability to specialize in the artistic styles of specific artists with limited available samples. Obviously this is a strong assumption given that it is a just an introductory, trivial and preliminary insight, which could lead to further investigation on the composition of the dataset employed by the DM.

Given the obtained results, for the final objective of this paper this last model will be employed for further investigations described in the next Section.

4.2 Learning a Metric for Mixed Artworks

While the goal of a single Resnet-18 was to learn a hierarchy of feature representations to solve the discriminative tasks between 5 classes (the five considered artists), a Siamese Neural Network can be exploited for weakly supervised metric learning tasks. Instead of taking single sample as input, the network takes

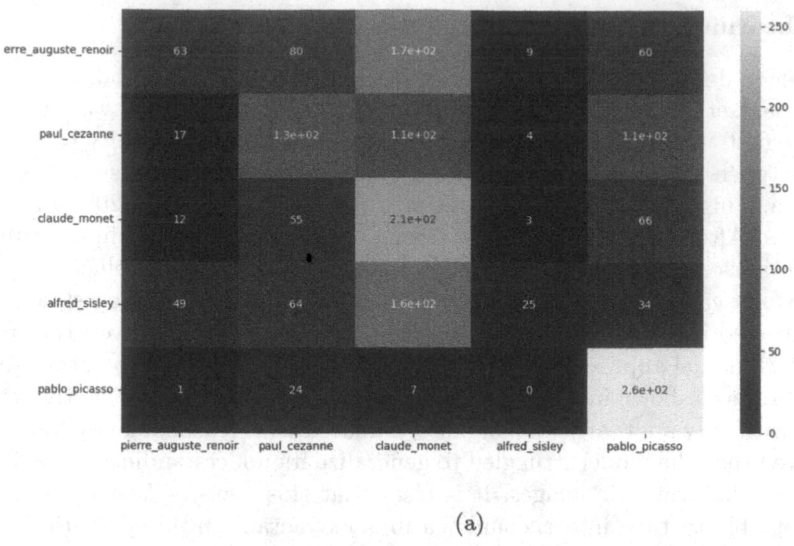

(a)

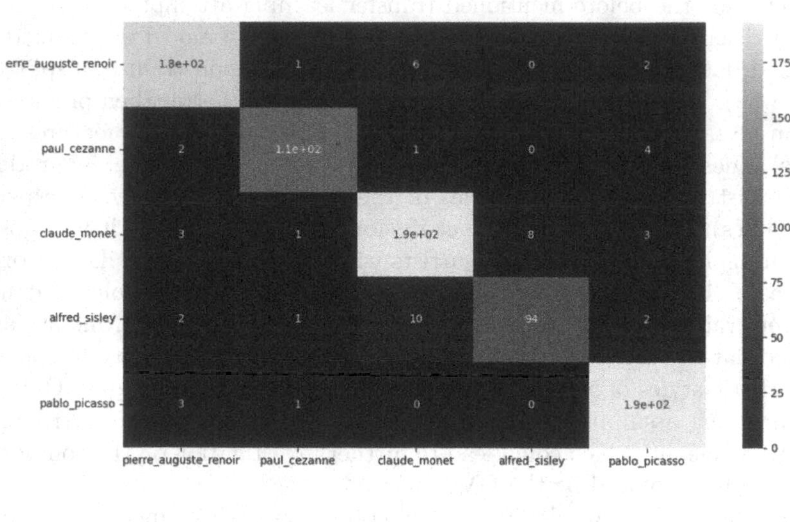

(b)

Fig. 2. (a) Transfer learning attempt. Results are shown as a confusion matrix computed on synthetic images with the Resnet-18 model trained only on original artworks. (b) Confusion matrix obtained on test set with the Resnet-18 model trained only on mixed dataset containing both original and synthetic images.

a pair of samples, and the loss functions are usually defined over pairs. In this case the loss function of a pair has the following form:

$$L(s_1, s_2, y) = (1 - y)\alpha D_w^2 + y\beta e^{\gamma D_w} \tag{1}$$

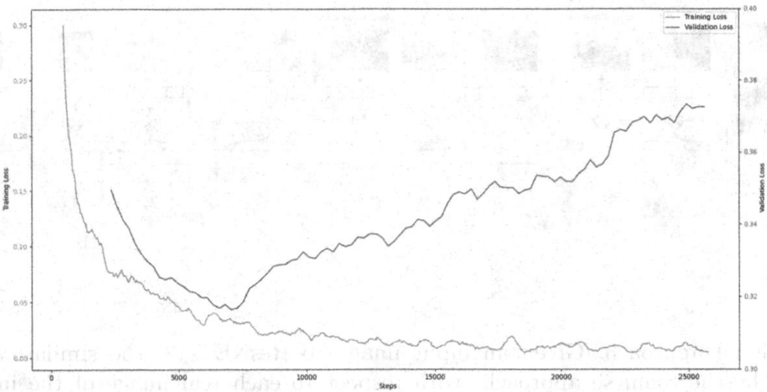

Fig. 3. Training losses of the siamese architecture. Orange shows the trend on the training set while blue is the validation one. (Color figure online)

where s_1 and s_2 are two pair of samples, $y \in 0, 1$ is the similarity label, and D_w is a distance function defined as in [21]. Parameters were set as follows $\alpha = 1/C_p$, $\beta = C_n$ and $\gamma = -2.77/C_n$ where $C_p = 0.2$ and $C_n = 10$.

Unlike methods that assign binary similarity labels to pairs, the network aims at bringing the output feature vectors closer for input pairs of the training labeled as similar, or push the feature vectors away if the input pairs are labeled as dissimilar.

Figure 1 shows the overall architecture showing as baselines the Resnet-18 models as trained and described in previous Section. The siamese network was trained employing the same training set used to previously train the single Resnet-18 models from the mixed dataset. Training and validation loss are shown in Fig. 3. Best model was considered for experiments with respect to the elbow of validation loss as shown in Fig. 3.

Given the properties of the loss L learned by the siamese network in a context of balances classes, the trained distance $\widehat{d}$ is a likelihood function describing the possibility to have image I being generated by one of the five authors' names took into account [4]. Thus, it is possible to define the probability that image I was generated by the tool DALL-E 2, by prompting a string containing an artist's name a between the five considered:

$$P(I|a) = 1 - min(\widehat{d}(I, s_a)) \quad (2)$$

where $\widehat{d}$ is the distance obtained by the trained siamese network, s is the set of images generated by DALL-E 2 with the corresponding author name a. In order to have $P \in [0, 1]$, Eq. 2 is evaluated only when $min(\widehat{d}(I, s_a)) \leq 1$. Figure 4 shows graphically the proposed final approach.

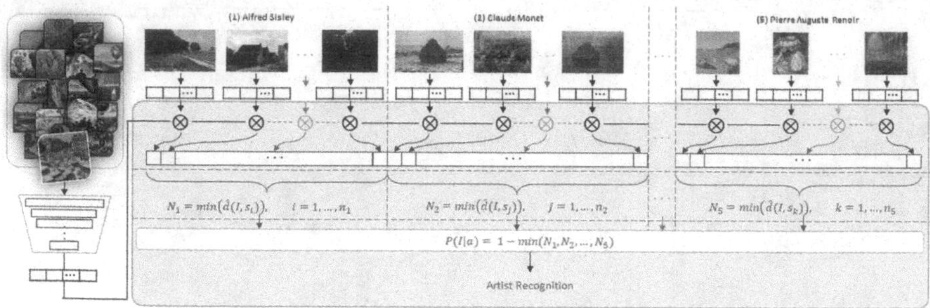

Fig. 4. Final approach. Given an input image to ResNET-18, the similarity score (defined by the Siamese approach) with respect to each real image of the involved artists is calculated. A voting process will define the number of images for each artist whose similarity score S exceeds a set threshold value T. Then, the test image will be assigned to a specific artist with respect to the highest voting score obtained in the previous step.

5 Experiments and Discussion

The siamese neural network with a baseline Resnet-18 model specialized in discriminating authors was developed for the purposes of this study as described in previous Sections. The Resnet-18 discriminative model was trained using the following parameters: ADAM optimization, 50 epochs, batch size of 32, weight decay and LR of 0.0001. The siamese configuration was trained with the loss as defined in Eq. 1, ADAM optimization, 250 epochs, batch size of 64, weight decay and lr of 0.0001 and contrastive loss margin of 2.

In order to evaluate the effectiveness of the method a retrieval test was carried out. The retrieval test is able to show how the metric is working in correcting detecting the author by means of comparison between query images and all original artworks in the test set of the mixed dataset D.

For each author a single generated image was selected as query and all the original artworks samples were employed to perform the retrieval test. The retrieval performance has been evaluated with the probability of the successful retrieval $P(n)$ in a number of test queries $P(n) = Q_n/Q$, where Q_n is the number of successful queries according to top-n criterion, i.e., the correct classification is among the first n retrieved images, and Q is the total number of queries. The average of $P(n)$ values with respect to all queries by considering only images achieving a distance $\widehat{d} \leq T$ where $T \in \{0.1, 0.2, 0.3, 0.4, 0.5\}$ is reported in Fig. 5 at varying of n.

It has to be noted that images that are very close in terms of $\widehat{d}$ are not similar in terms of aspect, style, contents and semantics. This empirically implies that $\widehat{d}$ is not working on those features but learned a sort of probability of having the author name as input prompt at generation time: the only thing that unites all the artworks of an artist.

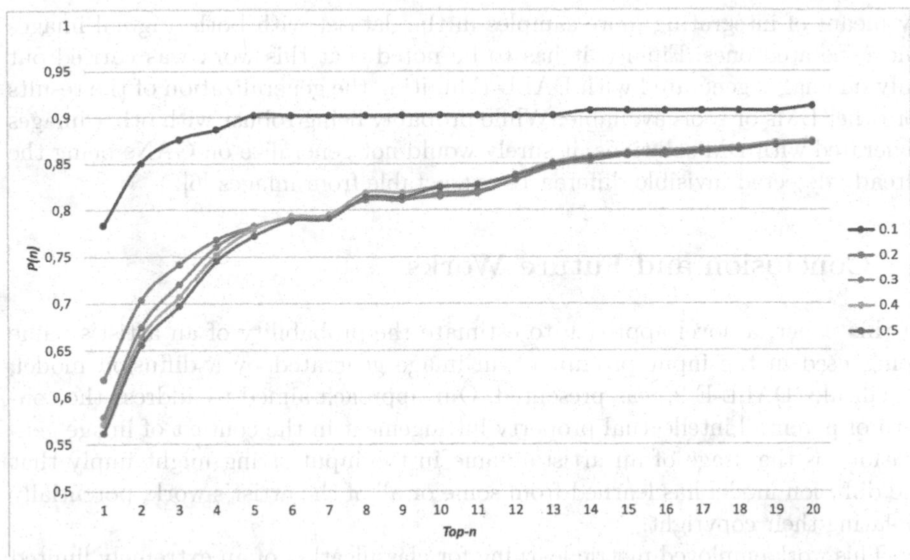

Fig. 5. Retrieval test results with different threshold values for the distance $\widehat{d}$ at varying of the top-n retrieved samples.

Fig. 6. An example with a query image I (the generated one) and the real closest image found by means of the learned metric. As an overall result, a probability is obtained that the name Picasso was used for the generation of image I.

The results obtained demonstrate that the proposed siamese approach could be employed to protect any kind of artists database given that a re-training phase is needed to a certain amount of images generated by prompting their name. Figure 6 shows a real case scenario. Generalization could be achieved only by means of integrating more samples in the dataset with both original images and generated ones. Finally, it has to be noted that this work was carried out only on images generated with DALL-E limiting the generalization of the results for other DMs or tools available. While probably being robust with other images generated with DM solutions, it surely would not generalize on GANs being the already detected invisible differences extractable from images [6].

6 Conclusion and Future Works

In this paper, a novel approach to estimate the probability of an artist's name being used in the input prompt of an image generated by a diffusion model, specifically DALL-E 2, was presented. Our approach aimed to address the concern of potential intellectual property infringement in the context of image generation, as the usage of an artist's name in the input string might imply that the diffusion model has learned from some or all of the artist's work, potentially violating their copyright.

This work employed metric learning for classification of an extremely limited number of authors but in future more sophisticated similarity measures, larger and more diverse datasets, and additional techniques to refine the estimation of the input string's content could be explored. Additionally, investigating the ethical implications and possible solutions to the challenges posed by AI-generated content in relation to copyright and intellectual property protection should be a priority for the research community.

References

1. Abbott, R.: Intellectual property and artificial intelligence: an introduction. In: Research Handbook on Intellectual Property and Artificial Intelligence, pp. 2–21. Edward Elgar Publishing (2022)
2. Battiato, S., Giudice, O., Guarnera, F., Puglisi, G.: CNN-based first quantization estimation of double compressed jpeg images. J. Vis. Commun. Image Representation **89**, 103635 (2022)
3. Battiato, S., Giudice, O., Paratore, A.: Multimedia forensics: discovering the history of multimedia contents. In: Proceedings of the 17th International Conference on Computer Systems and Technologies 2016, pp. 5–16 (2016)
4. Berlemont, S., Lefebvre, G., Duffner, S., Garcia, C.: Class-balanced Siamese neural networks. Neurocomputing **273**, 47–56 (2018)
5. Carlini, N., et al.: Extracting training data from diffusion models. arXiv preprint arXiv:2301.13188 (2023)
6. Corvi, R., Cozzolino, D., Zingarini, G., Poggi, G., Nagano, K., Verdoliva, L.: On the detection of synthetic images generated by diffusion models. arXiv preprint arXiv:2211.00680 (2022)
7. Dhariwal, P., Nichol, A.: Diffusion models beat GANs on image synthesis. Adv. Neural Inf. Process. Syst. **34**, 8780–8794 (2021)
8. Gillotte, J.L.: Copyright infringement in AI-generated artworks. UC Davis L. Rev. **53**, 2655 (2019)

9. Giudice, O., Guarnera, L., Battiato, S.: Fighting Deepfakes by detecting GAN DCT anomalies. J. Imaging **7**(8), 128 (2021). https://doi.org/10.3390/jimaging7080128, https://www.mdpi.com/2313-433X/7/8/128
10. Giudice, O., Guarnera, L., Paratore, A.B., Farinella, G.M., Battiato, S.: Siamese ballistics neural network. In: 2019 IEEE International Conference on Image Processing (ICIP), pp. 4045–4049. IEEE (2019)
11. Goodfellow, I., et al.: Generative adversarial nets. In: Advances in Neural Information Processing Systems, pp. 2672–2680 (2014)
12. Guarnera, F., Allegra, D., Giudice, O., Stanco, F., Battiato, S.: A new study on wood fibers textures: documents authentication through LBP fingerprint. In: 2019 IEEE International Conference on Image Processing (ICIP), pp. 4594–4598. IEEE (2019)
13. Guarnera, L., Giudice, O., Battiato, S.: Fighting Deepfake by exposing the convolutional traces on images. IEEE Access **8**, 165085–165098 (2020). https://doi.org/10.1109/ACCESS.2020.3023037
14. Guarnera, L., et al.: The face deepfake detection challenge. J. Imaging **8**(10), 263 (2022)
15. He, K., Zhang, X., Ren, S., Sun, J.: Deep residual learning for image recognition. In: Proceedings of the IEEE Conference on Computer Vision and Pattern Recognition, pp. 770–778 (2016)
16. Hristov, K.: Artificial intelligence and the copyright dilemma. Idea **57**, 431 (2016)
17. Ramesh, A., Dhariwal, P., Nichol, A., Chu, C., Chen, M.: Hierarchical text-conditional image generation with clip latents. arXiv preprint arXiv:2204.06125 (2022)
18. Sha, Z., Li, Z., Yu, N., Zhang, Y.: DE-FAKE: detection and attribution of fake images generated by text-to-image diffusion models. arXiv preprint arXiv:2210.06998 (2022)
19. Shan, S., Cryan, J., Wenger, E., Zheng, H., Hanocka, R., Zhao, B.Y.: Glaze: protecting artists from style mimicry by text-to-image models. arXiv preprint arXiv:2302.04222 (2023)
20. Vyas, N., Kakade, S., Barak, B.: Provable copyright protection for generative models. arXiv preprint arXiv:2302.10870 (2023)
21. Wang, F., Kang, L., Li, Y.: Sketch-based 3d shape retrieval using convolutional neural networks. In: Proceedings of the IEEE Conference on Computer Vision and Pattern Recognition, pp. 1875–1883 (2015)
22. Wang, R., et al.: Fakespotter: a simple yet robust baseline for spotting AI-synthesized fake faces. arXiv preprint arXiv:1909.06122 (2019)
23. Wang, S.Y., Wang, O., Zhang, R., Owens, A., Efros, A.A.: CNN-generated images are surprisingly easy to spot... for now. In: Proceedings of the IEEE/CVF Conference on Computer Vision and Pattern Recognition, pp. 8695–8704 (2020)
24. Zhang, X., Karaman, S., Chang, S.F.: Detecting and simulating artifacts in GAN fake images. In: 2019 IEEE International Workshop on Information Forensics and Security (WIFS), pp. 1–6. IEEE (2019)
25. Zirpoli, C.T.: Generative artificial intelligence and copyright law (2023)

Benchmarking of Blind Video Deblurring Methods on Long Exposure and Resource Poor Settings

Maria Ausilia Napoli Spatafora(✉), Massimo O. Spata, Luca Guarnera, Alessandro Ortis, and Sebastiano Battiato

Department of Mathematics and Computer Science, University of Catania, Viale A. Doria, 6, 95125 Catania, Italy
maria.napolispatafora@phd.unict.it,
{massimo.spata,luca.guarnera,alessandro.ortis,
sebastiano.battiato}@unict.it

Abstract. This paper presents a benchmark evaluation of blind video deblurring methods in specific challenging settings. The employed videos are affected by severe deblurring artifacts and acquisition conditions (e.g., low resolution, high exposure, camera motion, complex scene motion, etc.). An in depth state of the art investigation has been carried out. Then, a specific set of methods based on mathematical optimization with image priors has been involved in our benchmark evaluation. The selected methods have been evaluated quantitatively and qualitatively.

Keywords: Blind Video Deblurring · Blind Deconvolution · Long Exposure Artifacts · Video Restoration

1 Introduction and Motivations

This paper reports a benchmark evaluation of blind video deblurring methods in the context of a research project involving long exposure devices and resource poor settings. The aim of this research activity is to develop methods addressing the problem of video motion deblurring in such a specific setting. The input videos are recorded at 120 fps, long exposure and low resolution (i.e., lower than 240p). To set our benchmark, we acquired this dataset by taking advantage of a proprietary infrared camera, which is the focus of this research project. The acquired sequences come from a microbolometer-array camera which provides greyscale thermal images. This kind of camera is intrinsically more subject to motion blur than traditional cameras (i.e., CCD, CMOS) due to its always-integrating sensor and long thermal time constant (in the range of 10 ms). In this context, the blurring artifacts are originated by several factors: the camera is moving during the recording, camera shake, motion of different objects in the scene, each with its own motion pattern, long exposure acquisition. Figure 1 shows some examples of severe blurring concerning the presented study. The

G. L. Foresti et al. (Eds.): ICIAP 2023, LNCS 14233, pp. 376–386, 2023.
https://doi.org/10.1007/978-3-031-43148-7_32

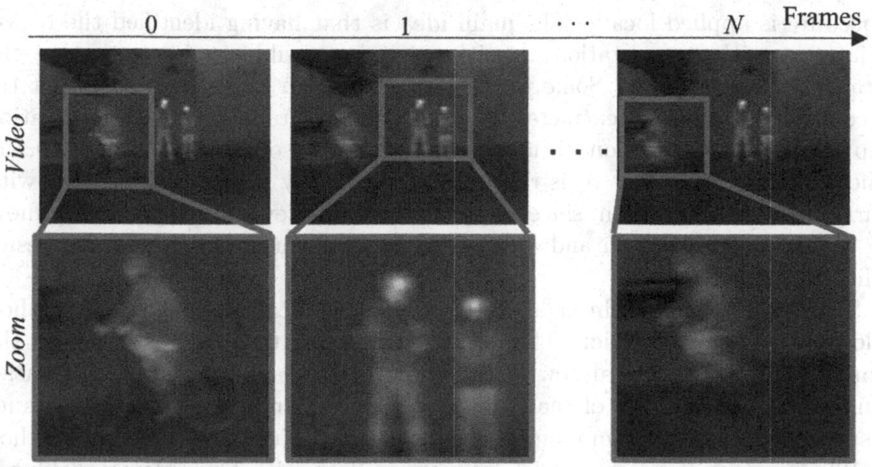

Fig. 1. Sample frames of the employed dataset (first row) and detail on severe blurred areas (second row).

objective is to produce videos in which all the elements in the scene are sharped. Other problem constraints are low computational resources and performances close to real-time computation, as one of the project's objective is the integration of the deblurring algorithm in the camera acquisition pipeline, as done by several acquisition artifact correction methods [2]. These last two constraints discard any modern ANN-based approach. The above described constraints and the problem settings drastically reduced the state-of-the-art methods to be included in the evaluation benchmark.

Video deblurring is an old and still open problem, for real-world scenarios. Moreover, due to the intrinsic difficulty of the general problem, most of research works focus on specific settings and constraints. Several works focus on camera motion, whereas others focus on subject deblurring. However, the underlined assumption is that the acquired scene is affected by only one Point Spread Function (PSF), which is often known, in addition to some random noise. This is actually very far from real scenarios. As well as from the types of blurring artifacts that the present study aims to address. Our preliminary literature study concluded that the problem has never been investigated previously with such a level of specificity on the scene settings. In particular, the main difference is that the state of the art approaches assume one or more of the following settings: the camera is stable, the whole image is affected by the same PSF, high resolution, the PSF is known.

As a first attempt, we selected works that aim to perform blind deblurring locally, by automatically detecting and reverting the local blur. These works often share a pipeline which is based on two main steps. First, the scene is segmented at pixel level, based on the local estimation of motion [6,8,19], planarity [10,19], entropy [11] or combinations of them. Then, specific deblurring procedure is applied locally. The main idea is that having identified the region of interest with segmentation algorithms, it is possible to try to restore the data under examination. Some works further add an integration step that try to combine information extracted from subsequent frames. For the integration step, most of them exploit variants of the super resolution technique [10,19], which requires sub-pixel registration that is usually difficult to achieve with blurry images. In addition, since the scene contains several independently moving objects, segmentation and area based registration for each segment result difficult.

Specifically, Yamaguchi et al. [19] proposed a local motion deblur method followed by super-resolution. The motion deblurring technique is based on the combination of the optical flow of the scene with scene segmentation. Considering an high frame-rate of the video and that variance of feature appearance in successive frames and motion of feature points are usually small, the method in [19] estimates scene geometries from video data with blur. Motion deblur is applied for each frame, and then, super resolution is applied to the deblurred image set. The authors also propose an adaptive super-resolution technique considering different defocus blur effects dependent on depth. Further, image quality is improved by considering the different defocus blur for each segment dependent on different depth with the adaptive super-resolution technique.

While in [11] Shapri et al. proposed a method based on image partitioning and entropy measurement feedback by analysing the randomness of grey level distribution in the image and the maximum value of entropy in subimages. Region of Interests (ROI) are defined based on entropy, then a PSF is estimated for each region and a proper deconvolution is applied.

Another approach is investigated by Hu et al. [8]. Based on optical flow and edge cues, they addressed the problems related to the fast motion, motion blur and occlusions by developing a novel saliency estimation technique as well as a novel neighbourhood graph for effective unsupervised foreground-background video segmentation.

Moreover, in the paper proposed by Fergus et al. [7], the authors apply a Bayesian approach to remove the blur produced by camera shake in digital images. The Bayesian approach allows to find the blur kernel implied by a distribution of probable images. Given that kernel, the image is then reconstructed using a standard deconvolution algorithm.

Finally, in Matsushita et al. [10] multiple moving objects are divided into segments representing independent objects. Then, deblurring and super-resolution techniques are applied to each segment which will be integrated to create the final result. The main difference with respect to [19] is related to the first phase (i.e., registration) which exploits feature points and clustering to segment the images based on planarity. Then, the segments tracked along a group of subsequent frames are integrated to get one deblurred image. To remove blurry noise in digital images, Delbracio et al. [6] proposed an approach based on the concept of aggregation. Specifically, given a burst of images, they take what is least blurry in each frame to construct a sharper, less noisy image of all the images in the burst. This can be done by performing a weighted average in the Fourier domain, with weights depending on the magnitude of the Fourier spectrum.

Although most of the above mentioned blind deblurring methods exhibit interesting performances on standard dataset or controlled experimental settings, our experimental investigation revealed that they can't be applied on the videos subject of our research. In particular, the videos under analysis presents severe blurring effects, originated by multiple factors. Moreover, the quality of the video itself further limits the information acquired by the camera, and hence the data that can be exploited for restoration. At this stage, starting from the insights of the first investigation, we decided to explore more solutions coming from different research areas. In particular, we investigated optimization methods based on priors, which revealed some interesting results on the filed of restoration of video/images acquired in long exposure settings. The paper is organized as follows. Prior based methods are described in Sect. 2. In Sect. 3 benchmark settings are reported. Experimental Results and Conclusions are exposed in Sect. 4.

2 Prior Based Methods

The blind deblurring methods detailed in the previous Section often have difficulties with severe blurring and saturated regions because these cases violate the linear blur model described in Eq. 1. Indeed, long exposure often results in heavy blur in the captured image.

$$B = I \otimes K + n \tag{1}$$

According to this formulation, besides the noise n, there are two main elements in this equation that we don't know. First we don't know the sharp image I, and second we don't know the scale of the filter K with which it was applied to obtain the blurred image B, since we don't know the depth (in case of defocus), nor the motion (in case of motion blurring). This problem is highly ill posed because both the sharp image and blur kernel are unknown. In order to make this problem well-posed, most existing methods exploits specific image information [3,16] to estimate a better kernel. One strategy may be trying to deconvolve the image with a set of predefined kernels at different scales, and somehow choose the best result. However, although this method is rough and performs several approximations, the two steps of this strategy are still quite challenging. First, deconvolution is known to be difficult, and existing algorithms result in significant ringing artifacts (e.g., Richardson-Lucy [9]). The second challenge is that there is not a simple good criterion to identify the correct scale. Indeed, scales larger than perfect one will look very wrong, whereas smallest scales will not be that bad because they don't cause evident artifacts as the larger ones. Therefore, it's not easy to decide which is the perfect scale by means of such an approach. The deconvolution problem can be defined as finding an image x, such that convolved with a filter f, will be equal to the blurred input y. This corresponds to minimizing the convolution error (see Eq. 2). Convolution is a linear operation, therefore we have linear constraints on x.

$$|f \otimes x - y|^2 \tag{2}$$

However, in many cases the blur filter erases some of the image information and we don't have a full rank system, which means that there may not be a unique solution. If we could computationally characterize typical images, we can bias algorithms to output 2D arrays that are more likely to look natural. This is a very powerful idea that is becoming popular for a number of problems in computer vision and image processing. One way to characterize images is to observe that the gradients are rare, and the normal image of the example has much fewer gradients comparing to the bad ones. Some approaches exploit this fact by penalizing potential outputs that have high gradients. Therefore we say that we want to find an image $\hat{I}$ that will minimize the convolution error, but also add the request that it will have fewer gradients. Or in other words, we try to minimize the sum of a gradient measure over all image pixels. Therefore despite the fact that more solutions have equal convolution error, we can prefer those that have fewer gradients. The following subsection briefly describes the methods based on such a strategy, that have been included in our benchmark study.

2.1 Hyper Laplacian

Bai et al. [1] proposed a graph-based blind image deblurring algorithm by interpreting an image patch as a signal on a weighted graph. The authors show that in the skeleton image patch, unlike blurry patches, the edge weights of a graph have a unique bi-modal distribution. For this reason, they proposed a graph-based image prior RGTV (reweighted graph total variation) that promotes a bi-modal weight distribution to reconstruct a skeleton patch from a blurry observation by considering the graph L1-Laplacian regularizer.

2.2 Dark Channel Prior

Pan et al. [14] proposed a blind image deblurring method based on the dark channel prior. The idea is based on the fact that most image patches in the clean image contain some dark pixels and these are not dark when averaged with neighbouring high intensity pixels during the blur process. The authors introduced a linear approximation of the min operator to compute the dark channel to blind deblurring on various scenarios.

2.3 Enhance Sparse Blind

Chen et al. [4] proposed new approach for improving the blind image deblurring task by introducing a new term that is better suited to fit the complex natural noise found in real-world images. The authors use a combination of a dense model (l2) and a newly-designed enhanced sparse model (le) to achieve this goal. Additionally, they suggest, using the le model to regularize image gradients, which helps to better preserve important image details.

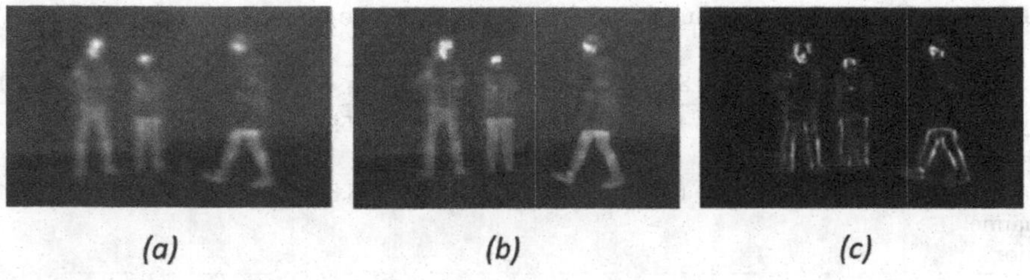

Fig. 2. A sample frame from our dataset depicting a severe blurred scene (a), the corresponding output generated by the Hyper Laplacian method (b) and the heatmap depicting the areas where the algorithm acts (c).

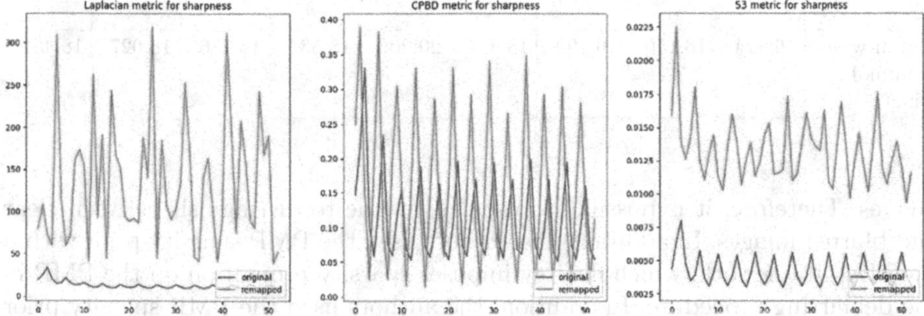

Fig. 3. An example of the benchmark metrics comparison between an original video (blue line) and the enhanced video (orange line) for the Hyper Laplacian method. (Color figure online)

2.4 Gradient Prior

An approach for deblurring text images was proposed by Pan et al. [13] that utilizes an L0-regularized prior that takes into account both the intensity and gradient of the image. This prior was motivated by the unique characteristics of text images. The approach also includes an efficient optimization method that generates reliable intermediate results for kernel estimation without the need for complex filtering strategies. To improve the final deblurred image, a simple method for removing artifacts was also proposed.

2.5 Patch Wise Minimal Pixel

Wen et al. [18] proposed a novel method for removing blur from images, by proposing a patch-wise minimal pixels (PMP) prior. PMP is a collection of local minimal pixels. The blur process tends to increase the intensity of a local minimal pixel. As a result, the PMP of clear images is much sparser than that of blurred

Table 1. Execution time (in seconds) for each sequence of each involved methods.

	Seq #1	Seq #2	Seq #3	Seq #4	Seq #5	Seq #6	Seq #7	Seq #8	Seq #9
Hyper Laplacian	49.432	48.337	47.105	23.605	23.767	23.594	23.654	22.665	23.648
Dark channel prior	179.961	139.112	124.139	121.917	134.175	198.394	213.421	230.833	236.334
Blind saturated	27.643	33.321	28.600	N.A.	N.A.	N.A.	N.A.	N.A.	N.A.
Enance sparse blind	13.342	12.161	20.552	20.191	23.186	21.930	22.706	21.793	24.106
Gradient prior	38.637	41.148	41.155	41.008	34.353	37.592	42.040	37.115	37.215
Patch wise minimal pixel	16.424	16.150	16.199	18.967	20.202	18.534	16.286	18.027	18.451

images. Therefore, it is possible to use this metric to distinguish between clear and blurred images. In addition, the authors used the PMP sparsity prior within the MAP framework, which flexibly imposes sparsity promotion on the PMP in the deblurring procedure. In addition, the authors used the PMP sparsity prior within the MAP framework, which flexibly imposes sparsity promotion on the PMP in the deblurring procedure.

2.6 Blind Saturated

Chen et al. [5] proposed a new method for images deblurring by introducing a new blurring model that takes both saturated and unsaturated pixels into account. Additionally, an efficient maximum a posteriori (MAP)-based optimization framework is developed based on this new blur model. In the following benchmark, this method has been evaluated just on a subset of the dataset, as the code provided by the authors of the paper raises an exception for some videos (N.A. in Table 2). Nevertheless, considering the relevance of [5] we decided to keep the results obtained on such subset of videos.

3 Benchmark Settings

Considering previous studies and literature on video deblurring, we selected three metrics to evaluate our benchmark: derivative of Laplacian [15]; the Cumulative Probability of Blur Detection CPBD [12] and S3 [17]. For all the metrics, the greater value denotes sharper images. The latter two (i.e., CPBD and S3) ranges from 0 to 1, whereas the derivative of Laplacian is unbounded. These metrics

Table 2. Experimental quantitative results of the presented benchmark. Baseline refers to the metrics computed on the original blurred sequences.

		Seq #1	Seq #2	Seq #3	Seq #4	Seq #5	Seq #6	Seq #7	Seq #8	Seq #9
Baseline	*Laplacian*	676.745	17.653	9.668	676.745	17.653	9.668	676.745	17.653	9.668
	CPBD	0.927	0.033	0.005	0.927	0.033	0.005	0.927	0.033	0.005
	S3	0.257	0.103	0.172	0.257	0.103	0.172	0.257	0.103	0.172
Hyper Laplacian	*Laplacian*	3186.528	79.723	26.106	1654.424	1419.341	1345.364	1554.359	1447.298	1554.359
	CPBD	0.963	0.306	0.115	0.139	0.246	0.542	0.246	0.384	0.246
	S3	0.716	0.150	0.121	0.910	0.887	0.891	0.905	0.903	0.904
Dark channel prior	*Laplacian*	8958.466	106.327	26.106	1654.424	1419.342	1345.364	1554.359	1498.005	1442.128
	CPBD	0.995	0.420	0.115	0.139	0.246	0.542	0.246	0.410	0.404
	S3	0.904	0.183	0.122	0.910	0.887	0.891	0.904	0.914	0.906
Blind saturated	*Laplacian*	1976.109	179.168	99.697	N.A.	N.A.	N.A.	N.A.	N.A.	N.A.
	CPBD	0.983	0.376	0.245	N.A.	N.A.	N.A.	N.A.	N.A.	N.A.
	S3	0.646	0.207	0.298	N.A.	N.A.	N.A.	N.A.	N.A.	N.A.
Enance sparse blind	*Laplacian*	5980.988	80.206	44.911	1963.080	1687.135	1760.431	1612.685	1910.105	1715.726
	CPBD	0.993	0.371	0.051	0.333	0.667	0.667	0.667	0.889	0.333
	S3	0.839	0.197	0.165	0.926	0.931	0.926	0.936	0.945	0.934
Gradient prior	*Laplacian*	6336.282	574.437	362.787	1308.01	1259.159	1259.159	1110.676	1746.564	1110.676
	CPBD	0.996	0.897	0.778	0.0	0.0	0.0	0.0	0.0	0.0
	S3	0.821	0.416	0.446	0.871	0.840	0.840	0.815	0.895	0.815
Patch wise minimal pixel	*Laplacian*	7583.493	131.651	95.318	1765.303	1464.855	1464.855	1572.094	1687.081	1572.094
	CPBD	0.997	0.627	0.485	0.0	0.0	0.0	0.0	0.333	0.0
	S3	0.879	0.250	0.238	0.921	0.859	0.859	0.891	0.908	0.891

are computed on a dataset which satisfies all the characteristics described in Sect. 1. Our dataset consists of 9 sequences, each with a different duration and depicting a different context. We conduct extensive experiments based on the prepared dataset. To demonstrate the accuracy of methods, we compare it also in terms of time performance metrics. All the evaluated methods are implemented in Matlab and run on a machine equipped with an Intel(R) Core(TM) i5 CPU, 16 GB memory and NVIDIA GTX 1080 GPU. Table 1 shows the detailed results for each sequence. Given the nature of the problem, we proposed a qualitative assessment to accompany the quantitative one just described. In particular, in addition to quantitative evaluation, our benchmark further produced an heatmap which highlights the areas where the algorithm acts (see Fig. 2)[1]. The metrics are computed for each frame and then averaged, moreover, in order to compare the improvement of each method on the single frames, beside evaluating the average behaviour for the entire video sequence we also considered the trend over time by plotting the metrics trends. Figure 3 shows the comparison of the metrics for one video example computed with the Hyper Laplacian method. The gap between the orange line and the blue line represents the improvement in terms of the specific metric.

1 More visual examples will be provided at conference time.

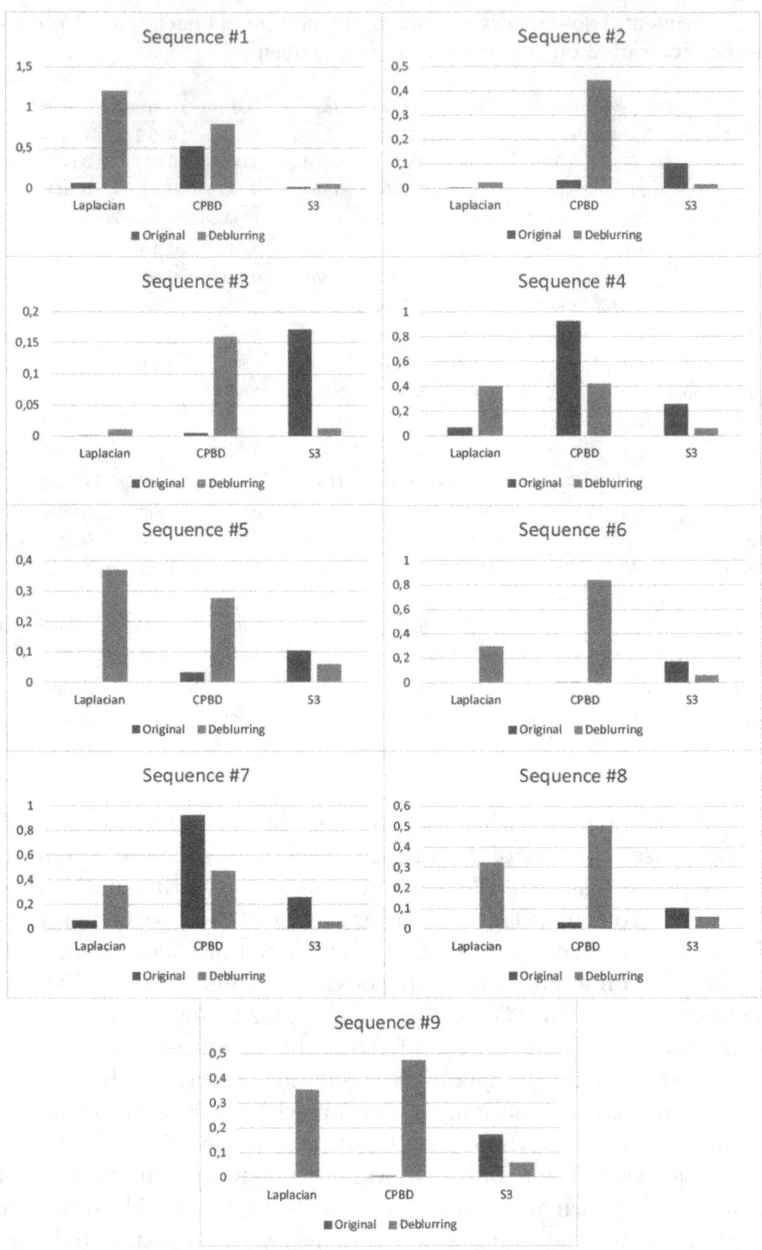

Fig. 4. Comparative histograms of the three metrics employed for the Hyper Laplacian method. The blue bins represent the baseline values computed on the original sequences, whereas the orange bins represent the values computed on the enhanced videos. (Color figure online)

4 Results and Conclusions

In this study, we applied several techniques to reduce the blurring caused by camera motion with long exposure. The evaluation includes a range of images taken under low level of lighting conditions and with different levels of camera motion. The results of our benchmark showed that the prior-based blind deblurring algorithms were successful in restoring the clarity of the images and improving the visual quality, in general. The most effective method are Hyper Laplacian Deconvolution and Dark Channel Prior techniques which are able to significantly reduce the amount of blurring artifacts. Table 2 shows the detailed results for each sequence. However, these values are computed considering the average performance on the whole video. Some methods achieved lower quantitative performances but are less prone to artifacts. By the other hand, methods such as the Dark channel prior algorithm introduces artifacts despite the high average performances.

Based on the performed benchmark, we concluded that the Hyper Laplacian method is the best trade off between quantitative and qualitative performances. Other methods introduce artifacts and shown higher computation times.

For this reason, we computed also an histogram of the three evaluation metrics for each sequence to assess robustly our benchmark. The Fig. 4 shows the histograms for the Hyper Laplacian method (see Sect. 2.1) that is the most effective. Since the derivative of Laplacian is unbounded, we rescaled these values by a factor of 10, 000 in order to make easier the comparison with respect to CPDB and S3 values.

There are several avenues for future research in this field. One possible direction is to investigate the use of multiple sensors, such as visible and infrared cameras, to improve the quality of the deblurred videos. Finally, the proposed method can be extended to handle other types of video degradation, such as noise and compression artifacts.

Acknowledgements. The research has been founded thanks to the collaboration with Huawei Technologies France, that supplied the employed dataset.

Acknowledge financial support from: PNRR MUR project PE0000013-FAIR.

References

1. Bai, Y., Cheung, G., Liu, X., Gao, W.: Graph-based blind image deblurring from a single photograph. IEEE Trans. Image Process. **28**(3), 1404–1418 (2018)
2. Battiato, S., Bosco, A., Castorina, A., Messina, G.: Automatic image enhancement by content dependent exposure correction. EURASIP J. Adv. Signal Process. **2004**, 1–12 (2004)
3. Cao, X., Ren, W., Zuo, W., Guo, X., Foroosh, H.: Scene text deblurring using text-specific multiscale dictionaries. IEEE Trans. Image Process. **24**(4), 1302–1314 (2015)

4. Chen, L., Fang, F., Lei, S., Li, F., Zhang, G.: Enhanced sparse model for blind deblurring. In: Vedaldi, A., Bischof, H., Brox, T., Frahm, J.-M. (eds.) ECCV 2020. LNCS, vol. 12370, pp. 631–646. Springer, Cham (2020). https://doi.org/10.1007/978-3-030-58595-2_38
5. Chen, L., Zhang, J., Lin, S., Fang, F., Ren, J.S.: Blind deblurring for saturated images. In: Proceedings of the IEEE/CVF Conference on Computer Vision and Pattern Recognition, pp. 6308–6316 (2021)
6. Delbracio, M., Sapiro, G.: Burst deblurring: removing camera shake through fourier burst accumulation. In: Proceedings of the IEEE Conference on Computer Vision and Pattern Recognition, pp. 2385–2393 (2015)
7. Fergus, R., Singh, B., Hertzmann, A., Roweis, S.T., Freeman, W.T.: Removing camera shake from a single photograph. In: ACM SIGGRAPH 2006 Papers, pp. 787–794 (2006)
8. Hu, Y.T., Huang, J.B., Schwing, A.G.: Unsupervised video object segmentation using motion saliency-guided spatio-temporal propagation. In: Proceedings of the European Conference on Computer Vision (ECCV), pp. 786–802 (2018)
9. Lanteri, H., Aime, C., Beaumont, H., Gaucherel, P.: Blind deconvolution using the richardson-lucy algorithm. In: Optics in Atmospheric Propagation and Random Phenomena, vol. 2312, pp. 182–192. SPIE (1994)
10. Matsushita, Y., Kawasaki, H., Ono, S., Ikeuchi, K.: Simultaneous deblur and super-resolution technique for video sequence captured by hand-held video camera. In: 2014 IEEE International Conference on Image Processing (ICIP), pp. 4562–4566. IEEE (2014)
11. Mohd Shapri, A.H., Abdullah, M.Z.: Accurate retrieval of region of interest for estimating point spread function and image deblurring. Imaging Sci. J. **65**(6), 327–348 (2017)
12. Narvekar, N.D., Karam, L.J.: A no-reference image blur metric based on the cumulative probability of blur detection (CPBD). IEEE Trans. Image Process. **20**(9), 2678–2683 (2011). https://doi.org/10.1109/TIP.2011.2131660
13. Pan, J., Hu, Z., Su, Z., Yang, M.H.: Deblurring text images via L0-regularized intensity and gradient prior. In: Proceedings of the IEEE Conference on Computer Vision and Pattern Recognition, pp. 2901–2908 (2014)
14. Pan, J., Sun, D., Pfister, H., Yang, M.H.: Blind image deblurring using dark channel prior. In: Proceedings of the IEEE Conference on Computer Vision and Pattern Recognition, pp. 1628–1636 (2016)
15. Pech-Pacheco, J., Cristobal, G., Chamorro-Martinez, J., Fernandez-Valdivia, J.: Diatom autofocusing in brightfield microscopy: a comparative study. In: Proceedings 15th International Conference on Pattern Recognition, ICPR-2000, vol. 3, pp. 314–317 (2000). https://doi.org/10.1109/ICPR.2000.903548
16. Varghese, N., MR, M.M., Rajagopalan, A.: Fast motion-deblurring of IR images. IEEE Signal Process. Lett. **29**, 459–463 (2022)
17. Vu, C.T., Phan, T.D., Chandler, D.M.: s_3: a spectral and spatial measure of local perceived sharpness in natural images. IEEE Trans. Image Process. **21**(3), 934–945 (2012). https://doi.org/10.1109/TIP.2011.2169974
18. Wen, F., Ying, R., Liu, P., Truong, T.K.: Blind image deblurring using patch-wise minimal pixels regularization. arXiv preprint arXiv:1906.06642 (2019)
19. Yamaguchi, T., Fukuda, H., Furukawa, R., Kawasaki, H., Sturm, P.: Video deblurring and super-resolution technique for multiple moving objects. In: Kimmel, R., Klette, R., Sugimoto, A. (eds.) ACCV 2010. LNCS, vol. 6495, pp. 127–140. Springer, Heidelberg (2011). https://doi.org/10.1007/978-3-642-19282-1_11

LieToMe: An LSTM-Based Method for Deception Detection by Hand Movements

Danilo Avola[1(✉)], Luigi Cinque[1], Maria De Marsico[1], Angelo Di Mambro[1], Alessio Fagioli[1], Gian Luca Foresti[2], Romeo Lanzino[1], and Francesco Scarcello[3]

[1] Department of Computer Science, Sapienza University, Via Salaria 113, 00198 Rome, Italy
{avola,cinque,demarsico,dimambro,fagioli,lanzino}@di.uniroma1.it

[2] Department of Mathematics, Computer Science and Physics, University of Udine, Via delle Scienze 206, 33100 Udine, Italy
gianluca.foresti@uniud.it

[3] Department of Computer Engineering, Modeling, Electronics and Systems Engineering, University of Calabria, Via Pietro Bucci, 87036 Rende, Italy
francesco.scarcello@unical.it

Abstract. The ability to detect lies is a crucial skill in essential situations like police interrogations and court trials. At present, several devices, such as polygraphs and magnetic resonance, can ease the deception detection task. However, the effectiveness of these tools can be compromised by intentional behavioral changes due to the subject awareness of such appliances, suggesting that alternative ways must be explored to detect lies without using physical devices. In this context, this paper presents an approach focused on the extraction of meaningful features from hand gestures. The latter provide cues on the person's behavior and are used to address the deception detection task in RGB videos of trials. Specifically, the proposed system extracts hands skeletons from an RGB video sequence and generates novel handcrafted features from the extrapolated keypoints to reflect the subject behavior through hand movements. Then, a long short-term memory (LSTM) neural network is used to classify these features and estimate whether the person is lying or not. Extensive experiments were performed to assess the quality of the derived features on a public collection of famous real-life trials. On this dataset, the proposed system sets new state-of-the-art performance on the unimodal hand-gesture deception detection task, demonstrating the effectiveness of the proposed approach and its handcrafted features.

Keywords: Deception detection · Hand gestures · LSTM

1 Introduction

Lying is a common aspect of our daily lives where people tell, on average, at least two lies per day [14]. Some of these lies are harmless, while others can have serious consequences, especially when considering essential situations like

G. L. Foresti et al. (Eds.): ICIAP 2023, LNCS 14233, pp. 387–398, 2023.
https://doi.org/10.1007/978-3-031-43148-7_33

police interrogations and court trials, where an innocent could be convicted or a culprit acquitted. What is more, studies indicate that humans can understand whether a person is lying or not with 54% accuracy only [11], suggesting that more sophisticated approaches are required to address this task. To this end, several procedures and devices can help discern deceptive behavior, e.g., electroencephalography (EEG), functional magnetic resonance imaging (fMRI), and polygraph. Although effective, these approaches employ bulky and expensive equipment, which can also hinder the ability to detect deception due to the person's awareness of the circumstances. For instance, it was demonstrated that trained subjects could deceive even the polygraph [16]. To overcome the aforementioned procedures, it is possible to exploit cameras. In this context, computer vision (CV) and artificial intelligence (AI) advances can be used as successful tools. In fact, these emerging technologies are being applied successfully to address heterogeneous tasks such as medical image analysis [2,6,28], object classification [7,8,32], and emotion recognition [4,9,15], also including deception detection itself [13]. On the latter, several studies demonstrated that different persons show common deception cues that can be observed in facial expressions, body posture, and hand movements [29,30]. Moreover, these cues can be captured through cameras in a possibly covert way, thus making the deception detection task worth exploring via CV and AI solutions [1].

In the existing deception detection research, methods either focus on specific modalities (e.g., video, voice) or explore multimodal methods to capture multiple cues simultaneously. Inspired by the results shown in [5], this work focuses on deception detection exclusively from hand movements and introduces novel handcrafted features to represent cues associated with hands. In detail, this paper explores various hand skeleton positions typical of people telling the truth or lying (e.g., relaxed hands and clenched fists) and classifies the extracted features via an LSTM architecture to consider the time evolution of a video sequence. To validate the effectiveness of the designed pipeline, extensive experiments were performed on a public collection of real-life trials, where the proposed method improved the deception detection from hand gestures by nearly 4% while also achieving performance close to multimodal approaches, fully highlighting the effectiveness of the designed procedure.

The remainder of this paper is organized as follows. Section 2 introduces related works addressing the deception detection task. Section 3 describes the proposed pipeline, with particular attention to the extracted handcrafted features. Section 4 reports the extensive experiments performed on a public collection of real-life trials. Finally, Sect. 5 concludes this paper.

2 Related Work

The deception detection approaches currently proposed in the literature can be broadly classified into non-verbal, verbal, and multimodal methods. Those in the first category focus on physiological measures. For instance, the authors of [20] examine EEG frequencies via fuzzy reasoning applied to spectral analysis to avoid human bias in data interpretation. Functional near-infrared spectroscopy (fNIRS) and functional magnetic resonance imaging (fMRI) are instead used in [21] and [27], respectively, to assess blood oxygen levels and brain activation areas associated with deception. While effective, these methods use bulky machinery that might compromise the test due to the person's awareness. Therefore, different approaches exploit alternative technologies such as cameras, which are usually tied to the eyes and, more in general, to facial features. For example, the scheme presented in [33] analyzes patterns in eye movements to identify gaze trajectories and fixation points, while the authors of [24] implement a neural network to detect lies from pupil dilation. Facial micro-expressions are explored, among others, in [3,10]. The former extracts emotions by applying local binary pattern (LBP) and histogram of optical flow (HOF) algorithms to images; the latter represents micro-expressions through additional procedures such as LBPs from three orthogonal planes (LBP-TOP) or improved dense trajectories (IDT). Finally, the solution presented in [5] describes exclusively hand movements via handcrafted features and analyzes them through a Fisher-LSTM. Differently from this last approach, the presented method focuses on a novel set of handcrafted features to recognize deception from hand motions which achieve higher performances with respect to [5], as reported in the experimental section.

The second category, which accounts for verbal methods, usually focuses on the analysis of speech via frequency-level examination or through linguistic characteristics. For example, the scheme in [23] analyzes mel-frequency cepstral coefficients (MFCCs) together with the voice pitch of a person and classifies the extracted features through a support vector machine (SVM). The method presented in [22], instead, is based on the linguistic inquiry and word count (LIWC) lexicon [25], which is an effective tool for deception detection and generates audio transcripts by computing dominant word classes present in a text.

The third and last category, i.e., multimodal approaches, includes the most performing models currently present in the state-of-the-art. These works address the deception detection task by exploiting features derived from various modalities, e.g., hand motion, facial micro-expression, and speech. For instance, the scheme proposed in [31] exploits video and audio via micro-expression and MFCC and performs a logistic regression to detect lies. The authors of [19] use video, audio, and micro-expression along with text features as input for improved performances on a multi-layer perceptron (MLP). The scheme described in [26] uses verbal and hand gesture features classified using the decision tree (DT) and random forest (RF) algorithms. Finally, the solution introduced in [18] leverages audio, video, and EEG-extracted features to detect lies through a custom convolutional neural network (CNN) architecture.

Multimodal approaches generally achieve higher performances with respect to their single-modality counterparts. However, exploring unimodal solutions is still relevant, especially when considering hand gestures are seldom used and usually only at a high level, e.g., hand presence mentioned in [26]. As a matter of fact, the authors of [5] have shown the effectiveness of some related handcrafted features, indicating that different deception cues can be recognized from hand movements, as is also demonstrated by the novel features exploited in this work.

3 Proposed Method

This section describes the proposed pipeline, which is summarized in Fig. 1. The chosen strategy extracts features from each input frame of a video sequence and classifies them as truthful or deceptive behavior via a neural network. These processes are described in greater detail in Sect. 3.1 and Sect. 3.2, respectively.

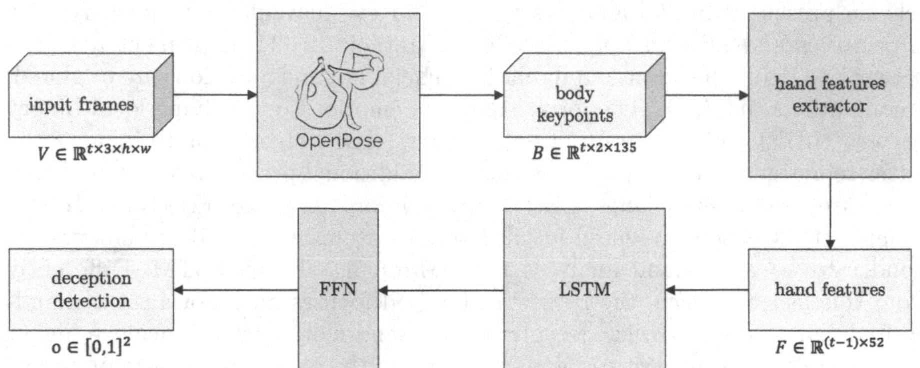

Fig. 1. Scheme of the proposed pipeline. Handcrafted feature extraction modules are in the top row, and the deception detection classifier is in the bottom row.

3.1 Feature Extraction

The first step in the presented pipeline entails the extraction of hand features. The system takes as input a video sequence $V \in \mathbb{R}^{t\times3\times h\times w}$ capturing hand movements in the form of t frames with a resolution of $h \times w$ pixels each over the three RGB channels. The system extracts from each frame $v \in V$ a set of features representing detailed hand keypoints. The latter are retrieved by OpenPose [12]. For each person present in the image, this multi-stage CNN outputs 135 2D coordinates corresponding to human skeleton and muscle keypoints. Out of this full set, the system only records the 42 keypoints that are associated with hand joints. Considering the full sequence of t frames, this produces a set of features $B \in \mathbb{R}^{t\times2\times135}$.

Starting from B, several handcrafted features $F \in \mathbb{R}^{(t-1)\times52}$ are generated to represent a deceitful behavior. A visual rendition of some of these features is shown in Fig. 2. Formally, let $m \in \{l, r\}$ indicate whether a feature belongs to the left or right hand. The first feature $\mathbf{f}_1 \in \mathbb{R}^{(t-1)\times2}$ is the index-thumb fingertip distance, shown in Fig. 2a, as these are the fingers that tend to move the most when gesturing during speech:

$$\mathbf{f}_1 = \bigoplus_{k=1}^{t-1}\left[\bigoplus_{m\in\{l,r\}} [\|\mathbf{i}^m_{4,k} - \mathbf{t}^m_{3,k}\|_2]\right], \tag{1}$$

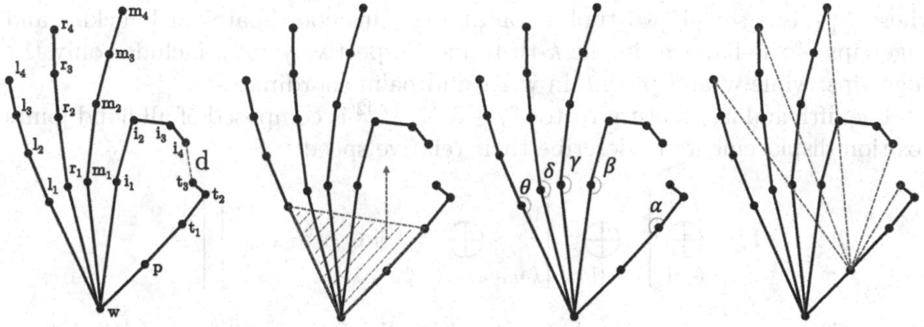

(a) Index-thumb fingertips distance. (b) Normal of palm surface. (c) Knuckles mean angles. (d) Palm-fingertips distances.

Fig. 2. Handcrafted feature extraction examples illustrating $\mathbf{f}_1$, $\mathbf{f}_2$, and $\mathbf{f}_4$.

where $\bigoplus$ indicates the concatenation operation; while $\mathbf{i}^m_{4,k}$ and $\mathbf{t}^m_{3,k}$ are 2D vectors representing, respectively, hand m index finger and thumb fingertips coordinates in the k-th frame.

The second feature $\mathbf{f}_2 \in \mathbb{R}^{(t-1)\times 2}$, represented in Fig. 2b, describes the surface normal of the triangle formed by the thumb base, wrist, and little finger proximal phalanx joints, as people tend to hide their palms when lying:

$$\mathbf{f}_2 = \bigoplus_{k=1}^{t-1} \left[\bigoplus_{m\in\{l,r\}} \left[(\mathbf{t}^m_{1,k} - \mathbf{w}^m_k) \times (\mathbf{l}^m_{1,k} - \mathbf{w}^m_k) \right] \right], \tag{2}$$

where $\mathbf{t}^m_{1,k}$, $\mathbf{l}^m_{1,k}$, and $\mathbf{w}^m_k$ indicate the 2D vectors relative to hand m thumb knuckle, little finger knuckle, and wrist joints in the k-th frame.

The third feature $\mathbf{f}_3 \in \mathbb{R}^{(t-1)\times 2}$ corresponds to the angle between the wrist and thumb proximal phalanx base in two consecutive frames, using the wrist as the vertex. This feature was chosen to detect a lying behavior based on the gestures a person makes when trying to describe something that does not exist, which often involves wrist rotation:

$$\mathbf{f}_3 = \bigoplus_{k=1}^{t-1} \left[\bigoplus_{m\in\{l,r\}} \left[\frac{(\mathbf{t}^m_{1,k} - \mathbf{w}^m_k) \cdot (\mathbf{t}^m_{1,k+1} - \mathbf{w}^m_k)}{|\mathbf{t}^m_{1,k} - \mathbf{w}^m_k| \cdot |\mathbf{t}^m_{1,k+1} - \mathbf{w}^m_k|} \right] \right], \tag{3}$$

where $\mathbf{t}^m_{1,k}$, $\mathbf{t}^m_{1,k+1}$, and $\mathbf{w}^m_k$ represent hand m 2D joint coordinates of the thumb knuckle in two consecutive frames and wrist in the k-th frame.

The fourth feature $\mathbf{f}_4 \in \mathbb{R}^{(t-1)\times 4}$, illustrated in Fig. 2c and 2d, contains the average angles and distances between finger joints and the palm, respectively, as people tend to stiffen and clench their hands when lying:

$$\mathbf{f}_4 = \bigoplus_{k=1}^{t-1} \left[\bigoplus_{m\in\{l,r\}} \left[\left(\frac{1}{5} \sum_{(\mathbf{a},\mathbf{b})\in S^m_{1,k}} \frac{(\mathbf{b}-\mathbf{a}) \cdot (\mathbf{w}-\mathbf{a})}{|\mathbf{b}-\mathbf{a}| \cdot |\mathbf{w}-\mathbf{a}|} \right), \left(\frac{1}{5} \sum_{\mathbf{a}\in S^m_{2,k}} \|\mathbf{a}-\mathbf{p}\|_2 \right) \right] \right], \tag{4}$$

where $S^m_{1,k}$ is a set of two tuples containing joint coordinates of knuckles and fingertips of the hand m in the k-th frame, respectively; $S^m_{2,k}$ includes only the fingertips; while $\mathbf{w}$ and $\mathbf{p}$ contain wrist and palm coordinates.

The fifth and last feature vector $\mathbf{f}_5 \in \mathbb{R}^{(t-1)\times 42}$ is composed of all hand joints position displacements to describe their relative speed:

$$\mathbf{f}_5 = \bigoplus_{k=1}^{t-1} \left[\bigoplus_{m \in \{l,r\}} \left[\bigoplus_{(\mathbf{a}_k, \mathbf{a}_{k+1}) \in S^m_{3,k}} \|\mathbf{a}_k - \mathbf{a}_{k+1}\|_2 \right] \right], \tag{5}$$

where $S^m_{3,k}$ is a set of two tuples containing all joint coordinates of hand m in two consecutive frames k and $k+1$. This feature captures the possible presence of cadenced movements in situations where the person's stress can increase. In fact, as noted by the authors of [5], hand speed tends to decrease when lying.

Finally, the handcrafted features are concatenated into a single vector F, used as input for the neural network model:

$$F = \mathbf{f}_1 \oplus \mathbf{f}_2 \oplus \mathbf{f}_3 \oplus \mathbf{f}_4 \oplus \mathbf{f}_5. \tag{6}$$

3.2 Deception Detection Classifier

To detect deception from the handcrafted features F, the proposed pipeline employs a combination of LSTM and feed-forward network (FFN).

An LSTM [17] is a recurrent neural network (RNN) that is designed to remember long-term dependencies in sequential data. It keeps track of both long- and short-term dependencies through cell states and hidden states, respectively. These networks use a series of gates to control the information flow, i.e., what data enters, is stored, and leaves the network. In particular, there are three kinds of gates: input, forget, and output. The input gate determines the information to be stored in the current state. The forget gate decides which parts of the cell state are useful, given the previous hidden state and new input data. The output gate controls what information to output from the current state, considering the previous and current states. To link the handcrafted features evolution through time, vector F is given as input to a single-layer LSTM returning a sequence with the same input size, using a hidden size equal to the number of features.

The last step to detect deception requires the classification of the LSTM output. This procedure is performed through an FFN that, differently from RNNs, contains one or multiple sequential layers without cyclic connections. The FFN adopted in the proposed method is composed of a linear layer followed by a ReLU activation function, a 50% dropout layer, and a softmax activation function to transform the result into a probability vector. Thus, once the input features F have been transformed by the LSTM, the resulting output is fed to the FFN to obtain the classification scores for the two classes $\mathbf{o} \in [0,1]^2$:

$$\mathbf{o} = \mathrm{FFN}(\mathrm{LSTM}(F)). \tag{7}$$

Fig. 3. Frames examples of OpenPose extracted skeletons.

4 Experiments

This section assesses the effectiveness of the proposed method in the deception detection task. In detail, Sect. 4.1 describes the dataset used to test it together with the employed hyper-parameters. Section 4.2 presents ablation studies on the pipeline components. Finally, Sect. 4.3 compares the proposed method with the existing literature.

4.1 Implementation Details

The proposed pipeline was tested on the public collection described in [26], which contains 121 real-life courtroom trials associated with 58 different persons. Since the presented method is wholly based on hands, which might be partially or fully occluded in some videos, as can be observed in Fig. 3, it uses the same subset proposed in [5] for a fair comparison. The resulting dataset comprises 77 recordings of 47 identities divided into 37 truthful and 40 deceptive videos. Moreover, all sequences containing both hands are treated as two distinct samples of the same class, reaching 119 sequences as previously described by the authors of [5].

Regarding the experimental settings, tests were performed using the same testing protocol employed by [5,31], i.e., a 10-fold cross-validation across subjects with random 80/20 training and test splits per fold, which guarantees that a given identity is not contained in both sets simultaneously. Each model was trained for 200 epochs using the Adam optimizer, with a learning rate set to 1e-3, batch size of 32, and dropout probability of 50%. Standard classification metrics, e.g., accuracy and F1-score, were used to evaluate all of the experiments.

Lastly, the presented pipeline was implemented using the Python programming language and Pytorch library. Furthermore, all experiments were conducted on an Intel i7 1.99 GHz CPU with 8 GB of RAM.

4.2 Ablation Studies

To thoroughly evaluate the presented pipeline, ablation studies were performed on the handcrafted features and possible recurrent architectures. Regarding the extracted hand features, tests were conducted to assess the effectiveness of different feature groups by defining several subsets. While many combinations were derived from the extracted features, three groups were identified that

Table 1. Ablation study on different feature groups.

Features	Accuracy	Precision	Recall	F1-score
G_1	88.00%	66.67%	93.13%	77.71%
G_2	92.00%	93.88%	80.71%	86.80%
G_3	94.00%	94.73%	90.01%	92.30%

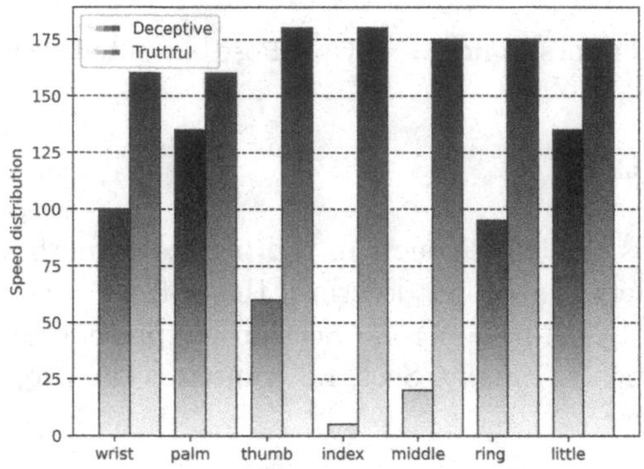

Fig. 4. Finger speed differences in truthful and deceptive videos.

provided the highest performances. In detail, the first group contains features associated with the palm surface normal, wrist rotation, and joint speed, i.e., $G_1 = \{\mathbf{f}_2, \mathbf{f}_3, \mathbf{f}_5\}$; the second group comprises features encompassing the index-thumb fingertip distance, the mean distance of fingers from the palm, and palm surface normal, i.e., $G_2 = \{\mathbf{f}_1, \mathbf{f}_2, \mathbf{f}_4\}$; while the third group comprehends all features described in Sect. 3.1, i.e., $G_3 = F$. The obtained results are reported in Table 1. As can be observed, group G_3 reaches the highest scores as it exploits all the extracted features. Nevertheless, G_1 and G_2 still manage to achieve high performances, primarily due to joint displacements ($\mathbf{f}_5$) that capture cadenced movements and distance between fingertips and palm ($\mathbf{f}_4$) that highlights possible hand contraptions, which are typical in lying persons. Regardless of these observations, G_2 performs worse with respect to the other two subsets as it relies on the palm surface normal ($\mathbf{f}_3$) and wrist rotation ($\mathbf{f}_3$), which can suffer from being partially occluded as they normally appear behind the hand. These findings indicate that speed and finger distances might be more useful for the deception detection task as they can capture deceptive behavior such as stiffening and subsequent decrease in movements in a more robust way. In fact, deceptive videos have a lower average finger, wrist, and palm speed, as illustrated in Fig. 4 and as it was also observed by the authors of [5].

Concerning possible recurrent models, an ablation study was carried out to investigate the contribution of different recurrent layers on the overall performance of the model presented in Sect. 3.2. The results obtained using feature group G_3 are summarized in Table 2. As can be observed, all recurrent layer

Table 2. Recurrent architecture ablation study.

Model	Layers	Accuracy	Precision	Recall	F1-score
GRU	1	88.76%	85.22%	79.37%	82.19%
BiLSTM	1	91.88%	89.45%	77.13%	82.83%
LSTM	2	92.68%	90.71%	90.09%	90.40%
LSTM	1	94.00%	94.73%	90.01%	92.30%

types achieve significantly high performances, with the gated recurrent units (GRU) falling slightly short of the other models. In any case, all configurations can capture temporal characteristics from the generated handcrafted features, suggesting that the extracted vector F well describes deception-related movements. On the other hand, more complex or advanced models do not necessarily perform better. For instance, the bidirectional LSTM (BiLSTM) obtained lower metrics with respect to the standard LSTM even though it analyzes the sequence in both directions. Similarly, using multiple LSTM layers results in performance degradation. This outcome is likely associated with the number of extracted hand features, i.e., 52, which is too small for more complex and advanced architectures that end up overfitting over the training set. Concluding, the deception detection task is best addressed via a single-layer LSTM when analyzing only hand movements through the designed handcrafted features F.

4.3 Performance Evaluation

The last step needed to evaluate the proposed pipeline requires a comparison with the state-of-the-art, which is shown in Table 3. In the reported results, models not focusing on hand gestures, i.e., [10,18,19,26], employ all 121 samples available in the dataset, whereas the work described by the authors of [5] and the presented approach make use of 77 videos, corresponding to 119 samples as mentioned in Sect. 4.1, since those are the only sequences where hands are visible. As it can be observed from the table, multimodal approaches tend to perform better than unimodal ones, which is an expected outcome due to their intrinsic ability to merge cues deriving from diverse sources such as video and audio. Despite this, the presented approach managed to achieve performances that are comparable to multimodal systems, even though it focuses exclusively on hand movements, indicating that the designed features are indeed able to describe and capture deceptive behavior. What is more, when compared to other unimodal models presented in [5], the proposed approach significantly improves upon their accuracy even if a simpler architecture is employed. In fact, when compared to the LSTM model used by [5], there is almost ≈10% accuracy increase, while it shows ≈4% performance boost when compared to a more advanced Fisher-LSTM that is able to analyze more advanced characteristics. This outcome corroborates, on the one hand, the efficacy of handcrafted features in depicting a deceptive behavior and, on the other hand, the quality of the extracted features themselves

Table 3. SOTA performance comparison on real-life trial dataset.

Method	Features	Accuracy
RF [26]	MicroExpr, Hand Gestures, Text	75.20%
RBF-SV [10]	MicroExpr (Action Units)	76.84%
MLP [19]	Audio, MicroExpr, Text, Video	90.99%
MLP [19]	Audio, MicroExpr, Text+, Video	96.14%
LieNet [18]	Audio, Video	97.00%
LSTM [5]	Hand Gestures	84.52%
Fisher-LSTM [5]	Hand Gestures	90.96%
Our work	Hand Gestures	94.00%

since the LSTM model presented in this work is similar to the one described in [5] but leverages completely different handcrafted features, fully demonstrating the effectiveness of the devised pipeline.

5 Conclusion

This paper presented a pipeline addressing the deception detection task. Specifically, novel handcrafted features were designed to describe hand movements associated with deceptive behavior. These features were classified via an FFM using the output of an LSTM to temporally link hand features extracted at each frame. The pipeline was tested on a public collection of real-life trials where the designed features enabled the architecture to achieve performances that are close to those of multimodal approaches. Furthermore, the presented approach set a new state-of-the-art for unimodal hand gesture deception detection, reaching a 94% accuracy on it and fully demonstrating the proposed pipeline effectiveness.

As possible future work, additional handcrafted features are going to be defined so that more articulated models can be exploited to address the deception detection task. Moreover, alternative multimodal approaches should also be explored so that hand features can be tied together with other modalities such as speech, text, or facial micro-expressions.

Acknowledgements. This work was supported by "Smart unmannEd AeRial vehiCles for Human likE monitoRing (SEARCHER)" project of the Italian Ministry of Defence (CIG: Z84333EA0D); and "A Brain Computer Interface (BCI) based System for Transferring Human Emotions inside Unmanned Aerial Vehicles (UAVs)" Sapienza Research Projects (Protocol number: RM1221816C1CF63B); and "DRrone Aerial imaGe SegmentatiOn System (DRAGONS)" (CIG: Z71379B4EA); and Departmental Strategic Plan (DSP) of the University of Udine - Interdepartmental Project on Artificial Intelligence (2020–25); and "A proactive counter-UAV system to protect army tanks and patrols in urban areas (PROACTIVE COUNTER-UAV)" project of the Italian Ministry of Defence (Number 2066/16.12.2019); and the MICS (Made in Italy - Circular and Sustainable) Extended Partnership and received funding from

Next-Generation EU (Italian PNRR - M4 C2, Invest 1.3 - D.D. 1551.11-10-2022, PE00000004). CUP MICS B53C22004130001.

References

1. Alaskar, H., Sbaï, Z., Khan, W., Hussain, A., Alrawais, A.: Intelligent techniques for deception detection: a survey and critical study. Soft. Comput. **27**(7), 3581–3600 (2023)
2. Avola, D., Bacciu, A., Cinque, L., Fagioli, A., Marini, M.R., Taiello, R.: Study on transfer learning capabilities for pneumonia classification in chest-x-rays images. Comput. Methods Programs Biomed. **221**, 106833 (2022)
3. Avola, D., Cascio, M., Cinque, L., Fagioli, A., Foresti, G.L.: LieToMe: an ensemble approach for deception detection from facial cues. Int. J. Neural Syst. **31**(02), 2050068 (2021)
4. Avola, D., Cascio, M., Cinque, L., Fagioli, A., Foresti, G.L.: Affective action and interaction recognition by multi-view representation learning from handcrafted low-level skeleton features. Int. J. Neural Syst. **32**, 2250040 (2022)
5. Avola, D., Cinque, L., De Marsico, M., Fagioli, A., Foresti, G.L.: LieToMe: preliminary study on hand gestures for deception detection via fisher-LSTM. Pattern Recogn. Lett. **138**, 455–461 (2020)
6. Avola, D., Cinque, L., Fagioli, A., Filetti, S., Grani, G., Rodolà, E.: Multimodal feature fusion and knowledge-driven learning via experts consult for thyroid nodule classification. IEEE Trans. Circuits Syst. Video Technol. **32**(5), 2527–2534 (2021)
7. Avola, D., Cinque, L., Fagioli, A., Foresti, G.L.: Sire-networks: convolutional neural networks architectural extension for information preservation via skip/residual connections and interlaced auto-encoders. Neural Netw. **153**, 386–398 (2022)
8. Avola, D., et al.: Medicinal boxes recognition on a deep transfer learning augmented reality mobile application. In: Proceedings of the International Conference on Image Analysis and Processing (ICIAP), pp. 489–499 (2022)
9. Avola, D., Cinque, L., Fagioli, A., Foresti, G.L., Massaroni, C.: Deep temporal analysis for non-acted body affect recognition. IEEE Trans. Affect. Comput. **13**(3), 1366–1377 (2020)
10. Avola, D., Cinque, L., Foresti, G.L., Pannone, D.: Automatic deception detection in RGB videos using facial action units. In: Proceedings of the International Conference on Distributed Smart Cameras (ICDSC), pp. 1–6 (2019)
11. Bond, C.F., Jr., DePaulo, B.M.: Accuracy of deception judgments. Pers. Soc. Psychol. Rev. **10**(3), 214–234 (2006)
12. Cao, Z., Hidalgo, G., Simon, T., Wei, S.E., Sheikh, Y.: OpenPose: realtime multi-person 2d pose estimation using part affinity fields. IEEE Trans. Pattern Anal. Mach. Intell. **43**(1), 172–186 (2021)
13. Constâncio, A.S., Tsunoda, D.F., Silva, H.d.F.N., Silveira, J.M.d., Carvalho, D.R.: Deception detection with machine learning: a systematic review and statistical analysis. Plos One **18**(2), e0281323 (2023)
14. DePaulo, B.M., Kashy, D.A., Kirkendol, S.E., Wyer, M.M., Epstein, J.A.: Lying in everyday life. J. Pers. Soc. Psychol. **70**(5), 979 (1996)
15. Dzedzickis, A., Kaklauskas, A., Bucinskas, V.: Human emotion recognition: review of sensors and methods. Sensors **20**(3), 592 (2020)
16. Gannon, T.A., Beech, A.R., Ward, T.: Risk assessment and the polygraph. The Use of the Polygraph in Assessing, Treating and Supervising Sex Offenders: A Practitioner's Guide, pp. 129–154 (2009)

17. Hochreiter, S., Schmidhuber, J.: Long short-term memory. Neural Comput. **9**(8), 1735–1780 (1997)
18. Karnati, M., Seal, A., Yazidi, A., Krejcar, O.: LieNet: a deep convolution neural network framework for detecting deception. IEEE Trans. Cogn. Develop. Syst. **14**(3), 971–984 (2021)
19. Krishnamurthy, G., Majumder, N., Poria, S., Cambria, E.: A deep learning approach for multimodal deception detection. In: Proceedings of the International Conference on Computational Linguistics and Intelligent Text Processing (CICLing), pp. 87–96 (2023)
20. Lai, Y.F., Chen, M.Y., Chiang, H.S.: Constructing the lie detection system with fuzzy reasoning approach. Granul. Comput. **3**, 169–176 (2018)
21. Li, F., Yang, W., Liu, X., Sun, G., Liu, J.: Using high-resolution UAV-borne thermal infrared imagery to detect coal fires in Majiliang mine, Datong coalfield, Northern China. Remote Sens. Lett. **9**(1), 71–80 (2018)
22. Mihalcea, R., Pulman, S.: Linguistic ethnography: identifying dominant word classes in text. In: Proceedings of the International Conference on Computational Linguistics and Intelligent Text Processing (CICLing), pp. 594–602 (2009)
23. Nasri, H., Ouarda, W., Alimi, A.M.: ReLiDSS: novel lie detection system from speech signal. In: Proceedings of the IEEE/ACS International Conference of Computer Systems and Applications (AICCSA), pp. 1–8 (2016)
24. Nurçin, F.V., Imanov, E., Işın, A., Ozsahin, D.U.: Lie detection on pupil size by back propagation neural network. Procedia Comput. Sci. **120**, 417–421 (2017)
25. Pennebaker, J.W., Francis, M.E., Booth, R.J.: Linguistic inquiry and word count: LIWC 2001. Mahway Lawrence Erlbaum Associates **71**(2001), 2001 (2001)
26. Pérez-Rosas, V., Abouelenien, M., Mihalcea, R., Burzo, M.: Deception detection using real-life trial data. In: Proceedings of the ACM on International Conference on Multimodal Interaction (ICMI), pp. 59–66 (2015)
27. Rusconi, E., Mitchener-Nissen, T.: Prospects of functional magnetic resonance imaging as lie detector. Front. Hum. Neurosci. **7**, 594 (2013)
28. Van der Velden, B.H., Kuijf, H.J., Gilhuijs, K.G., Viergever, M.A.: Explainable artificial intelligence (XAI) in deep learning-based medical image analysis. Med. Image Anal. **79**, 102470 (2022)
29. Vrij, A., Edward, K., Roberts, K.P., Bull, R.: Detecting deceit via analysis of verbal and nonverbal behavior. J. Nonverbal Behav. **24**, 239–263 (2000)
30. Vrij, A., Semin, G.R.: Lie experts' beliefs about nonverbal indicators of deception. J. Nonverbal Behav. **20**, 65–80 (1996)
31. Wu, Z., Singh, B., Davis, L., Subrahmanian, V.: Deception detection in videos. In: Proceedings of the AAAI Conference on Artificial Intelligence (AAAI), vol. 32 (2018)
32. Zaidi, S.S.A., Ansari, M.S., Aslam, A., Kanwal, N., Asghar, M., Lee, B.: A survey of modern deep learning based object detection models. Digit. Signal Process. **126**, 103514 (2022)
33. Zuo, J., Gedeon, T., Qin, Z.: Your eyes say you're lying: an eye movement pattern analysis for face familiarity and deceptive cognition. In: Proceedings of the International Joint Conference on Neural Networks (IJCNN), pp. 1–8 (2019)

Spatial Transformer Generative Adversarial Network for Image Super-Resolution

Pantelis Rempakos[1], Michalis Vrigkas[2(✉)], Marina E. Plissiti[1], and Christophoros Nikou[1]

[1] Department of Computer Science and Engineering, University of Ioannina, 45110 Ioannina, Greece
{cs04279,marina,cnikou}@uoi.gr

[2] Department of Communication and Digital Media, University of Western Macedonia, 52100 Kastoria, Greece
mvrigkas@uowm.gr

Abstract. High-resolution images play an essential role in the performance of image analysis and pattern recognition methods. However, the expensive setup required to generate them and the inherent limitations of the sensors in optics manufacturing technology leads to the restricted availability of these images. In this work, we exploit the information retrieved in feature maps using the notable VGG networks and apply a transformer network to address spatial rigid affine transformation invariances, such as translation, scaling, and rotation. To evaluate and compare the performance of the model, three publicly available datasets were used. The model achieved very gratifying and accurate performance in terms of image PSNR and SSIM metrics against the baseline method.

Keywords: Image super-resolution · Spatial transformer · VGG · SRGAN

1 Introduction

The prospect of obtaining detailed digital high-resolution (HR) images from a set of low-resolution (LR) observations has been a topic of great interest in both the fields of signal and image processing [15,21]. In the last few decades, recent developments in convolutional neural networks and advancements in GPU technology have further lowered the barrier to accumulating high-resolution images and videos. In addition, state-of-the-art machine learning models have made major breakthroughs in conventional computer vision tasks [12,17].

Despite the aforementioned improvements, the methods of image and video super-resolution (SR) available today can be unsatisfying, in the sense that they fail to match expectations in perceptual quality and computational efficiency [11,14]. Nonetheless, the need for high-quality images has remained at large an

G. L. Foresti et al. (Eds.): ICIAP 2023, LNCS 14233, pp. 399–411, 2023.
https://doi.org/10.1007/978-3-031-43148-7_34

essential need for human interpretation of information and machine perception. In this work, we seek to achieve a feature fine, realistic, and computationally efficient super-resolution method that brings about quality image enhancement.

One of the main challenges when it comes to SR methods is the task of generating high-quality images that are realistic and sensible to human perception. This problem is difficult in the sense that the quality of the generated images is affected by multiple factors such as the quality of the input image taken by the image extracting sensor type, the image spectral and spatial resolution, and light variations. As a consequence, images generally tend to appear distorted, and blurry, with noise which SR models seek to adverse. In our approach, we aspire to develop a robust model that reliably understands this problem and generates HR images that are invariant to large geometric aberrations.

More specifically, in our approach, we construct a novel robust model that reliably generates high-resolution images that are invariant to these geometric aberrations. We call this model, the ST-SRGAN model, which essentially accounts for the fusion of spatial transformer networks (STN) alongside a super-resolution generative adversarial network (SRGAN) [11]. Our model exhibits remarkable performance in common publicly available datasets, as was verified by the experimental results.

2 Related Work

In the past, the consensus was to approach the SR problem with either statistical, prediction, or patch-based methods [16,19,20]. We can briefly summarize the related super-resolution methods in two major categories namely (i) learning-based and (ii) reconstruction-based methods.

Learning-based methods approximate HR images by using neighbor embedding, sparse coding, pixel-based, and example-based methods [6,7,24]. On the other hand, reconstruction-based methods use prior retrieved information to determine the HR limitations such as edge sharpening, regularization, and deconvolution methods [1,3,18]. Nowadays, researchers have substantially suppressed the limitations of the SR problem, leading to the development of cost-effective systems that allow researchers to make better use of big data.

Learning-Based Methods. A case in point is the super-resolution convolutional neural network (SRCNN) [4], which is regarded as the earliest CNN super-resolution model. The model structure consists of three parts. The first part relates to the extraction of data from the LR image. The second part implements non-linear mapping, a dimension-reducing method that attempts to retain the distances between data points as well as possible. Finally, in the third part, the model super resolves the image and reconstructs its high-resolution counterpart.

Kim *et al.* [9] showed that increasing network depth resulted in significant improvements in model accuracy. The VDSR network architecture consists of 20 weight layers which is much deeper than its SRCNN counterpart which only has three layers. By cascading small filters many times in a deep network structure, contextual information over large image regions is exploited efficiently. However using very deep networks, convergence speed would become a critical issue during training. To counter this problem the model would learn only residuals and use extremely high learning rates (104 times higher than SRCNN) enabled by adjustable gradient clipping.

The limitation of these methods is that increasing the resolution of LR images before the image-enhancing step may lead to high computational complexity. This is an especially problematic state for CNNs, where processing speed depends directly on the resolution of the input image. Furthermore, interpolation methods that are typically used to accomplish this task (e.g. bicubic interpolation) do not bring the additional information required to tackle the ill-posed nature of the SR reconstruction problem.

Reconstruction-Based Methods. Dong *et al.* [5] proposed an improvement to the SRCNN model that uses a post-upsampling reconstruction technique called FSRCNN. In this approach, feature extraction is performed in the low-resolution space. In addition, FSRCNN also uses a 1×1 convolutional layer after feature extraction to reduce the computational complexity cost by reducing the number of channels required. FSRCNN has a relatively shallow network which makes it easier to learn about the effect of each component. This model is even faster with better-reconstructed image quality than the previous SRCNN.

Nonetheless, for problems where LR images need to be upscaled by large factors (i.e., $8\times$), regardless of whether the upsampling is complete before or after passing through the deep SR network, the results are bound to be suboptimal. It makes more sense to progressively upscale the LR image in such cases to finally meet the spatial dimension criteria of the HR output rather than upscaling by $8\times$ in one shot. To this end, Lai *et al.* [10] proposed that the sub-band residuals of HR images can progressively be reconstructed. Sub-band residuals refer to the differences between the upsampled image and the ground truth HR image at the respective level of the network.

3 Method

Image SR methods aim to reconstruct high-resolution images given a set of low-resolution observations. In this section, we elaborate the details of the proposed ST-SRGAN method and describe the architectural units and training objectives.

3.1 Image Model Formulation

Since digital imaging systems are subject to hardware limitations, images are often degraded due to these limitations. The captured image may often be distorted by motion blur or additive noise because of the limited time window during the sensor of the image-capturing system is open. This problem is posed in its linear form as:

$$\mathbf{y}_k = \mathbf{W}_k \mathbf{z} + \mathbf{n}\,, \tag{1}$$

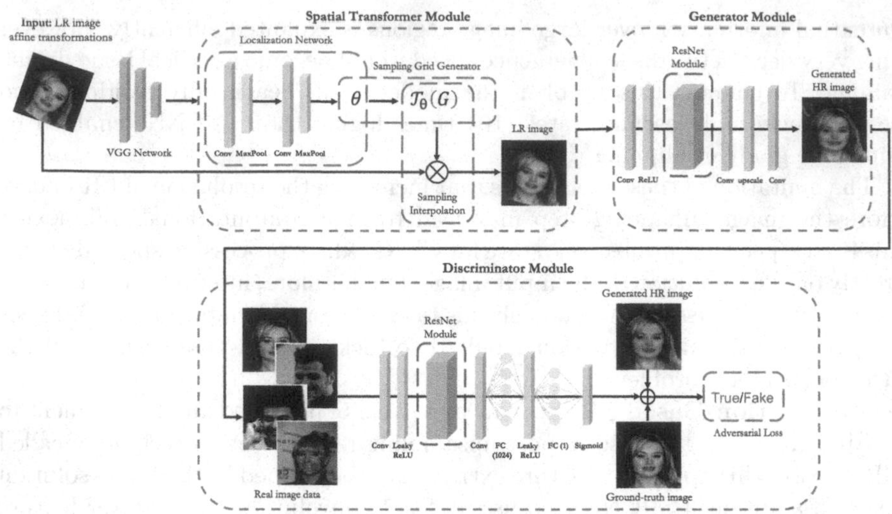

Fig. 1. Overview of the proposed ST-SRGAN architecture. First, the LR images are passed over the VGG network to extract a feature map and a spatial-transformer network is used to assess the affine transformations. Once the images have been aligned correctly, the generator module generates the estimated HR image. Finally, the generated HR image is detected as real or fake by the discriminator network based on real high-resolution images.

where $\mathbf{y}_k \in \mathbb{R}^{M \times N}$ is the k-th LR image, with $k = 1, \ldots, p$. The desired HR image $\mathbf{z} \in \mathbb{R}^{r_1 M \times r_2 N}$, where r_1 and r_2 are the up-scale factors in the horizontal and vertical directions, respectively. The degradation matrix $\mathbf{W}_k = \mathbf{D}_k \mathbf{B}_k \mathbf{M}_k$ for the k-th frame performs the operations of (i) a motion matrix $\mathbf{M}$ that includes rigid transformation parameters such as rotation angle and translation vector, (ii) a blurring matrix $\mathbf{B}$, and (iii) a sub-sampling matrix $\mathbf{D}$. Finally, $\mathbf{n}$ is additive Gaussian noise. Note that all images are ordered lexicographical order.

3.2 Proposed Approach

We propose a spatial transformer-enhanced SRGAN network (ST-SRGAN) to address the aforementioned limitations of CNN-based methods. More specifically, we design a generator network to model the relationship among different views of an LR image to capture the relevant perceptual information in a robust and geometrically invariant way. Compared to traditional CNN-based methods, this model can discriminately incorporate information from multiple angular views, noisy, and in general spatially distorted images. The transformer works as an added self-attention mechanism to the generator network. The architecture of the proposed ST-SRGAN network is depicted in Fig. 1.

The LR images are first examined by the VGG network which extracts high-quality feature maps from LR images, that may be used to generate their high-resolution counterparts. The localization network takes the input maps retrieved by the VGG network and outputs the parameters of the affine transformations that should be applied to the feature maps. Following this procedure, the spatial-transformer processes these feature maps. Then, the grid generator proceeds to generate a grid of (x, y) coordinates using the parameters of the affine transformation that correspond to a set of points where the input feature map should be sampled to produce the transformed output feature map followed by a bilinear interpolation. The correctly aligned images are then fed into the generator, which then generates the estimated HR image. Using real high-resolution images and the generated HR images, the discriminator network predicts whether an image given by the generator is real or fake.

The generator and discriminator units work similarly to the original SRGAN architecture. However, compared to traditional GAN approaches it should be noted that the capabilities of the proposed discriminator are hampered by adding Gaussian noise layers in between the original layers of the model. This improves the performance of the model as a strong discriminator model is proven to work as a step function that hinders the result by producing no useful gradients to update the generator. Finally, dropout layers are also used.

Spatial-Transformer Module. Existing super-resolution methods do not consider that, when image transmission is over noisy channels, the effect of any possible geometric transformations could incur significant quality loss and distortions. To address this problem, the proposed model is formulated as a fusion of the SRGAN [11] and the spatial-transformer network [8]. This allows the development of a robust, spatially-transformed deep learning framework that is able to simultaneously perform both geometric transformations and image super-resolution. The reason for using the spatial-transformer network is that it provides model invariance when it comes to spatial transformations of LR images such as rotation, translation, and scaling.

More specifically, the spatial-transformer module consists of three main components, namely, (i) the localization network, (ii) the sampling grid generator, and (iii) bilinear interpolation. The localization network input corresponds to a $4D$ tensor representation of a batch of LR images $\mathbf{y}_k \in \mathbb{R}^{M \times N \times C}$, where C is the number of channels. The network contains a few convolutional layers and a few dense layers. Its output prediction consists of the parameters of transformation matrix $\mathbf{W}_k$. These parameters are used to determine the input feature map transformations that the network must estimate, such as the rotation angle of the input LR images, the amount of translation, and the scaling factor required to focus on the region of interest in the input feature map.

Then, the sampling grid generator predicts the transformation parameters which are in turn used in the form of an affine transformation matrix of size 2×3 for each LR image in the batch. Thus, we obtain a sampling grid of transformed indices:

$$\begin{pmatrix} x_i^s \\ y_i^s \end{pmatrix} = \mathcal{T}_\theta (G_i) = \mathrm{A}_\theta \begin{pmatrix} x_i^t \\ y_i^t \\ 1 \end{pmatrix} = \begin{bmatrix} \theta_{11} & \theta_{12} & \theta_{13} \\ \theta_{21} & \theta_{22} & \theta_{23} \end{bmatrix} \begin{pmatrix} x_i^t \\ y_i^t \\ 1 \end{pmatrix}, \tag{2}$$

where $\mathcal{T}_\theta(G_i)$ represents the transformation of grid $G_i = (x_i^t, y_i^t)$ of the target coordinates of the regular grid in the output feature map and (x_i^s, y_i^s) are the source coordinates in the input feature map. Matrix A_θ corresponds to the affine transformation and the parameters θ represent the rotation angles, translation and scale parameters of each LR image in the batch. Finally, a bilinear interpolation on transformed indices is performed to estimate the pixel value at the transformed point (x_i^t, y_i^t) using the four nearest pixel values. For example, a point $(1, 1)$ after a counter clockwise 45° rotation of its axes, becomes $(2, 0)$.

Residual Network Module. In our approach, layers are reformulated as learning residual functions with reference to the layer inputs instead of learning unreferenced functions. Each residual block can be expressed as a sequence of the following equations:

$$y_l = h(x_l) + F(x_l, W_l), \tag{3}$$

$$x_{l+1} = f(y_l), \tag{4}$$

where x_l and x_{l+1} are the input and the output of the l-th block, F is a residual function, $h(x_l)$ is an identity mapping function and f is a ReLU function. The main idea behind this sequence is to learn the additive residual function F with respect to the $h(x_l)$, taking advantage of an identity mapping function $h(x_l) = x_l$. To formulate an identity mapping $f(y_l) = y_l$, activation functions ReLU and batch normalization are considered as the "preactivation" of the weight layers, while traditional techniques considered them as "post-activation".

Generator and Discriminator Module. The generator contains the residual network module, instead of deep convolution networks because residual networks are easy to train and allow to be substantially deeper to generate better results. During training, an HR image is down-sampled to a LR image. The generator unit then tries to up-sample the image from low to high resolution. After the image is passed into the discriminator, the latter tries to distinguish between a ground-truth super-resolution and the estimated HR image and generates the adversarial loss which is then backpropagated into the generator unit.

The discriminator unit implements LeakyReLU as activation. The network contains eight convolutional layers with of 3×3 filter kernels, increasing by a factor of two from 64 to 512 kernels. Strided convolutions are used to reduce the image resolution each time the number of features is doubled. The resulting 512 feature maps are followed by two dense layers with a leakyReLU applied between those two layers and a sigmoid activation function is used to obtain a probability for sample classification.

3.3 Training Objective

Content Loss: In this work, we used two types of content losses. The first one is the pixel-wise MSE loss $\mathcal{L}_{MSE}^{SR}$ of the residual network module, which is the most common MSE loss for image SR. However, MSE loss is not able to deal with high-frequency content in the image which resulted in producing overly smoother images.

$$\mathcal{L}_{MSE}^{SR} = \frac{1}{r_1 M r_2 N} \sum_{i=1}^{r_1 M} \sum_{j=1}^{r_2 N} (z_{i,j} - G_{\theta G}(y_{i,j}))^2 . \tag{5}$$

The second loss is the VGG loss, which is based on the ReLU activation layer of the pre-trained VGG-19 network. Here the VGG network works as a feature extractor and the feature map $\phi(\cdot)$ that is extracted is used in the loss function.

$$\mathcal{L}_{VGG}^{SR} = \frac{1}{MN} \sum_{i=1}^{M} \sum_{j=1}^{N} \left(\phi(z)_{i,j} - \phi(G_{\theta G}(y))_{i,j} \right)^2 . \tag{6}$$

Adversarial Loss: The adversarial loss forces the generator to generate an image more similar to the HR image by using the discriminator to differentiate between ground-truth and the estimated HR image.

$$\mathcal{L}_{G}^{SR} = \sum_{i=1}^{N} -\log D_{\theta D}(G_{\theta G}(y)) . \tag{7}$$

The total loss is computed as the sum of all the individual losses:

$$\mathcal{L}_{tot}^{SR} = \mathcal{L}_{MSE}^{SR} + \mathcal{L}_{VGG}^{SR} + \lambda \mathcal{L}_{G}^{SR} . \tag{8}$$

where $\lambda = 10^{-3}$ controls the importance of the $\mathcal{L}_{G}^{SR}$ term in the total loss. This loss is preferred over the mean-squared error loss because we do not care about the pixel-by-pixel comparison of the images. We are mostly concerned about the improvement in the quality of the images. Hence, by using this loss function in the ST-SRGAN model, we are able to achieve high-quality results.

4 Experiments

Data Selection. In this study, the CelebA Dataset is used [13] for training purposes. This dataset is a large-scale face attributes collection with more than 200K celebrity photographs, each with 40 attribute annotations. The images cover background clutter and large pose variations. Furthermore, Set5 [2] and Set14 [23] datasets were also used for testing.

Evaluation Metrics. To evaluate the SR reconstruction results, (i) the peak-signal-to-noise-ratio (PSNR), (ii) the mean squared error (MSE) of the original image and the degraded image, and (iii) the structural similarity index (SSIM) [22] were used.

Training Details and Parameters. The network was trained on an NVIDIA Titan XP GPU using a random sample of 10K images from the CelebA dataset. These images are distinct from the testing images. To obtain the 64×64 LR and 256×256 HR images, we down-sampled and up-sampled respectively using the OpenCV library. The same applies to different upscaling factors (i.e., $2\times$, $3\times$, and $4\times$) used for the evaluation of the proposed method.

The LR and HR input image range is scaled to $[-1, 1]$ because we are using the *tanh* activation function. The MSE loss was thus calculated on images of intensity range $[1, 1]$. Note that the VGG-22 network is used to extract the feature maps of the dataset images. Also, batch sizes of 1, 4, and 8 were used and Adam optimizer is used with a learning step of 2×10^{-4} and decay rates $\beta 1 = 0.5$ and $\beta 2 = 0.999$, respectively.

Table 1. Experimental results for *Config 1* variants (i.e., *c1-a*, *c1-b*, and *c1-c* with different up-scaling factors).

Up-scale factor	4×			3×			2×		
Configuration	*c1-a*	*c1-b*	*c1-c*	*c1-a*	*c1-b*	*c1-c*	*c1-a*	*c1-b*	*c1-c*
Batch size	$B = 1$	$B = 4$	$B = 8$	$B = 1$	$B = 4$	$B = 8$	$B = 1$	$B = 4$	$B = 8$
PSNR ↑	22.43	27.32	29.46	24.36	25.83	28.81	19.98	22.81	25.94
SSIM ↑	0.88	0.92	0.94	0.89	0.91	0.93	0.87	0.88	0.90
MSE ↓	0.28	0.24	0.22	0.38	0.32	0.23	0.23	0.21	0.16

The generator network is comprised of 16 residual blocks. In addition, dropout and Gaussian noise layers were added to the discriminator to avoid the vanishing gradient problem. The spatial-transformer module was applied prior to the convolutional layer of the generator and after receiving the extracted feature maps using the VGG-22 architecture. The localization network to the identity transformation of the spatial-transformer module was initialized before the training process and while building the generator network. Note that the MSE-based SRResNet network was employed as initialization to the generator network to avoid undesired local optima. The variant configurations were trained with 5×100 iterations in which the generator and discriminator have been trained alternatively between iterations.

4.1 Experimental Results

Image reconstruction measurements are accomplished via the PSNR, MSE (%), and SSIM (%) metrics. Parameter B represents the batch size of the experiment. In the tables below, we affirm the experimental results of various configurations made to evaluate our model i.e., namely *Config 1* and *Config 2* and their variations. These values are obtained through 5 random realizations of the experiment in each case.

- *Config 1*. This configuration consists of three variants namely, *c1-a*, *c1-b*, and *c1-c* that represent the behavior of the model when the input images suffer from blurring and additive Gaussian white noise for batch sizes of 1, 4, and 8. Finally, for each configuration, different up-scaling factors of 4×, 3×, and 2× were applied, respectively. The performance of the different configurations is shown in Table 1.

- *Config 2*. This configuration comprises also three variants namely, *c2-a*, *c2-b*, and *c2-c* that represent the behavior of the model when the input images suffer from blurring, additive Gaussian white noise, and spatial translations and rotations with batch sizes of 1, 4, and 8. For each configuration, different up-scaling factors of 4×, 3×, and 2× were employed, respectively. Quantitative results are summarized in Table 2.

Table 2. Experimental results for *Config 2* variants (i.e., *c2-a*, *c2-b*, and *c2-c* with different up-scaling factors).

Up-scale factor	4×			3×			2×		
Configuration	*c2-a*	*c2-b*	*c2-c*	*c2-a*	*c2-b*	*c2-c*	*c2-a*	*c2-b*	*c2-c*
Batch size	$B = 1$	$B = 4$	$B = 8$	$B = 1$	$B = 4$	$B = 8$	$B = 1$	$B = 4$	$B = 8$
PSNR ↑	17.85	18.38	20.84	18.28	19.84	22.74	19.01	21.85	21.75
SSIM ↑	0.86	0.87	0.90	0.85	0.92	0.92	0.87	0.92	0.92
MSE ↓	0.28	0.26	0.21	0.28	0.24	0.19	0.19	0.15	0.16

Table 3 shows a comparison of the nearest neighbor, bicubic, and SRGAN with the proposed method on benchmark data. The results confirm that the proposed ST-SRGAN methods outperform all reference methods regarding to the evaluation metrics. The values in bold indicate the best-quality reconstructed images. However, it is worth noting that visual inspection remains the main method to perform assessment for SR methods. Finally, visual results of the reconstructed HR images with 4× up-scaling are depicted in Fig. 2. As it can be observed, from LR images (first column), our method produces high-quality images (last column), which are a reliable and accurate approximation of the original image (middle column). Furthermore, our method overcomes both the limitations of blurring/noising and spatial translations and rotations.

Table 3. Comparison of NN, bicubic, and SRGAN [11] with 4× up-scaling.

	Config 1				*Config 2*			
	NN	Bicubic	SRGAN [11]	ST-SRGAN	NN	Bicubic	SRGAN [11]	ST-SRGAN
PSNR↑	26.26	28.43	29.40	**29.46**	17.14	19.01	20.80	**20.84**
SSIM ↑	0.76	0.82	0.85	**0.94**	0.67	0.72	0.90	**0.90**

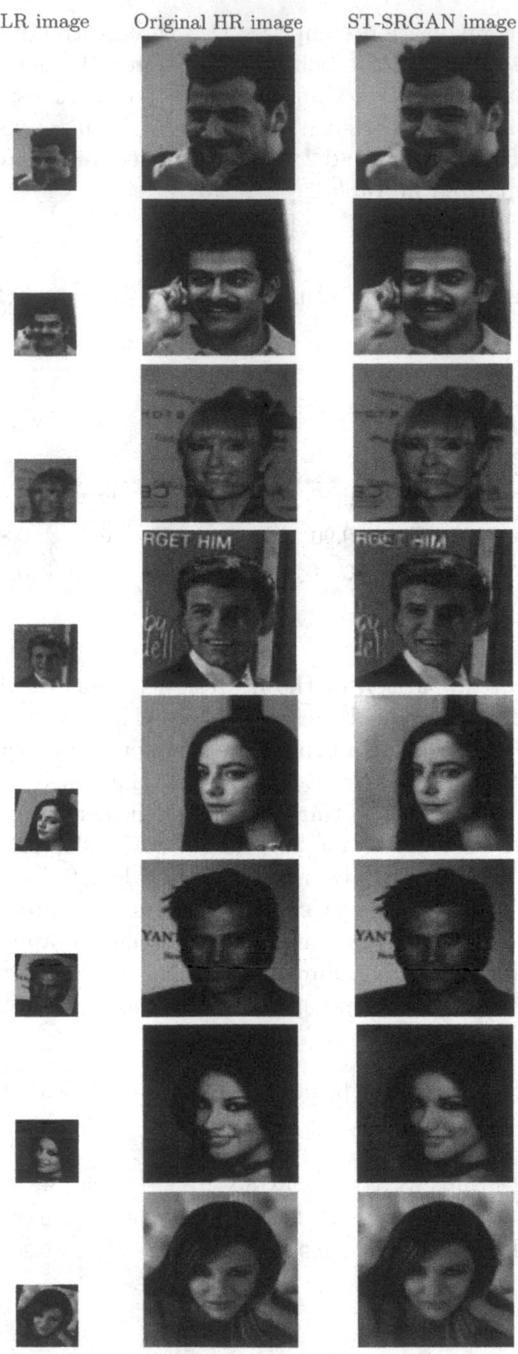

Fig. 2. Reconstructed HR images with 4× up-scaling.

5 Conclusion

We introduced a novel spatial transformer GAN network to solve the problem of image super-resolution. In this work, we aimed at showing the robustness of pairing a SR network with the spatial transformer network for estimating transformation parameters between LR images. CNN-based SR algorithms, such as the SRGAN can be massively improved when paired with spatial transformers. The spatial transformations applied to three publicly available image datasets were successfully learned by the network and this is further evaluated by performance metrics. In our experiments, 64×64 distorted face images were up-sampled in various degrees of up-scaling (e.g., 2×, 3×, and 4×). Moreover, the proposed method is not limited to its use in face SR but other image datasets. The robustness of the proposed model in spatial invariance is the reason behind its supremacy compared to baseline and state-of-the-art methods in super-resolution.

Furthermore, the cost of adding a spatial transformer model to our network is negligible. There are almost no extra computational costs in time and the size of the information required to process their trainable variables. The spatial transformer module has proved to be very powerful and very useful and its total potential is yet to be realised. In future work, we intend to extend the model for video super-resolution and exploit the spatiotemporal information found in between concurrent frames.

Acknowledgments. The authors gratefully acknowledge the support of NVIDIA Corporation with the donation of the Titan XP GPU used for this research. We acknowledge support of this work by the projects "Dioni: Computing Infrastructure for Big-Data Processing and Analysis" (MIS No. 5047222), which is implemented under the Action "Reinforcement of the Research and Innovation Infrastructure", and "Bessarion" (T6YB -00214), which is implemented under the call "Open Innovation in Culture", both funded by the Operational Program "Competitiveness, Entrepreneurship and Innovation" (NSRF 2014–2020) and co-financed by Greece and the European Union (European Regional Development Fund). All statements of fact, opinion or conclusions contained herein are those of the authors and should not be construed as representing the official views or policies of the sponsors.

References

1. Aly, H.A., Dubois, E.: Image up-sampling using total-variation regularization with a new observation model. IEEE Trans. Image Process. **14**(10), 1647–1659 (2005)
2. Bevilacqua, M., Roumy, A., Guillemot, C., Morel, M.L.A.: Low-complexity single-image super-resolution based on nonnegative neighbor embedding. In: Proceedings of the British Machine Vision Conference, No. 135. British Machine Vision Association (2012)
3. Dai, S., Han, M., Xu, W., Wu, Y., Gong, Y.: Soft edge smoothness prior for alpha channel super resolution. In: Proceedings of the IEEE/CVF Conference on Computer Vision and Pattern Recognition. IEEE, June 2007
4. Dong, C., Loy, C.C., He, K., Tang, X.: Image super-resolution using deep convolutional networks. IEEE Trans. Pattern Anal. Mach. Intell. **38**(2), 295–307 (2016)

5. Dong, C., Loy, C.C., Tang, X.: Accelerating the super-resolution convolutional neural network. In: Leibe, B., Matas, J., Sebe, N., Welling, M. (eds.) ECCV 2016. LNCS, vol. 9906, pp. 391–407. Springer, Cham (2016). https://doi.org/10.1007/978-3-319-46475-6_25
6. Freeman, W., Jones, T., Pasztor, E.: Example-based super-resolution. IEEE Comput. Graph. Appl. **22**(2), 56–65 (2002). https://doi.org/10.1109/38.988747
7. Gao, X., Zhang, K., Tao, D., Li, X.: Image super-resolution with sparse neighbor embedding. IEEE Trans. Image Process. **21**(7), 3194–3205 (2012)
8. Jaderberg, M., Simonyan, K., Zisserman, A., Kavukcuoglu, K.: Spatial transformer networks. arXiv, June 2015
9. Kim, J., Lee, J., Lee, K.: Accurate image super-resolution using very deep convolutional networks. In: Proceedings of the IEEE Conference on Computer Vision and Pattern Recognition, pp. 1646–1654. IEEE Computer Society, June 2016. https://doi.org/10.1109/CVPR.2016.182
10. Lai, W.S., Huang, J.B., Ahuja, N., Yang, M.H.: Deep Laplacian pyramid networks for fast and accurate super-resolution. In: Proceedings of the IEEE/CVF Conference on Computer Vision and Pattern Recognition, pp. 5835–5843 (2017). https://doi.org/10.1109/CVPR.2017.618
11. Ledig, C., et al.: Photo-realistic single image super-resolution using a generative adversarial network. arXiv abs/1609.04802 (2016). arxiv.org/abs/1609.04802
12. Liang, J., Zeng, H., Zhang, L.: Details or artifacts: a locally discriminative learning approach to realistic image super-resolution. In: Proceeding of the IEEE/CVF Conference on Computer Vision and Pattern Recognition, pp. 5657–5666, June 2022
13. Liu, Z., Luo, P., Wang, X., Tang, X.: Deep learning face attributes in the wild. arXiv abs/1411.7766 (2014). arxiv.org/abs/1411.7766
14. Lu, Z., Li, J., Liu, H., Huang, C., Zhang, L., Zeng, T.: Transformer for single image super-resolution. In: Proceedings of the IEEE/CVF Conference on Computer Vision and Pattern Recognition Workshops, pp. 457–466, June 2022
15. Maral, B.C.: Single image super-resolution methods: a survey. arXiv (2022). https://doi.org/1048550/ARXIV.2202.11763
16. Michaeli, T., Irani, M.: Nonparametric blind super-resolution. In: Proceedings of the IEEE International Conference on Computer Vision, pp. 945–952 (2013). https://doi.org/10.1109/ICCV.2013.121
17. Pesavento, M., Volino, M., Hilton, A.: Attention-based multi-reference learning for image super-resolution. In: Proceedings of the IEEE/CVF International Conference on Computer Vision, pp. 14697–14706, October 2021
18. Shan, Q., Li, Z., Jia, J., Tang, C.K.: Fast image/video upsampling. ACM Trans. Graph. **27**(5) (2008). https://doi.org/10.1145/1409060.1409106
19. Vrigkas, M., Nikou, C., Kondi, L.P.: On the improvement of image registration for high accuracy super-resolution. In: Proceedings of the IEEE International Conference on Acoustics, Speech and Signal Processing, Prague, Czech Republic, pp. 981–984, May 2011
20. Vrigkas, M., Nikou, C., Kondi, L.P.: A fully robust framework for map image super-resolution. In: Proceedings of the IEEE International Conference on Image Processing, Orlando, FL, pp. 2225–2228, September 2012
21. Wang, Y., Zhao, L., Liu, L., Hu, H., Tao, W.: URNet: a U-shaped residual network for lightweight image super-resolution. Remote Sensing (Basel) **13**(19), 3848 (2021)
22. Wang, Z., Bovik, A., Sheikh, H., Simoncelli, E.: Image quality assessment: from error visibility to structural similarity. IEEE Trans. Image Process. **13**(4), 600–612 (2004). https://doi.org/10.1109/TIP.2003.819861

23. Zeyde, R., Elad, M., Protter, M.: On single image scale-up using sparse-representations. In: Boissonnat, J.-D., et al. (eds.) Curves and Surfaces 2010. LNCS, vol. 6920, pp. 711–730. Springer, Heidelberg (2012). https://doi.org/10.1007/978-3-642-27413-8_47
24. Zhang, K., Gao, X., Tao, D., Li, X.: Single image super-resolution with non-local means and steering kernel regression. IEEE Trans. Image Process. **21**(11), 4544–4556 (2012)

Real-Time GAN-Based Model for Underwater Image Enhancement

Danilo Avola[1(✉)], Irene Cannistraci[1], Marco Cascio[1,2], Luigi Cinque[1], Anxhelo Diko[1], Damiano Distante[2], Gian Luca Foresti[3], Alessio Mecca[1], and Ivan Scagnetto[3]

[1] Department of Computer Science, Sapienza University of Rome, Via Salaria 113, 00198 Rome, Italy
{avola,cannistraci,cascio,cinque,diko,mecca}@di.uniroma1.it

[2] Department of Law and Economics, University of Rome UnitelmaSapienza, Piazza Sassari 4, 00161 Rome, Italy
{marco.cascio,damiano.distante}@unitelmasapienza.it

[3] Department of Mathematics, Computer Science and Physics, University of Udine, Via delle Scienze 206, 33100 Udine, Italy
{gianluca.foresti,ivan.scagnetto}@uniud.it

Abstract. Enhancing image quality is crucial for achieving an accurate and reliable image analysis in vision-based automated tasks. Underwater imaging encounters several challenges that can negatively impact image quality, including limited visibility, color distortion, contrast sensitivity issues, and blurriness. Among these, depending on how the water filters out the different light colors at different depths, the color distortion results in a loss of color information and a blue or green tint to the overall image, making it difficult to identify different underwater organisms or structures accurately. Improved underwater image quality can be crucial in marine biology, oceanography, and oceanic exploration. Therefore, this paper proposes a novel Generative Adversarial Network (GAN) architecture for underwater image enhancement, restoring good perceptual quality to obtain a more precise and detailed image. The effectiveness of the proposed method is evaluated on the EUVP dataset, which comprises underwater image samples of various visibility conditions, achieving remarkable results. Moreover, the trained network is run on the RPi4B as an embedded system to measure the time required to enhance the images with limited computational resources, simulating a practical underwater investigation setting. The outcome demonstrates the presented method applicability in real-world underwater exploration scenarios.

Keywords: Underwater image enhancement · GAN · Underwater exploration

1 Introduction

In the last few years, visual information analyzing tools have become increasingly attractive for solving heterogeneous perception-based tasks such as video surveil-

G. L. Foresti et al. (Eds.): ICIAP 2023, LNCS 14233, pp. 412–423, 2023.
https://doi.org/10.1007/978-3-031-43148-7_35

lance [2,8,14,31], biomedical imaging [6,9,32], environmental modeling [1,19], gesture recognition [5,11], or human behavioral analysis [3,4,7,13,27]. Among the others, underwater imaging helps to develop new technologies and techniques to discover underwater environments. However, this process is still challenging due to limited visibility and light-related phenomena, i.e., absorption, scattering, and refraction [17]. Indeed, underwater image processing is characterized by degradation issues, including loss of color information and other optical artifacts in captured images leading to decreasing performance of visual-related tasks, e.g., segmentation, detection, or classification. Despite the promising results in the literature on terrestrial imagery, there are still plenty of chances to enhance the perceptual quality of underwater images. Therefore, quality enhancement and detailed restoration strategies are crucial for better understanding the underwater environment, identifying new species, and monitoring marine ecosystems. To this end, the literature proposes several physics- and learning-based strategies [10,12,15,28]. The former is appropriate for color correction and dehazing; however, other than being computationally expensive, it requires scene depth and assessing water quality based on the interaction of light with water, which is not always available in automated applications. Instead, the latter is appropriate to enhance the overall perceptual image quality from large data collection. This paper presents an underwater image enhancement strategy based on a novel GAN-based architecture to improve the visual perception of the given poor-quality image. Motivated by the good results in [18], the proposed model leverages U-Net [26] and PatchGAN [18] networks for learning the mapping between poor- and good-quality underwater images. The effectiveness of the proposed network architecture is evaluated on the Enhancing Underwater Visual Perception (EUVP) [18] dataset, an underwater image collection suitable for adversarial model training. Finally, to simulate and test the feasibility of the presented method in real-world underwater exploration scenarios, usually characterized by inspection robots with limited computational resources, the trained model is run on an embedded system suitable for the industrial sector. This simulation measures the time required to enhance the perceptual quality of underwater images, which is crucial for practical applications. In summary, the main contributions of this paper are as follows:

- Designing an innovative GAN model for underwater image quality enhancement, suitable for oceanic explorations and human-robot collaborative experiments;
- Achieving of State-Of-the-Art (SOTA) performance on the EUVP benchmark, a real-world underwater image dataset containing paired collections of poor- and good-quality images for supervised learning;
- Performing tests on an embedded system to measure the time required for image quality enhancement, simulating a generic and vision-based practical underwater investigation setting with limited computational resources.

2 Related Work

Underwater image enhancement is gaining ever-increasing interest in literature thanks to the wide range of applications in developing, exploring, and protecting ocean resources. According to the literature, proposed solutions can be classified into physics- and learning-based approaches.

2.1 Physics-Based Approaches

Traditional physics-based strategies comprise the definition of formal models, including mathematical or simplified underwater image formation models. In fact, [25,29] proposed methods relying on the Beer-Lambert law to recover better details of underwater images considering the light attenuation related to the material properties through which the light travels. Instead, the authors in [22] manipulate poor-quality image color channels to dehaze an image that is later combined with an enhanced version, obtained through a color correction algorithm, to achieve the final good-quality result. Also, [24] presents an underwater image dehazing method manipulating the RGB color space. First, the authors estimate the background light using the quad-tree subdivision iteration algorithm. Afterward, the RGB color space dimensionality prior is compressed to the UV color space by clustering the pixels into a hundred haze-lines and setting a haze-free boundary to compute the dehazed version of the image accurately. In [10], the authors propose a method considering the water type in the enhancement process. Given the low-quality image, they first estimate the veiling-light, and the color restoration strategy is performed for multiple water types having different properties. Finally, the best enhanced version is selected automatically based on the gray-world assumption. However, previous methods require prior knowledge that may not always provide a reliable solution relying on local and global color distribution. Therefore, [12] proposes an approach for terrestrial and underwater image quality restoration without any prior information. A multi-band decomposition solution extracts the base and detail layers for intensity and Laplacian modules involved in restoring the image.

2.2 Learning-Based Approaches

Most recently, Deep Learning (DL) techniques have been applied to underwater image enhancement thanks to their capability of automatically learning the mappings between two domains from large data collection. In literature, the GAN model proved effective in improving the perceptual quality of underwater images. In [28], the authors propose the Class-conditional Attention GAN (CA-GAN) for underwater image enhancement in which the class label guides the generation of the good-quality image version. Differently, in [15], a CycleGAN-based [33] architecture is used to create a paired dataset of underwater poor- and good-quality images through the style transfer property; therefore, a fully convolutional encoder-decoder is trained for underwater image enhancement on such synthesized data. Instead, in [20], the CycleGAN model is used as the backbone for a network architecture learning the cross-domain mapping between underwater and terrestrial images. Also, in [23], the authors exploit a different GAN model that uses terrestrial images and depth maps to learn the color correction of monocular underwater images in an unsupervised fashion. In [30] is proposed a stacked conditional GAN consisting of a haze detection sub-network and color correction sub-network. In detail, the former produces a hazing detection mask from the underwater image given as input, while the latter corrects the image color by exploiting the previously predicted mask. Instead, [16] proposes a multiscale dense GAN combining residual learning, dense concatenation, and multiscale operation to correct color casts and restore image details. In [18], the authors propose a fully-convolutional conditional GAN-based model with a multi-modal objective function to evaluate the perceptual quality of the given underwater image considering global content, color, local texture, and style information. Finally, in [21], multiple inputs and a GAN architecture are combined to solve color casts, low contrast, and haze-like issues. Specifically, the model generator component comprises main and auxiliary sub-networks. Initially, the main module extracts the features from raw underwater images, whereas the auxiliary component extracts the features from the fusion-based enhanced version obtained through SOTA methods. Afterward, the two sub-networks outputs are merged to decode the restored image.

3 Method

To enhance underwater images by increasing their perceptual quality, the GAN-based architecture depicted in Fig. 1 was designed to find the non-linear mapping between poor- and good-quality underwater images. The proposed model expands the traditional GAN network by leveraging a Convolutional Neural Network (CNN) architecture to handle visual data by extracting the low-dimensional feature representation of underwater images used to improve the input images visually. To this end, the network training follows a supervised fashion, allowing to map the underwater poor-quality image subspace to the ground truth good-quality image subspace. With such a supervised training paradigm, the underwater image structural information and scene details are maintained while the network learns to generate its restored version.

3.1 Underwater Image Enhancement

Recently, learning-based approaches have shown to be effective for underwater image enhancement achieving promising results. The GAN-based models can successfully approximate a mapping function of a given input data distribution to a target data distribution by generating fake samples as if they were drawn from the target distribution itself. Indeed, the proposed network relies on this property using a CNN-based GAN architecture comprising generator (G) and discriminator (D) components, where the former produces fake images

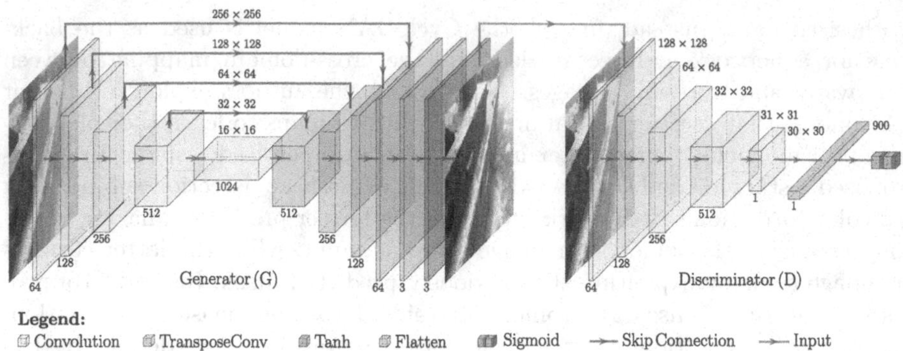

Fig. 1. The proposed GAN-based architecture. Given the poor-quality underwater image as input, the network is trained to restore good perceptual quality, obtaining a more precise and detailed image.

trying to learn the target distribution of the enhanced images while the latter discriminates against all real and fake enhanced images. Specifically, motivated by the success of skip connections in generative adversarial models for image-to-image mapping and quality enhancement, G is a customized version of the fully-convolutional U-Net architecture. In detail, the encoder part consists of five convolutional layers with 3×3 filter size and a stride of 2, each followed by batch normalization and Leaky Rectified Linear Unit (LeakyReLU) activation function. Alongside, the decoder component follows a reverse structure of the encoder with transposed convolutional layers rather than convolutions and Rectified Linear Unit (ReLU) instead of LeakyReLU activation functions to stabilize the training process. Finally, after the last transposed convolution, the decoder uses the hyperbolic tangent function to reconstruct the restored images. Precisely, given as input the RGB underwater image, with size $256 \times 256 \times 3$, the encoding part of the network learns feature maps of shape $16 \times 16 \times 1024$, representing its low-dimensional feature representation. Afterward, the decoder utilizes these latent features to generate an enhanced version of the image with the final shape $256 \times 256 \times 3$. In addition, to enable G for accurately reproducing features such as local texture and style, the discriminator is a PatchGAN-based network discriminating on patch-level information. Specifically, D comprises five padded convolutional layers with 4×4 filters and a stride of 2, except for the last layer set to 1, followed by batch normalization and LeakyReLU activation. Given the RGB image as input, the discriminator extracts feature maps of size 30×30 on which a linear layer and the sigmoid function are applied for discriminating between real or fake good-quality images. Following this adversarial training strategy, the generator G learns the significant low-dimensional features to maintain from underwater data; indeed, when D classifies both original and generated enhanced images as reals, it implies that the low-dimensional representation is very informative to the point of fooling the discriminator with the restored image. Formally, the underwater-to-enhance image mapping leverages the min-max game between G and D models, defined as:

$$L_{GAN}(D,G) = \min_{G}\max_{D}(E_Y[logD(Y)] + E_{X,Y}[log(1-(D(G(X))))]), \quad (1)$$

where X represents the underwater images as input, Y the ground truth good-quality images, and $G(X)$ the restored image from the input, $D(Y)$ and $D(G(X))$ indicate the estimated probabilities of given good-quality and enhanced images being real. Since the L_{GAN} is designed for learning to approximate the target distribution, it does not enforce the generator G to maintain important aspects like global content, color, style, and local structures. To this end, in order for G to generate an enhanced version of the underwater input image that is consistent with the corresponding good-quality ground truth, the Mean Absolute Error (MAE) and Mean Standard Error (MSE) between the restored and original ground truth images are defined as:

$$MAE_G = \frac{1}{(W \times H)} \sum_{i=1}^{W} \sum_{j=1}^{H} [X(i,j) - Y(i,j)], \quad (2)$$

$$MSE_G = \frac{1}{(W \times H)} \sum_{i=1}^{W} \sum_{j=1}^{H} [X(i,j) - Y(i,j)]^2, \quad (3)$$

where $X(i,j)$ and $Y(i,j)$ indicate a pixel in the enhanced and original good-quality images, respectively. Thus, the training objective of the overall network is to minimize the loss function, defined as follows:

$$L(D,G) = w_{GAN}L_{GAN}(D,G) + w_{MAE}MAE_G + w_{MSE}MSE_G. \quad (4)$$

where w_{GAN}, w_{MAE}, and w_{MSE} are weighting parameters adjusting the impact of individual losses.

4 Experiments

This section presents a comprehensive evaluation of the proposed GAN-based model on the EUVP dataset for the underwater image enhancement task. Specifically, it describes significant implementation details and quantitative and qualitative evaluations. Moreover, it also reports the image enhancement time of the trained network running on the Raspberry Pi 4 model B (RPi4B) as an embedded system, simulating a generic practical underwater investigation setting with limited computational resources.

4.1 EUVP Dataset

The EUVP dataset [18] is the publicly available benchmark focused on underwater image enhancement that contains poor and good perceptual quality underwater image samples suitable for models trained in a supervised fashion. Specifically, it contains 8670 images with a size of 256 × 256 comprising 3700 training

Fig. 2. EUVP dataset paired image samples. In the first row GT good-quality images; in the second row the corresponding underwater poor-quality images.

pairs and 1270 validation samples. Data augmentation is applied to training data, including horizontal flip, vertical flip, and random crop, increasing to 10882 the training pairs. For paired data, the ground truth good-quality images were generated by using the method proposed in [15]. Figure 2 illustrates some paired training image samples.

4.2 Implementation Details

The proposed architecture design has been developed using the PyTorch framework. The experiments were performed on two GPUs, i.e., ×2 NVIDIA GeForce RTX 2080 Ti with 11 GB of RAM. In detail, the network was trained following a supervised setting for 50 epochs by exploiting the original EUVP data split and training protocol. For the model training, Adam was used as the optimizer with a learning rate set to 0.0002, an ϵ parameter of 1e−8, a weight decay set to 1e−2, a first β_1 and second β_2 momentum initial decay rate of 0.5 and 0.999, respectively. Finally, the weight parameters adjusting the loss functions within Eq. (4) were set to $w_{GAN} = w_{MSE} = 1$ and $w_{MAE} = 10$. For the proposed method evaluation, the standard metrics Peak Signal to Noise Ratio (PSNR) and Structural Similarity Index Measure (SSIM) were used to compare the enhanced images with the corresponding ground truth quantitatively. Precisely, the PSNR estimates the reconstruction quality of the restored image, while the SSIM compares the two images considering three properties: luminance, contrast, and structure.

4.3 Underwater Image Enhancement Evaluation

Regarding the underwater image enhancement, the quantitative results are reported measuring the PSNR and the SSIM metrics between the proposed model enhanced image version and the available respective ground truth. With the aim to approximate the image reconstruction accuracy of the restored image, the PSNR is defined as follows:

$$PSNR(X,Y) = 10\log_{10}[255^2/MSE(X,Y)], \tag{5}$$

Table 1. Performance evaluation on the EUVP dataset and comparison with key literature physics- and learning-based methods.

Model	Type	$PSNR\uparrow$	$SSIM\uparrow$
Uw-HL [10]	Physics-based	18.85 ± 1.76	0.7722 ± 0.066
Mband-En [12]	Physics-based	12.11 ± 2.55	0.4565 ± 0.097
Res-GAN [18]	Learning-based	14.75 ± 2.22	0.4685 ± 0.122
Res-WGAN [18]	Learning-based	16.46 ± 1.80	0.5762 ± 0.014
LS-GAN [18]	Learning-based	17.83 ± 2.88	0.6725 ± 0.062
Pix2Pix [18]	Learning-based	20.27 ± 2.66	0.7081 ± 0.069
UGAN-P [15]	Learning-based	19.59 ± 2.54	0.6685 ± 0.075
CycleGAN [18]	Learning-based	17.14 ± 2.17	0.6400 ± 0.080
FUnIE-GAN-UP [18]	Learning-based	21.36 ± 2.17	0.8164 ± 0.046
FUnIE-GAN [18]	Learning-based	21.92 ± 1.07	0.8876 ± 0.068
Ours	**Learning-based**	$\mathbf{22.09 \pm 1.02}$	$\mathbf{0.9002 \pm 0.059}$

where X and Y are the enhanced and ground truth images, respectively. Instead, to measure the perceptual restored image quality, the SSIM is defined as:

$$SSIM(X,Y) = \left(\frac{2\mu_X\mu_Y + c_1}{\mu_X^2 + \mu_Y^2 + c_1}\right)\left(\frac{2\sigma_{XY} + c_2}{\sigma_X^2 + \sigma_Y^2 + c_2}\right) \tag{6}$$

where X and Y are always the enhanced and ground truth images being compared, μ_X and μ_Y are the pixel sample means of X and Y, respectively, σ_X and σ_Y are the standard deviations, σ_{XY} is the covariance of x and y; finally, c_1 and c_2 are constants used to stabilize the division when the denominator is close to zero. Table 1 reports the obtained results on test images and summarizes comparisons with key literature physics- and learning-based works. As can be observed, the learning-based methods generally perform significantly better than physics-based approaches. However, the proposed GAN-based network achieves remarkable and increased performances in reconstruction quality, i.e., PSNR, and restoring perceptual quality, i.e., SSIM. Regarding the qualitative evaluation, the obtained results are depicted in Fig. 3. As can be observed, the presented approach generates enhanced underwater images, restoring optical quality and successfully removing the typical blue or green tint in underwater images. Notice that, in some cases, the enhanced color in the restored images generated by the proposed method is closer to the true color rather than in the ground truths.

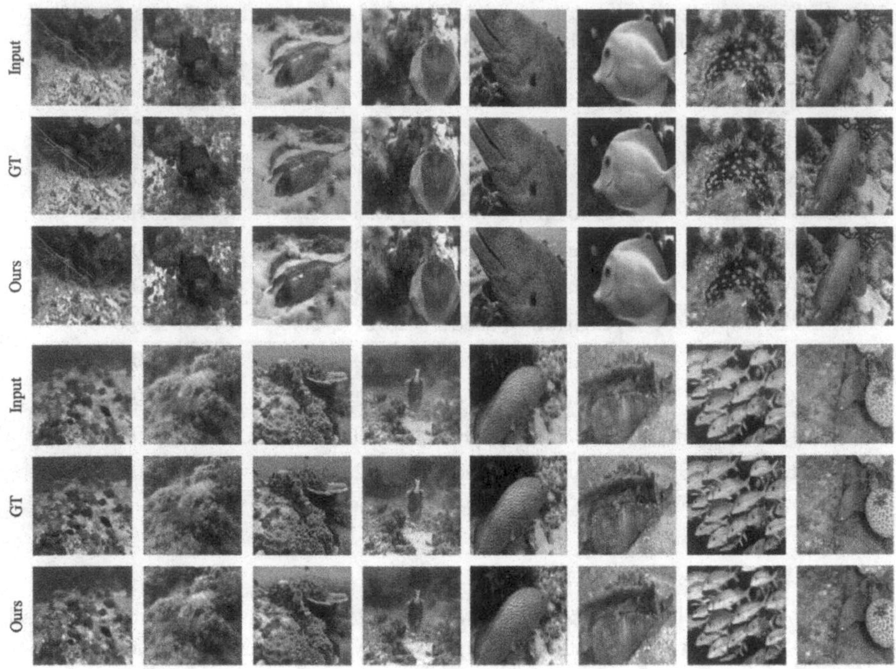

Fig. 3. Qualitative results for underwater image enhancement. In the first and fourth rows, the input underwater images; in the second and fifth rows, the corresponding good-quality GT images; in the third and sixth rows, the proposed network enhanced images.

4.4 Embedded System Evaluation for Practical Underwater Exploration Simulation

Concluding the proposed method evaluation, the trained model is run on commodity hardware with limited computational resources to simulate a practical underwater investigation setting. A generic vision-based oceanic exploration process is characterized by a visual sensor mounted on the automated underwater vehicle capturing poor-quality images. Therefore, an embedded system (e.g., RPi4B) can be used to enhance the captured image quality for underwater image processing tasks in real-time, e.g., during inspections of pipelines, dams, and offshore platforms, to enable safer and more efficient maintenance and repair operations in the industrial sector. In many applications, time plays a crucial role. The proposed neural model is tested on an RPi4B, a system based on a Quad-core Cortex-A72 (ARM v8) 64-bit SoC @1.5 GHz processor with 2 GB of RAM board, to simulate the described investigation setting. Since we use trained weights for the model, image enhancement performance does not change with respect to the reported in Sect. 4.3. Therefore, the critical factor is examining the time required to obtain a good-quality image. In detail, given a 256×256 image as input, the prediction time for a single image enhancement is about 0.029 s, thus handling roughly 34FPS in real-time. This result is also confirmed by evaluating the entire test set. Therefore, the model can be executed on commodity hardware at a considerable amount of FPS, thus demonstrating its applicability in real-world scenarios.

5 Conclusions

This paper presents a novel GAN-based architecture for underwater image enhancement to obtain a precise and detailed scene, restoring good perceptual quality. By learning the low-dimensional feature representation of underwater images, the proposed network maintains the structural information and details while discovering the mapping between poor- and good-quality images following a supervised training fashion. More importantly, the proposed method achieved state-of-the-art performance on a real-world underwater image dataset containing paired collections of poor- and good-quality images and demonstrated real-time capabilities on commodity hardware for up to 34FPS.

Acknowledgements. This work was supported by "Smart unmannEd AeRial vehiCles for Human likE monitoRing (SEARCHER)" project of the Italian Ministry of Defence (CIG: Z84333EA0D); and "A Brain Computer Interface (BCI) based System for Transferring Human Emotions inside Unmanned Aerial Vehicles (UAVs)" Sapienza Research Projects (Protocol number: RM1221816C1CF63B); and "DRrone Aerial imaGe SegmentatiOn System (DRAGONS)" (CIG: Z71379B4EA); and Departmental Strategic Plan (DSP) of the University of Udine - Interdepartmental Project on Artificial Intelligence (2020–25); and "An Integrated Platform For Autonomous Agents For Maritime Situational Awareness (ARGOS)" project of the Italian Ministry of Defence (PNRM 2022); and the MICS (Made in Italy - Circular and Sustainable) Extended Partnership and received funding from Next-Generation EU (Italian PNRR - M4 C2, Invest 1.3 - D.D. 1551.11-10-2022, PE00000004). CUP MICS B53C22004130001.

References

1. Avola, D., Bernardi, M., Cinque, L., Foresti, G.L., Massaroni, C.: Adaptive bootstrapping management by keypoint clustering for background initialization. Pattern Recognit. Lett. **100**, 110–116 (2017). https://doi.org/10.1016/j.patrec.2017.10.029
2. Avola, D., et al.: A novel GAN-based anomaly detection and localization method for aerial video surveillance at low altitude. Remote Sens. **14**(16), 4110 (2022). https://doi.org/10.3390/rs14164110
3. Avola, D., Cascio, M., Cinque, L., Fagioli, A., Foresti, G.L.: Affective action and interaction recognition by multi-view representation learning from handcrafted low-level skeleton features. Int. J. Neural Syst. 2250040 (2022). https://doi.org/10.1142/s012906572250040x
4. Avola, D., Cinque, L., De Marsico, M., Fagioli, A., Foresti, G.L.: LieToMe: preliminary study on hand gestures for deception detection via fisher-LSTM. Pattern Recognit. Lett. **138**, 455–461 (2020). https://doi.org/10.1016/j.patrec.2020.08.014

5. Avola, D., Cinque, L., Fagioli, A., Foresti, G.L., Fragomeni, A., Pannone, D.: 3d hand pose and shape estimation from RGB images for keypoint-based hand gesture recognition. Pattern Recognit. **129**, 108762 (2022). https://doi.org/10.1016/j.patrec.2017.10.029
6. Avola, D., Cinque, L., Fagioli, A., Foresti, G., Mecca, A.: Ultrasound medical imaging techniques: a survey. ACM Comput. Surv. **54**(3), 1–38 (2021). https://doi.org/10.1145/3447243
7. Avola, D., Cinque, L., Foresti, G.L., Pannone, D.: Automatic deception detection in RGB videos using facial action units. In: International Conference on Distributed Smart Cameras, pp. 1–6 (2019). https://doi.org/10.1145/3349801.3349806
8. Avola, D., Foresti, G.L., Martinel, N., Micheloni, C., Pannone, D., Piciarelli, C.: Real-time incremental and geo-referenced mosaicking by small-scale UAVs. In: Battiato, S., Gallo, G., Schettini, R., Stanco, F. (eds.) ICIAP 2017. LNCS, vol. 10484, pp. 694–705. Springer, Cham (2017). https://doi.org/10.1007/978-3-319-68560-1_62
9. Avola, D., Petracca, A., Placidi, G.: Design of a framework for personalised 3d modelling from medical images. Comput. Methods Biomech. Biomed. Eng. Imaging Vis. **3**(2), 76–83 (2015). https://doi.org/10.1080/21681163.2013.853622
10. Berman, D., Levy, D., Avidan, S., Treibitz, T.: Underwater single image color restoration using haze-lines and a new quantitative dataset. IEEE Trans. Pattern Anal. Mach. Intell. **43**(8), 2822–2837 (2021). https://doi.org/10.1109/TPAMI.2020.2977624
11. Budzan, S., et al.: Using gesture recognition for AGV control: preliminary research. Sensors **23**(6), 3109 (2023). https://doi.org/10.3390/s23063109
12. Cho, Y., Jeong, J., Kim, A.: Model-assisted multiband fusion for single image enhancement and applications to robot vision. IEEE Robot. Autom. Lett. **3**(4), 2822–2829 (2018). https://doi.org/10.1109/LRA.2018.2843127
13. Dzedzickis, A., Kaklauskas, A., Bucinskas, V.: Human emotion recognition: review of sensors and methods. Sensors **20**(3), 592 (2020). https://doi.org/10.3390/s20030592
14. Elhoseny, M.: Multi-object detection and tracking (MODT) machine learning model for real-time video surveillance systems. Circuits Syst. Signal Process. **39**, 611–630 (2020)
15. Fabbri, C., Islam, M.J., Sattar, J.: Enhancing underwater imagery using generative adversarial networks. In: IEEE International Conference on Robotics and Automation, pp. 7159–7165 (2018). https://doi.org/10.1109/ICRA.2018.8460552
16. Guo, Y., Li, H., Zhuang, P.: Underwater image enhancement using a multiscale dense generative adversarial network. IEEE J. Ocean. Eng. **45**(3), 862–870 (2020). https://doi.org/10.1109/JOE.2019.2911447
17. Islam, M.J., Ho, M., Sattar, J.: Understanding human motion and gestures for underwater human-robot collaboration. J. Field Robot. **36**(5), 851–873 (2019). https://doi.org/10.1002/rob.21837
18. Islam, M.J., Xia, Y., Sattar, J.: Fast underwater image enhancement for improved visual perception. IEEE Robot. Autom. Lett. **5**(2), 3227–3234 (2020). https://doi.org/10.1109/LRA.2020.2974710
19. Kang, Z., Yang, J., Yang, Z., Cheng, S.: A review of techniques for 3d reconstruction of indoor environments. ISPRS Int. J. Geo-Inf. **9**(5), 330 (2020). https://doi.org/10.3390/ijgi9050330
20. Li, C., Guo, J., Guo, C.: Emerging from water: underwater image color correction based on weakly supervised color transfer. IEEE Signal Process. Lett. **25**(3), 323–327 (2018). https://doi.org/10.1109/LSP.2018.2792050

21. Li, H., Zhuang, P.: DewaterNet: a fusion adversarial real underwater image enhancement network. Signal Process. Image Commun. **95**, 116248 (2021). https://doi.org/10.1016/j.image.2021.116248
22. Li, H., Zhuang, P., Wei, W., Li, J.: Underwater image enhancement based on dehazing and color correction. In: IEEE International Conference on Parallel Distributed Processing with Applications, Big Data Cloud Computing, Sustainable Computing Communications, Social Computer and Networking, pp. 1365–1370 (2019). https://doi.org/10.1109/ISPA-BDCloud-SustainCom-SocialCom48970.2019.00196
23. Li, J., Skinner, K.A., Eustice, R.M., Johnson-Roberson, M.: WaterGAN: unsupervised generative network to enable real-time color correction of monocular underwater images. IEEE Robot. Autom. Lett. **3**(1), 387–394 (2018). https://doi.org/10.1109/LRA.2017.2730363
24. Liu, Y., Rong, S., Cao, X., Li, T., He, B.: Underwater single image dehazing using the color space dimensionality reduction prior. IEEE Access **8**, 91116–91128 (2020). https://doi.org/10.1109/ACCESS.2020.2994614
25. Petit, F., Capelle-Laize, A.S., Carre, P.: Underwater image enhancement by attenuation inversion with quaternions. In: IEEE International Conference on Acoustic Speech Signal Process, pp. 1177–1180 (2009). https://doi.org/10.1109/ICASSP.2009.4959799
26. Ronneberger, O., Fischer, P., Brox, T.: U-Net: convolutional networks for biomedical image segmentation. In: Navab, N., Hornegger, J., Wells, W.M., Frangi, A.F. (eds.) MICCAI 2015. LNCS, vol. 9351, pp. 234–241. Springer, Cham (2015). https://doi.org/10.1007/978-3-319-24574-4_28
27. Sharma, P., et al.: Student engagement detection using emotion analysis, eye tracking and head movement with machine learning. In: Reis, A., Barroso, J., Martins, P., Jimoyiannis, A., Huang, R.YM., Henriques, R. (eds.) TECH-EDU 2022. CCIS, vol. 1720, pp. 52–68. Springer, Cham (2023). https://doi.org/10.1007/9783031229183_5
28. Wang, J., et al.: CA-GAN: class-condition attention GAN for underwater image enhancement. IEEE Access **8**, 130719–130728 (2020). https://doi.org/10.1109/ACCESS.2020.3003351
29. Xiong, J., Zhuang, P., Zhang, Y.: An efficient underwater image enhancement model with extensive Beer-Lambert law. In: IEEE International Conference on Image Processing (ICIP), pp. 893–897 (2020). https://doi.org/10.1109/ICIP40778.2020.9191131
30. Ye, X., Xu, H., Ji, X., Xu, R.: Underwater image enhancement using stacked generative adversarial networks. In: Pacific Rim Conference on Multimedia (PCM), pp. 514–524 (2018). https://doi.org/10.1007/9783030007645_47
31. Zhou, J.T., Du, J., Zhu, H., Peng, X., Liu, Y., Goh, R.S.M.: AnomalyNet: an anomaly detection network for video surveillance. IEEE Trans. Inf. Forensics Secur. **14**(10), 2537–2550 (2019). https://doi.org/10.1109/TIFS.2019.2900907
32. Zhou, S.K., et al.: A review of deep learning in medical imaging: imaging traits, technology trends, case studies with progress highlights, and future promises. Proc. IEEE **109**(5), 820–838 (2021). https://doi.org/10.1109/JPROC.2021.3054390
33. Zhu, J.Y., Park, T., Isola, P., Efros, A.A.: Unpaired image-to-image translation using cycle-consistent adversarial networks. In: IEEE International Conference on Computer Vision (ICCV), pp. 2223–2232 (2017)

HERO: A Multi-modal Approach on Mobile Devices for Visual-Aware Conversational Assistance in Industrial Domains

Claudia Bonanno[1(✉)], Francesco Ragusa[1,2], Antonino Furnari[1,2], and Giovanni Maria Farinella[1,2,3]

[1] FPV@IPLAB, DMI, University of Catania, Catania, Italy
claudia.bonanno@phd.unict.it,
{francesco.ragusa,antonino.furnari,giovanni.farinella}@unict.it
[2] Next Vision s.r.l. - Spinoff of the University of Catania, Catania, Italy
[3] Cognitive Robotics and Social Sensing Laboratory, ICAR-CNR, Palermo, Italy

Abstract. We present HERO, an artificial assistant designed to communicate with users with both natural language and images to aid them carrying out procedures in industrial contexts. Our system is composed of five modules: 1) the input module retrieves user utterances and collects raw data, such as text and images, 2) the Natural Language Processing module processes text from user utterances, 3) the object detector module extracts entities by analyzing images captured by the user, 4) the Question Answering module generates responses to users' specific questions on procedures, and 5) the output module selects the final response to give to the user. We deployed and evaluated the system in an industrial laboratory furnished with different tools and equipment for carrying out repair and test operations on electrical boards. In this setting, the HERO system allows the user to retrieve information on tools, equipment, procedures, and safety rules. Experiments on domain-specific labeled data, as well as a user study suggest that the design of our system is robust and that its use can be beneficial for users over classic methods for retrieving information and guide workers, such as printed manuals.

Keywords: First Person Vision · Conversational Agents

1 Introduction

In recent years, there has been a significant increase of interest in developing virtual assistants capable of communicating with humans using natural language, both in the research community and in industry. Artificial intelligent assistants like Alexa, Siri, and Google Assistant are becoming increasingly popular in everyday life as they can assist users in simple tasks such as setting

---

Supplementary Information The online version contains supplementary material available at https://doi.org/10.1007/978-3-031-43148-7_36.

G. L. Foresti et al. (Eds.): ICIAP 2023, LNCS 14233, pp. 424–436, 2023.
https://doi.org/10.1007/978-3-031-43148-7_36

and reminding calendar appointments, controlling and automating smart home devices, playing music or audiobooks, and communicating with humans through natural language. Also the recently well-known ChatGPT conversational agent can be used as an assistant, as shown in its integration into Bing's search engine. The exploitation of such assistants can bring several quality of life improvements, helping users to access information and solve problems more easily. In fact, dialogue-based assistants are already used in customer care to support users in solving specific tasks such as making bookings and reservations, knowing latest expenses, and receiving updates about latest orders and product recommendations.

While the use of these assistants brings many advantages, current systems are not yet able to process different kinds of information besides natural language in the form of audio or text. Indeed, an assistant capable of communicating with users via natural language and analyzing images and videos acquired from the user's point of view with wearable and mobile devices such as smart glasses and smartphones, could be useful in several contexts including daily life (supporting people inside their kitchens or while doing housework) or work scenarios (assisting an operator working on a maintenance procedure). In industrial contexts, in particular, an assistant able to see and understand in real-time what the user is doing could warn them when they're doing something wrong by saying: "Stop! You're doing *this* wrong. Do *this* instead.". Exploiting images or videos collected from the user's point of view allows to gain more contextual information and provide useful answers. With this kind of support, an industrial operator may work even in a little known context with ease and retrieve specific instructions or details about an object or a procedure just by asking questions like "What is the object I am looking at? How do I use that?" or "What am I supposed to do now?". While recent efforts have focused on developing assistants that can analyze a user's surroundings and provide human-like assistance, such approaches have been primarily explored in consumer contexts (e.g., Facebook's SIMMC challenges [11], and Aye-saac [16]), rather than in industrial environments.

This paper presents HERO, an assistive visual dialogue agent that interacts with users in a conversational fashion within an industrial environment (see Fig. 1). The system is composed of five modules: 1) the input module retrieves user utterances and extracts raw data, such as text and images, 2) the Natural Language Processing (NLP) module processes text from user utterances, 3) the object detector module extracts entities by analyzing images captured by the user, 4) the Question Answering (QA) module generates responses about users' specific questions on procedures, and 5) the output module selects the final response to be given to the user. The developed agent is capable of understanding natural utterances made by a user, keeping track of the state of the conversation and giving useful answers. The system uses both textual cues obtained from the utterance and visual cues obtained from the images captured from the user's point of view, which allows to make the interaction with the user more natural as compared to those of systems relying solely on natural language. Furthermore, the aforementioned vision capabilities enable users to ask questions about objects within an industrial laboratory, regardless of whether they know the name of the

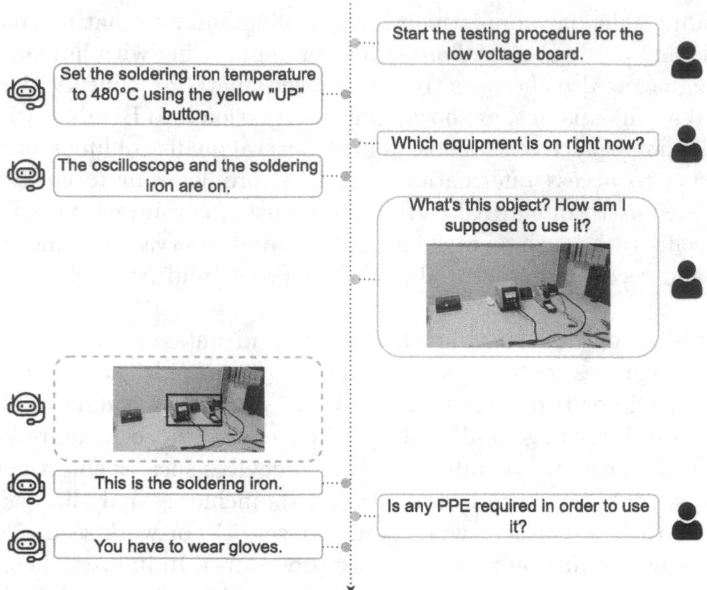

Fig. 1. Concept of the proposed assistant, HERO. HERO can support users to achieve their goals in the considered industrial setting. The user (right) can ask for instructions on a given procedure, retrieve information on the state of specific objects ("Which equipment is on"), ask visually-grounded questions ("What's this object?"), and obtain general information about safety instructions ("Is any PPE required?"). HERO replies using natural language (left), while an object detection module is used to answer visually-grounded questions (e.g., by recognizing the soldering iron).

object or not. For example, users can ask questions about the objects present in the environment ("How am I supposed to use this object?"), the laboratory itself ("What equipment does this laboratory provide?") or the procedures that can be carried out ("I'm in front of panel A. What should I do next?").

We measured the performance of the NLP and Object Detection modules and validated the whole system with a user study based on a group of 11 participants, who were asked to perform an industrial procedure supported by HERO and one by a standard paper-based manual. After the trial, subjects were asked to fill out a questionnaire in order to evaluate different aspects of the system. Results show that users generally prefer interacting with HERO over consulting a printed manual when seeking information about objects, procedures, and the laboratory itself. However, the study also suggests that there is still room for improvement in the system's performance and user experience.

The main contribution of our work lies in integrating vision capabilities in a classic conversational assistance paradigm, in a way that can be applied in several industrial use cases. This gives our system the ability to provide a conversational interface specifically designed for industrial contexts that incorporates visual intelligence capabilities. Users can submit images along with their queries to facilitate more natural and intuitive communication. Even people not familiar with a laboratory environment could complete the unknown procedures with ease. This approach not only makes it easier for users to articulate their questions and obtain relevant information, but also provides a more efficient and accurate mean of communication within industrial laboratory settings.

2 Related Work

2.1 Dialogue State Tracking

Dialogue State Tracking (DST) is the task of keeping track of the current belief state of a dialogue, understanding the context, the goals of the user, and the current state of the system given the entire previous conversation. The authors of [19] perform dialogue state tracking using natural language descriptions for intents and slots, whereas the authors of [6] approach DST as a reading comprehension task to answer the question "What is the state of the current dialog?". The authors of [10] tackle DST by generating values for slots using an encoder-decoder architecture. Lastly, the authors of [7,9] work on a fixed ontology with open vocabulary prediction.

Recently, Dialogue State Tracking systems have been extended to exploit multiple signals simultaneously (Multimodal Dialogue State Tracking), such as text, speech, images, videos, and gestures. Different works exploited additional signals such as images [3] and 3D models of the objects present in the environment [11] to perform Multimodal Dialogue State Tracking (MM-DST) in the domain of retail. Furthermore, the authors of [13] approached the task on the CLEVR diagnostic dataset, exploiting videos instead of images.

Our system integrates DST capabilities into a classic Rasa chatbot architecture [2] by incorporating custom actions. To tackle MM-DST, we include an object detection module in the architecture that exploits visual cues.

2.2 Visual Question Answering

Visual Question Answering (VQA) is the task of generating a natural language answer given an image and a question about the image. This task requires to process both visual and textual information and reason about their relationship to produce accurate answers. VQA has numerous applications in areas such as image captioning, automated image retrieval, and interactive systems. Earlier approaches to this task were presented in [1,8].

Dialogue context can be used as a useful cue for VQA. Indeed, the model must understand both the image and the dialogue history in order to generate a response to a question asked by the user. Several works explored the visual dialogue task [4]. The use of reinforcement learning has been explored in [5,15], while in [12] visual coreference resolution is used. Recently, a wide variety of language models performing vision-and-language connections [14,18] have been designed to solve several downstream tasks, including VQA.

Our system offers similar functionalities to VQA, answering questions on a given set of objects that can be found in the target context in which the system

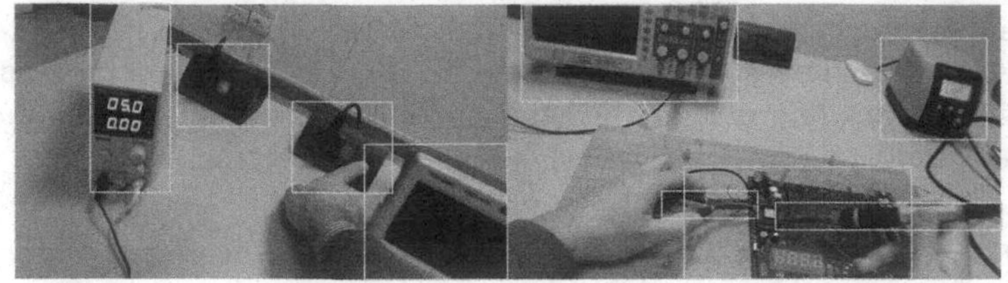

Fig. 2. Sample images of the industrial laboratory with bounding box annotations around the objects present in the scene.

is deployed, allowing users to leverage multimodal signals into conversations, such as images.

3 Industrial Laboratory

We tested HERO in an industrial laboratory including tools and equipment to carry out test and repair operations on electrical boards. Specifically, the laboratory contains 23 objects such as electrical boards, an oscilloscope and an electric screwdriver and allows to perform four procedures, namely *high voltage board repair, low voltage board repair, high voltage board test* and *low voltage board test*. Based on this environment, we collected two sets of data for training the NLP module and the object recognition module respectively, as detailed in the following sections. Figure 2 shows some images depicting users performing the procedures in the considered industrial laboratory, along with bounding box annotations suitable for object recognition.

3.1 Dataset for the NLP Module

We constructed a dataset of utterances in order to train our NLP module considering the industrial laboratory, the contained objects, and the procedures that can be performed. During the dataset creation phase, we structured our data using the intent-entity paradigm. The intent of an utterance represents what the user wants to communicate when posing a question, while entities are crucial pieces of information that can be extracted from the utterances to determine the conversation state. Taking into account domain semantics, human actions, and objects, we determined four types of entities, namely *objects, electronic boards, components, procedures* and 24 different intents such as *greet, procedure_tutorial* and *object_warnings*.

To train the intent and entity classification model needed in the NLP module, for each intent, we provided several examples representing different utterances that a user might make while conveying that intent. Each utterance is in a structured format in which entities are marked with their class. An example of such structured data is "How am I supposed to use the [oscilloscope](object)?". In the example, "[oscilloscope]" indicates that the word "oscilloscope" represents an entity, while "(object)" indicates that the entity is "object" type. This utterance has "object_warnings" intent and contains the "oscilloscope" entity of type "object". Ultimately, we created 151 examples related to the 24 intents.

3.2 Dataset for the Object Detection Module

To obtain data suitable for training the object recognition module, we collected 42 videos of users performing test and repair operations in the considered laboratory using the Microsoft Hololens 2 device. For each of these videos, we extracted the first frame where a hand touches an object and the subsequent frame where the same hand releases the object. For each frame, we annotated all objects with (x, y, w, h, c) tuples, where c represents the object class among the 23 object categories, and (x, y, w, h) denotes the bounding box coordinates. With this procedure, we labeled 16K images including 90K objects belonging to the 23 different categories. Figure 2 shows some examples of the annotated frames.

4 Proposed System

Figure 3 illustrates a high-level view of the HERO architecture. It comprises five main modules: 1) input module, 2) NLP module, 3) object detector module, 4) QA module, and 5) output module. The architecture modules are described in detail in the following sections. The current system was deployed on Facebook Messenger, using Rasa's channel connector and can be accessed through a smartphone. Thanks to the modular framework provided by Rasa, implementations in other platforms such as wearable devices is possible.

Input Module. The input module is responsible for retrieving user utterances and extracting coarse data, such as text and images. Text is sent to the NLP and QA modules, while images are processed by the object detector module.

NLP Module. This module, based on the Rasa framework [2], is responsible for processing text from user utterances. Our NLP pipeline differs from Rasa's provided standard pipeline and relies on the *SpacyNLP, SpacyTokenizer, CountVectorsFeaturizer, SpacyFeaturizer, DIETClassifier, EntitySynonymMapper,* and *ResponseSelector* components provided by Rasa[1]. These components are trained on the collected training utterances for predicting intents and extracting entities. The NLP module predicts the intent and the entities from each utterance, which are then sent to the Output module. Entities may be absent for certain intents or due to the structure of the user's utterance. If the system detects that an object entity is required but missing from text, the image is sent to the object detector module. If the predicted user intent is to receive information on a specific procedure, the QA module is used to process the final response.

[1] https://rasa.com/docs/rasa/components.

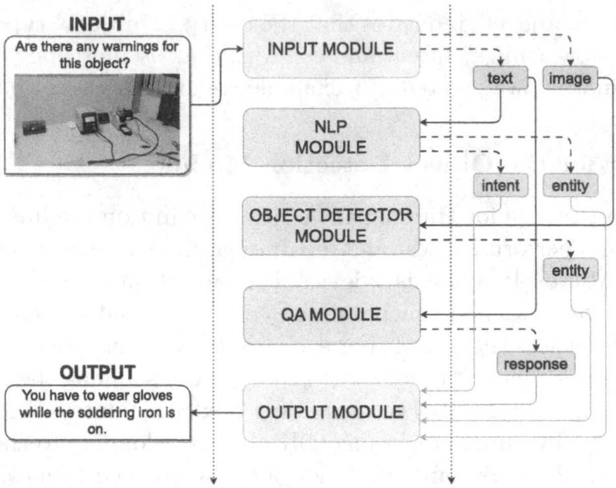

Fig. 3. The architecture of the proposed HERO system. The left column lists two utterances (input/output) exchanged by the user and the system, while the central column lists the main modules of HERO. Yellow boxes are used for modules that extract/assemble data, while blue boxes represent modules that process raw data to obtain higher level information. The right column shows extracted and processed data, where red boxes denote raw data and green and purple boxes indicate output high level information. Given an utterance comprised of an image and some text, the input module extracts raw data such as text and the images. The NLP module receives text from the input module, predicts an intent and extracts entities if present. The object detector module extracts the class of the object closest to the center of the image sent by the input module as an entity. If the intent of the user is to obtain information on a procedure, the QA module generates a custom response from text sent by the input module. The output module selects the final response based on high level information received by the NLP, object detector, and QA modules. (Color figure online)

Object Detector Module. This module utilizes the two-stage object detector Faster R-CNN [17] trained to recognize the objects of interest and predict a (x, y, w, h, c) tuple for each object in the image received by the input module. As humans typically direct their attention towards the center of their field of view while interacting with an object, we consider only the object closest to the center when more that one object are detected. The predicted class that represents an entity is sent to the Output module.

QA Module. This module utilizes the OpenAI APIs[2] to prompt the *text-davinci-003* model with the content of a specific procedure and the user's question about this procedure in order to generate a response, which is then sent to the Output module.

[2] https://openai.com/blog/openai-api.

Table 1. Intent and entity classification results obtained by the NLP module.

	Precision	Accuracy	F_1-score
Entity classification	0.944	0.959	0.866
Intent Classification	0.729	0.735	0.700

Output Module. This module selects the final response from a lookup table. The choice is based on the intent and entities detected and aggregated by the previous modules. If the intent is to ask a question on a procedure, the output of the QA module is forwarded to the user.

5 Experiments and Results

In this section, we analyze the performance of the NLP and Object Detection modules and report the results of our user study aimed to assess the overall usefulness of the developed assistant.

5.1 Performance of the NLP Module

We split our dataset in training and test sets using an 80:20 ratio. Considering the small size of our dataset, we find the optimal parameters performing a 5-fold cross-validation on our training set. We trained the final model on the whole training set considering the mean of the parameters for each split. The model has been trained on an Intel Core i5 CPU for 100 epochs with learning rate 0.001 and a variable batch size that linearly increases from 64 to 256 during training. We evaluate our model using the precision, accuracy and F_1-score measures. Table 1 reports the results obtained for entity classification (first row) and intent classification (last row). Our pipeline recognizes entities with a precision of 0.944, an accuracy of 0.959 and an F_1-score of 0.866. The intent classification obtains a precision of 0.729, an accuracy of 0.735 and F_1-score of 0.700. These results may be attributed to the small size of our dataset.

5.2 Performance of the Object Detection Module

We split the object detection dataset into training, validation and test sets with a 2:1:1 ratio and being careful to include all the frames extracted from a video into a single split. For our object detector module, we used the two-stages Faster R-CNN with a ResNet-101 backbone, as implemented in the Detectron2 framework. The experiments were conducted on a Nvidia V100 GPU. We trained our model for 30,000 iterations with a batch size of 2, a learning rate of 0.001 and 1000 learning rate warm-up iterations. We decreased the learning rate by a factor of 10 after 15,000 and 20,000 iterations. We evaluated the model using the COCO mean Average Precision (mAP) metric with an Intersection over Union (IoU) of 0.5 (mAP@50). The achieved mAP of 82.57% indicates that the module is able to correctly localize and recognize the objects present in the images.

Table 2. List of questions included in the questionnaire aimed to evaluate the subjects' degree of satisfaction with our proposed system.

ID	Question
1.1	How satisfied are you overall with the experience in a range from 1 to 5? 1-definitely not satisfied, 5-definitely satisfied
1.2	How natural did you find the interaction with the app in a range from 1 to 5? 1-definitely not natural, 5-definitely natural
1.3	How often did you use the photo sending feature to communicate with the bot? a-never, b-once, c-more than once
1.4	How natural did you find this feature (if you didn't use this feature, you can skip this question) in a range from 1 to 5? 1-definitely not natural, 5-definitely natural
1.5	How helpful do you think the technology demonstrated in this application prototype can be in a range from 1 to 5? 1-definitely not helpful, 5-definitely helpful
1.6	Do you think the technology demonstrated in this prototype can be used in other contexts besides the industrial context? a-yes, b-no
1.7	How often did the system correctly recognize the intent of your questions in a range from 1 to 5? 1-never, 5-each time
1.8	How useful do you think the information received from the application is in a range from 1 to 5? 1-definitely not useful, 5-definitely useful
1.9	How clear do you think the information received from the application is in a range from 1 to 5? 1-definitely not clear, 5-definitely clear
1.10	How satisfied are you with the system response time in a range from 1 to 5? 1-definitely not satisfied, 5-definitely satisfied
1.11	How useful do you think it is for the application to be available on the phone rather than another device (wearable devices, tablets, fixed screens) in a range from 1 to 5? 1-I'd prefer a different device, 5-I prefer a mobile device
1.12	Would you prefer a version with voice dictation? a-yes, b-no

5.3 User Study

We recruited a total of 11 participants and asked them to perform two different procedures consecutively in the considered industrial laboratory. In the execution of the first procedure, participants were asked to use HERO to receive instructions on the next steps to follow and whenever they had a question about how to proceed. In the execution of the second procedure, participants were asked to use a standard paper-based manual. The manual contains the same information on the laboratory, its objects and procedures that can be accessed by interacting with HERO, and is structured in four different sections: *boards, objects, Personal Protection Equipment (PPE)*, and *procedures*. Each of the two procedures is composed of 10 instructions such as "Set the soldering iron temperature to 480°C using the yellow UP button" or "Fix the electronic board to the working area using the electric screwdriver" and there is no overlap between

Table 3. List of questions included in the questionnaire aimed to compare the two experiences.

ID	Question
2.1	Which experience satisfied you the most in a range from 1 to 5? 1-definitely the application, 5-definitely the paper-based manual
2.2	How convenient did you find the use of the paper-based manual in a range from 1 to 5? 1-definitely not convenient, 5-definitely convenient
2.3	How much do you think the technology demonstrated in this application prototype could support you, compared to the use of the paper-based manual in a range from 1 to 5? 1-I found the manual more supportive, 5-I found the application more supportive
2.4	Which tool allowed you to complete the instructions more quickly in a range from 1 to 5? 1-I found the manual as the quickest tool, 5-I found the application as the quickest tool
2.5	How useful do you think the information received from the application is compared to the information obtained through the paper-based manual in a range from 1 to 5? 1-I found the manual instructions more useful, 5-I found the application instructions more useful
2.6	Which tool provided clearer instructions in a range from 1 to 5? 1-I found the manual instructions clearer, 5-I found the application instructions clearer
2.7	Which experience did you prefer overall? a-the use of the application, b-the use of the paper-based manual

the objects required to complete the two procedures. After completing each of the two procedures, the participants were asked to fill out a questionnaire. The first questionnaire aimed to evaluate the subjects' degree of satisfaction with our proposed system, as well as its utility and usability. The second questionnaire aimed to compare the two experiences when using HERO or the paper-based manual. We randomize the order in which each participant performs the two procedures, but let the user rely on HERO to carry out the first of the two performed procedures and the paper-based manual for the second one. This way, each procedure is performed about 50% of the time with either the two systems. Tables 2 and 3 provide the list of questions included in the two questionnaires.

Results. We visualize the distribution of the responses to the questionnaires through boxplots in Fig. 4. We include only questions that require to express a score in a range from 1 to 5. As shown by the boxplots, the participants expressed satisfaction with the overall experience and response time (questions 1.1 and 1.10). However, the natural language component was perceived less natural as compared to the vision component (questions 1.2 and 1.4). Some participants reported difficulty in formulating their queries in a way that the system could understand (question 1.7, which has a median score of 3), but found the retrieved

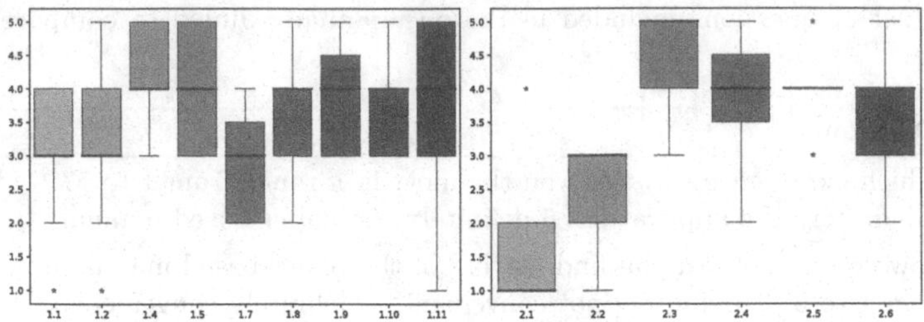

Fig. 4. Boxplot of the results for each question that required a discrete answer in the first questionnaire (left) and in the second questionnaire (right).

information useful and easy to understand (questions 1.8 and 1.9, which have a median score of 4). Even though a few participants expressed a preference for a different type of device (e.g., wearable device), most of the participants appear to be satisfied with the mobile application (question 1.11). The participants generally believed that this technology could be useful in various contexts besides the industrial one (we obtained 100% "yes" preferences in question 1.6). Most users noted that voice dictation can be useful (81.8% "yes" preferences on question 1.12). Despite the participants' pre-existing bias and familiarity with the industrial laboratory, when completing a procedure using a paper-based manual after completing the first procedure using the app, they still preferred our system over the manual. The participants found our system more satisfying, convenient, and quicker to use than the manual (questions 2.1, 2.2, 2.3, and 2.4 in Fig. 4 (right)). Although the information contained in the paper-based manual and in the chatbot's responses was identical, the participants perceived our system's responses as more useful and clearer (questions 2.5 and 2.6). Overall, the participants preferred the chatbot experience (90.9% users replied "the use of the application" to question 2.7).

6 Conclusion

In this paper, we presented a multi-modal conversational assistant designed to support users carrying out procedures within industrial laboratories. In order to improve the system's helpfulness and the naturalness of the interactions, we incorporated object recognition capabilities into the system. Experiments highlight the good performance in the considered industrial laboratory on intent-entity prediction for both text and visual inputs. To evaluate the system qualitatively, we conducted a user study in which 11 volunteers reported on the execution of two procedures in two different settings: testing the HERO system and consulting a standard paper-based manual. The results suggest that the approach presented in this work can be useful for users in industrial contexts. Future work could focus on extending HERO including a speech-to-text and text-to-speech module, as well as developing a wearable device version of the system in order to provide a hands-free experience.

Acknowledgements. This research is supported by Next Vision (Next Vision: https://www.nextvisionlab.it/.) s.r.l. and by the project Future Artificial Intelligence Research (FAIR) - PNRR MUR Cod. PE0000013 - CUP: E63C2200194000.

References

1. Anderson, P., et al.: Bottom-up and top-down attention for image captioning and visual question answering. In: 2018 IEEE/CVF Conference on Computer Vision and Pattern Recognition, pp. 6077–6086. IEEE (2018)
2. Bocklisch, T., Faulkner, J., Pawlowski, N., Nichol, A.: Rasa: open source language understanding and dialogue management. arXiv preprint arXiv:1712.05181 (2017)
3. Cui, C., Wang, W., Song, X., Huang, M., Xu, X.S., Nie, L.: User attention-guided multimodal dialog systems. In: Proceedings of the 42nd International ACM SIGIR Conference on Research and Development in Information Retrieval. Association for Computing Machinery (2019)
4. Das, A., et al.: Visual dialog. In: Proceedings of the IEEE Conference on Computer Vision and Pattern Recognition (CVPR) (2017)
5. Das, A., Kottur, S., Moura, J.M., Lee, S., Batra, D.: Learning cooperative visual dialog agents with deep reinforcement learning. In: Proceedings of the IEEE International Conference on Computer Vision (ICCV) (2017)
6. Gao, S., Sethi, A., Agarwal, S., Chung, T., Hakkani-Tur, D.: Dialog state tracking: a neural reading comprehension approach. In: Proceedings of the 20th Annual SIGdial Meeting on Discourse and Dialogue, pp. 264–273 (2019)
7. Goel, R., Paul, S., Hakkani-Tür, D.: HyST: a hybrid approach for flexible and accurate dialogue state tracking. arXiv preprint arXiv:1907.00883 (2019)
8. Goyal, Y., Khot, T., Summers-Stay, D., Batra, D., Parikh, D.: Making the V in VQA matter: elevating the role of image understanding in visual question answering. In: Proceedings of the IEEE Conference on Computer Vision and Pattern Recognition, pp. 6904–6913 (2017)
9. Heck, M., et al.: TripPy: a triple copy strategy for value independent neural dialog state tracking. In: Proceedings of the 21th Annual Meeting of the Special Interest Group on Discourse and Dialogue, pp. 35–44 (2020)
10. Hosseini-Asl, E., McCann, B., Wu, C.S., Yavuz, S., Socher, R.: A simple language model for task-oriented dialogue. In: Advances in Neural Information Processing Systems, pp. 20179–20191 (2020)
11. Kottur, S., Moon, S., Geramifard, A., Damavandi, B.: SIMMC 2.0: a task-oriented dialog dataset for immersive multimodal conversations. In: Conference on Empirical Methods in Natural Language Processing, pp. 4903–4912 (2021)
12. Kottur, S., Moura, J.M.F., Parikh, D., Batra, D., Rohrbach, M.: Visual coreference resolution in visual dialog using neural module networks. In: Ferrari, V., Hebert, M., Sminchisescu, C., Weiss, Y. (eds.) ECCV 2018. LNCS, vol. 11219, pp. 160–178. Springer, Cham (2018). https://doi.org/10.1007/978-3-030-01267-0_10
13. Le, H., Chen, N., Hoi, S.: Multimodal dialogue state tracking. In: Proceedings of the 2022 Conference of the North American Chapter of the Association for Computational Linguistics: Human Language Technologies, pp. 3394–3415 (2022)
14. Lu, J., Batra, D., Parikh, D., Lee, S.: ViLBERT: pretraining task-agnostic visiolinguistic representations for vision-and-language tasks. In: Advances in Neural Information Processing Systems (2019)

15. Murahari, V., Chattopadhyay, P., Batra, D., Parikh, D., Das, A.: Improving generative visual dialog by answering diverse questions. In: Proceedings of the Conference on Empirical Methods in Natural Language Processing (EMNLP) (2019)
16. Ramil Brick, E., et al.: Am i allergic to this? Assisting sight impaired people in the kitchen. In: Conference on Multimodal Interaction (2021)
17. Ren, S., He, K., Girshick, R., Sun, J.: Faster R-CNN: towards real-time object detection with region proposal networks. In: Advances in Neural Information Processing Systems (2015)
18. Tan, H., Bansal, M.: LXMERT: learning cross-modality encoder representations from transformers. In: Proceedings of the 2019 Conference on Empirical Methods in Natural Language Processing and the 9th International Joint Conference on Natural Language Processing (EMNLP-IJCNLP), pp. 5100–5111 (2019)
19. Zhao, J., et al.: Description-driven task-oriented dialog modeling. arXiv preprint arXiv:2201.08904 (2022)

A Computer Vision-Based Water Level Monitoring System for Touchless and Sustainable Water Dispensing

Andrea Felicetti[1], Marina Paolanti[2(✉)], Rocco Pietrini[1], Adriano Mancini[1], Primo Zingaretti[1], and Emanuele Frontoni[2]

[1] VRAI - Vision Robotics and Artificial Intelligence Lab, Dipartimento di Ingegneria dell'Informazione, Università Politecnica delle Marche, 60131 Ancona, Italy
{a.felicetti,r.pietrini,a.mancini,p.zingaretti}@univpm.it

[2] Department of Political Sciences, Communication and International Relations, University of Macerata, 62100 Macerata, Italy
{marina.paolanti,emanuele.frontoni}@unimc.it

Abstract. In recent years, the need for contactless and sustainable systems has become increasingly relevant. The traditional water dispensers, which require contact with the dispenser and often involve single-use plastic cups or bottles, are not only unhygienic but also contribute to environmental pollution. This paper presents a touchless water dispenser system that uses artificial intelligence (AI) to control the dispensing of water or any liquid beverage. The system is designed to fill a container under the nozzle, dispense water when the container is aligned with the flow, and stop dispensing when the container is full, all without requiring any physical contact. This approach ensures compliance with hygiene regulations and promotes environmental sustainability by eliminating the need for plastic bottles or cups, making it a "plastic-free" and "zero waste" system. The prototype is based on a computer vision approach that employs an RGB camera and a Raspberry Pi board, which allows for real-time image processing and machine learning operations. The system uses image processing techniques to detect the presence of a container under the nozzle and then utilizes AI algorithms to control the flow of liquid. The system is trained using machine learning models and optimized to ensure accuracy and efficiency. We discuss the development and implementation of the touchless water dispenser system, including the hardware and software components used, the algorithms employed, and the testing and evaluation of the system. The results of our experiments show that the touchless water dispenser system is highly accurate and efficient, and it offers a safe and sustainable alternative to traditional water dispensers. The system has the potential to be used in a variety of settings, including public spaces, hospitals, schools, and offices, where hygiene and sustainability are of utmost importance.

Keywords: Touchless Water Dispenser · Computer Vision · Artificial Intelligence · Sustainability

G. L. Foresti et al. (Eds.): ICIAP 2023, LNCS 14233, pp. 437–449, 2023.
https://doi.org/10.1007/978-3-031-43148-7_37

1 Introduzione

The COVID-19 pandemic has prompted a reconsideration of our daily habits and behaviours. One of the most significant changes has been an increased focus on hygiene and the need to avoid direct contact with shared surfaces. This has shifted towards touchless or contactless technology in various industries, including hospitality, retail, and public services [4]. At the same time, the issue of plastic pollution has become more pressing than ever before. The plastic waste produced by single-use bottles and cups is a significant contributor to ocean pollution and poses a threat to marine life. As a result, there is a growing interest in finding sustainable solutions to reduce the use of single-use plastics [13].

To address these challenges, the concept of redesigning commonly shared devices with touchless technology (e.g., public toilets, supermarket checkouts, vending machines for food and beverages, etc.) is very popular today [6,11]. Moreover, the rapid growth in the technological field of artificial intelligence (AI) to the point where it can be integrated into every electronic device makes it essential to design new intelligent devices reduced waste, and enhanced convenience [10,15,16].

A water dispenser is a typical example of the challenges faced by institutions in maintaining and monitoring their drinking water supply [9]. In facilities such as hospitals, universities, and office buildings, water dispensers are commonly installed in various locations to meet the drinking water needs on-site [3,14]. Traditional dispensers require users to press buttons or touch surfaces to fill their containers, increasing the risk of spreading germs and viruses [12]. In this context, the development of smart water dispensers that use touchless technology to fill bottles or containers has become an increasingly popular option. To tackle these issues, a touchless water dispenser that operates based on computer vision has been proposed. This dispenser is capable of detecting and filling the container under the nozzle, dispensing water only when the container is aligned with the flow, and automatically stopping the dispensing process when the container is full. The system's control is based on AI, ensuring accurate and efficient dispensing. The touchless feature of the dispenser makes it fully compliant with hygiene regulations, promoting a healthier and safer environment. Additionally, the reuse of the same container instead of dispensing plastic cups or bottles makes it a plastic-free system, promoting environmental sustainability [1,5]. Finally, the dispenser's ability to stop dispensing once the container is full, makes it a zero-waste system, avoiding unnecessary water waste. This paper describes the design and implementation of this touchless water dispenser system, highlighting its benefits and features, as well as its potential for wider adoption in various settings.

The main contributions of this paper can be summarized as follows: i) development of a touchless water dispenser system that uses AI to fill and dispense liquid without the need for physical contact. This system complies with hygiene regulations and promotes the principles of environmental sustainability by minimizing plastic waste and being "zero waste"; ii) the use of computer vision techniques to monitor the water level in the container and control the dispensing process. This allows for precise and efficient dispensing, reducing waste and ensuring a positive user experience; iii) the potential for this system to be expanded beyond water to other liquid beverages, making it a versatile and adaptable solution for various settings and applications.

The paper is structured as follows: Sect. 2 describes the architecture implemented as well as the dataset used to evaluate its performance; Sect. 3 presents the results of our experimental phase and finally, in Sect. 4, conclusions are drawn after analyzing the experimental results.

2 Materials and Methods

In this section, we introduce the prototype system design, which is schematically is depicted in Fig. 1.

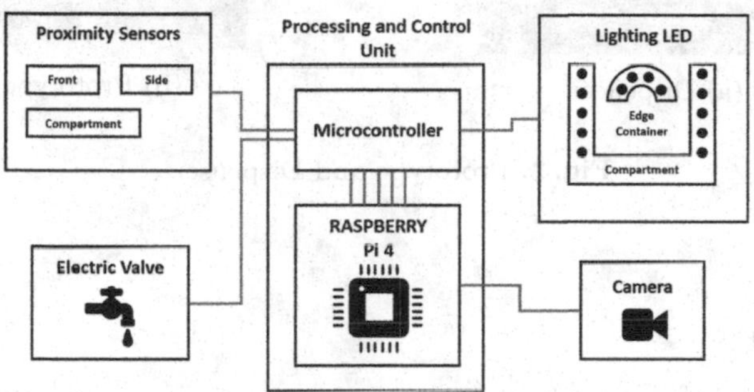

Fig. 1. Prototype System Design

The prototype system consists of a low-resolution camera connected to a Raspberry Pi 4 processing unit and an electric valve to control the flow, all integrated inside the dispenser. The camera is positioned in the upper part of the dispensing compartment, with a top-down view and offset and misalignment with respect to the axis where the liquid flows. During development, the same prototype has been revised and integrated several times by adding hardware components to simplify the software control system. In each of these updates, design, functionality, and cost constraints have always been respected. The final model includes the prototype system composed of a camera, electric valve, and Raspberry processing unit. LED lights have been added to illuminate the dispenser compartment, proximity sensors to manually command the flow, control the presence of objects in the compartment, and recognize the front presence of a person, microcontroller to manage the input of proximity sensors and adjust the LED light intensity, and control the electric valve. The microcontroller and Raspberry Pi 4 communicate through an ad hoc interface. A 3.5-in front display connected to the Raspberry completes the hardware. The hardware is entirely enclosed in the dispenser. The lighting system illuminates the interior and edge of the container, and two lateral arrays illuminate the entire compartment. A proximity sensor on the side of the dispensing column is used to control the water flow if the dispenser is configured for manual operation. Two front proximity sensors, symmetrically arranged under the display, detect the front presence of a person and enable the LED lights to illuminate the compartment. The dispenser is depicted in Fig. 2a.

(a) Dispencer

(b) Prototype

Fig. 2. Prototype and Dispenser

2.1 Dataset

After defining and creating the final model, a dataset was collected for software tuning. Seven water bottles were randomly selected, and videos were recorded for each of them simulating the filling process. In this phase, manual control of the dispenser was necessary to control the water flow. The videos start recording from an empty dispenser state. The container, held in hand, enters the dispenser and appears in the camera's field of view. Once positioned in the centre of the dispenser (resting on the grid or held in hand) in a way that allows the water to fall inside the container, the dispenser is activated. The dispenser continues until the container is full, and water spills out of the container. At that point, the dispenser is deactivated, and the full container is removed from the dispenser. The recording ends with an empty dispenser state. The recordings were performed under different environmental conditions to recreate various lighting conditions, which greatly affect the luminance and chrominance of the image and/or video frames captured by the camera. For each water bottle and lighting condition, two videos were recorded, one for training and one for testing. All videos were recorded with a resolution of 480 × 640 pixels, a frame rate of 30 fps, and saved in AVI format. All videos were manually labelled, and the timestamps of significant events during the recording were annotated for each of them. For each of them, the temporal instants related to significant events during the recording have been annotated. The annotated events are described in Table 1. Some of these events can occur multiple times during the video.

Table 1. Labelled Time Events

Event	Description
InsertionTimeStart	The container appears at the edge of the video frame, carried in the hand towards the centre of the compartment
BottleVisibleStart	The container, carried in the hand, appears at the edge of the ROI
BottleCenteredStart	The container is aligned with the axis where the liquid flows
InsertionTimeEnd	The hand carrying the container goes out of the camera's field of vision. Only for containers left lying in the compartment
FlushTimeStart	Start of delivery
OverBoardTime	The liquid leaks out of the mouth of the full container
FlushTimeEnd	End of delivery
RemovalTimeStart	The hand that picks up the container enters the camera's field of vision. Only for containers left lying in the compartment
BottleCenteredEnd	The container is misaligned to the axis where the liquid flows
BottleVisibleEnd	The container, carried in the hand, emerges from the edges of the ROI
RemovalTimeEnd	The container emerges from the edges of the video frame, carried in the hand out of the compartment

Additional videos were acquired to fine-tune the software in undesired situations (contrary to the correct use of the device) to which the system can be forced. These situations are the insertion of unwanted objects into the compartment and containers placed in a parallax condition. Unwanted objects are defined as objects that are unsuitable or in such a condition that they cannot contain fluids (e.g., corked or already full flasks, various objects, and empty hands). For the problem of unwanted objects, various objects were randomly selected. In the selection, we took care to equally select round objects (rolls of tape, bottle caps, glasses, coins, etc.), general objects (key rings, smartphones, cigarette packets, etc.), and 'empty hands' (even if wearing finger rings and/or wristwatches). The recorded videos, under different ambient light conditions, were annotated using some of the same events used in the water bottle filling videos. Specifically, the events recorded are: InsertionTimeStart, BottleVisibleStart, InsertionTimeEnd, RemovalTimeStart, BottleCenteredEnd, RemovalTimeEnd. In this case, however, it is not the water bottle that is the object in the video, and no dispensing is enabled.

2.2 Algorithm

As already mentioned for the hardware, the software was also developed and fine-tuned in several stages, in line with the hardware additions implemented to overcome the limitations of the software. In addition, the use of machine learning algorithms and models, which require large amounts of annotated data to learn a task (detection or classification), were improved using annotated data from the same software with the previous model version. The final version of the software implementing the vision method consists of two macro-tasks. The main task handles the recognition of the container's edge, verifies the alignment condition with the axis where the water flows and sends the command to start dispensing, recognises when the container is full (anticipating water spillage) and sends the command to end dispensing.

Calibration. The mechanical configuration of the dispenser, in particular, the precision in the arrangement of the vision system and axis where the liquid flows is a fundamental requirement. The position, and mutual alignment, of the camera must meet defined constraints unless there is a certain tolerance to factory deviations, as well as deviations due to shock after assembly and vibration during operation. Barring gross hardware deviations that would require mechanical recalibration, a calibration is also required on the software side to make the camera's viewing window, at least in the ROI (Region Of Interest), independent from dispencer to dispencer; in addition to developing algorithms that are as independent as possible from these causes of disturbance.

Following, the contour of the water flow as seen above is marked by noting the coordinates of the four vertices of the trapezoid delimiting it. The major base of the trapezoid lies on the right edge of the frame and corresponds to the greatest visible height of the water flow, while the minor base coincides approximately with the diameter of the hole where the flow touches the bottom of the compartment. The flow rate is regulated and stabilised by a mechanical valve at 14 s per litre.

Edge Detection and Full Detection. The main task handles the recognition of the container's edge, verifies the alignment condition with the axis where the water flows and sends the command to start dispensing, recognises when the container is full (anticipating water spillage) and sends the command to end dispensing. Recognition of the container edge at the same time as recognition of the "container full" state was approached as an image-based detection task, using a state-of-the-art neural architecture. In particular, we used the SSD-MobileNetv3 (Single Shot Detection) model, a "lightweight" model, optimised to run on mobile devices and/or with limited computing resources [8]. The dataset of collected videos related to water bottle filling (centred and off-centred) were preprocessed by obtaining sets of labelled RGB frames to train and test the detection model. Each video was segmented into frame sequences based on the temporal annotation of events. The key events considered in this phase are:

bottleCentredStart, flushInTheMiddle, fullTime, overBoardTime. The fullTime event was defined as the time instant 2 s after the overboadTime. Considering that the videos are recorded at 30 fps, there are 60 frames between fullTime and overBoardTime. The event flushInTheMiddle was defined as the time instant in the middle of the fill, i.e. in the middle between flushStart and overBoardTime. For each extracted frame, the bounding box was annotated by hand. The frames between bottleCenteredStart and flushInTheMiddle were labelled as NoFull, while the frames between fullTime and overBoardTime are labelled as Full; considering that the videos are recorded at 30 fps, there are 60 full frames per video. Frames beyond the overBoardTime were not considered in the training of the model because in the correct functioning of the model they represent an undesirable situation. Frames between flushInTheMiddle and fullTime were not considered in order to have a clear separation margin between the two classes. Frames prior to bottleCenteredStart were also not considered in order to force the model not to recognise container edges beyond the centred condition. This strategy inherently helps to avoid parallax conditions for containers seen during training. For each extracted frame (full and noFull) the bounding box of the container edge was annotated. The box annotation was performed semi-automatically.

In a first step, a reduced set of frames was created to train the same detection model (SSD-MobileNetv3) to detect the container edge regardless of class (full or noFull). Preliminary tests verified the robustness of the model in detecting container edges with good accuracy. Therefore, four frames corresponding to the events: insertionTimeStart, bottleCenteredStart, flushInTheMiddle, and overBoardTime were extracted from each video. In the second step, the trained model in the first step was used to annotate the container edge on all frames extracted from the videos of the bottle-filling dataset. Next, a manual selection of all correctly annotated frames was performed, discarding the frames the container edge detection failed. Thus, we have the set of frames to train and test the detection model. Overall, the number of frames extracted is shown in Table 2.

Table 2. Filling Dataset. Cardinality of frames across training and test set.

class	training		test	
	centred	decentralised	centred	decentralised
full	3837	5412	952	1368
noFull	3568	5484	900	1356
total	7405	10897	1852	2724

The extracted frames were cropped by selecting the ROI. Consequently, the annotation of the boxes in each frame was also adapted to the cropping. The annotations of all images were stored in two text files (one for training and the other for validation) according to the COCO annotation standard. The model, consisting of a feature extractor with a MobileNetV3 small architecture, is pre-trained on the COCO14 dataset, a benchmark dataset for the image detection task [7]. The model was set up to detect a maximum of one object per image (after all, one container can be filled at a time in the proper operation of the dispenser), with a threshold of 0.6 on the IoU (intersection over Union) metric for the suppression of non-maximums. The input size was set to $200 \times 200 \times 3$, so the input RGB images are rescaled to 200×200 pixels before being processed by the model. For the training process, the optimisation algorithm used is SGD (stochastic gradient descent) with batch size set to 128 and a total of 10000 iterations. The loss function to be minimised is a combination of the localisation and classification losses. The initial learning rate was set to 0.05 with cosine decay as iterations increase. Geometric (vertical flip and jittering) and colorimetric (luminosity variation, hue, saturation, and contrast) augmentation operations were used to generalise to conditions not seen during training. Among the geometric operations, we have no random crop, horizontal flip rotations, as there is no great variation in the viewing area between distributor and distributor. On the other hand, jittering (set plus or minus 5 pixels) and vertical flips are used to harden the model to small variations in the viewing area, residual of the calibration, as the vertical symmetry allows this. All colourimetric operations, on the other hand, help to harden the model to different brightness conditions of the installation environment.

Classification of Unwanted Objects. Among the control tasks, the detection of unwanted objects introduced in the dispenser compartment was approached as an image classification task. In particular, we used the mobileNetv3 neural architecture (same architecture used for detection but structured for the classification task). Specifically, we use a mobileNetV3 small feature extractor to which a final convolutional classifier is added. The dataset of collected videos related to water bottle filling and generic objects were preprocessed by obtaining sets of labelled RGB frames to train and test the classification model. Each video was segmented into sequences of frames based on the temporal annotation of events. For each video, all frames between the bottleVisibleStart and bottleVisibleEnd events are labelled bottle, generic object, or empty hand; while all frames preceding insertionTimeStart and those following removalTimeEnd are labelled empty. The frame set obtained shows a high bias in favour of the majority class. In order to train the classification model avoiding bias towards the majority class, it was necessary to balance the classes, truncating the numerosity of all of them to the numerosity of the minority class. Since both were undesirable, the generic object class, and empty hand were merged into the object class. The numerosity of the frame set obtained to train and test the unwanted object classification model is shown in Table 3. The extracted frames were cropped by selecting the ROI.

The mobileNetV3 model, with feature extractor pre-trained on the imagenet dataset [2], a benchmark dataset for the image classification task, was configured to classify three classes: empty, bottle, object. The alpha parameter, which takes

Table 3. Unwanted Object Dataset. Cardinality of frames.

class	unbalanced			balanced			
	training	valid+seen	unseen	training	validation	seen	unseen
empty	19893	15164	2163	14682	5163	5163	2163
bottle	84959	80890	6161	14682	5163	5163	2163
object	10486	6330	7508	14682	5163	5163	2163
hands	4196	3996	0				
total	119534	106380	15832	44046	15489	15489	6489

into account the model size (number of trainable parameters), was set to 0.75. The input size was set to $200 \times 200 \times 3$, so the input RGB images are rescaled to 200×200 pixels before being processed by the model. The input RGB image is preprocessed using a bipolar scaling that compresses the range of values from -1 to 1.

For the training process, the optimisation algorithm used is ADAM with batch size set to 128 and a total of 60 epochs. The loss function to be minimised is the categorical crossentropy. The initial learning rate was set to 0.01 and reduced by a factor of 10 from epochs 20 (0.001) and 40 (0.0001). To generalise to conditions not seen during training, augmentation operations were used, the same as those used to train the detection model.

Control Task: Circle-Flow Alignment Check and State Management. From the container's edge box, the parameters of the inscribed ellipse are extracted. The alignment condition (seen from above), a necessary but not sufficient condition as it is subject to parallax error, occurs when the ellipse intersects both oblique sides of the trapezoid outlining the water flow in at least one point. Up to this point, we have dealt with "instantaneous" tasks (single frame processing). In order to control the delivery of the flow, ensuring a more accurate and continuous operation, it is necessary to consider the readings not only at the current frame but also at the previous ones. For this, a status register was defined containing "strong" flags, to which the combinatorial function that commands the delivery is linked. Another status register was defined to count the succession of "weak" flags, obtained from the processing of the single frame. The "strong" flags depend on the sequence of "weak" flags at the current and previous frames. In particular, the (strong and weak) flags considered indicate the scene state (empty, bottle, object), the alignment state (true, false), and the full state (true, false). The hysteresis parameters (for how many consecutive frames to check the output of the models to activate or deactivate a strong flag) were set and refined after careful observation of the behaviour on live tests, finding the right compromise between accuracy, continuity and responsiveness. Higher hysteresis parameters guarantee more continuous behaviour, avoiding discontinuities due to isolated errors during the processing of a frame. Furthermore, consensus over several consecutive frames guarantees greater accuracy in state transitions. On the other hand, the higher the hysteresis parameters, the less responsive the system is. Activation of the Bottle scene flag occurs when the scene classification model detects a bottle. Activation of the Empty scene flag occurs when the edge and full detection model does not detect circles for at least 2 consecutive frames. Activation of the Object scene flag occurs when the scene classification model detects object for at least 2 consecutive frames. Activation of the alignment flag occurs when the alignment condition is verified for at least 2 consecutive frames. Deactivation of the alignment flag occurs when the alignment condition is not verified for at least 2 consecutive frames.

3 Results and Discussions

Tests of the detection model on the videos in the dataset (same flasks seen during training) confirm excellent accuracy in both edge detection and full/no full state recognition, as reported in Table 4 and in Table 5.

Table 4. Fullness classification results over Centred Filling Dataset

	precision	recall	f1-score	accuracy	
Full	0.967	0.991	0.979		952
NoFull	0.990	0.964	0.977		900
overall	0.978	0.978	0.978	0.978	1852

Table 5. Fullness classification results over Decentralised Filling Dataset

	precision	recall	f1-score	accuracy	support
full	0.942	0.956	0.949		1368
noFull	0.955	0.941	0.948		1356
overall	0.949	0.949	0.949	0.949	2724

Live tests confirm a high robustness in detecting the bounding box of generic flasks. The most critical part is the correct and stable detection (between successive frames) of the full/no full status, especially for unseen containers during the training phase. It is detected very well in the dispenser known in the dataset when resting on the compartment. In some cases, when the dispenser is held in the hand, full is not detected, especially if the dispenser is too close to the nozzle. However, the right setting of the control parameters ensures good operation even on unfamiliar flasks. The most frequently encountered residual problem remains intermittent dispensing in some cases. While we also accept a slightly early delivery STOP. Full sensing would seem to be robust to changes in ambient light conditions.

Tests of the scene classification model confirm very good accuracy on the test data. To verify the generalisation to objects not seen during the training phase, two separate analyses were performed. The model trained on frames extracted from training videos of the dataset, was tested on frames extracted from test videos of the same SEEN dataset (different videos but same objects seen during training), and on frames extracted from videos of the UNSEEN test-only dataset (videos of objects not seen during training) respectively reported in Table 6 and in Table 7.

Table 6. Scene classification results over Unwanted Object Dataset. Seen scenario.

	precision	recall	f1-score	accuracy	support
empty	0.97	0.94	0.95		5163
bottle	0.93	0.99	0.96		5163
object	0.99	0.95	0.96		5163
overall	0.96	0.96	0.96	0.96	15489

Table 7. Scene classification results over Unwanted Object Dataset. Unseen scenario.

	precision	recall	f1-score	accuracy	support
empty	0.95	0.96	0.95		2163
bottle	0.95	0.96	0.95		2163
object	0.98	0.96	0.97		2163
overall	0.96	0.96	0.96	0.96	6489

4 Conclusions and Future Works

The touchless water dispenser system presented in this paper provides a promising solution for addressing the challenges of hygiene and sustainability in water dispensing. The use of AI and computer vision algorithms, combined with the hardware and software components, ensures that the system is efficient, accurate, and easy to use. The system offers several advantages over traditional water dispensers, including eliminating the need for physical contact, promoting the reuse of containers, and reducing plastic waste. Our experiments showed that the touchless water dispenser system is highly accurate and efficient. Additionally, the system has the potential to be used in a variety of settings where hygiene and sustainability are a top priority, such as public spaces, hospitals, schools, and offices. The touchless water dispenser system represents a significant step towards achieving a more sustainable and hygienic future. By eliminating the need for physical contact and promoting the reuse of containers, the system can contribute to reducing plastic waste and promoting environmental sustainability. We believe that the touchless water dispenser system can serve as a basis for further research and development in this field, with the potential to revolutionize water dispensing and contribute to a more sustainable and hygienic world.

References

1. Coelho, P.M., Corona, B., ten Klooster, R., Worrell, E.: Sustainability of reusable packaging-current situation and trends. Resour. Conserv. Recycl. X **6**, 100037 (2020)
2. Deng, J., Dong, W., Socher, R., Li, L.J., Li, K., Fei-Fei, L.: ImageNet: a large-scale hierarchical image database. In: 2009 IEEE Conference on Computer Vision and Pattern Recognition, pp. 248–255. IEEE (2009)
3. Dhanasekar, S., Nageshwar, S.S., Ranjani, S.S., Vidhya, S.S.S., Prakash, C.S., Arunkumar, N.: A survey on IoT-based hand hygiene dispenser with temperature and level monitoring systems. In: 2022 8th International Conference on Advanced Computing and Communication Systems (ICACCS), vol. 1, pp. 01–05. IEEE (2022)
4. Erjavec, J., Manfreda, A.: Online shopping adoption during Covid-19 and social isolation: extending the UTAUT model with herd behavior. J. Retail. Consum. Serv. **65**, 102867 (2022)
5. Evode, N., Qamar, S.A., Bilal, M., Barceló, D., Iqbal, H.M.: Plastic waste and its management strategies for environmental sustainability. Case Stud. Chem. Environ. Eng. **4**, 100142 (2021)
6. Ighalo, J.O., Adeniyi, A.G., Marques, G.: Internet of things for water quality monitoring and assessment: a comprehensive review. In: Artificial Intelligence for Sustainable Development: Theory, Practice and Future Applications, pp. 245–259 (2021)
7. Lin, T.-Y., et al.: Microsoft COCO: common objects in context. In: Fleet, D., Pajdla, T., Schiele, B., Tuytelaars, T. (eds.) ECCV 2014, Part V. LNCS, vol. 8693, pp. 740–755. Springer, Cham (2014). https://doi.org/10.1007/978-3-319-10602-1_48
8. Liu, W., et al.: SSD: single shot MultiBox detector. In: Leibe, B., Matas, J., Sebe, N., Welling, M. (eds.) ECCV 2016, Part I. LNCS, vol. 9905, pp. 21–37. Springer, Cham (2016). https://doi.org/10.1007/978-3-319-46448-0_2
9. Madana, A.L., Sadath, L.: IoT applications in automated water level detections. In: 2020 International Conference on Intelligent Engineering and Management (ICIEM), pp. 401–407. IEEE (2020)
10. Mastaneh, Z., Mouseli, A.: Technology and its solutions in the era of Covid-19 crisis: a review of literature. Evid. Based Health Policy Manage. Econ. **4**, 138–149 (2020)
11. Sahoo, A.K., Udgata, S.K.: A novel ANN-based adaptive ultrasonic measurement system for accurate water level monitoring. IEEE Trans. Instrum. Meas. **69**(6), 3359–3369 (2019)

12. Seneviratne, S., Koggalage, R., Rasanjana, K.H., Srimal, H.: Design of automatic sanitizer for door handles and push buttons. University of Vocational Technology
13. Silva, A.L.P., et al.: Rethinking and optimising plastic waste management under Covid-19 pandemic: policy solutions based on redesign and reduction of single-use plastics and personal protective equipment. Sci. Total Environ. **742**, 140565 (2020)
14. Tadikonda, C., et al.: Smart sanitizer disperser with level monitoring. Turk. J. Comput. Math. Educ. (TURCOMAT) **12**(12), 994–999 (2021)
15. Wu, J., Wang, X., Dang, Y., Lv, Z.: Digital twins and artificial intelligence in transportation infrastructure: classification, application, and future research directions. Comput. Electr. Eng. **101**, 107983 (2022)
16. Zhang, Z., Wen, F., Sun, Z., Guo, X., He, T., Lee, C.: Artificial intelligence-enabled sensing technologies in the 5g/internet of things era: from virtual reality/augmented reality to the digital twin. Adv. Intell. Syst. **4**(7), 2100228 (2022)

Smoothing and Transition Matrices Estimation to Learn with Noisy Labels

Simone Ricci[1(✉)], Tiberio Uricchio[2], and Alberto Del Bimbo[1]

[1] Università degli Studi di Firenze, Florence, Italy
{simone.ricci,alberto.bimbo}@unifi.it
[2] Università degli Studi di Macerata, Macerata, Italy
tiberio.uricchio@unimc.it

Abstract. In recent years, there has been impressive progress in learning with noisy labels, particularly in leveraging a small set of clean data. Meta-learning-based label correction techniques have further advanced performance by correcting noisy labels during training. However, these methods require multiple back-propagation steps, which considerably slows down the training process. Alternatively, some researchers have attempted to estimate the label transition matrix on-the-fly to address the issue of noisy labels. These approaches are more robust and faster than meta-learning-based techniques. The use of the transition matrix makes the classifier skeptical about all corrected samples, thereby mitigating the problem of label noise. We propose a novel three-head architecture that can efficiently estimate the label transition matrix and two new label smoothing matrices at each iteration. Our approach enables the estimated matrices to closely follow the shifting noise and reduce over-confidence on classes during classifier model training. We report extensive experiments on synthetic and real world noisy datasets, achieving state of the art performance on synthetic variants of CIFAR-10/100 and on the challenging Clothing1M datasets. Code at https://github.com/z3n0e/STM.

Keywords: Learning with noisy labels · Label correction · Label smoothing · Matrix estimation

1 Introduction

Supervised learning has achieved great success in various classification tasks such as image classification [9] or face recognition [27] by utilizing large annotated datasets [4]. However, when annotations are obtained from coarse-grained sources, noisy labels can arise and lead to degraded performance [3], despite the availability of a large amount of annotated data. Recently, several methods have been proposed to build robust classifiers that are insensitive to noisy labels [1,10,19,23,29]. Unlike traditional methods [1,19] that consider all labels to be potentially corrupted, recent approaches utilize a small set of clean data to improve performance further. Loss correction methods [10,29] modify loss functions based on the clean data set, reducing the influence of noisy labels. Re-weighting methods [22,23] penalize noisy samples during training. Recent label correction methods [23,30], which leverage model-agnostic meta-learning, achieve remarkable performance by relabeling noisy labels to directly reduce noise levels.

G. L. Foresti et al. (Eds.): ICIAP 2023, LNCS 14233, pp. 450–462, 2023.
https://doi.org/10.1007/978-3-031-43148-7_38

In this paper, we propose a novel and efficient method for learning with noisy labels, which estimates a transition matrix and two smoothing matrices to correct noisy labels in real-time during training, improving generalization with the estimated matrices closely follows the shifting noise distribution induced by label correction. To estimate all three matrices in an efficient manner, we propose a three-head architecture with a shared feature extractor. The noisy corrected classifier head estimates the label transition matrix, while the pure noisy classifier head is responsible for the estimation of the two smoothing matrices. Finally, the third head is the main clean classifier, trained to be statistically consistent using all the estimated matrices.

We demonstrate the efficacy of our proposed method on synthetic and real-world noisy label datasets, specifically, CIFAR-10/100 [14] variants with artificially added noise, and Clothing1M [32], respectively. Our contribution in this paper is threefold: (1) We propose a novel method for learning with noisy labels by estimating three noise correction matrices, a transition matrix to relabel noisy samples and two smoothing matrices that mitigate overconfidence and enhances generalization. (2) We propose a three-head architecture that optimizes the matrices estimation with a single back-propagation step. (3) We provide extensive experimental results that validate the effectiveness of our proposed method in terms of predictive performance.

2 Related Work

2.1 Learning with Noisy Labels

Learning with noisy labels assumes that labels of the training samples could be potentially corrupted. Various methods have been proposed, such as using different loss functions [19,28], regularizations [11], re-weighting training samples [17,22,30], and correcting noisy labels [7,8]. However, different losses or regularizations often yield inferior performance to state-of-the-art methods [11,17,36], and re-weighting methods can filter out noisy but helpful samples for extracting features, leading to sub-optimal performance [25,30]. Label correction methods can circumvent these issues by relabeling noisy samples, but they are prone to propagate errors when miscorrected labels are continuously accumulated [17]. Further methods estimate a label transition matrix [20,23] using meta-learning, but have a limitation in that they need multiple training stages. In [15], class label only noise is estimated by directly evaluating a transition matrix with an auxiliary head. In contrast to such works we propose a three-head architecture to directly estimate a label transition matrix in conjunction with two new smoothing matrices that mitigate overconfidence in the transition matrix and enhances generalization for instance based noise.

2.2 Noise Estimation via Small Clean Dataset

Several recent studies argue that a small clean dataset is easily obtained by techniques such as image retrieval [21]. Many studies have successfully adapted this idea and shown massive performance improvement compared to traditional methods [2,10,37]. Early methods required multiple training stages, which hindered training efficiency

[10]. Recent studies widely adopt meta-learning [5,23] to various strategies discussed above, such as sample re-weighting [13,22], label correction [30,38], and label transition matrix estimation [29]. These approaches perform a virtual update with the noisy dataset, find optimal parameters using the clean dataset, and then update the actual parameters by the newly found parameters. However, this virtual update process requires three back-propagation steps per iteration, leading to at least three times the computational cost. Moreover, the clean set is not directly used for training the main model. In contrast, our proposed method estimate the label transition matrix and two smoothing matrices, exploiting directly the learning of the model on the clean samples without multiple back-propagation passes.

3 Methods

Label correction techniques are utilized to enhance the accuracy of machine learning models by identifying and correcting mislabeled data points. However, existing methods for label correction may incorrectly classify data points as either clean or noisy, which can lead to erroneous model training. It is well-known that the estimation of an accurate label transition matrix can address this miscorrection issue [10,15,20,26,31, 35]. Each element T_{ij} of the label transition matrix $T \in \mathbb{R}^{N \times N}$ is defined as the probability of a clean label i to be corrupted as a noisy label j, i.e. $T_{ij} = p(\overline{y} = j | y = i)$. We introduce two estimated smoothing matrix in order to enhance the generalization capacity of machine learning models reducing the over-confidence on classes. Our proposed method is illustrated in Fig. 1 and its effectiveness is demonstrated through extensive experiments.

3.1 Batch Creation

The estimation of matrices can be done using a clean batch of examples. To ensure effective estimation, we formulate the clean batch such that it contains the same number of samples per class. To compose the clean batch $d = (x, y)$ of size M, we randomly select samples for the entire N classes in the clean dataset D, ensuring that there is the same number of clean samples for each class. Instead the noisy batch $\hat{d} = (\hat{x}, \hat{y})$, is generated randomly sample from the noisy dataset $\hat{D}$. From the previously noisy batch $\hat{d}$, we generate a new corrected batch $\overline{d} = (\overline{x}, \overline{y})$ where a direct correction with a head (see Sect. 3.3) is applied to the label $\hat{y}$ obtaining the corrected label $\overline{y}$.

3.2 Transition Matrix

Several methods have been proposed to estimate the label transition matrix accurately utilizing a clean dataset directly [10,15,31,35].

The strategy that we use employs a feature extractor Φ and an auxiliary noisy linear classifier $\overline{\Theta}$, that is trained with first noisy and then corrected labels, to estimate the transition probability $p(\overline{y}|y)$ using clean samples $(x, y) \in d$. The definition of T is given by the following equation:

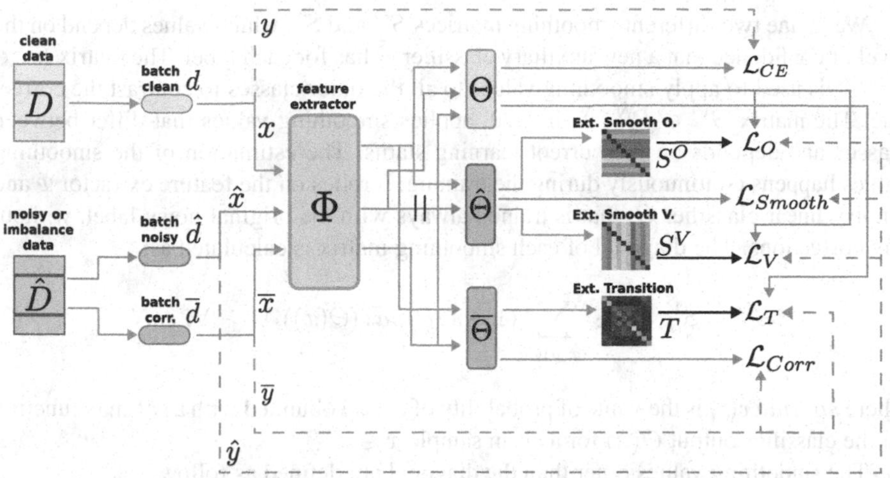

Fig. 1. Overview of our proposed method which involves using two datasets: a clean dataset D and a noisy dataset $\hat{D}$. At each iteration, we prepare three batches of data: a batch of clean data d, a batch of noisy data $\hat{d}$, and a batch of corrected noisy data $\overline{d}$, which is obtained by using the correction classifier $\overline{\Theta}$. We extract features from all three batches using an extractor Φ, and then classify the output using three different heads: Θ, $\hat{\Theta}$, and $\overline{\Theta}$. These heads are responsible for classifying the sample x, estimating two smooth matrices, and estimating the transition matrix, respectively, along with a corrected label $\overline{y}$ that is kept for the next iterations.

$$T = \frac{1}{M} \sum_{(x,y) \in d} softmax(\overline{\Theta}(x)) \quad (1)$$

Each element T_{ij} of the label transition matrix $T \in \mathbb{R}^{N \times N}$ is defined as the probability of a label i to be predicted as an another label j. However, the estimation of the transition matrix using only a limited number of clean samples within a single batch may result in a slightly inaccurate estimation. The transition matrix T is designed to handle the problem of noisy labels, but it is not well suited to dynamically decrease model's overconfidence on learned classes.

3.3 Smoothing Matrices

During the training process of a model, especially in the final stages when the network tends to adapt to the training data, identification of noisy labels can pose a challenge to the transition matrix. However, label smoothing can mitigate this issue by assigning higher probabilities to classes with lower confidence and lower probabilities to those with higher confidence. The application of label smoothing not only prevents overfitting but also enhances the generalization of the model. Moreover, it can also improve the estimation of the transition matrix T, thereby augmenting the model's capacity to handle noisy labels present in the data.

We define two different smoothing matrices S^O and S^V which values depend on the level of confidence that a new auxiliary classifier $\hat{\Theta}$ has for each label. The matrix $S^O \in R^{N \times N}$ is used to apply smoothing values to all the other classes to contrast the correct one. The matrix $S^V \in R^{N \times N}$, instead, applies smoothing values that differ between classes and depends on the current learning status. The estimation of the smoothing values happens continuously during the training. It relies on the feature extractor Φ and a noisy linear classifier $\hat{\Theta}$, that is trained always with the original noisy label, without any corrections. The diagonal of each smoothing matrix is calculated as:

$$S^O_{ii} = S^V_{ii} = \frac{1}{M} \sum_{(x,y) \in d} (1 - softmax(\hat{\Theta}(x))_i) \qquad \forall i \in N \tag{2}$$

where $softmax()_i$ is the value of probability of class i obtained with a softmax function on the classifier output $\hat{\Theta}(x)$ for a clean sample $x \in d$.

The smoothing values other than the diagonal are defined as follow:

$$S^O_{ij} = \frac{1}{M} \sum_{(x,y) \in d} \frac{softmax(\hat{\Theta}(x))_i}{N} \qquad \forall j \neq i \in N \tag{3}$$

$$S^V_{ji} = \frac{1}{M} \sum_{(x,y) \in d} \frac{softmax(\hat{\Theta}(x))_i}{N} \qquad \forall j \neq i \in N \tag{4}$$

The matrix S^O is used to compensate the overconfidence on a class applying the same value of smoothing to all the others categories. Instead, the matrix S^V provide a value of smoothing that depends on the confidence of each single class.

3.4 Learning with Estimated Matrices

A clean classifier Θ is trained with all the three estimated matrices T, S^O and S^V. To effectively capture the expected interaction of our matrices with the output of individual model heads, we propose the definition of distinct loss functions for each of them. The formulation of these loss allows for a more targeted optimization of the label transitioning and smoothing. The first loss $\mathcal{L}_T$ is computed as:

$$\mathcal{L}_T = \sum_{(\overline{x},\overline{y}) \in \overline{d}} \mathcal{L}^{CE}(T^\top \Theta(\overline{x}), \overline{y}) \tag{5}$$

where $\mathcal{L}^{CE}$ is the cross-entropy loss. We introduce a second and third loss that rely on the smoothing matrices S^O and S^V respectively.

$$\mathcal{L}_O = \sum_{(\hat{x},\hat{y}) \in \hat{d}} \mathcal{L}^{CE}(gumbel(\Theta(\hat{x})) \cdot diag(S^O) + (S^O - diag(S^O))), \hat{y}) \tag{6}$$

$$\mathcal{L}_V = \sum_{(\hat{x},\hat{y}) \in \hat{d}} \mathcal{L}^{CE}(gumbel(\Theta(\hat{x})) \cdot diag(S^V) + (S^V - diag(S^V))), \hat{y}) \tag{7}$$

The *gumbel*() is the Gumbel-Softmax function [12], and *diag*() function gives the diagonal value of a matrix. Gumbel-Softmax is a reparameterization technique that allows the generation of samples from a categorical distribution with a continuous relaxation, as such, it can be optimized using gradient-based methods. We use the Gumbel-Softmax function in order to sample discrete categorical variables in a differentiable way. This function returns one-hot label on which we can apply our smoothing technique directly. The formulation of the final loss for the clean classifier of our model is a combination of a cross-entropy on a clean batch and all the previous defined three losses:

$$\mathcal{L}_{Clean} = \sum_{(x,y)\in d} \mathcal{L}_{CE}(\Theta(x), y) + \mathcal{L}_T + \mathcal{L}_O + \mathcal{L}_V \tag{8}$$

The estimation of the three matrices relies on the capability of the two auxiliary heads of the model, $\overline{\Theta}$ and $\hat{\Theta}$. We train the $\overline{\Theta}$ head used to estimate the transition matrix T with the corrected data optimizing the cross-entropy loss:

$$\mathcal{L}_{Corr} = \sum_{(\overline{x},\overline{y})\in \overline{d}} \mathcal{L}_{CE}(\overline{\Theta}(\overline{x}), \overline{y}) \tag{9}$$

The last $\hat{\Theta}$ head, employed for the purpose of inferring the two smoothing matrices S^O and S^V, is trained using directly the noisy data:

$$\mathcal{L}_{Smooth} = \sum_{(\hat{x},\hat{y})\in \hat{d}} \mathcal{L}_{CE}(\hat{\Theta}(\hat{x}), \hat{y}) \tag{10}$$

The whole model is trained end-to-end with a weighted sum of the three losses:

$$\mathcal{L}_{Tot} = \lambda \cdot \mathcal{L}_{Clean} + \frac{\lambda}{2} \cdot \mathcal{L}_{Corr} + \frac{\lambda}{2} \cdot \mathcal{L}_{Smooth} \tag{11}$$

where λ is empirically set to 1. Because the $\hat{\Theta}$ head is trained only with noisy data without any correction strategy, we disallow the back-propagation of the gradient to the model's backbone from the $\hat{\Theta}$ head. We observed that this improves the performance of the final trained image classification model.

4 Experiments on Noisy Labels

In this section, we will assess the predictive performance of our proposed learning approach during a training with noisy labels, which are artificially generated or belong to real-world noisy label distribution.

4.1 Baselines

To evaluate our method, we have specifically chosen baselines that utilize a small, clean dataset to learn with noisy labels. These baselines can be categorized into three types:

- Re-weighting: The L2RW [22] algorithm assigns weights to training samples based on their gradients. The MW-Net [24] model trains an explicit weighting function with the training samples. The MFRW-MES [23] method learns how to mask visual feature and to weight the calculated gradient in order to address noise label problem.
- Label transition matrix estimation: The GLC [10] algorithm estimates the label transition matrix using the small clean dataset. The MLoC [29] method considers the label transition matrix as trainable parameters to be obtained through meta-learning. FasTEN [15] makes use of an efficient estimated transition matrix.
- Label correction: The MLaC [38] algorithm trains a label correction network as a meta-process to provide corrected labels. The MSLC [30] method uses soft labels with loss balancing weight through meta-gradient descent step under the guidance of the clean dataset.

4.2 Experiments on CIFAR-10/100

The first experiment is performed on variants of the CIFAR-10/100 datasets, where controlled noise is added.

Dataset and Settings. The CIFAR-10/100 datasets have become widely used as benchmarks for evaluating the robustness of machine learning algorithms to noisy labels. To introduce noise into these datasets, synthetic label manipulation techniques have been employed. Specifically, two types of noise have been injected: symmetric and asymmetric. In the case of symmetric noise, the labels are randomly flipped with a uniform distribution. In contrast, for asymmetric noise, the labels are flipped with a distribution that depends on the class, following the exact evaluation protocol described in [20,34]. The strength of the noise is regulated by a parameter p, that indicates the percentage of noise injected in the labels.

Table 1. Test accuracy on CIFAR10 and CIFAR100 dataset with symmetric label noise. The backbone used is a ResNet-34. p denotes the different levels of noise. The results for the cited methods are reported directly from their original papers. Instead, † indicates the results obtained by our implementation. The best result is marked in bold.

Dataset	Sym CIFAR-10				Sym CIFAR-100			
Noise p	0.2	0.4	0.6	0.8	0.2	0.4	0.6	0.8
L2RW [22]	88.26	83.76	74.54	42.60	57.79	44.82	30.01	10.71
MW-Net [24]	89.76	86.52	81.68	56.56	66.73	59.44	49.19	19.04
GLC [10]	89.66	85.30	80.34	67.44	60.99	49.00	33.38	20.38
MLoC [29]	90.50	87.20	81.95	54.64	68.16	62.09	54.49	20.23
MLaC [38]	89.75	86.63	82.20	71.94	49.81	35.15	20.15	12.85
MSLC [30]	90.94	88.36	83.93	64.90	68.62	63.30	53.83	21.07
MFRW-MES† [23]	91.38	88.07	83.51	20.52	69.90	62.90	53.65	16.07
FasTEN [15]	91.94	90.07	86.78	79.52	68.75	63.82	55.22	37.36
STM (Ours)	**92.79**	**90.79**	**89.14**	**85.57**	**72.43**	**67.38**	**59.73**	**46.32**

We extracted the clean dataset consisting of 1K samples from a training set containing 50K samples. We utilized ResNet-34 as the backbone network for all of our experiments on these datasets.

Table 2. Test accuracy on CIFAR10 and CIFAR100 dataset with asymmetric label noise. The backbone used is a ResNet-34. p denotes the different levels of noise. The results for the cited methods are reported directly from their original papers. Instead, † indicates the results obtained by our implementation. The best result is marked in bold.

Dataset	Asym CIFAR-10			Asym CIFAR-100		
Noise p	0.2	0.4	0.6	0.2	0.4	0.6
L2RW [22]	88.79	85.86	–	59.11	55.12	–
MW-Net [24]	91.31	88.69	–	67.90	64.50	–
GLC [10]	91.56	89.76	–	64.43	54.20	–
MLoC [29]	91.15	89.35	–	69.20	66.48	–
MLaC [38]	91.45	90.26	–	56.46	49.20	–
MSLC [30]	91.45	89.26	–	70.86	66.99	–
MFRW-MES† [23]	92.11	91.15	84.63	69.31	64.01	52.04
FasTEN [15]	92.29	90.43	86.70	70.35	67.93	64.8
STM (Ours)	**92.81**	**91.34**	**86.75**	**73.49**	**70.67**	**66.68**

Results Comparison. Table 1 presents a summary of the evaluation outcomes obtained for CIFAR-10/100 datasets exposed to symmetric noise. On the other hand, Table 2 displays the performance results obtained for asymmetric noise. Our proposed approach has attained the state-of-the-art performance on all noise levels for both CIFAR-10/-100 datasets. Our method has demonstrated superior performance over the baselines, particularly under high noise levels ($p = 0.8$ sym and $p = 0.6$ asym). These results establish that our proposed methodology effectively copes with the challenges posed by noisy labels during the learning process.

Qualitative Results. We show in Fig. 2 the estimated matrices T, S^O, and S^V at distinct training epochs. During the initial phase, the transition matrix properly captures the label noise, which is evident from Figs. 2a and 2b. However, as the training progresses towards the final stages, the transition matrix T starts transforming into a diagonal matrix, resulting in a decline in its ability to correct labels. Conversely, the smoothing matrices S^O and S^V exhibit an opposite trend. In the early stages of training, the diagonal elements of the smoothing matrices possess higher values than the other matrix elements. In contrast, as the training approaches completion, there is a broader distribution of higher values across all matrix elements, as illustrated in Fig. 2. This means that when the model is less confident on a class (initial phase of the training) a small amount of smoothness is applied to the other labels. Instead, when the model becomes more confident on a class (last phases of the training process) the smoothing value added to the others is higher.

4.3 Experiments on Clothing1M

Dataset and Settings. The Clothing1M dataset [32] is a large-scale real-world dataset containing one million samples, out of which 14K samples have been manually annotated for clean labels. The dataset exhibits a significant degree of noise, which renders it a challenging benchmark for the task of image classification. To conduct a rigorous evaluation of the performance of deep neural networks on this dataset, we adopt the original division of clean and noisy data. In order to ensure fairness and comparability with prior works, we employ the ResNet-50 architecture, which has been pre-trained on the ImageNet dataset [4], as the initial backbone architecture for our experiments.

Table 3. Comparison with state-of-the-art methods in test accuracy (%) on Clothing1M dataset with real-world noise. Results are reported from the original papers. The best result is marked in bold.

Method	Accuracy (%)
Baseline (CE) [24]	68.94
F-correction [20]	69.84
S-adaptation [6]	70.36
MLoC [29]	71.10
L2RW [22]	72.04
GLC [10]	73.69
MW-Net [24]	73.72
MSLC [30]	74.02
FaMUS [33]	74.43
DivideMix [16]	74.76
AugDesc [18]	75.11
MLaC [38]	75.78
MFRW-MES [23]	77.44
FasTEN [15]	77.83
STM (Ours)	**78.23**

Baselines. In addition to the baselines introduced in the previous experiments, we compare DivideMix [16], and AugDesc [18] that utilize semi-supervised learning techniques and diverse data augmentation strategies. F-correction [20], employs transition matrix estimation that leverage selected data points, without the need for labeled data.

Results Comparison. Table 3 presents the evaluation results of our proposed method on the Clothing1M dataset, which comprises class and instance-dependent noisy labels. Our method outperforms the baselines by a substantial margin, attaining remarkable performance. These findings suggest that our proposed strategy is well-suited to addressing real-world problems that often involve label corruption.

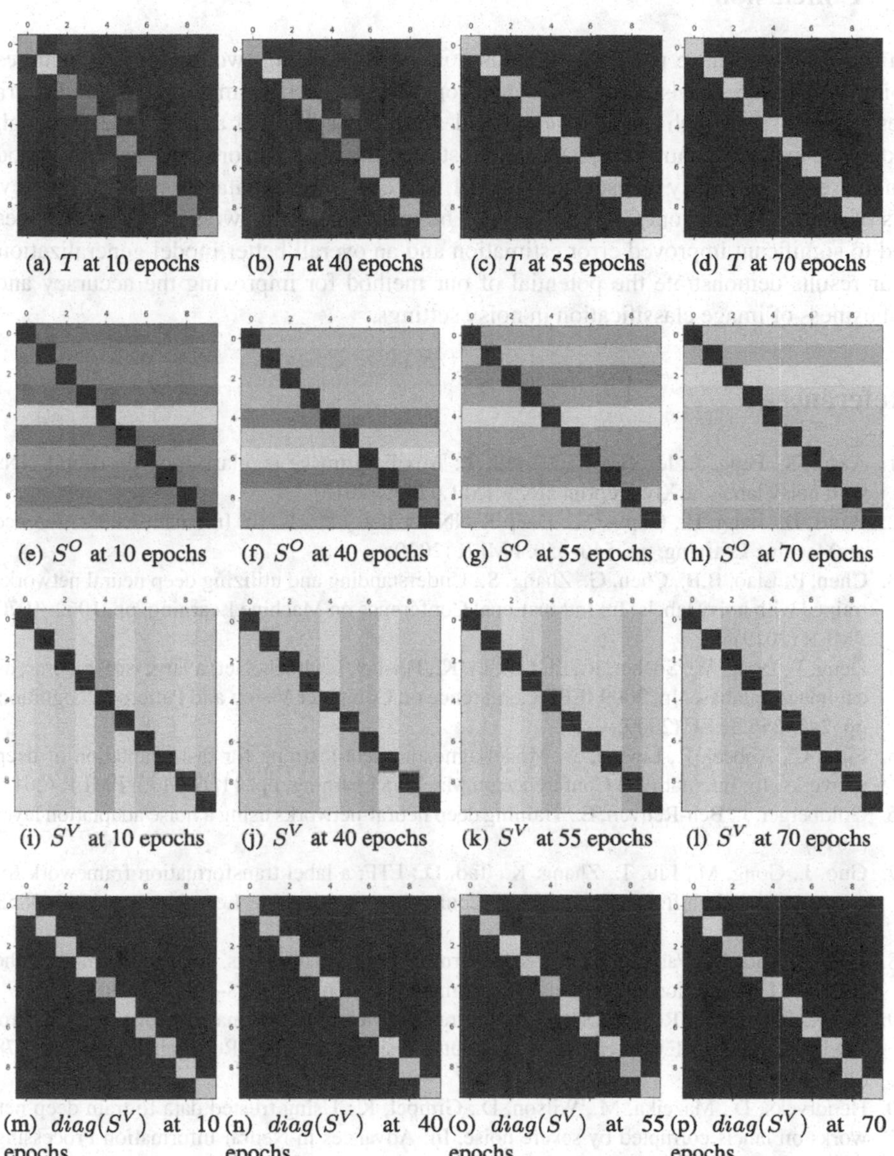

(a) T at 10 epochs (b) T at 40 epochs (c) T at 55 epochs (d) T at 70 epochs

(e) S^O at 10 epochs (f) S^O at 40 epochs (g) S^O at 55 epochs (h) S^O at 70 epochs

(i) S^V at 10 epochs (j) S^V at 40 epochs (k) S^V at 55 epochs (l) S^V at 70 epochs

(m) $diag(S^V)$ at 10 epochs (n) $diag(S^V)$ at 40 epochs (o) $diag(S^V)$ at 55 epochs (p) $diag(S^V)$ at 70 epochs

Fig. 2. Sequence of estimated matrices calculated during training on CIFAR10 with sym noise ($p = 0.8$) at 10, 40, 55, 70 epochs, respectively from left to right. On first row there are the transition matrices T learned at the different epochs. In the second row the horizontal smoothing matrix S^O without the diagonal. In the third row the vertical smoothing matrix S^V without the diagonal. In the last row the diagonal of S^O and S^V (the diagonal is the same for the two matrices). The value of the matrices are displayed with the standard colors that the matplotlib library uses to visualize a 2D matrix as color-coded image.

5 Conclusion

In this work, we have presented a methodology for learning two smoothing matrices jointly with a label transition matrix. Our proposed approach estimates these three matrices using a small high-quality dataset and a three-head model architecture. Through extensive experimentation, we have demonstrated the great performance of our method on the real-world noisy dataset Clothing1M, and on synthetic datasets with varying levels of noise derived from CIFAR-10/100. The inclusion of the two smoothing matrices led to significant improved error estimation and an overall better model generalization. Our results demonstrate the potential of our method for improving the accuracy and robustness of image classification in noisy settings.

References

1. Azadi, S., Feng, J., Jegelka, S., Darrell, T.: Auxiliary image regularization for deep CNNs with noisy labels. arXiv preprint arXiv:1511.07069 (2015)
2. Bahri, D., Jiang, H., Gupta, M.: Deep K-NN for noisy labels. In: International Conference on Machine Learning, pp. 540–550. PMLR (2020)
3. Chen, P., Liao, B.B., Chen, G., Zhang, S.: Understanding and utilizing deep neural networks trained with noisy labels. In: International Conference on Machine Learning, pp. 1062–1070. PMLR (2019)
4. Deng, J., Dong, W., Socher, R., Li, L.J., Li, K., Fei-Fei, L.: ImageNet: a large-scale hierarchical image database. In: 2009 IEEE Conference on Computer Vision and Pattern Recognition, pp. 248–255. IEEE (2009)
5. Finn, C., Abbeel, P., Levine, S.: Model-agnostic meta-learning for fast adaptation of deep networks. In: International Conference on Machine Learning, pp. 1126–1135. PMLR (2017)
6. Goldberger, J., Ben-Reuven, E.: Training deep neural-networks using a noise adaptation layer (2016)
7. Guo, J., Gong, M., Liu, T., Zhang, K., Tao, D.: LTF: a label transformation framework for correcting label shift. In: International Conference on Machine Learning, pp. 3843–3853. PMLR (2020)
8. Han, J., Luo, P., Wang, X.: Deep self-learning from noisy labels. In: Proceedings of the IEEE/CVF International Conference on Computer Vision, pp. 5138–5147 (2019)
9. He, K., Zhang, X., Ren, S., Sun, J.: Deep residual learning for image recognition. In: Proceedings of the IEEE Conference on Computer Vision and Pattern Recognition, pp. 770–778 (2016)
10. Hendrycks, D., Mazeika, M., Wilson, D., Gimpel, K.: Using trusted data to train deep networks on labels corrupted by severe noise. In: Advances in Neural Information Processing Systems, vol. 31 (2018)
11. Hu, W., Li, Z., Yu, D.: Simple and effective regularization methods for training on noisily labeled data with generalization guarantee. arXiv preprint arXiv:1905.11368 (2019)
12. Jang, E., Gu, S., Poole, B.: Categorical reparameterization with Gumbel-Softmax. arXiv preprint arXiv:1611.01144 (2016)
13. Jiang, L., Zhou, Z., Leung, T., Lif, L.J., Fei-Fei, L.: MentorNet: learning data-driven curriculum for very deep neural networks on corrupted labels. In: International Conference on Machine Learning, pp. 2304–2313. PMLR (2018)
14. Krizhevsky, A., Hinton, G., et al.: Learning multiple layers of features from tiny images (2009)

15. Kye, S.M., Choi, K., Yi, J., Chang, B.: Learning with noisy labels by efficient transition matrix estimation to combat label miscorrection. In: Avidan, S., Brostow, G., Cissé, M., Farinella, G.M., Hassner, T. (eds.) Computer Vision-ECCV 2022: 17th European Conference, Tel Aviv, Israel, 23–27 October 2022, Proceedings, Part XXV, vol. 13685, pp. 717–738. Springer, Cham (2022). https://doi.org/10.1007/978-3-031-19806-9_41
16. Li, J., Socher, R., Hoi, S.C.: DivideMix: learning with noisy labels as semi-supervised learning. arXiv preprint arXiv:2002.07394 (2020)
17. Mirzasoleiman, B., Cao, K., Leskovec, J.: Coresets for robust training of deep neural networks against noisy labels. In: Advances in Neural Information Processing Systems, vol. 33 (2020)
18. Nishi, K., Ding, Y., Rich, A., Höllerer, T.: Augmentation strategies for learning with noisy labels. arXiv preprint arXiv:2103.02130 (2021)
19. Patrini, G., Nielsen, F., Nock, R., Carioni, M.: Loss factorization, weakly supervised learning and label noise robustness. In: International Conference on Machine Learning, pp. 708–717. PMLR (2016)
20. Patrini, G., Rozza, A., Krishna Menon, A., Nock, R., Qu, L.: Making deep neural networks robust to label noise: a loss correction approach. In: Proceedings of the IEEE Conference on Computer Vision and Pattern Recognition, pp. 1944–1952 (2017)
21. Radford, A., et al.: Learning transferable visual models from natural language supervision. In: International Conference on Machine Learning, pp. 8748–8763. PMLR (2021)
22. Ren, M., Zeng, W., Yang, B., Urtasun, R.: Learning to reweight examples for robust deep learning. In: International Conference on Machine Learning, pp. 4334–4343. PMLR (2018)
23. Ricci, S., Uricchio, T., Bimbo, A.D.: Meta-learning advisor networks for long-tail and noisy labels in social image classification. ACM Trans. Multimedia Comput. Commun. Appl. **19**(5s), 1–23 (2023)
24. Shu, J., et al.: Meta-weight-net: learning an explicit mapping for sample weighting. arXiv preprint arXiv:1902.07379 (2019)
25. Song, H., Kim, M., Lee, J.G.: SELFIE: refurbishing unclean samples for robust deep learning. In: International Conference on Machine Learning, pp. 5907–5915. PMLR (2019)
26. Sukhbaatar, S., Bruna, J., Paluri, M., Bourdev, L., Fergus, R.: Training convolutional networks with noisy labels. arXiv preprint arXiv:1406.2080 (2014)
27. Taigman, Y., Yang, M., Ranzato, M., Wolf, L.: DeepFace: closing the gap to human-level performance in face verification. In: Proceedings of the IEEE Conference on Computer Vision and Pattern Recognition, pp. 1701–1708 (2014)
28. Wang, Y., Ma, X., Chen, Z., Luo, Y., Yi, J., Bailey, J.: Symmetric cross entropy for robust learning with noisy labels. In: Proceedings of the IEEE/CVF International Conference on Computer Vision, pp. 322–330 (2019)
29. Wang, Z., Hu, G., Hu, Q.: Training noise-robust deep neural networks via meta-learning. In: Proceedings of the IEEE/CVF Conference on Computer Vision and Pattern Recognition, pp. 4524–4533 (2020)
30. Wu, Y., Shu, J., Xie, Q., Zhao, Q., Meng, D.: Learning to purify noisy labels via meta soft label corrector. In: Proceedings of the AAAI Conference on Artificial Intelligence, vol. 35, pp. 10388–10396 (2021)
31. Xia, X., et al.: Are anchor points really indispensable in label-noise learning? arXiv preprint arXiv:1906.00189 (2019)
32. Xiao, T., Xia, T., Yang, Y., Huang, C., Wang, X.: Learning from massive noisy labeled data for image classification. In: Proceedings of the IEEE Conference on Computer Vision and Pattern Recognition, pp. 2691–2699 (2015)
33. Xu, Y., Zhu, L., Jiang, L., Yang, Y.: Faster meta update strategy for noise-robust deep learning. In: Proceedings of the IEEE/CVF Conference on Computer Vision and Pattern Recognition, pp. 144–153 (2021)

34. Yao, J., Wu, H., Zhang, Y., Tsang, I.W., Sun, J.: Safeguarded dynamic label regression for noisy supervision. In: Proceedings of the AAAI Conference on Artificial Intelligence, vol. 33, pp. 9103–9110 (2019)
35. Yao, Y., et al.: Dual T: reducing estimation error for transition matrix in label-noise learning. arXiv preprint arXiv:2006.07805 (2020)
36. Zhang, H., Cisse, M., Dauphin, Y.N., Lopez-Paz, D.: Mixup: beyond empirical risk minimization. arXiv preprint arXiv:1710.09412 (2017)
37. Zhang, X., Wu, X., Chen, F., Zhao, L., Lu, C.T.: Self-paced robust learning for leveraging clean labels in noisy data. In: Proceedings of the AAAI Conference on Artificial Intelligence, vol. 34, pp. 6853–6860 (2020)
38. Zheng, G., Awadallah, A.H., Dumais, S.: Meta label correction for noisy label learning. In: Proceedings of the 35th AAAI Conference on Artificial Intelligence (2021)

Semi-supervised Classification for Remote Sensing Datasets

Itza Hernandez-Sequeira[1(✉)], Ruben Fernandez-Beltran[2], Yonghao Xu[3], Pedram Ghamisi[3,4], and Filiberto Pla[1]

[1] Institute of New Imaging Technologies, University Jaume I, Castellón de la Plana, Spain
{isequeir,pla}@uji.es

[2] Department of Computer Science and Systems, University of Murcia, Murcia, Spain
rufernan@um.es

[3] Institute of Advanced Research in Artificial Intelligence, Vienna, Austria
yonghao.xu@iarai.ac.at

[4] Helmholtz-Zentrum Dresden-Rossendorf, Dresden, Germany
p.ghamisi@hzdr.de

Abstract. Deep semi-supervised learning (DSSL) is a rapidly-growing field that takes advantage of a limited number of labeled examples to leverage massive amounts of unlabeled data. The underlying idea is that training on small yet well-selected examples can perform as effectively as a predictor trained on a larger number chosen at random [14]. In this study, we explore the most relevant approaches in DSSL literature like FixMatch [19], CoMatch [13], and, the class aware contrastive SSL (CCSSL) [25]. Our objective is to perform an initial comparative study of these methods and assess them on two remote sensing (RS) datasets: UCM [27] and AID [22]. The performance of these methods was determined based on their accuracy in comparison to a supervised benchmark. The results highlight that the CoMatch framework achieves the highest accuracy for both the UCM and AID datasets, with accuracies of 95.52% and 93.88% respectively. Importantly, all DSSL algorithms outperform the supervised benchmark, emphasizing their effectiveness in leveraging a limited number of labeled examples to enhance classification accuracy for remote sensing scene classification tasks. The code used in this study was adapted from CCSSL [25] and the detailed implementation will be accessible at https://github.com/itzahs/SSL-for-RS.

Keywords: semisupervised learning · deep learning · reduced labels · remote sensing

1 Introduction

Remote sensing (RS) is an Earth Observation (EO) technology that utilizes a range of sensors on satellites, aircraft, and terrestrial platforms to collect non-

This work was supported by the Ministry of Science and Innovation of Spain (PID2021-128794OB-I00), and the University Jaume I (PREDOC/2020/50 and Becas de Estancia de Investigación E-2022-30).

G. L. Foresti et al. (Eds.): ICIAP 2023, LNCS 14233, pp. 463–474, 2023.
https://doi.org/10.1007/978-3-031-43148-7_39

intrusive EO data [6]. These sensors enable diverse applications such as agriculture monitoring, urban planning, and natural disaster response [20]. However, RS datasets often suffer from the scarcity of labeled data, which is costly and time-consuming to acquire, particularly for large-scale datasets.

In the field of RS and geoscience, deep learning (DL) algorithms have gained prominence due to their autonomous learning capability, scalability on GPU/TPU architectures [3], and versatility in tasks such as land cover classification and object detection [15]. Moreover, deep semi-supervised learning (DSSL) algorithms leverage both labeled and unlabeled data to improve model performance. By integrating DSSL algorithms into RS, the challenges associated with limited labeled data can be addressed, as these algorithms effectively utilize the available labeled data alongside a larger pool of unlabeled data.

DSSL models employ various techniques, including generative, consistency regularization, graph-based, pseudo-labeling, and hybrid methods [26]. Generative modeling refers to creating artificial instances from a dataset but retaining similar characteristics of the original dataset, for example, semi-supervised Generative Adversarial Networks (GANs) [5] and Variational AutoEncoders (VAEs) [9]. In terms of consistency regularization, the model is fed two of the same unlabeled samples with different perturbations like additive noise [17], random augmentation [23], or adversarial training [16]. Then, a regularization term applied to the final loss (e.g. cross-entropy) forces the model to output similar predictions. Graph-based methods implement deep embedding methods such as AutoEncoders [10] and Graph Neural Networks (GNN) [11] to create representations of the training samples (nodes/vertices) and the relationships between the nodes (edges). Then it proceeds to place new unlabeled samples within the context of the graph. Finally, pseudo-labeling [12] uses the model's own confident predictions to generate high-confidence pseudo-labels, which can be added to the training dataset as labeled examples.

Hybrid methods are a combination of various methodologies and other components used to improve performance [26]. FixMatch [19] uses pseudo labeling on weakly-augmented unlabeled data. If the prediction is above the threshold of 0.95, it is converted to a hard "one-hot" pseudo-label. Then, via consistency regularization, it guides the strongly-augmented unlabeled data toward the same label. For CoMatch [13], similar to FixMatch, uses the weakly-augmented images to create pseudo-labels that are then used as targets to train the class predictions on strongly-augmented ones. However, it introduces contrastive learning to use graph-based feature representations to create smooth constraints on pseudo-labels.

Currently, there are unified benchmarks of the SSL methods [21] however these focus on the performance with few labeled samples and do not consider tasks such as imbalanced or out-of-distribution datasets. In this work, aside from the hybrid methods like FixMatch and CoMatch, we include a Class Aware Contrastive SSL (CCSSL) [27]. CCSSL introduces the idea that raw data contains in-distribution (known classes and balanced datasets) and out-of-distribution (unknown classes or unlabeled distribution) data. A threshold divides the unlabeled images: for highly possible in-distribution data it applies class-aware clustering in the feature space and for out-of-distribution data, it follows a contrastive learning mechanism.

In this study, our focus is on utilizing deep semi-supervised learning (DSSL) algorithms for remote sensing datasets (RS). In Sect. 2, we introduce the specific RS datasets used to evaluate the considered models. Then we provide details on the algorithms used, and the experimental configurations, including the choice of backbone architecture, augmentation techniques employed, and other relevant hyperparameters. In Sect. 3, we present the comparative results of the models based on their best-reported accuracy. Additionally, we provide a comprehensive computational cost analysis that takes into account various factors, including GPU memory usage, total training time, and data loading time. Finally, in Sect. 4, we offer conclusions, discuss any limitations encountered, and propose potential directions for future research.

2 Data and Methods

In this section, we provide an overview of the remote sensing datasets utilized in the context of semi-supervised learning (SSL), specifically focusing on the number of labels available for each dataset. Additionally, we present the deep semi-supervised learning (DSSL) algorithms employed in our study and provide details regarding their code base and the necessary configuration for conducting the experiments.

2.1 Datasets Splits for Supervised and Semi-supervised

We used two aerial image-based datasets, the UCM dataset [27] containing 21 classes with a total of 100 images of 256×256 pixels for each class, and the AID dataset [22] with 30 classes of about 200 to 400 samples per class with a size of 600×600 pixels in each class.

In fully supervised scenarios, the chosen train, and test splits are 50% both for UCM and AID datasets [2]. The algorithm then uses 1,050 labeled samples for UCM and 5,000 for AID for both training and testing. As for semi-supervised learning, [19] provided the benchmark for evaluating SSL algorithm in computer vision (CV) with 4, 25, and 100 labeled samples per class. The correspondence in the remote sensing datasets is summarized in Table 1, we used 4/25/40 labeled examples per class. Using 4 labeled samples per class corresponds to a total of 84 labeled samples in UCM versus the 1,050 labeled samples required in a

Table 1. Remote sensing datasets and the number of labels per class for SSL classification.

Dataset	Class	# Labels per class	# Training data	# Testing data	Train/Test split
UCM [27]	21	4/25/40	84/525/840	1, 050	50%/50%
AID [22]	30	4/25/40	120/750/1200	5, 000	50%/50%

supervised manner. This is the smallest amount of labeled data used during training, as the goal is for the algorithm to perform better with fewer labeled samples.

2.2 Experiments Configuration

For the experiments, we used the CVPR2022 PyTorch implementation of the Class Aware Contrastive Semi-Supervised Learning (CCSSL) [25]. Taking into consideration the computational requirements, we customized the available configuration files with the selected algorithms, datasets, backbone, augmentations, and training settings.

Algorithms. The repository contains the FixMatch, CoMatch, and CCSSL algorithms. FixMatch (Fig. 1) combines consistency regularization and pseudo-labeling. The pseudo-labeling process generates artificial labels based on weakly-augmented images and enforces the loss against strongly-augmented versions [19]. However, a drawback of FixMatch is that it relies solely on the model's output to justify its predictions, which can lead to confirmation bias [1].

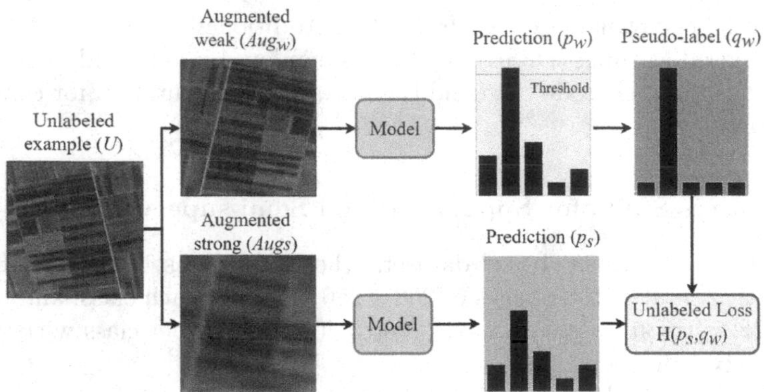

Fig. 1. Diagram of FixMatch (adapted from [19]).

CoMatch (Fig. 2) aims to address some of FixMatch limitations by incorporating ideas from contrastive learning, and graph-based learning. It utilizes a co-training framework where class probabilities from the classification head and low-dimensional embeddings from the projection head interact and co-evolve [13]. CCSSL, builds on CoMatch and introduces class-aware contrastive learning to semi-supervised training. In CCSSL, the unlabeled data is separated into in-distribution and out-of-distribution based on the likelihood of their maximum prediction confidence. For in-distribution data, class-aware clustering is applied in the feature space to enhance compatibility with downstream tasks. For out-of-distribution data, contrastive learning is employed using different views of the same image as positive samples [25].

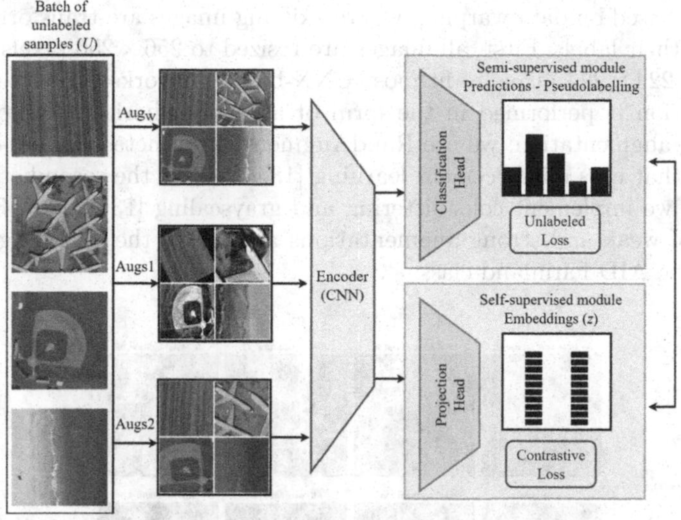

Fig. 2. Diagram of CoMatch (adapted from [13]).

Backbone. Following thorough experimentation, we chose the WideResNet-28-2 [28] network architecture as the backbone for our study. This choice was motivated by its superior classification performance when compared to the other available backbones, including WRN-28-8 [28], ResNet18 [7], and RN-50 [7]. The WideResNet is known for its ability to capture more diverse and complex features due to its wider layers compared to standard ResNet architectures. This is accomplished by augmenting the number of channels or filters within each layer [28]. However, this enhancement comes at the cost of slower training, as the addition of more features increased the computational resources and time needed. Although WRN initially achieved higher accuracies than ResNet in our calculations, it requires three times more GPU memory and seven times more training time per epoch than ResNet-18 [13].

The WRN-28-2 architecture was primarily used for most experiments with a parameter count of 1.47M. However, for the CoMatch and FixMatch+CCSSL techniques, an additional projection head was incorporated, resulting in a parameter count of 1.49M. The projection head consisted of a 2-layer Multilayer Perceptron (MLP) that generated 64-dimensional embeddings for the implementation of the contrastive loss.

Data Augmentation. The codebase includes a class constructor for custom datasets, for which we have provided the train and test text files. We obtained these files from the UAE-RS repository [24]. A key component of the DSSL algorithms training is the data augmentation. In the case of FixMatch, it contains a weak Aug_w and a strong augmentation. However, both CoMatch and CCSSL require one weak and two strong augmentations Aug_{s1} and Aug_{s2}. All augmentations are based on data warping, where existing images are transformed while preserving their labels. First, all images are resized to 256×256 pixels and then cropped to 224×224 pixels to fit most CNN-based networks. Then a geometric transformation is performed in the form of a random horizontal flip. For the first strong augmentation, we use RandAugment [4], a meta-learning data augmentation that uses reinforcement learning [18]. And for the second strong augmentation, we implement color jittering and grayscaling [13]. Figure 3 presents examples of weak and strong augmentations applied to the UCM Agricultural class and the AID Farmland class.

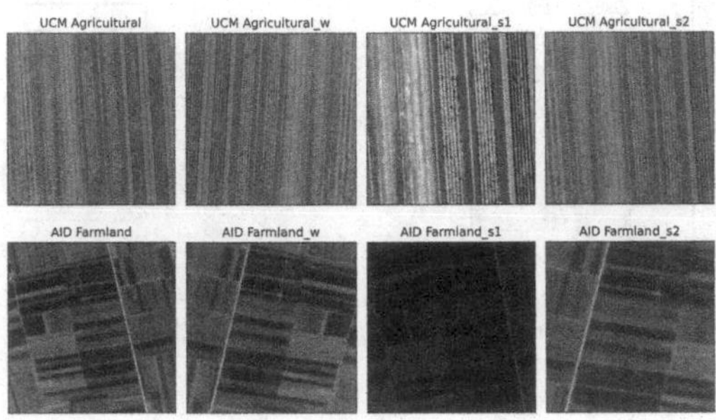

Fig. 3. Weak and strong augmentation examples for UCM (upper) and AID (lower).

Training Setting. The remaining hyperparameters were kept as [19], these being the weight of the unlabeled loss $\lambda_u = 1$, initial learning rate $\eta = 0.03$, the momentum SGD $\beta = 0.9$, the pseudo-label and the confidence threshold $\tau = 0.95$. For the CCSSL model, we ran it without the separation of in-out distribution (tpush $= 0$) because both UCM and AID are balanced datasets. Tpush is used for noisier datasets and improves the classification when out-of-distribution data is present [25].

For this initial study, we conducted a range of experiments to evaluate different approaches. Firstly, we employed fully supervised learning, utilizing 50% of the available training data. This involved using half of the labeled data for training the model and holding out the remaining half for evaluation purposes. Additionally, we explored supervised learning scenarios with limited labeled samples per class. Specifically, we experimented training the model with 4, 25, and 40 labeled samples per class. Lastly, we investigated the performance of semi-supervised learning algorithms, including FixMatch, CoMatch, and FixMatch+CCSSL with 4, 25, and 40 labeled samples per class. During each step of the training process, the batch consisted of an equal number of labeled and unlabeled examples. The ratio of unlabeled to labeled data was 7:1 ($\mu = 7$), which implies that for every batch containing 8 labeled examples, there were 56 unlabeled examples.

Computational Requirements. The experiments in this study were conducted using two different GPUs: an NVIDIA A5000 with 24 GB of VRAM and an NVIDIA A100 SXM4 with 40 GB of VRAM. To achieve a balance between computational resources and convergence, we used a batch size of 8 and 200 epochs as the standard configuration for training. However, there were exceptions in which a batch size of 4 was employed. These exceptions occurred for the CoMatch and the FixMatch+CCSSL algorithms applied to both the UCM and AID datasets with 40 labeled examples. When using 40 labeled examples, the model size exceeded the capacity of the GPU, as further computational details can be found in the results section of the paper.

To enhance future research in deep semi-supervised learning, it is recommended to increase the batch sizes to 16, 32, or 64 and extend the training durations. To ensure more robust results and consider the significant variability in training with limited labeled samples, it is advised to conduct experiments using five different random seeds, as suggested by [19].

It is worth noting that the FixMatch paper utilized a TPU with 32 cores, which is roughly equivalent to 32 NVIDIA V100 in terms of computing power [19]. This high computational power allowed them to train with larger batch sizes, aiming to improve overall performance. Similarly, the CoMatch paper [13] conducted their experiments using 8 NVIDIA V100, and the CCSSL paper [25] employed 4 NVIDIA V100 for their experiments. The use of multiple GPUs allowed for parallel processing and efficient utilization of computational resources.

3 Experimentation Results

This section presents the results obtained from the experimentation with various SSL methods for remote sensing scene classification. We begin by comparing the accuracy of the SSL methods to their supervised baseline and then to that reported in previous works. Then, we discuss the computational costs to train these models.

3.1 Accuracy Comparison

The results shown in Table 2 correspond to the best accuracy obtained on the testing dataset for 200 epochs using supervised and SSL methods on UCM and AID datasets. The table shows that the CoMatch framework achieved the best performance with 95.52% and 93.88% accuracy for UCM and AID datasets, respectively. CoMatch proved to be a robust SSL algorithm that can generalize well to different datasets and label settings. In general, all the accuracies reported for the SSL algorithms outperformed their supervised benchmark. If we compare FixMatch, when using 40 labels the accuracy increases in 1% for UCM and 2% for AID reaching 94.48% and 92.8% respectively, when compared to the supervised approach.

Table 2. Results for 200 epochs for Fully Supervised, Supervised, and SSL methods.

Dataset	UCM [27]			AID [22]		
# Labels per class	4	25	40	4	25	40
# Total samples	84	525	840	120	750	1200
Supervised	90.00	90.86	93.52	85.34	85.90	89.94
FixMatch [19]	94.38	92.00	94.48	92.66	87.30	92.8
CoMatch [13]	95.52	94.76	93.81	93.04	93.88	89.26
FixMatch+CCSSL [25]	94.00	94.67	94.67	90.94	91.28	92.00
# Total samples	1, 050			5, 000		
Fully Supervised	93.71			90.2		

Our experiments with supervised classification using 25 samples on the UCM dataset were in line with previous findings. Specifically, the literature reported accuracy rates of over 86% when utilizing 10 samples on the same dataset [8]. Overall, the accuracy of the classification increased with the number of labeled samples which suggests that a large number of labeled samples is required to improve the accuracy in supervised RS scene classification. However, it is important to note that the supervised literature has reported higher accuracies using a CNN-CapsNet, with values of 97.59% and 96.81% for AID and UCM datasets, respectively [29].

In terms of the number of labeled samples per class, our results indicate that with as little as 4 labeled samples per class, the SSL algorithms were able to outperform the supervised learning algorithm, which required 50% of the training dataset. For instance, when using 4 labels per class, UCM reported 90% on the supervised baseline compared to 94.38% on FixMatch or 95.52% on CoMatch. This suggests that SSL can reduce the labeling cost and enable the training of accurate classifiers with very few labeled samples.

When applying the CCSSL framework to FixMatch with 25 labels per class, it improved the FixMatch by about 2% for UCM and 4% for AID. Interestingly, adding the CCSSL framework generally improves results even without threshold separation. In contrast to CoMatch, FixMatch+CCSSL demonstrated lower accuracy across all label settings and datasets. It is important to note that although CoMatch+CCSSL was available as an option, it was not included in the conducted experiments due to its high computational demands.

3.2 Analysis of Computational Costs

This section provides a detailed computational cost analysis for different trainers. Table 3 summarizes the findings, including factors such as the number of labels per class used, the batch size, GPU memory usage, total training time, the time per batch, and data loading time.

Table 3. Overview of computational costs for the trainers.

Trainer	N°	UCM					AID				
	Label	B	GPU	Ttime	Btime	DLoad	B	GPU	Ttime	Btime	DLoad
Fully Supervised	50%	8	<12	50:53	0.89	0.72	8	<12	177:25 h	2.94	2.74
Supervised	4	8	<12	12:39	0.22	0.03	8	<12	13:01	0.21	0.03
Supervised	25	8	<12	6:37	0.11	0.02	8	<12	12:49	0.21	0.12
Supervised	40	8	<12	66:06	1.15	0.97	8	<12	177:23	2.94	2.74
FixMatch	4	8	12–24	68:17	1.2	0.25	8	12–24	123:56	2.11	1.19
FixMatch	25	8	12–24	50:33	0.88	0.03	8	12–24	51:04	0.88	0.02
FixMatch	40	8	12–24	67:39	1.19	0.23	8	12–24	210:09	3.53	2.55
CoMatch	4	8	24–40	71:08	1.24	0.04	8	24–40	70:41	1.22	0.03
CoMatch	25	8	24–40	71:31	1.26	0.04	8	24–40	336:38	1.25	0.03
CoMatch	40	4	12–24	59:11	1.04	0.29	4	12–24	90:43	1.51	0.76
FM+CCSSL	4	8	24–40	70:40	1.23	0.04	8	24–40	71:21	1.22	0.03
FM+CCSSL	25	8	24–40	70:39	1.26	0.04	8	24–40	71:37	1.25	0.03
FM+CCSSL	40	4	12–24	59:49	1.04	0.31	4	12–24	46:17	0.78	0.02

Note: B denotes batch size, GPU indicates usage in GiB, Ttime represents training time (hours), Btime is time per batch, and DLoad is data loading time (seconds).

The training times vary significantly across the different methods. Fully supervised demonstrates relatively shorter training times, ranging from 50 h and 53 min to 177 h and 25 min. In contrast, FixMatch, CoMatch, and FM+CCSSL exhibit longer training durations, spanning from 68 h and 17 min to 336 h and 38 min.

The GPU usage remains consistent across most methods, with values less than 12 GiB for supervised methods, ranging from 12–24 GiB for FixMatch with varying label ratios, and from 24–40 GiB for CoMatch and FixMatch+CCSSL. It is worth noting that for CoMatch and FixMatch+CCSSL with 40 labels, the batch size had to be reduced to accommodate GPU limitations.

Both UCM and AID images are resized to 224 for the model, with UCM containing 21 classes and AID having 30 classes. On average, the processing time for each training batch is higher for AID, ranging from 0.21 to 3.53 s, compared to UCM, which ranges from 0.11 to 1.26 s.

FixMatch emerges as the SSL method with the fastest training time across different label ratios, making it an efficient choice. However, it is crucial to note that CoMatch achieved the best classification accuracy among the methods. However, CoMatch may face challenges when applied to large-scale datasets with a significantly higher number of classes. Constructing a meaningful pseudo-label graph requires a sufficient number of samples from each class in the unlabeled batch, which can be challenging when there are memory limitations. Therefore, the scalability of CoMatch should be carefully considered when dealing with datasets containing a large number of classes.

4 Conclusions

In this study, we compared the performance of several deep semi-supervised (DSSL) learning algorithms on two remote sensing datasets, UCM and AID, using different numbers of labeled examples per class. The results showed that the hybrid methods, especially CoMatch, outperformed other algorithms in both datasets. Even in datasets with low noise, including the CCSSL framework to separate in and out of distribution data proved useful for improving the accuracy obtained with FixMatch. Additionally, our experiments suggest that fully supervised learning requires a large number of labeled samples to achieve higher accuracy. Overall, our results demonstrate that DSSL models can achieve better accuracy with a small number of labeled samples, which is significant for practical applications where obtaining labeled data is costly and time-consuming. To optimize computational costs, it is crucial to consider alternative backbones, given the computational demands associated with using WideResNet (WRN). In situations where GPU availability is limited, reducing image size and adjusting the labeled-to-unlabeled data ratio can potentially enable an increase in batch size, contributing to improved efficiency.

References

1. Arazo, E., Ortego, D., Albert, P., O'Connor, N.E., McGuinness, K.: Pseudo-labeling and confirmation bias in deep semi-supervised learning. CoRR abs/1908.02983 (2019). arxiv.org/abs/1908.02983
2. Cheng, G., Xie, X., Han, J., Guo, L., Xia, G.S.: Remote sensing image scene classification meets deep learning: challenges, methods, benchmarks, and opportunities. IEEE J. Sel. Topics Appl. Earth Observ. Remote Sens. **13**, 3735–3756 (2020). https://doi.org/10.1109/JSTARS.2020.3005403
3. Chollet, F.: Deep Learning with Python, 2nd edn. Manning Publications, New York (2021). https://www.manning.com/books/deep-learning-with-python-second-edition
4. Cubuk, E.D., Zoph, B., Shlens, J., Le, Q.: RandAugment: practical automated data augmentation with a reduced search space. In: Larochelle, H., Ranzato, M., Hadsell, R., Balcan, M., Lin, H. (eds.) Advances in Neural Information Processing Systems, vol. 33, pp. 18613–18624. Curran Associates, Inc. (2020)
5. Goodfellow, I., et al.: Generative adversarial nets. In: Ghahramani, Z., Welling, M., Cortes, C., Lawrence, N., Weinberger, K. (eds.) Advances in Neural Information Processing Systems, vol. 27. Curran Associates, Inc. (2014). https://proceedings.neurips.cc/paper/2014/file/5ca3e9b122f61f8f06494c97b1afccf3-Paper.pdf
6. Guo, H., Goodchild, M.F., Annoni, A. (eds.): Manual of Digital Earth. Springer, Singapore (2020). https://doi.org/10.1007/978-981-32-9915-3
7. He, K., Zhang, X., Ren, S., Sun, J.: Deep residual learning for image recognition. In: 2016 IEEE Conference on Computer Vision and Pattern Recognition (CVPR), pp. 770–778 (2016). https://doi.org/10.1109/CVPR.2016.90
8. Hu, F., Xia, G.S., Hu, J., Zhang, L.: Transferring deep convolutional neural networks for the scene classification of high-resolution remote sensing imagery. Remote Sen. **7**(11), 14680–14707 (2015). https://doi.org/10.3390/rs71114680

9. Kingma, D.P., Welling, M.: Auto-encoding variational Bayes. In: Bengio, Y., LeCun, Y. (eds.) 2nd International Conference on Learning Representations, ICLR 2014, Banff, AB, Canada, 14–16 April 2014, Conference Track Proceedings (2014). arxiv.org/abs/1312.6114
10. Kipf, T.N., Welling, M.: Variational graph auto-encoders (2016). https://doi.org/10.48550/ARXIV.1611.07308, arxiv.org/abs/1611.07308
11. Kipf, T.N., Welling, M.: Semi-supervised classification with graph convolutional networks. In: 5th International Conference on Learning Representations, ICLR 2017, Toulon, France, 24–26 April 2017, Conference Track Proceedings. OpenReview.net (2017). https://openreview.net/forum?id=SJU4ayYgl
12. Lee, D.H.: Pseudo-label : the simple and efficient semi-supervised learning method for deep neural networks. In: Workshop on Challenges in Representation Learning ICML, vol. 3, no. 2 (2013)
13. Li, J., Xiong, C., Hoi, S.C.H.: CoMatch: semi-supervised learning with contrastive graph regularization. In: 2021 IEEE/CVF International Conference on Computer Vision (ICCV), pp. 9455–9464 (2021). https://doi.org/10.1109/ICCV48922.2021.00934
14. Liu, Q., Liao, X., Carin, L.: Detection of unexploded ordnance via efficient semisupervised and active learning. IEEE Trans. Geosci. Remote Sens. **46**(9), 2558–2567 (2008). https://doi.org/10.1109/TGRS.2008.920468
15. Ma, L., Liu, Y., Zhang, X., Ye, Y., Yin, G., Johnson, B.A.: Deep learning in remote sensing applications: a meta-analysis and review. ISPRS J. Photogramm. Remote Sens. **152**, 166–177 (2019). https://doi.org/10.1016/j.isprsjprs.2019.04.015, https://www.sciencedirect.com/science/article/pii/S0924271619301108
16. Miyato, T., Maeda, S., Koyama, M., Ishii, S.: Virtual adversarial training: a regularization method for supervised and semi-supervised learning. IEEE Trans. Pattern Anal. Mach. Intell. **41**(8), 1979–1993 (2019). https://doi.org/10.1109/TPAMI.2018.2858821
17. Rasmus, A., Berglund, M., Honkala, M., Valpola, H., Raiko, T.: Semi-supervised learning with ladder networks. In: Cortes, C., Lawrence, N., Lee, D., Sugiyama, M., Garnett, R. (eds.) Advances in Neural Information Processing Systems, vol. 28. Curran Associates, Inc. (2015). https://proceedings.neurips.cc/paper/2015/file/378a063b8fdb1db941e34f4bde584c7d-Paper.pdf
18. Shorten, C., Khoshgoftaar, T.M.: A survey on image data augmentation for deep learning. J. Big Data **6**(1), 60 (2019). https://doi.org/10.1186/s40537-019-0197-0
19. Sohn, K., et al.: FixMatch: simplifying semi-supervised learning with consistency and confidence. In: Larochelle, H., Ranzato, M., Hadsell, R., Balcan, M., Lin, H. (eds.) Advances in Neural Information Processing Systems, vol. 33, pp. 596–608. Curran Associates, Inc. (2020). https://proceedings.neurips.cc/paper/2020/file/06964dce9addb1c5cb5d6e3d9838f733-Paper.pdf
20. Srivastava, P.K.K., Kumar, A., Mall, R.K., Saikia, P.: Earth Observation in Urban Monitoring: Techniques and Challenges, 1st edn. Elsevier, New York (2023)
21. Wang, Y., et al.: USB: a unified semi-supervised learning benchmark for classification (2022). https://doi.org/10.48550/ARXIV.2208.07204, arxiv.org/abs/2208.07204
22. Xia, G.S., et al.: AID: a benchmark data set for performance evaluation of aerial scene classification. IEEE Trans. Geosci. Remote Sens. **55**(7), 3965–3981 (2017). https://doi.org/10.1109/TGRS.2017.2685945
23. Xie, Q., Dai, Z., Hovy, E., Luong, T., Le, Q.: Unsupervised data augmentation for consistency training. In: Larochelle, H., Ranzato, M., Hadsell, R., Balcan, M.,

Lin, H. (eds.) Advances in Neural Information Processing Systems, vol. 33, pp. 6256–6268. Curran Associates, Inc. (2020). https://proceedings.neurips.cc/paper/2020/file/44feb0096faa8326192570788b38c1d1-Paper.pdf

24. Xu, Y., Ghamisi, P.: Universal adversarial examples in remote sensing: methodology and benchmark. IEEE Trans. Geosci. Remote Sens. **60**, 1–15 (2022). https://doi.org/10.1109/TGRS.2022.3156392
25. Yang, F., et al.: Class-aware contrastive semi-supervised learning. In: 2022 IEEE/CVF Conference on Computer Vision and Pattern Recognition (CVPR), pp. 14401–14410 (2022). https://doi.org/10.1109/CVPR52688.2022.01402
26. Yang, X., Song, Z., King, I., Xu, Z.: A survey on deep semi-supervised learning. IEEE Trans. Knowl. Data Eng. **35**(9), 1–20 (2022). https://doi.org/10.1109/TKDE.2022.3220219
27. Yang, Y., Newsam, S.: Bag-of-visual-words and spatial extensions for land-use classification. In: Proceedings of the 18th SIGSPATIAL International Conference on Advances in Geographic Information Systems, GIS 2010, pp. 270–279. Association for Computing Machinery, New York, NY, USA (2010). https://doi.org/10.1145/1869790.1869829
28. Zagoruyko, S., Komodakis, N.: Wide residual networks. In: Wilson, R.C., Hancock, E.R., Smith, W.A.P. (eds.) Proceedings of the British Machine Vision Conference (BMVC), pp. 87.1–87.12. BMVA Press, September 2016. https://doi.org/10.5244/C.30.87
29. Zhang, W., Tang, P., Zhao, L.: Remote sensing image scene classification using CNN-CapsNet. Remote Sens. **11**(5), 494 (2019). https://doi.org/10.3390/rs11050494, https://www.mdpi.com/2072-4292/11/5/494

Exploiting Exif Data to Improve Image Classification Using Convolutional Neural Networks

Ralf Lederer, Martin Bullin(✉), and Andreas Henrich

University of Bamberg, 96047 Bamberg, Germany

martin.bullin@uni-bamberg.de

Abstract. In addition to photo data, many digital cameras and smartphones capture Exif metadata which contain information about the camera parameters used when a photo was captured. While most semantic image recognition approaches only use pixel data for classification decisions, this work aims to examine whether Exif data can improve image classification performed by Convolutional Neural Networks (CNNs). We compare the classification performance and training time of fusion models that use both, image data and Exif metadata, for image classification in contrast to models that use only image data. The most promising result was obtained with a fusion model which was able to increase the classification accuracy for the selected target concepts by 7.5% compared to the baseline, while the average total training time of all fusion models was reduced by 7.9%.

Keywords: Convolutional Neural Networks · fusion networks · Exif metadata · semantic scene classification

1 Introduction

Due to the widespread availability and use of smartphones, it has never been so easy to take a quick photo in almost any situation in life. Therefore, countless digital photos are captured every day, which are then for example uploaded to social media platforms, shared with friends via instant messaging services, or stored in private or public photo libraries [8]. In order to extract knowledge from such a large amount of data, automatic image recognition approaches are used, which are able to identify objects in photos and classify photos into different meaningful categories [13]. Since image classification is usually performed using only raw pixel data, the Exif metadata captured by many digital cameras mostly remain unconsidered for the classification task. Exif data is a special kind of photo metadata, which amongst others provides information about the camera parameters used when a photo was captured, such as focal length, exposure time and ISO value. The Exif specification describes over 150 tags [3]. Depending on the camera used, different tags are recorded. While certain Exif tags have been

G. L. Foresti et al. (Eds.): ICIAP 2023, LNCS 14233, pp. 475–486, 2023.
https://doi.org/10.1007/978-3-031-43148-7_40

used successfully to improve the classification of selected target concepts, their usefulness in conjunction with deep learning approaches has not been further explored. The main contribution of this work is to evaluate whether a combination of EXIF metadata and raw image data can provide benefits in terms of classification accuracy and training time when using deep CNNs. We use different state-of-the-art CNN architectures to create baseline models that classify selected concepts based on pure image data. In addition, we create fusion models by combining the used CNNs with a multilayer perceptron (MLP) that classifies the same concepts based on Exif metadata.

The paper is organized as follows. Section 2 gives an overview of previous work in the area. In Sect. 3, the selected classification tasks "indoor-outdoor", "moving-static" and "object-landscape" are described. In addition, we describe the structure of the used training data which was exported from the image portal Flickr[1]. Subsequently, we highlight our approach to incorporate Exif data into the image classification process. We discuss the selection of Exif tags in the context of the chosen classification tasks and describe the architectures and the training process of the models used to classify images based on Exif metadata, image data and a combination of both. The experimental results are evaluated in Sect. 4. Section 5 concludes the paper.

2 Background and Related Work

The idea of using Exif metadata for image classification is not new. Boutell et al. integrate selected Exif tags into a Bayesian network to combine low-level image features with metadata cues [14]. They selected different Exif tags for each of the target concepts indoor-outdoor, sunset and manmade-natural, based on their classification power [2]. They then fuse the classification of a Support Vector Machine (SVM) based on color histograms and textual image features with the selected Exif data. The authors were able to increase the classification accuracy for all selected concepts, compared to the SVM baseline models. They used a dataset with a total of 24,000 photos. Ghazali et al. use different combinations of selected Exif tags to classify photos based solely on metadata [7]. They use a SVM and a k-Nearest Neighbor classifier (k-NN) to perform scenery-object classification and show that the Exif tags "FNumber" and "Flash" in combination with the image dimensions mainly contribute to the classification task. The best k-NN classifier reaches an accuracy of 75.15% for a dataset of 500 images. Ku et al. provide the idea of using Exif data to classify photos into different scene modes [12]. Scene modes provide presets for camera parameters used to capture certain types of scenes, such as portraits, scenery, and night shots. The authors use a SVM to decide whether an image was captured with the configuration of a firework scene mode, based on 20 Exif tags. They use a total of 1178 photos for evaluation and reach a classification accuracy of 80.8% for firework and 98.25% for non-firework photos. Safonov et al. use Exif data to detect defects, such as photo artefacts, noise, exposure problems, and red eyes, in order to reduce the computation time of an automatic image enhancement pipeline [17].

[1] https://www.flickr.com/.

Finally, it has to be noted that the GPS attribute information is also part of Exif. This specific information and its classification power has already been addressed in [1]. Therefore, we do not consider the GPS information here.

The approach of Boutell et al. [2,14] is the most similar compared to ours as the authors also use a combination of image data and Exif metadata to perform image classification. However, in contrast to our work, the authors do not use deep learning methods for their experiment. Our approach is also different in that we use more Exif tags for classification and investigate the classification power of Exif data in the context of super-concepts and sub-concepts. In addition, we investigate the influence of Exif data when using different and particularly small image sizes.

3 Experimental Design

The experiment is conducted using three individual classification tasks. *Indoor-outdoor* classification is the task of deciding whether a photo was taken indoors or outdoors. While outdoor photos are usually captured in daylight, indoor photos are often taken with artificial lighting. *Moving-static* classification describes the task of deciding whether the dominant object in the photo is in motion or stationary. We mainly use pictures of moving vehicles, like cars or boats, but also photos of people and other objects. The classification task *object-landscape* describes the task of deciding whether a photo shows a landscape or an object. Objects can be all kinds of everyday objects, such as bicycles or toys. We also consider living things such as animals or plants as objects, since the central task is to distinguish wide-angle scenes from close-ups. For example, a photo of a skyline should be classified as landscape and not as object. It does not matter whether the photo was captured indoors or outdoors. To investigate whether Exif data help to further separate photos that belong to the same *super-concept*, each classification task is divided into various *sub-concepts*. For example, we divide indoor photos into different types of rooms, such as kitchen or living room and outdoor photos into different types of landscapes, such as river or mountain. Table 1 provides an overview of all used super-concepts and sub-concepts. Each image is assigned one or two labels, with the super-concept label always present and one optional sub-concept label.

3.1 Training Data

Each instance in our dataset consists of a photo and the associated Exif data. The data was exported from the image portal Flickr using the Flickr API [6]. It is divided into three disjoint datasets, one for each classification task. To collect the data for individual concepts, we primarily relied on topic-based photo pools, so-called Flickr groups. These are used by Flickr users to collect photos on specific topics and are thus a way to obtain training data without the need for manual labeling [16]. Additionally, we collected photos and Exif data using Flickr free text search. When a free text search is performed, the Flickr API

Table 1. Considered concepts with the number of images for each concept. For non-unique sub-concepts the distribution to the two super-concepts is given.

Super-concepts	Sub-concepts
indoor (37887), outdoor (36473)	bathroom (2016), bedroom (2028), corridor (2041), kitchen (1997), office (2020), beach (2191), forest (2325), mountain (2116), river (2230), urban (2027), plant (5412|4897), dog (2963|3736), furniture (2388|2405), cat (3026|3183), portrait (2896|3511), sport (4387|4954)
moving (30386), static (28281)	boat (4022|3318), plane (1786|2837), motorcycle (2291|3169), car (3653|902)
object (23591), landscape (19395)	food (5000), furniture (2388), toys (6203), vehicle (5000), beach (2191), forest (2325), mountain (2116), skyline (7763)

returns photos whose title, description or user tags contain the corresponding search term. User tags are keywords used to succinctly describe the content of a photo. They can be added by the image authors. In addition, user tags are automatically determined and added by Flickr robots using semantic image recognition approaches [21]. Since Flickr does not define strict rules for image titles, descriptions and user tags, and because the way Flickr determines the relevance of photos for search terms is unknown, training data collected via free text search contains a certain amount of noise. The same problem exists for Flickr groups if users upload photos that don't comply with the group's content rules. We chose not to manually clean the collected training data in order to conduct the experiment in the context of user data, where noise is common. However, we manually checked the applicability of the groups and search terms we used and verified the training data on a random basis. We collected a total of 176,013 images, 74,360 for indoor-outdoor, 58,667 for moving-static and 42,986 for object-landscape.[2]

3.2 Exif Tag Selection

The Exif specification defines over 150 standard tags of which only some appear useful for the classification of the chosen concepts. Our selection is primarily based on camera settings, which are adjusted depending on the type of scene being captured. For more information we refer to [15]. We use a total of 14 tags for our experiment.

Focal Length : The focal length is measured in millimeters (mm) and describes the distance between the optical center of the camera lens and the photo sensor. In combination with the used sensor, it determines the camera's field of view and magnification. Short focal lengths are typically used for landscape photography

[2] The dataset containing references to the Flickr images has been published with the DOI "10.48564/unibafd-vmm4f-m4y33". The source code associated with this publication has also been published with the DOI "10.48564/unibafd-q73v9-wz721". The corresponding GitHub repository can be accessed through the provided references.

because they capture a wide field of view and make objects appear smaller. In contrast, long focal lengths are usually used to take close-up pictures of distant objects, as they reduce the field of view and magnify objects. In addition to the Exif tag "FocalLength", we use the Exif tag "FocalLengthIn35mmFormat", which specifies the focal length relative to a 35mm film camera and makes it easier to compare values for cameras with different hardware.

Exposure Time and ISO: The exposure time determines how long the camera sensor is exposed to light. In combination with the ISO value, which controls the sensitivity of the camera sensor, the brightness of a photo can be influenced. The more light there is in the environment, the shorter the chosen exposure time can be to take an adequately exposed photo. To take a sharp photo and to avoid blur when shooting fast-moving objects, very fast exposure times are necessary. In situations where only little light is available, or fast moving objects are to be captured, it can be helpful to increase the ISO value. However, a rule of thumb is to keep the ISO value as low as possible, since higher values lead to more noise.

F-Number: The f-number is the ratio between the focal length and the diameter of the aperture. It determines how much light reaches the photo sensor. The aperture can be adjusted mechanically on many cameras, usually allowing a pre-defined range of selected f-numbers to be used. Larger apertures can be particularly advantageous when taking pictures of moving objects under poor lighting conditions, as they allow shorter exposure times to be used. Therefore, in addition to the Exif tag "FNumber", we also use "MaxApertureValue" which states the maximum possible aperture diameter of the used camera.

Flash and Light Source: The Exif tag "Flash" indicates whether a flash was used when a photo was captured, which should mainly be the case for indoor photos. In addition, we use the tag "LightSource" which provides the kind of light source used to perform the white balance of the camera.

Additional Exif Tags: To integrate further background knowledge about the ambient lighting conditions, we use the tags "BrightnessValue" and "ExposureCompensation" which describe the measured average luminance of the captured scene. In addition, we use the tags "Orientation", "Contrast", "Sharpness" and "Saturation". While orientation describes the rotation of the camera, the other tags provide information about post-processing procedures performed by the camera after the photo was captured.
Due to the heterogeneity of the used training data, not all Exif tags exist for every photo. The distribution is as follows:
"FocalLength", "ExposureTime", "Flash", "ISO", "FNumber" (100%), "ExposureCompensation" (98%), "MaxApertureValue" (78%), "Orientation" (69%), "LightSource", "Sharpness" (55%), "FocalLengthIn35mmFormat" (52%), "Saturation", "Contrast" (50%), "BrightnessValue" (19%).

3.3 Model Training

For the classification of Exif data, a multilayer perceptron with four hidden layers is used. We conducted a hyper-parameter search to find adequate values for learning rate, regularization parameters and model architecture [22]. Before the training, missing Exif tags are imputed. We use the most frequent value for categorical and binary tags, and the mean value for numerical tags [4]. The image-only baseline models are created using the four state-of-the-art CNN architectures MobileNetV2, EfficientNetB0, EfficientNetB4 and ResNet50V2 which differ in depth, width and the used network blocks [9,18,20]. For creating fusion models, we combine the baseline models with the multilayer perceptron using concatenation layers. Additionally, we add fully connected layers (FC) before and after the concatenation layers to increase the model accuracy. To facilitate and shorten the training process of baseline and fusion models, transfer learning is applied. We initialize the used CNNs with pre-trained weights based on the ImageNet [11] dataset and add a new classification head consisting of a global average pooling layer (GAP), a dropout layer (DO) and a fully connected output layer. For EfficientNets, we include a batch normalization layer (BN) after the pooling layer. In fusion models, the output layer is added on top of the fully connected layer after the concatenation layer. The architecture of fusion models is sketched in Fig. 1.

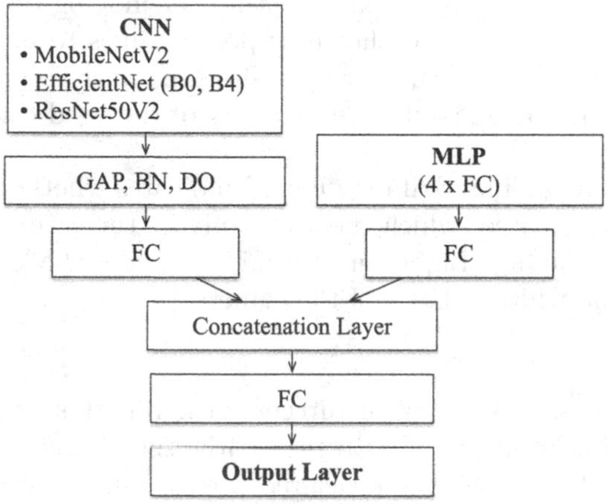

Fig. 1. Network architecture of fusion models

During early training, the weights of the classification head are updated without changing the pre-trained weights of the layers of the CNN. In the subsequent fine-tuning phase, the weights of the top layers of the CNN are unlocked and adjusted using a low learning rate in order to learn high-level image features that are important for the classification task. The learning rate is selected based on the current epoch. We use discrete learning rates from 0.001 to 0.00001 for the first 30 epochs and then gradually decrease the learning rate in each epoch. In the fine-tuning phase, which starts after 15 epochs, early stopping is applied to terminate the training process as soon as the loss no longer decreases appropriately. In addition, we use the image data augmentation techniques rotation, flipping, and translation to increase the robustness of the models [19]. The classification of super-concepts and sub-concepts is conducted individually for each classification task. When classifying super-concepts, each photo only has one ground truth label. Hence, we define the classification task as multi-class problem and use a softmax activation function in the output layer of the used model. In contrast, for the classification of sub-concepts, the learning task is defined as a multi-label classification problem. Here, each photo can have up to two ground truth labels, one for the super-concept and one for the sub-concept. Thus, we use a sigmoid activation function in the output layer of the used model. It is likely that Exif data has a greater impact on the training process when using small image sizes. Smaller images contain less information and thus it is more difficult to detect relevant features which are important for the classification task [10]. To test our assumption, we used two different image resolutions for training, 50×50 pixels and 150×150 pixels. We built a total of 16 image-only baseline models, 16 fusion models, and 2 Exif-only models for each problem scenario. The dataset was divided into training (70%), validation (20%) and test (10%) sets.

4 Evaluation

Figure 2 provides the macro F1-scores achieved on the test set for all created models. The classification accuracy could be increased for all super-concepts. In sub-concept classification, the classification accuracy could be increased in 22 of 24 cases. When considering individual problem scenarios, the accuracy increased the most in moving-static classification and the least in object-landscape classification. In the case of indoor-outdoor classification, good results are obtained

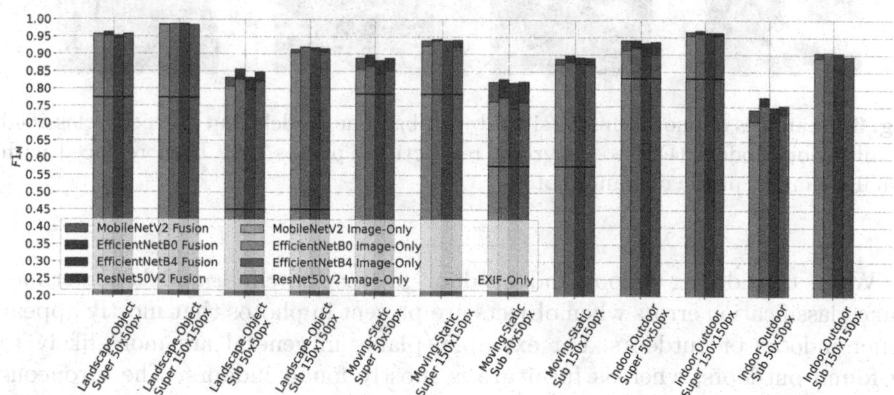

Fig. 2. Macro F1-scores of fusion models and baseline models for each problem scenario and image resolution

classifying super-concepts. However, when classifying sub-concepts, the accuracy decreased slightly for the two EfficientNet variants compared to the baseline, which can be attributed to the poor classifiability of the selected sub-concepts using Exif data only. In particular, it is difficult to distinguish between different types of rooms solely on the basis of Exif data. In contrast, the individual objects in moving-static classification could be distinguished with a higher accuracy based on Exif data. One reason could be that the backgrounds in the images showing these objects differ significantly (water, sky, roads), which could be reflected in the Exif data.

4.1 Classification Examples

Since the usefulness of Exif data seems to decrease with increasing image resolution (see Fig. 2), we focus our discussion here on cases in which Exif data may support image classification independent of the image size used. Figure 3 shows photos misclassified by all baseline models, but correctly classified by all fusion models.

Fig. 3. Examples of photos misclassified by all baseline models, but correctly classified by all fusion models. (Due to copyright restrictions, photos have been replaced with visually similar, public domain photos.)

When considering indoor and outdoor photos, baseline models sometimes make classification errors when objects are present in photos that mostly appear either indoors or outdoors. For example, plants in general are more likely to be found outdoors, whereas furniture is mostly found indoors. The erroneous results of the baseline models in these cases could be due to the fact that this tendency is embedded in the pre-trained network weights based on the ImageNet dataset. Thus, Exif data seem to be helpful for determining whether a photo was taken inside or outside, regardless of the objects present in the photo. This can be helpful in cases where the background of the photo doesn't allow for a clear decision. In moving-static classification Exif data is most supportive in situations where it is hard even for humans to decide whether the objects present in the photo are moving. For example, it is almost impossible to recognize that the fan in Fig. 3 is rotating without further knowledge. The same problem occurs for the photo showing the aircraft. For object-landscape classification, Exif data is least supportive. Nevertheless, in ambiguous cases (Fig. 3, right side) the image-only models seem to have problems when wide-angle photos contain further dominant objects. In such cases, Exif data can support the classification. In more obvious cases, however, it seems to be unproblematic to distinguish wide-angle shots from close-ups based on image data only.

4.2 Exif Tag Importance

To asses the importance of individual Exif tags for different problem scenarios, we first randomly permuted the corresponding Exif tag values in the test data [5]. We then compared the F1-scores obtained on the permuted test data with those obtained on the unchanged test data. We performed a total of 50 permutations for each Exif tag and measured the average F1-score drop of Exif-only models. The results for the super-concepts are shown in Table 2.

Table 2. The 5 most important Exif tags in super-concept classification, based on the average F1-score drop of randomly permuted data, FN = f-number, ET = exposure time, FL = focal length, FL35 = focal length in 35mm format, MA = maximum aperture value, EC = exposure compensation

Rank	indoor-outdoor	moving-static	object-landscape
1	ISO (−11%)	FL (−11.7%)	FN (−13.8%)
2	FN (−8.3%)	FL35 (−8.5%)	ISO (−6.2%)
3	ET (−7.6%)	MA (−7.6%)	FL35 (−5%)
4	FL (−6.7%)	FN (−6.7%)	FL (−4.8%)
5	Flash (−5%)	EC (−2.8%)	ET (−3%)

The five most important Exif tags for indoor-outdoor classification contain three camera parameters related to lighting conditions. This seems plausible, as indoor lighting conditions are usually worse than outdoor lightning conditions, requiring the camera parameters to be adjusted accordingly. For example, a flash is probably mostly used indoors. Outdoors, on the other hand, shorter exposure times and lower ISO values can be used. In the case of moving-static classification, the properties of the camera lens have a major impact on the classification result. The maximum lens aperture seems to be an important factor. One explanation for this could be that lenses with high focusing speeds are required when photographing fast moving objects. Usually, the smaller the maximum f-number of the lens, the faster it can change the focus. The classification result of object-landscape classification is mainly influenced by the F-Number and the focal length, which seems plausible as they strongly affect the camera's field of view. The importance of the ISO value could be explained by the fact that most wide-angle scenes are probably shot outdoors. In general, focal length, F-Number and ISO value seem to be important features in image classification based on Exif metadata.

4.3 Training Time

Figure 4 highlights the differences in average training time of the fusion models compared to the baseline models, summarized for all problem scenarios. The training time decreased the most with EfficientNet-based models. One reason for this could be that Exif data may guide the learning of individual image channel importances, which is performed by squeeze-and-excitation blocks that are not present in the other network architectures used. This in turn could lead to faster convergence. The average training time of baseline models was 52 minutes for an image size of 50×50 px and 48.5 min for an image size of 150×150px. Fusion models needed 45.6 and 45 minutes, respectively. Exif-only models needed 2.7 min, which is 94.6% less compared to image-data based models.

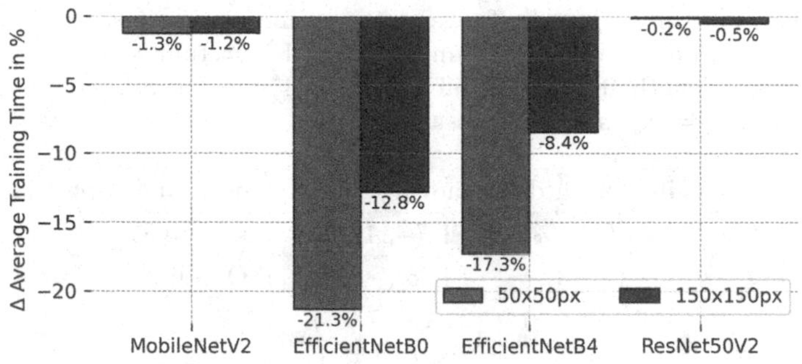

Fig. 4. Total average training time delta of fusion models compared to baseline models

5 Conclusion

We mixed image data with Exif metadata to improve image classification in the context of three problem scenarios. The results show that the classification accuracy of the created fusion models could be increased by an average of 3.2% for images with a resolution of 50×50 pixels and by an average of 0.8% for images with 150×150 pixels, while the training time could be reduced by an average of 10% and 5.7%. When considering super-concepts, the classification accuracy could be increased in all cases. However, in sub-concept classification, the accuracy slightly decreased in 2 of 24 cases compared to the baseline. Therefore, the usefulness of incorporating Exif data into the image classification process highly depends on the used target concepts and seems to be particularly helpful in cases where Exif data can help to identify certain properties of objects in photos that are also difficult for humans to recognize, such as whether an object in the photo is moving. Moreover, Exif data can be useful to shorten the required training time of the created models, which in turn leads to less energy consumption. Future considerations should include use cases in which image classification can be based exclusively on Exif data. This appears to be promising in particular for economic and ecological reasons, as the training times are significantly shorter compared to image-data based approaches. Also, training data with more instances, less noise and complete Exif tags should be used. Furthermore, it would be of interest to identify further target concepts that provide good classifiability using only Exif metadata.

References

1. Arbinger, C., Bullin, M., Henrich, A.: Exploiting geodata to improve image recognition with deep learning. In: Companion Proceedings of the Web Conference 2022, WWW 2022, pp. 648–655. ACM, New York, NY, USA (2022). https://doi.org/10.1145/3487553.3524645
2. Boutell, M., Luo, J.: Beyond pixels: exploiting camera metadata for photo classification. Pattern Recogn. **38**(6), 935–946 (2005). https://doi.org/10.1016/j.patcog.2004.11.013, https://www.sciencedirect.com/science/article/pii/S0031320304003978
3. Electronics, J., Information Technology Industries Association, J.: Exchangeable image file format for digital still cameras: Exif version 2.32. Technical report, Camera & Imaging Products Association, May 2019. https://www.cipa.jp/std/documents/download_e.html?DC-008-Translation-2019-E
4. Emmanuel, T., Maupong, T., Mpoeleng, D., Semong, T., Mphago, B., Tabona, O.: A survey on missing data in machine learning. J. Big Data **8**(1), 140 (2021). https://doi.org/10.1186/s40537-021-00516-9
5. Fisher, A., Rudin, C., Dominici, F.: All models are wrong, but many are useful: learning a variable's importance by studying an entire class of prediction models simultaneously. J. Mach. Learn. Res. JMLR **20**, 177 (2019)
6. Flickr: The app garden, API Documentation. www.flickr.com/services/api/. Accessed 13 Dec 2022
7. Ghazali, J., et al.: Image classification using EXIF metadata. Int. J. Eng. Trends Technol. **1**, 69–73 (2020). https://doi.org/10.14445/22315381/CATI3P211
8. Hand, M.: Ubiquitous photography. Polity (2012)
9. He, K., Zhang, X., Ren, S., Sun, J.: Identity mappings in deep residual networks. In: Leibe, B., Matas, J., Sebe, N., Welling, M. (eds.) ECCV 2016. LNCS, vol. 9908, pp. 630–645. Springer, Cham (2016). https://doi.org/10.1007/978-3-319-46493-0_38
10. Kannojia, S., Jaiswal, G.: Effects of varying resolution on performance of CNN based image classification an experimental study. Int. J. Comput. Sci. Eng. **6**, 451–456 (2018). https://doi.org/10.26438/ijcse/v6i9.451456

11. Krizhevsky, A., Sutskever, I., Hinton, G.E.: ImageNet classification with deep convolutional neural networks. Commun. ACM **60**(6), 84–90 (2017). https://doi.org/10.1145/3065386
12. Ku, W., Kankanhalli, M.S., Lim, J.-H.: Using camera settings templates ("Scene Modes") for image scene classification of photographs taken on manual/expert settings. In: Ip, H.H.-S., Au, O.C., Leung, H., Sun, M.-T., Ma, W.-Y., Hu, S.-M. (eds.) PCM 2007. LNCS, vol. 4810, pp. 10–17. Springer, Heidelberg (2007). https://doi.org/10.1007/978-3-540-77255-2_2
13. Lazebnik, S.: Computer Vision: A Reference Guide, Object Class Recognition (Categorization), pp. 533–536. Springer, Cham (2014). https://doi.org/10.1007/978-0-387-31439-6_337
14. Luo, J., Boutell, M., Brown, C.: Pictures are not taken in a vacuum - an overview of exploiting context for semantic scene content understanding. IEEE Signal Process. Mag. **23**(2), 101–114 (2006). https://doi.org/10.1109/MSP.2006.1598086
15. Maître, H.: From Photon to Pixel: The Digital Camera Handbook. Wiley, New York (2017). https://doi.org/10.1002/9781119402442, https://onlinelibrary.wiley.com/doi/abs/10.1002/9781119402442.ch1
16. Negoescu, R.A., Gatica-Perez, D.: Analyzing Flickr groups. In: Proceedings of the 2008 international Conference on Content-Based Image and Video Retrieval, CIVR 2008, pp. 417–426. ACM, New York, NY, USA (2008). https://doi.org/10.1145/1386352.1386406
17. Safonov, I.V., Kurilin, I.V., Rychagov, M.N., Tolstaya, E.V.: Image enhancement pipeline based on EXIF metadata. In: Adaptive Image Processing Algorithms for Printing. SCT, pp. 65–83. Springer, Singapore (2018). https://doi.org/10.1007/978-981-10-6931-4_3
18. Sandler, M., Howard, A., Zhu, M., Zhmoginov, A., Chen, L.C.: MobileNetV2: inverted residuals and linear bottlenecks. In: Proceedings of the IEEE Conference on Computer Vision and Pattern Recognition, pp. 4510–4520, June 2018. https://doi.org/10.1109/CVPR.2018.00474
19. Shorten, C., Khoshgoftaar, T.M.: A survey on image data augmentation for deep learning. J. Big Data **6**(1), 60 (2019). https://doi.org/10.1186/s40537-019-0197-0
20. Tan, M., Le, Q.V.: EfficientNet: rethinking model scaling for convolutional neural networks. In: International Conference on Machine Learning, pp. 6105–6114 (2019)
21. Thread, F.H.F.O.: Updates on tags, May 2015. https://www.flickr.com/help/forum/en-us/72157652019487118/. Accessed 8 Nov 2022
22. Yu, T., Zhu, H.: Hyper-parameter optimization: a review of algorithms and applications. ArXiv arXiv:abs/2003.05689 (2020)

Weak Segmentation-Guided GAN for Realistic Color Edition

Vincent Auriau[1(✉)], Emmanuel Malherbe[1,2], and Matthieu Perrot[1]

[1] L'Oréal AI Research, Clichy, France
{vincent.auriau,matthieu.perrot}@loreal.com
[2] Artefact Research Center, Paris, France
emmanuel.malherbe@artefact.com

Abstract. Editing the color of images in a realistic way finds many applications such as changing the perception of an image, data augmentation or film post processing. The design of an automatic tool is a complex and long addressed challenge. In particular, two properties are difficult to meet altogether: generating realistic results without artifacts and the possibility to precisely choose the future color. Conventional methods using segmentation and histogram matching maximize the controllability but also introduce a lack of realism or are complex to automate. On the contrary, GANs that specialize in realism are difficult to control. To overcome these challenges, we propose a novel GAN architecture leveraging any differentiable segmentation model. We demonstrate the genericness of our framework that presents state of the art results on different use cases. It generates images that look realistic while offering a precise color control.

Keywords: color edition · GAN · generative · segmentation

1 Introduction

Color edition control is the process of modifying an object's color in a graphical representation. A user can control the edition by specifying desired colors. It is a great tool for images post processing. By altering the color palette, it is possible to modify the global atmosphere or perception of the image [24]. One might want to be more precise and only adjust the color of a single area: a lighter blue sky in a landscape photograph or changing an actor's hair color in a movie. Simulating several color variations of the same car or garment can help to choose the right color among the millions of possibilities before any material realization. It can also be used to help customers select the right make up or hair coloration they desire. While physical renderers handle colors very well [25], it is still a real challenge to work with existing photographic pictures. Color transfer [14] and colorization [27] are two other applications that can be seen as specific cases of color edition. Most recent methodologies use neural networks and particularly GANs [28], however they still meet challenges with some non satisfying results.

E. Malherbe—Part the job was done while employed at L'Oréal.

G. L. Foresti et al. (Eds.): ICIAP 2023, LNCS 14233, pp. 487–499, 2023.
https://doi.org/10.1007/978-3-031-43148-7_41

Fig. 1. Example results of our framework on three different use cases: coloring hair and lips, clothing and cats and dogs.

Our goal is to realistically and precisely change an object's color in an image. We consider the general color perception of an object in the image, characterizing the object by a single color. In order to keep the human control simple, interaction of light with the object's geometry or local color variations coming from the textures are to be handled automatically.

To better define the objective, we enumerate four properties to evaluate the performances of a color edition model:

(i) **Colorization accuracy:** The perceived resulting color of the object is as close as possible as the desired one
(ii) **Realism:** The generated image is realistic to the human eye
(iii) **Stability:** The geometry of the considered object and local color variations are kept unchanged
(iv) **Restraint:** Objects and areas that should not be modified, or "background", are left untouched

We describe our robust method for color edition that overcomes the two main difficulties of existing techniques: producing realistic results while reaching a precise color control. Our main contributions are:

- A generic model architecture and training strategy for editing human perceived color while ensuring realism, stability and constraint
- A versatile object-aware color edition tool based on segmentation that addresses various use-cases and applications: color edition, color transfer and colorization

- Experimental comparison on various datasets and use cases that show how our approach achieves the best trade-off compared to the state-of-the art models

2 Related Work

Color manipulation in images is a long-addressed problem. Different semi-automatic or fully automatic methods have been developed, with different advantages and drawbacks. We specifically focus on color edition on the core of this paper, however inspirational technologies have been developed for color transfer or colorization.

Color Transfer consists of extracting the right color features from a reference image, and to transfer them to the content image. In [20], a simple but effective function for color transfer is given. The color reference image X and the content image $\tilde{X}$ are described respectively by (μ, σ) and $(\tilde{\mu}, \tilde{\sigma})$, the mean and standard deviation of the color distributions of each image in the L*a*b* space. The color transfer function is then defined as:

$$g(\tilde{X}_{i,j}) = \mu + \frac{\sigma}{\tilde{\sigma}}(\tilde{X}_{i,j} - \tilde{\mu}) \tag{1}$$

More complex transformation manipulate the whole color histogram [17] [26]. It enables modifications of the color palettes that would not be possible with a simple linear translation. In return, the needed computation is much greater. Most state of the art are today produced by deep learning. In [5] a neural color transfer method between images is proposed. It uses a trained VGG 19 to match semantically meaningful entities between the reference and content images, before applying Eq. 1. While having an image to transfer the color from is of great help, it might be long or complicated to find the very example image that perfectly fits the user's intention. It also limits creativity, forcing the use of already existing color palettes.

Convolution neural network have also shown the first good results on automatic **Gray Image Colorization** [6]. The use of conditional GAN with the pix2pix model [7] have led to more realistic results. These methods do not allow any color controlability as the resulting color cannot be decided by the user. Such control is introduced in [23] with colored drawn strokes. No measure of the color precision is developed and background is often impacted, creating undesired artifacts.

In **Color Edition Control**, a user can modify at will the colors of an object in an image. Pixels to be edited can be selected using a segmentation model [19], which was more recently performed using neural segmentation [13,26]. The desired color is defined by the user in the form of a histogram or color look up tables. Respectively histogram matching algorithms and interpolation are then applied to modify the selected pixels. While providing good results, these methods suffer in terms of realism, particularly around the edges of the computed mask where artifacts tend to appear. Moreover the color formats are complicated to manipulate for a user.

Another strategy that overcomes those limitations is to use a conditional GAN architecture [16] such as the one of StarGAN [2], with the colors as condition variables. CAGAN [10] proposes to automatically extract colors and let the generator indirectly focus on the right area without further explicit supervision. The major persisting challenges are generated images color precision and the inherent modifications of background pixels, that create inconsistencies. To overcome this issue, MagGAN [29] suggests to feed the model with manually labelled masks to improve the focus on a specific zone. However it is not designed for continuous color space and requires mask labels as input.

In this paper, we aim at bridging the different approaches presented above, GANs for realism and segmentation for object selection. It leads us to create an accurate, controllable and versatile coloring tool for edition.

3 Proposed Model

Our approach consists of Generator G that produces the transformed image $\hat{X} = G(X, c)$ from the input image X, with the considered object colorized as described by the color vector c. We also define S, a generic and differentiable segmentation model such that $S(X)$ is the probability for each pixel to represent the object: $S(X) \in [0;1]^{W \times H}$ where $S(X)_{i,j} = \mathbb{P}(x_{i,j} \in object)$. In practice we use a neural network for S that is trained separately from the rest of the framework and frozen, but it can be generalized to any differentiable function.

G is a convolutional neural network with an encoder-decoder architecture. The decoder takes the image features created by the encoder and the desired color as inputs and generates the colorized image. G is trained to optimize four modules that ensure the required properties: C for colorization accuracy, D for realism, R for stability and finally OF for restraint. The general organisation is illustrated Fig. 2, and the modules described in the following sections.

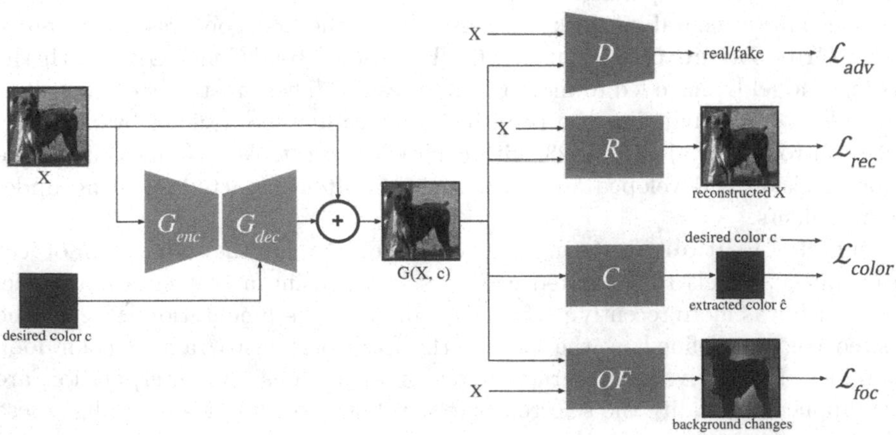

Fig. 2. General organization of our framework. Generated images are evaluated through the four defined modules that design G's training: C, D, OF and R. G actually predicts a difference that is applied to the input image.

3.1 Color Estimation Module C

We need a robust color estimation function that should be differentiable for later optimization purposes. We describe an object's color as an average RGB value over its representing pixels. This estimator characterizes well the general color of a uniformly colored object and can be easily represented and understood by a user. The weighted mean color of an object $C(X)$ is then:

$$C(X) = \mathbb{E}_{object}[X] = \frac{\sum_{i,j} S(X)_{i,j} * x_{i,j}}{\sum_{i,j} S(X)_{i,j}} \tag{2}$$

with $x_{i,j}$ the color vector at pixel (i, j). It is illustrated in Fig. 3.

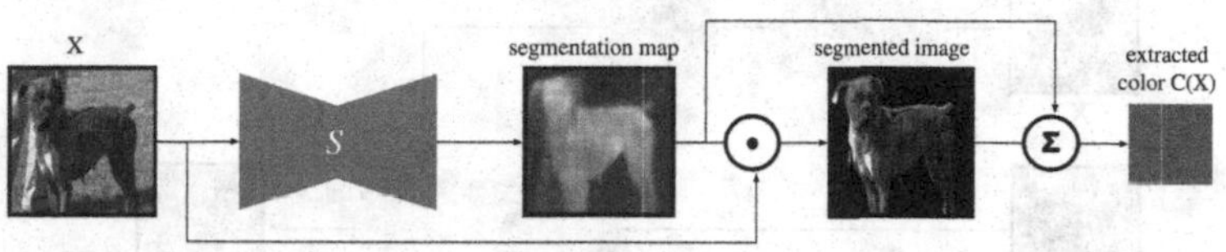

Fig. 3. Color estimation module C: from the segmentation map given by the segmentation model S, the weighted mean $C(X)$ is computed to represent the color of the object.

This color estimator is used to estimate the color of our object in the generated image $\hat{X}$ and compare it to the desired color c. The color loss is a L_2 loss computed in the $CIEL^*a^*b^*_{76}$ color space as it better matches human perception [10,21].

$$\mathcal{L}_{color} = ||C(\hat{X}) - c||_{L^*a^*b^*_{76}} \tag{3}$$

Thanks to the use of S, we are sure to extract and therefore control the color of the object itself.

3.2 Discriminator Module D

The GAN structure ensures that the main coloring model G produces realistic results by adversarially training a discriminator neural network, D. It tries to distinguish real images from the one edited by G. Our adversarial loss is the standard Wasserstein GAN loss with gradient penalty [1]:

$$\mathcal{L}_{adv} = D(\hat{X}) - D(X) + \lambda_{gp} \nabla_{\tilde{X}} D(\tilde{X}) \tag{4}$$

with $\tilde{X}$ randomly obtained for each batch by interpolating between real images X and generated images $\hat{X}$.

3.3 Reconstruction Module R

In order to ensure stability, we use the popular cycle consistency loss introduced in CycleGAN [31] for image to image translation. The model is trained to be able to recreate the original image from its own generation. It ensures that G maintains the objects geometry and that no pixel or information is lost through the process. The reconstruction loss is defined as:

$$\mathcal{L}_{rec} = ||X - G(G(X,c), C(X))||^2 \tag{5}$$

The reconstruction module is illustrated in Fig. 4.

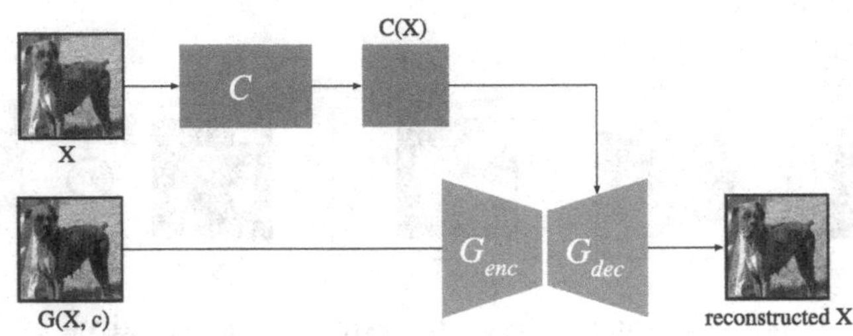

Fig. 4. Reconstruction module R: The edition model G is asked to reconstruct an original image from one of its own edition.

3.4 Object Focus Module OF

With our frozen segmentation model S, we can also ensure restraint, meaning that G keeps background pixels unchanged. We only want pixels that relate to the object's color change to be edited by the model. These background alterations is a common artifact produced by GAN based approaches. The Object Focus loss $\mathcal{L}_{foc}$ penalizes any change operated outside of the segmentation mask representing our object:

$$\mathcal{L}_{foc} = ||(X - \hat{X})) \odot (\mathbf{1} - S(X))||^2 \tag{6}$$

with $\mathbf{1}$ the matrix of same shape as $S(X)$ and filled with ones and $\odot$ the element wise product on the R, G and B channels. The resulting object focus module is illustrated in Fig. 5.

3.5 General Objective

We summarize the final objective to train G and D as follows:

$$\mathcal{L}_G = \mathcal{L}_{adv} + \lambda_{color}\mathcal{L}_{color} + \lambda_{foc}\mathcal{L}_{foc} + \lambda_{rec}\mathcal{L}_{rec} \tag{7}$$

$$\mathcal{L}_D = -\mathcal{L}_{adv} \tag{8}$$

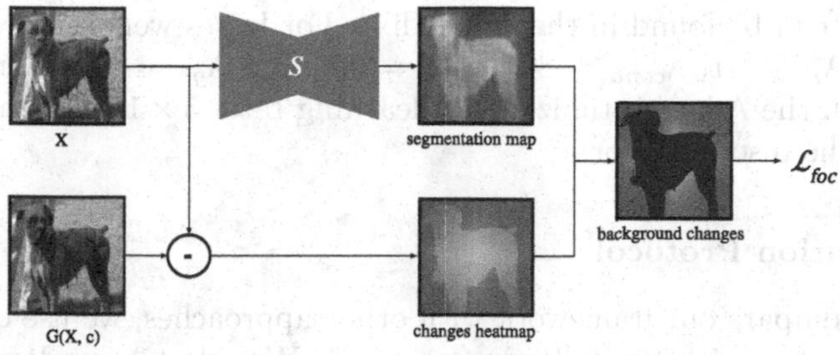

Fig. 5. Object focus module OF: Generated images are compared to the original ones and all pixels modified outside the segmentation maps are penalized. This helps to keep a background consistency between original and generated images.

where the different λ are used to set up and control the importance of the different losses in the final objective. The general workflow of the model is illustrated in Fig. 2. Models G and D are trained by this process, S is trained beforehand separately.

We have presented our framework to edit a single object's color. It can be easily extended to any number of objects. One only needs to sum the losses over the different objects.

4 Experiments

4.1 Datasets

We have tested our framework on three different use cases supported by open-source datasets designed to train the segmentation model. First, we used CelebA-HQ dataset [12] to train our hair and lips segmentation network. It was then completed with the FFHQ dataset [8] to improve the data diversity. Secondly, the cat and dogs dataset [15] was used for coloring pets. For this use case, we used an already trained segmentation network from [4]. We plugged this model to our framework that works without further need of adaptation. Finally, the clothing dataset [11] was used to train the segmentation model as well as the color edition model for the garnments use-case. Example of results can be found in Fig. 1. The results presented in this part are all obtained using the first datasets. Additional material and examples are shared in the appendix[1].

4.2 Implementation

For the following experiments, the parameters described in this part were used, unless it is specifically specified otherwise. The models were trained on 256×256 pixels images resized by bicubic interpolation. For the trained segmentation models, we used a U-Net segmentation network [22]. The generator and discriminator

[1] Appendix can be found at: https://github.com/VincentAuriau/wsg-gan-color-edition/blob/main/pdf_files/appendix.pdf.

architectures can be found in the Appendix[3]. For losses weighting, we used the parameters: $\lambda_{rec} = 1$, $\lambda_{cond} = 80$, $\lambda_{foc} = 600$ and $\lambda_{gp} = 6$. We trained both networks with the Adam optimizer with learning rates 5×10^5 for the generator and 10^5 for the discriminator.

4.3 Evaluation Protocol

In order to compare our framework with other approaches, we use quantitative metrics computed with the following protocol: We select a random subset $\mathcal{X}_1$ composed of 10k images that we keep for testing. We select a second subset $\mathcal{X}_2$, of same size as $\mathcal{X}_1$. We generate $G(X_1^i, C(X_2^i))$ for each $(X_1^i, X_2^i) \in (\mathcal{X}_1, \mathcal{X}_2)$ to build the dataset $\mathcal{X}_{1\to 2}$.

Metrics. For realism quantification, we use the Fréchet Inception Distance, or FID [18]. We compare our generated images $\mathcal{X}_{1\to 2}$ with $\mathcal{X}_2$: $FID_{1\to 2} = FID(\mathcal{X}_{1\to 2}, \mathcal{X}_2)$. For color precision estimation, we use $L_{1\to 2}^{color} = \frac{1}{n}\sum_{i=1}^{n} ||C(X_{1\to 2}^i) - C(X_2^i)||_{L^*a^*b^*_{76}}$. We also compute a modified version of $\mathcal{L}_{foc}$ to quantify how well the different models perform on the task of focusing on the right pixels. L_{bg}^{avg} is defined as:

$$L_{bg}^{avg} = \frac{1}{3n}\sum_{i=1}^{n}\frac{||(X_{1\to 2}^i - X_1^i)\odot(\mathbf{1} - M_1^i)||_1}{||\mathbf{1} - M_1^i||_1} \tag{9}$$

with $(X_1^i)_n$ being all the images of X_1, M_1^i the manually annotated mask corresponding to X_1^i, $X_{1\to 2}^i$ the corresponding generated image and $\mathbf{1}$ the matrix filled with ones. L_{bg}^{avg} represents the average modification of a background pixel's channel. The $\mathcal{L}_1$ norm is taken to better express human perception in RGB [3].

User Test. We have performed a user study for human quantitative evaluation of the different approaches. We have asked five annotators which was the most realistic edition among the one generated by the different approaches. We have repeated the experience with 30 random triplets of editions, totalling 150 annotations.

4.4 Comparison with Other Approaches

We compare our framework with other approaches that aim at coloring objects in images. The first approach, alpha blending [13,17] uses segmentation paired with RGB shift. For fair comparison, the area to edit is selected using S, then Eq. 1 is applied for the average color of segmented pixels to match the desired one.

The second one, CAGAN [10], is a straightforward adaptation of StarGAN [2] for color control. A third model is trained besides D for color extraction. It uses labels, without any guidance on the object. For fair comparison we extracted color labels using S. This third model induces noise in the global color precision of the editions.

Finally, we also benchmark our model with one of the most performing GAN, StyleGAN [9]. As it is not built specifically for coloring, we developed a straight forward strategy on the style space [30] which is detailed in the Appendix[3]. This simple adaptation enables to perform color edition using StyleGAN. More complex adaptation could be designed, but it is beyond the scope of this paper.

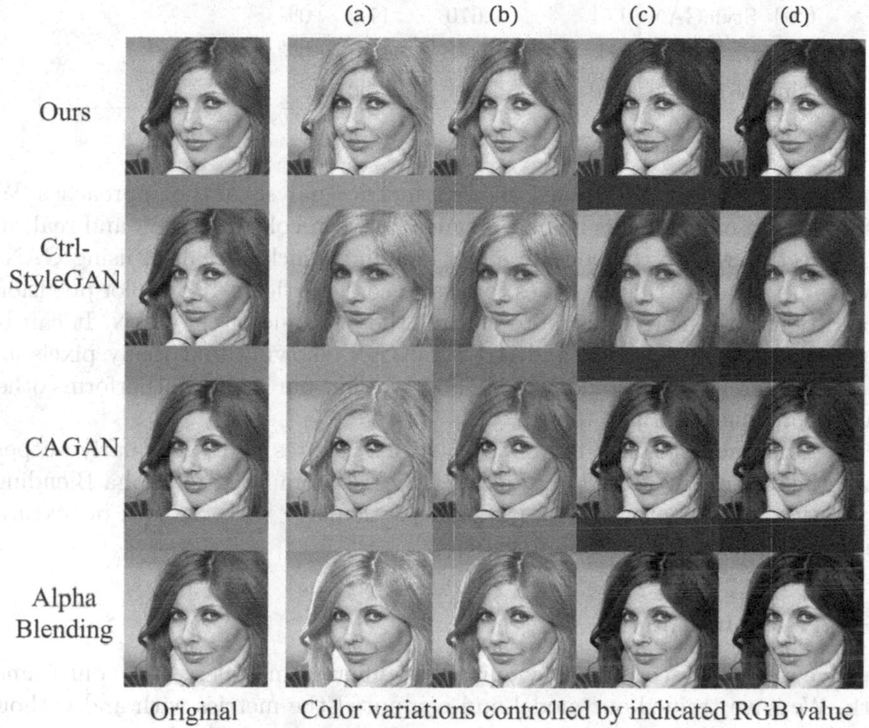

Fig. 6. Comparison of different methods for hair coloring. The desired hair color is illustrated on each column between rows.

Qualitative Evaluation. Figure 6 shows examples of results for hair colorization. We observed the main defaults of compared approaches. Alpha Blending creates artifacts on the edges of the segmented mask (see (a)). CAGAN tends to edit pixels outside hair pixels. It is particularly obvious on (a) with the face color that is different and on (d) where the background is affected. Finally, StyleGAN which is not built for image-to-image translation changes the faces attributes, the projection failing at perfectly reproducing the original image.

Table 1. Quantitative results of the color edition frameworks. We show that our approach outperforms alpha blending in terms of FID and other approaches on all metrics. Our model also outperforms others on the users votes. For the first three metrics, lower is better. Best results are in blue and second best in green.

Model	$L_{bg}^{avg} \times 10^3$	$FID_{1\rightarrow 2}$	$L_{1\rightarrow 2}^{color}$	User % Most Realistic
Ours	0.92	0.050	5.03	61%
Ctrl- StyleGAN	9.51	0.670	17.2	0%
CAGAN	121	0.096	11.3	0%
Alpha Blending	0.00	1.301	0.00	34%

Quantitative Evaluation. Table 1 shows a metric analysis of the approaches. We observed that our model is a good balance between color precision and realism. Our objective was to compensate alpha blending's lack of realism using GANs. However, typical GAN approaches come at a cost, particularly on color precision. In addition, it is more difficult to control the outcome of the GAN. It can be seen with L_{bg}^{avg} for CAGAN and ctrl-StyleGAN, showing that many pixels are modified whereas they should not. We observe that our model outperforms other GAN approaches on these topics.

Our approach obtained the majority of users votes and particularly outperforms the other GAN-based approach, CAGAN. Compared to Alpha Blending, our method seems to have better results particularly with the lack of textures in images with darker colored objects.

4.5 Ablation Study

We have also analysed the behaviour of the different modules within our framework. We have trained our model and computed the metrics with and without the reconstruction and object focus losses. The results can be found in Table 2.

Table 2. Ablation study of our proposed model. The Object Focus loss improves realism and background preservation while reconstruction positively affects all metrics. For all metrics, lower is better. Best results are in blue and second best in green, identity excluded

Model	$L_{bg}^{avg} \times 10^3$	$FID_{1\rightarrow 2}$	$L_{1\rightarrow 2}^{color}$
Identity	*0.00*	*0.045*	*25.2*
$\lambda_{foc} = 0, \lambda_{rec} = 0$	8.38	0.051	4.46
$\lambda_{foc} = 0, \lambda_{rec} = 1$	8.00	0.060	4.80
$\lambda_{foc} = 600, \lambda_{rec} = 0$	1.01	0.056	5.99
Ours	0.92	0.050	5.03
$GTmask_{C+OF}$	3.40	0.082	8.51
$GTmask_{OF}$	1.79	0.052	5.20

In order to emphasize the importance of S, we also train $GTmasks$, an equivalent of our framework with losses computed with ground truth masks instead of S's predictions. We test this strategy for Object Focus, $GTmask_{OF}$, and combined with Color Estimation, $GTmask_{C+OF}$. $GTmask_{C+OF}$ is thus optimized without a fully differentiable $C(X)$.

We observe that optimising the object focus loss fulfills the objective of limiting the modification of pixels outside of the colored object. Combined with reconstruction loss, it further improves all metrics. Finally, the use of a differentiable S function considerably improves the results.

5 Conclusion

We presented our framework for generic objects colorizing. Our new object-aware colorization model based on segmentation, introduces specific losses to improve the realism and the color precision of the results. It can be used on any type of image of objects and only needs a pretrained and differentiable segmentation model to work. We have shown state of the art results on several datasets, with the best balance between realism of generated images, accurate color of transformed object as well as maintained geometry and background. Our framework can be adapted for different applications such as color edition, color transfer or colorization, as shown through the experiments on several use-cases. As future work, we will focus on spatial distribution of the color. In order to let the user manipulate more than one general color of the object while keeping the simplicity of use.

References

1. Arjovsky, M., Chintala, S., Bottou, L.: Wasserstein gan (2017)
2. Choi, Y., Choi, M., Kim, M., Ha, J., Kim, S., Choo, J.: Stargan: unified generative adversarial networks for multi-domain image-to-image translation. CoRR abs/1711.09020 (2017). https://arxiv.org/abs/1711.09020
3. Concha, A., Civera, J.: An evaluation of robust cost functions for rgb direct mapping. In: 2015 European Conference on Mobile Robots (ECMR) (2015)
4. Zakirov, B.E.: Keras implementation of deeplabv3+ (2019)
5. He, M., Liao, J., Yuan, L., Sander, P.V.: Neural color transfer between images. CoRR abs/1710.00756 (2017). https://arxiv.org/abs/1710.00756
6. He, Z., Zuo, W., Kan, M., Shan, S., Chen, X.: Attgan: facial attribute editing by only changing what you want (2018)
7. Isola, P., Zhu, J., Zhou, T., Efros, A.A.: Image-to-image translation with conditional adversarial networks. CoRR abs/1611.07004 (2016). https://arxiv.org/abs/1611.07004
8. Karras, T., Laine, S., Aila, T.: A style-based generator architecture for generative adversarial networks. In: IEEE CVPR 2019 (2019)
9. Karras, T., Laine, S., Aittala, M., Hellsten, J., Lehtinen, J., Aila, T.: Analyzing and improving the image quality of stylegan. CoRR abs/1912.04958 (2019). https://arxiv.org/abs/1912.04958

10. Kips, R., Gori, P., Perrot,M., Bloch, I.: CA-GAN: weakly supervised color aware GAN for controllable makeup transfer. CoRR abs/2008.10298 (2020). http://arxiv.org/2008.10298
11. Lakshmanamoorthy, R.: People clothing segmentation (2021). www.kaggle.com/rajkumarl/people-clothing-segmentation
12. Lee, C.H., Liu, Z., Wu, L., Luo, P.: Maskgan: towards diverse and interactive facial image manipulation. In: IEEE Conference on Computer Vision and Pattern Recognition (CVPR) (2020)
13. Levinshtein, A., Chang, C., Phung, E., Kezele, I., Guo, W., Aarabi, P.: Real-time deep hair matting on mobile devices (2018)
14. Liu, S.: An overview of color transfer and style transfer for images and videos (2022)
15. Microsoft: Cats and dogs: a dataset for kaggle challenge (2017). www.microsoft.com/en-us/download/details.aspx?id=54765/
16. Mirza, M., Osindero, S.: Conditional generative adversarial nets. CoRR abs/1411.1784 (2014). http://arxiv.org/1411.1784
17. Neumann, L., Neumann, A.: Color style transfer techniques using hue, lightness and saturation histogram matching. In: Proceedings of the First Eurographics Conference on Computational Aesthetics in Graphics, Visualization and Imaging, pp. 111–122. Computational, Eurographics Association, Goslar, DEU (2005)
18. Nunn, E.J., Khadivi, P., Samavi, S.: Compound frechet inception distance for quality assessment of GAN created images. CoRR abs/2106.08575 (2021). https://arxiv.org/abs/2106.08575
19. Pal, N.R., Pal, S.K.: A review on image segmentation techniques. Pattern Recogn. **26**(9), 1277–1294 (1993)
20. Reinhard, E., Ashikhmin, M., Gooch, B., Shirley, P.: Color transfer between images. IEEE Comput. Graph. Appl. **21**, 34–41 (2001)
21. Reinhard, E., Pouli, T.: Colour spaces for colour transfer. In: Schettini, R., Tominaga, S., Trémeau, A. (eds.) CCIW 2011. LNCS, vol. 6626, pp. 1–15. Springer, Heidelberg (2011). https://doi.org/10.1007/978-3-642-20404-3_1
22. Ronneberger, O., Fischer, P., Brox, T.: U-net: convolutional networks for biomedical image segmentation. CoRR abs/1505.04597 (2015). https://arxiv.org/abs/1505.04597
23. Sangkloy, P., Lu, J., Fang, C., Yu, F., Hays, J.: Scribbler: controlling deep image synthesis with sketch and color (2016). https://doi.org/10.48550/ARXIV.1612.00835
24. Seifi, H., DiPaola, S., Enns, J.T.: Exploring the effect of color palette in painterly rendered character sequences. In: Computational Aesthetics in Graphics, Visualization, and Imaging (2012)
25. Tewari, A., et al.: State of the art on neural rendering. In: Computer Graphics Forum (2020)
26. Tkachenka, A., et al.: Real-time hair segmentation and recoloring on mobile gpus. CoRR abs/1907.06740 (2019). https://arxiv.org/abs/1907.06740
27. Varga, D., Szirányi, T.: Fully automatic image colorization based on convolutional neural network. In: 23rd ICPR (2016)
28. Vitoria, P., Raad, L., Ballester, C.: Chromagan: adversarial picture colorization with semantic class distribution. In: The IEEE Winter Conference on Applications of Computer Vision, pp. 2445–2454 (2020)
29. Wei, Y., et al.: Maggan: high-resolution face attribute editing with mask-guided generative adversarial network. CoRR abs/2010.01424 (2020). https://arxiv.org/abs/2010.01424

30. Wu, Z., Lischinski, D., Shechtman, E.: Stylespace analysis: disentangled controls for stylegan image generation. CoRR abs/2011.12799 (2020). https://arxiv.org/abs/2011.12799
31. Zhu, J.Y., Park, T., Isola, P., Efros, A.A.: Unpaired image-to-image translation using cycle-consistent adversarial networks (2020)

Hand Gesture Recognition Exploiting Handcrafted Features and LSTM

Danilo Avola[1(✉)], Luigi Cinque[1], Emad Emam[1], Federico Fontana[1], Gian Luca Foresti[2], Marco Raoul Marini[1], and Daniele Pannone[1]

[1] Department of Computer Science, Sapienza University, Via Salaria 113, 00198 Rome, Italy
{avola,cinque,emam,fontana.f,marini,pannone}@di.uniroma1.it
[2] Department of Mathematics, Computer Science and Physics, University of Udine, Via delle Scienze 206, 33100 Udine, Italy
gianluca.foresti@uniud.it

Abstract. Hand gesture recognition finds application in several heterogeneous fields, such as Human-Computer Interaction, serious games, sign language interpretation, and more. Modern recognition approaches use Deep Learning methods due to their ability in extracting features without human intervention. The drawback of this approach is the need for huge datasets which, depending on the task, are not always available. In some cases, handcrafted features increase the capability of a model in achieving the proposed task, and usually require fewer data with respect to Deep Learning approaches. In this paper, we propose a method that synergistically makes use of handcrafted features and Deep Learning for performing hand gesture recognition. Concerning the features, they are engineered from hand joints, while for Deep Learning, a simple LSTM together with a multilayer perceptron is used. The tests were performed on the DHG dataset, comparing the proposed method with both state-of-the-art methods that use handcrafted features and methods that use learned features. Our approach overcomes the state-of-the-art handcrafted features methods in both 14 and 28 gestures recognition tests, while we overcome the state-of-the-art learned features methods for the 14 gesture recognition test, proving that it is possible to use a simpler model with well engineered features.

Keywords: Handcrafted Feature · Deep Learning · LSTM

1 Introduction

Nowadays, hardware and software improvements in acquisition devices allow obtaining impressive results in several fields, including robotics [8,13,24,28,34], environmental analysis [3,4,21,23,30], and Human Computer Interaction (HCI) [1,2,6,10,19]. By focusing on the latter, natural interaction revolutionized a plethora of contexts, ranging from cultural heritage to medical applications [11,12].

G. L. Foresti et al. (Eds.): ICIAP 2023, LNCS 14233, pp. 500–511, 2023.
https://doi.org/10.1007/978-3-031-43148-7_42

A hot topic in natural interaction regards the Hand Gesture Recognition (HGR) [9,32], in which a sequence of hand poses is analyzed. Modern approaches for HGR involve Deep Learning (DL) techniques, mainly due to the characteristics of such techniques to automatically extract relevant features from the data. However, in order to have the most generalized features, DL models are usually comprised of a massive amount of layers, which leads to long training times. In addition, there is the need of a large amount of memory for handling the computation and, lastly, the need of datasets having thousands or even millions of samples.

For solving these problems, it is possible to exploit handcrafted features, which are specifically engineered for the task that must be solved. In this paper, we focused on the design of specific features for the recognition of dynamic gestures in a video stream. Such features, extracted from the joints of the hand, are then fed to a vanilla Long-Short Term Memory (LSTM), which output is classified by a multilayer perceptron (MLP). Such model has been deliberately kept simple to prove the goodness of the proposed features. Tests performed on the state-of-the-art DHG Dataset [17], highlight the superiority of our method with respect to other state-of-the-art handcrafted feature-based approaches, as well as with respect to automatic feature-based approaches.

The following are the strengths of the proposal:

- We proposed a handcrafted set of feature specifically designed for dynamic hand gesture recognition;
- The reliability of the engineered features allows keeping the DL model simple;
- Our method overcome the current state-of-the art methods that use handcrafted features, in both 14 and 28 gesture recognition tasks, while we overcome the current state-of-the art methods that use automatic extracted features in the 14 gesture recognition task.

The rest of the paper is structured as follows. In Sect. 2, an overview of the literature regarding hand gesture recognition techniques is provided. In Sect. 3, the proposed handcrafted features are described in detail. Section 4, presents the performed experiments and the comparison with the current state-of-the-art approaches. Finally, Sect. 5, the conclusions are provided, highlighting the effectiveness of the proposal and its possible future improvements.

2 Related Work

This section presents the most recent works in both handcrafted feature extraction and DL methods for hand gesture recognition. By considering feature extraction, authors in [17] proposed three descriptors for coding spatial information of a gesture, namely, Shape of Connected Joints (SoCJ), Histogram of Wrist Rotations (HoWR), and Histogram of Hand Directions (HoHD). SoCJ is a set of relevant hand joints, while the HoWR and the HoHD are histograms where each direction and rotation are respectively localized at a unique bin. Then, the temporal domain is managed with a Temporal Pyramid (TP), and the classification is performed with a Support Vector Machine (SVM). The TP is a valuable strategy in this application field, and its effectiveness is a popular choice for many works [14,18,33]. In [27], the 3D spatial information is retrieved with a Leap Motion Controller (LMC), and the following information is extracted: palm direction, palm normal, palm position, and fingertips positions. Subsequently, a feature vector is created by with data computed by using the above mentioned information, e.g. the Euclidean distance among fingertips, fingertip angle, adjacent fingertip-angle, the absolute angle between adjacent fingertips, and more. Then, vectors are classified by using a Hidden Conditional Neural Field. Authors in [16], proposed a hybrid approach. Firstly, hand crafted features such as shape of connected joints, intra/inter finger relative distance, and global relative translation are computed. Then, temporal dependencies are learned by exploiting LSTM cells.

Regarding DL methods in HGR, authors in [5] used an LMC for extracting hand joints and then fed them to a Deep LSTM (DLSTM). Authors in [25], instead, managed the structure of the hand as a graph, thanks to their proposed Hand Gesture Graph Convolutional Network (HG-GCN). In [29], a different type of neural network is involved: the Spatial-Temporal and Temporal-Spatial Hand Gesture Recognition Network (ST-TS-HGR-NET). The main idea deals with Symmetric Positive Definite (SPD) Matrix Learning and Classification Sub-Network (SPDC-NET). It learns an SPD matrix from a set of SPD matrices and maps such matrix to a Euclidean space for performing classification. Authors in [31], proposed a method in which the 3d data about skeleton and hands is processed by a Convolutional Neural Network (CNN), and the extracted features are then fed to a LSTM to classify gestures. Authors in [7], instead, exploited LSTM and Fisher vector encoding of gesture for detecting deceptions through gestures.

With respect to the works reported in this section, our approach is to use synergistically handcrafted features, designed by us, with a vanilla LSTM, to keep the complexity of the model as low as possible.

3 Method

In this section, the handcrafted features designed by us, together with the used DL model, are presented.

3.1 Handcrafted Features

The aim of designing handcrafted features is to have the best data that allows to discriminate as much as possible the different gestures. To achieve this result, we engineered features that take in consideration both spatial and temporal properties of the hand joints through time.

The joints are represented as a tuple of 3 elements which are, namely, the X, Y, and Z coordinate of the joint. Formally:

$$J_i = (X_i, Y_i, Z_i), i \in [0, 21] \tag{1}$$

In the subsequent sections, the features are described in detail.

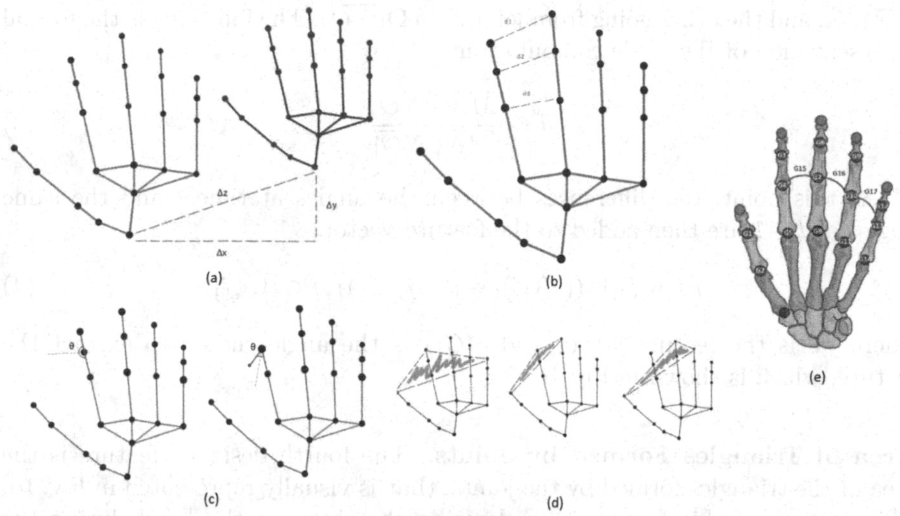

Fig. 1. A visual representation of the designed handcrafted features: (a) The position of the joints over time, (b) Average distance between equivalent joints in adjacent fingers, (c) Angular difference of time, (d) Area of defined triangles, (e) The selected angles for angular difference over time.

Joints Position over Time. One of the most relevant characteristic used in the analysis of skeleton data is the displacement of the joints over time. As depicted in Fig. 1a, this feature captures the movement of the hand during the gesture and reflects the speed at which the hand is moving:

$$fv = X_{i_t} - X_{i_{t-1}}, Yi_t - Yi_{t-1}, Zi_t - Zi_{t-1}, i \in [0, 21] \tag{2}$$

where fv is the feature vector containing all the handcrafted features, X_i, Y_i and Z_i are the X, Y, and Z coordinates of the *i-th* joint, t is the current frame, and $t-1$ is the frame related to the previous time instant.

Average Distance Between Adjacent Joints. The second proposed features consists in the average distance between adjacent joints, which is shown in Fig. 1b. This feature describes the spatial arrangement of the hand and reflects the overall shape and size of the hand during a gesture. To compute this feature, the distances between adjacent joints of each finger are calculated, and then the average distance is computed over all the fingers.

Angular Difference over Time. As the third feature, we have considered the angular difference between adjacent joints over time, that is shown in Fig. 1c. Let us consider two joints of the hand, namely, N and M. The angle at joint N is computed as the angle between two segments, one going from joint M to

N($\overline{MN}$), and the other going from joint N to O($\overline{NO}$). The following is the formal representation of the angle computation:

$$\theta = \frac{\overline{MN} \cdot \overline{NO}}{|\overline{MN}||\overline{NO}|} \tag{3}$$

At this point, the differences between the angles at time t and the same angles at $t-1$ are then added to the feature vector:

$$fv = fv \cup \left((\theta(Gi)_t - \theta(Gi)_{t-1}) \right), i \in [1, 17] \tag{4}$$

where fv is the feature vector and $\theta(Gi)$ is the angle chosen to extract the feature, which is shown in Fig. 1e.

Area of Triangles Formed by Joints. The fourth designed feature is the area of the triangles formed by the joints, that is visually represented in Fig. 1d. The purpose of this feature is to describe the shape of the hand during the gesture, and to reflect the configuration of the fingers. In details, it provides information on how much the hand is closed, allowing to discriminate better similar gestures. For the computation, three triangles are considered, formed by the following joints:

- Thumb, index, and pinky fingertips;
- Thumb, index, and middle fingertips;
- Thumb, middle, and ring fingertips.

For each triangle, the lengths of the three sides $(s1, s2, s3)$ are computed, and then the areas are found as follows:

$$p = (l1 + l2 + l3)/2 \tag{5}$$

$$A = \sqrt{p * (p - l1) * (p - l2) * (p - l3)} \tag{6}$$

Subsequently, also these features are added to fv.

Other Features. In addition to our defined features, fv contains features that are well-known in the field of gesture recognition. These features are:

- Joints speed;
- 3D space distance between the thumb and index fingertip;
- 3D space distance between the thumb fingertip and the center of the hand, indicating how closed the thumb is;
- Average of the distances between the center of the hand and respectively: the index, middle, ring, and little fingertip, indicating how closed is the hand overall;
- Ratio between the sum of the angles of the thumb and index finger divided by the sum of the angles of the middle and ring fingers, to discriminate gestures performed with one finger or with all fingers;

- Difference between the three coordinates of x, y, and z from the wrist joint to the center of the hand in the same frame, to indicate the hand's rotation position.

Despite the used dataset includes depth images of the gestures, we focused only on the joints due to the following reasons. Firstly, the joints features lead to have fewer parameters with respect to an image input. Secondly, we wanted to keep the model as simple as possible to highlight the impact of well-designed handcrafted features.

3.2 LSTM

The DL approach chosen for analyzing the proposed feature is the LSTM. The choice fell on this model due to its capabilities in handling sequential data. The implemented LSTM consists of two stacked LSTM nodes, has a hidden size of 130, and a dropout layer was inserted between the two layers for regularization. The LSTM takes as input a sequence of T vectors X $= \{x_0, x_1, \ldots, x_T\}$, each consisting of the features defined in the previous section. The sequence is 130 frames per 118 features for each gesture, and outputs a vector of 118 elements that is used by a Multilayer Perceptron (MLP) for classifying the gesture. Below is a summary of the equations that define the LSTM:

$$\begin{aligned} i_t &= \sigma\left(W_{xi} x_t + W_{hi} h_{t-1} + b_i\right) \\ f_t &= \sigma\left(W_{xf} x_t + W_{hf} h_{t-1} + b_f\right) \\ c_t = f_t \odot c_{t-1} + i_t &\odot \tanh\left(W_{xc} x_t + W_{hc} h_{t-1} + b_c\right) \\ o_t &= \sigma\left(W_{xo} x_t + W_{ho} h_{t-1} + b_o\right) \\ h_t &= o_t \odot \tanh\left(c_t\right) \end{aligned} \tag{7}$$

Here, $\mathrm{i}, \mathrm{f}, c$, and o represent the input gate, forget gate, cell state, and output gate, respectively. Moreover, h_t is the internal state at time t (also seen as short-term memory), $W_{xi}, W_{xf}, W_{xo}, W_{xc}$ are the weights between the input layer and, respectively, the input gate, forget gate, output gate, and cell state; $W_{hi}, W_{hf}, W_{ho}, W_{hc}$ are the weights between the hidden layer and, respectively, the input gate, forget gate, output gate, and cell state; b_i, b_f, b_c, b_o are the bias weights of the input gate, forget gate, cell state, and output gate, respectively. Finally, σ indicates the sigmoid function, and $\odot$ indicates the Hadamard product between vectors.

4 Experiments

In this section, the dataset together with the used hardware and software are described. Then, the performed experiments, together with the comparison with methods at the state-of-the-art, are shown.

Table 1. Comparison of our approach with state-of-the-art methods that use automatic features extraction. The used metric is accuracy.

Method	14 gestures	28 gestures
HG-GCN [25]	92.8%	88.3%
STA-GCN: [35]	91.5%	87.7%
ST-TS-HGR-NET [29]	94.2%	89.4%
CNN+LSTM [31]	85.6%	81.1%
HPEV [26]	92.5%	88.8%
Devineau et al. [20]	91.2%	84.3%
STA-Res-TCN [22]	89.2%	85.0%
Our proposed method	**94.6%**	**89.0%**

4.1 Dataset

Experiments are conducted on the DHG 14/28 [17] dataset. DHG 14/28 contains various sequences of 14 hand gestures performed with 2 different finger configurations (in total 14 + 14). Each gesture is performed 5 times by 20 participants in 2 ways, resulting in 2800 sequences. All participants are right-handed. The sequences are labeled according to their gesture, e.g. grab, tab, expand, pinch, rotation, swipe, shake, the number of fingers used, the performer and the trial number. Each sequence frame contains a depth image, the coordinates of 22 joints in both 2D depth image space and 3D world space, forming a complete hand skeleton. The depth images and hand skeletons were captured at 30 frames per second, with a depth image resolution of 640×480. The length of each sample gesture varies from 20 to 50 frames.

4.2 Implementation Details

After pre-processing and feature extraction, the dataset was split into two parts, the training set (75%) and the testing set (25%). We train our model using Pytorch on hardware consisting of a 3.4 GHz R5 2600 processor with 16 GB of RAM and NVidia RTX2080 GPU. The training has been performed for 200 epochs with a batch size of 32. The initial learning rate is 0.001 and is later updated by the Adam optimizer. The test methodology used is the 10-Fold-Cross-Validation.

4.3 Results and Comparison

Table 1 and Table 2 show the comparison with the works proposed by the state-of-the-art models that use the DHG 14/28 dataset.

The comparison with the state of the art is, in part, dependent on the test protocol adopted. Table 1, which shows the state-of-the-art methods that automatically extract the features on the DHG-14 dataset, shows that our proposed

Table 2. Comparison with state-of-the-art results which use handcrafted methods for feature extraction. The used metric is accuracy.

Method	14 gestures	28 gestures
De Smedt et al. [18]	86.8%	84.2%
Boulahia et al. [15]	90.4%	80.4%
Boutaleb et al. [16]	95.21%	–
Our proposed method	**94.6%**	**89.0%**

method outperforms them all when testing on 14 gestures. The confusion matrix of 14 classes is shown in Fig. 2a to further illustrate the performance of our proposed model. Furthermore, testing on 28 gestures, our approach also produces better results than nearly all the state-of-the-art methods except for [29]. Hence, by using a relatively simple model, the features we have designed manage to accurately capture the expression of a hand during any gesture. Moreover, these features are stable and well-conceived in that they can be compared with methods that are not only more complex but also extract features in a more substantial way using automatic methods. Instead, Table 2 shows that our approach also outperforms the majority of state-of-the-art based only on handcrafted features, further demonstrating the effectiveness of our model. As it is possible to see, the method presented in [16] has better results on the 14 gesture classification. This is mainly due to a more complex developed model with respect to our simple LSTM. In addition, no information regarding the 28 gestures recognition is provided, thus a complete comparison cannot be done.

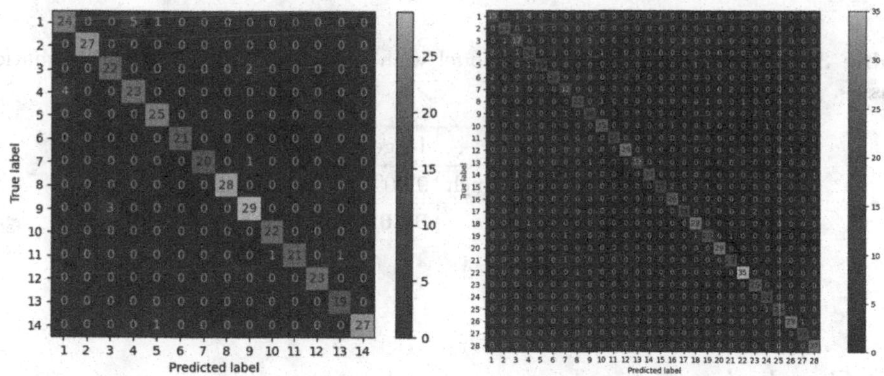

Fig. 2. The confusion matrix for a) 14 gestures, and b) 28 gestures.

As can be seen from the confusion matrix in Fig. 2a, a large part of the error occurs by confusing class 1 gestures (grab) with class 4 gestures (pinch). In Figs. 3a and 3b, we have samples of frames from the dataset taken from the grab and pinch sequences performed by the same individual. The high similarity of these two gestures, along with other variables such as involuntary variations in the gestures performed that causes some samples in the data to be not entirely visually distinguishable, leads the model to be incorrect when making a distinction between grab and pinch.

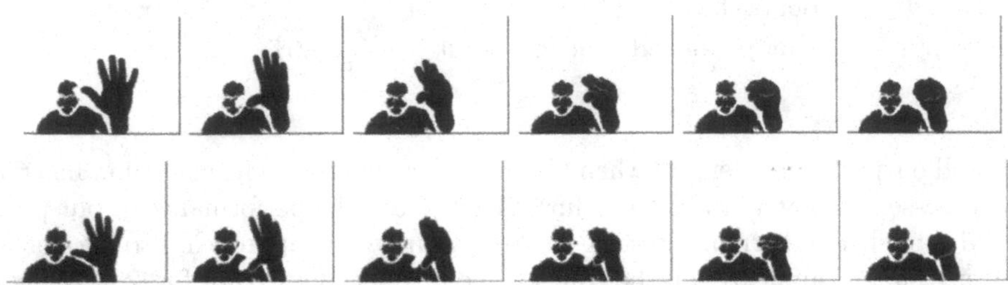

Fig. 3. Sample frames from sequence a) Gesture 1 - Subject 1 - Attempt 1 (Grab), and b) Gesture 4 - Subject 1 - Attempt 1 (Pinch).

The same confusion problem was encountered between the Grab and Pinch gesture classes with the 28 gestures. The confusion matrix of 28 classes is shown in Fig. 2b, where the four classes that include the two gestures performed with both finger configurations showed generated confusion in the model. Therefore, an attempt to train the model was performed on the dataset, excluding the two or four gesture classes that created ambiguity in the case of 14 or 28 gestures, respectively. The results obtained, shown in Table 3, shows significantly greater accuracy.

Table 3. Comparison of the original model with a model that excludes problematic classes.

Method	14 gestures	28 gestures
Excluding Grab and Pinch	97.0%	92.5%
With all gestures	**94.6%**	**89.0%**

5 Conclusion

Hand gesture recognition has found applications in various heterogeneous fields, including Human-Computer Interaction, serious games, and sign language interpretation, among others. Modern approaches to recognition utilize Deep Learning techniques, owing to their ability to extract features without the need for human intervention. However, a major drawback of this approach is its reliance on vast datasets that may not always be available. For some tasks, handcrafted features can enhance a model capability in achieving the desired result, and generally require less data compared to Deep Learning approaches.

This paper proposes a method that synergistically combines handcrafted features and Deep Learning for hand gesture recognition. The features are engineered from hand joints, while a simple Long Short-Term Memory (LSTM) model is employed for Deep Learning. The proposed method is evaluated on the DHG dataset, and is compared to state-of-the-art methods that use handcrafted features, as well as methods that use learned features.

The results of our experiments indicate that our approach outperforms state-of-the-art methods that use handcrafted features in both the 14 and 28 gesture recognition tests. Furthermore, our approach surpasses state-of-the-art methods that use learned features in the 14 gesture recognition test. These findings demonstrate that it is possible to employ a simpler model with well-engineered features to achieve high performance in hand gesture recognition.

Acknowledgement. This work was supported by "Smart unmannEd AeRial vehiCles for Human likE monitoRing (SEARCHER)" project of the Italian Ministry of Defence (CIG: Z84333EA0D); and "A Brain Computer Interface (BCI) based System for Transferring Human Emotions inside Unmanned Aerial Vehicles (UAVs)" Sapienza Research Projects (Protocol number: RM1221816C1CF63B); and "DRrone Aerial imaGe SegmentatiOn System (DRAGONS)" (CIG: Z71379B4EA); and Departmental Strategic Plan (DSP) of the University of Udine - Interdepartmental Project on Artificial Intelligence (2020-25); and "A proactive counter-UAV system to protect army tanks and patrols in urban areas (PROACTIVE COUNTER-UAV)" project of the Italian Ministry of Defence (Number 2066/16.12.2019), and the MICS (Made in Italy - Circular and Sustainable) Extended Partnership and received funding from Next-Generation EU (Italian PNRR - M4 C2, Invest 1.3 - D.D. 1551.11-10-2022, PE00000004). CUP MICS B53C22004130001

References

1. Alrowais, F., et al.: Modified earthworm optimization with deep learning assisted emotion recognition for human computer interface. IEEE Access **11**, 35089–35096 (2023)
2. Au, S., Dilworth, P., Herr, H.: An ankle-foot emulation system for the study of human walking biomechanics. In: Proceedings of the IEEE International Conference on Robotics and Automation (ICRA), pp. 2939–2945 (2006)
3. Avola, D., Cinque, L., Foresti, G.L., Martinel, N., Pannone, D., Piciarelli, C.: Low-Level Feature Detectors and Descriptors for Smart Image and Video Analysis: A Comparative Study, pp. 7–29 (2018)
4. Avola, D., Bernardi, M., Cinque, L., Foresti, G.L., Massaroni, C.: Adaptive bootstrapping management by keypoint clustering for background initialization. Pattern Recogn. Lett. **100**, 110–116 (2017)
5. Avola, D., Bernardi, M., Cinque, L., Foresti, G.L., Massaroni, C.: Exploiting recurrent neural networks and leap motion controller for the recognition of sign language and semaphoric hand gestures. IEEE Trans. Multimedia **21**(1), 234–245 (2019)

6. Avola, D., Caschera, M.C., Ferri, F., Grifoni, P.: Ambiguities in sketch-based interfaces. In: 40th Annual Hawaii International Conference on System Sciences (HICSS'07), p. 290b (2007)
7. Avola, D., Cinque, L., De Marsico, M., Fagioli, A., Foresti, G.L.: Lietome: preliminary study on hand gestures for deception detection via fisher-LSTM. Pattern Recogn. Lett. **138**, 455–461 (2020)
8. Avola, D., et al.: Low-altitude aerial video surveillance via one-class SVM anomaly detection from textural features in UAV images. Information **13**(1) (2022)
9. Avola, D., Cinque, L., Fagioli, A., Foresti, G.L., Fragomeni, A., Pannone, D.: 3D hand pose and shape estimation from RGB images for keypoint-based hand gesture recognition. Pattern Recogn. **129**, 108762 (2022)
10. Avola, D., et al.: Medicinal boxes recognition on a deep transfer learning augmented reality mobile application. In: Sclaroff, S., Distante, C., Leo, M., Farinella, G.M., Tombari, F. (eds.) Image Analysis and Processing - ICIAP 2022, pp. 489–499 (2022)
11. Avola, D., Cinque, L., Foresti, G.L., Marini, M.R.: An interactive and low-cost full body rehabilitation framework based on 3D immersive serious games. J. Biomed. Inform. **89**, 81–100 (2019)
12. Avola, D., Cinque, L., Foresti, G.L., Marini, M.R., Pannone, D.: VRheab: a fully immersive motor rehabilitation system based on recurrent neural network. Multimedia Tools Appl. **77**(19), 24955–24982 (2018)
13. Avola, D., Foresti, G.L., Cinque, L., Massaroni, C., Vitale, G., Lombardi, L.: A multipurpose autonomous robot for target recognition in unknown environments. In: 14th International Conference on Industrial Informatics (INDIN), pp. 766–771 (2016)
14. Bai, X., Li, C., Tian, L., Song, H.: Dynamic hand gesture recognition based on depth information. In: International Conference on Control, Automation and Information Sciences), pp. 216–221 (2018)
15. Boulahia, S.Y., Anquetil, E., Multon, F., Kulpa, R.: Dynamic hand gesture recognition based on 3D pattern assembled trajectories. In: Seventh International Conference on Image Processing Theory, Tools and Applications, pp. 1–6 (2017)
16. Boutaleb., Y., Soladie., C., Duong., N., Kacete., A., Royan., J., Seguier., R.: Efficient multi-stream temporal learning and post-fusion strategy for 3D skeleton-based hand activity recognition. In: Proceedings of the 16th International Joint Conference on Computer Vision, Imaging and Computer Graphics Theory and Applications (VISIGRAPP 2021) - Volume 4: VISAPP, pp. 293–302 (2021)
17. De Smedt, Q., Wannous, H., Vandeborre, J.P.: Skeleton-based dynamic hand gesture recognition. In: IEEE Conference on Computer Vision and Pattern Recognition Workshops, pp. 1206–1214 (2016)
18. De Smedt, Q., Wannous, H., Vandeborre, J.P.: Heterogeneous hand gesture recognition using 3d dynamic skeletal data. Comput. Vis. Image Underst. **181**, 60–72 (2019)
19. Deng, Z., Gao, Q., Ju, Z., Yu, X.: Skeleton-based multifeatures and multistream network for real-time action recognition. IEEE Sens. J. **23**(7), 7397–7409 (2023)
20. Devineau, G., Moutarde, F., Xi, W., Yang, J.: Deep learning for hand gesture recognition on skeletal data. In: 13th IEEE International Conference on Automatic Face and Gesture Recognition, pp. 106–113 (2018)
21. Fernando, T., Fookes, C., Gammulle, H., Denman, S., Sridharan, S.: Toward onboard panoptic segmentation of multispectral satellite images. IEEE Trans. Geosci. Remote Sens. **61**, 1–12 (2023)

22. Hou, J., Wang, G., Chen, X., Xue, J.-H., Zhu, R., Yang, H.: Spatial-temporal attention Res-TCN for skeleton-based dynamic hand gesture recognition. In: Leal-Taixé, L., Roth, S. (eds.) ECCV 2018. LNCS, vol. 11134, pp. 273–286. Springer, Cham (2019). https://doi.org/10.1007/978-3-030-11024-6_18
23. Khokher, M.R., et al.: Early yield estimation in viticulture based on grapevine inflorescence detection and counting in videos. IEEE Access **11**, 37790–37808 (2023)
24. Lee, J., Olsman, W., Triebel, R.: Learning fluid flow visualizations from in-flight images with tufts. IEEE Robot. Autom. Lett. **8**(6), 3677–3684 (2023)
25. Li, Y., He, Z., Ye, X., He, Z., Han, K.: Spatial temporal graph convolutional networks for skeleton-based dynamic hand gesture recognition. EURASIP J. Image Video Process. **2019**(1), 78 (2019)
26. Liu, J., Liu, Y., Wang, Y., Prinet, V., Xiang, S., Pan, C.: Decoupled representation learning for skeleton-based gesture recognition. In: IEEE/CVF Conference on Computer Vision and Pattern Recognition, pp. 5750–5759 (2020)
27. Lu, W., Tong, Z., Chu, J.: Dynamic hand gesture recognition with leap motion controller. IEEE Sig. Process. Lett. **23**(9), 1188–1192 (2016)
28. Luna, M.A., et al.: Spiral coverage path planning for multi-UAV photovoltaic panel inspection applications. In: International Conference on Unmanned Aircraft Systems (ICUAS), pp. 679–686 (2023)
29. Nguyen, X.S., Brun, L., Lézoray, O., Bougleux, S.: A neural network based on SPD manifold learning for skeleton-based hand gesture recognition. In: IEEE/CVF Conference on Computer Vision and Pattern Recognition, pp. 12028–12037 (2019)
30. Nian, B., Jiang, B., Shi, H., Zhang, Y.: Local contrast attention guide network for detecting infrared small targets. IEEE Trans. Geosci. Remote Sens. **61**, 1–13 (2023)
31. Núñez, J.C., Cabido, R., Pantrigo, J.J., Montemayor, A.S., Vélez, J.F.: Convolutional neural networks and long short-term memory for skeleton-based human activity and hand gesture recognition. Pattern Recogn. **76**, 80–94 (2018)
32. Oudah, M., Al-Naji, A., Chahl, J.: Hand gesture recognition based on computer vision: a review of techniques. J. Imaging **6**(8) (2020)
33. Yang, K., Li, R., Qiao, P., Wang, Q., Li, D., Dou, Y.: Temporal pyramid relation network for video-based gesture recognition. In: 25th IEEE International Conference on Image Processing, pp. 3104–3108 (2018)
34. Yao, L., Fu, C., Li, S., Zheng, G., Ye, J.: SGDViT: saliency-guided dynamic vision transformer for UAV tracking. In: IEEE International Conference on Robotics and Automation (ICRA), pp. 3353–3359 (2023)
35. Zhang, W., Lin, Z., Cheng, J., Ma, C., Deng, X., Wang, H.: STA-GCN: two-stream graph convolutional network with spatial-temporal attention for hand gesture recognition. Vis. Comput. **36**(10), 2433–2444 (2020)

An Optimized Pipeline for Image-Based Localization in Museums from Egocentric Images

Nicola Messina[1(✉)], Fabrizio Falchi[1], Antonino Furnari[2], Claudio Gennaro[1], and Giovanni Maria Farinella[2]

[1] ISTI-CNR, via G. Moruzzi 1, 56017 Pisa, Italy
nicola.messina@isti.cnr.it

[2] University of Catania, Viale A. Doria 6, 95125 Catania, Italy

Abstract. With the increasing interest in augmented and virtual reality, visual localization is acquiring a key role in many downstream applications requiring a real-time estimate of the user location only from visual streams. In this paper, we propose an optimized hierarchical localization pipeline by specifically tackling cultural heritage sites with specific applications in museums. Specifically, we propose to enhance the Structure from Motion (SfM) pipeline for constructing the sparse 3D point cloud by a-priori filtering blurred and near-duplicated images. We also study an improved inference pipeline that merges similarity-based localization with geometric pose estimation to effectively mitigate the effect of strong outliers. We show that the proposed optimized pipeline obtains the lowest localization error on the challenging Bellomo dataset [11]. Our proposed approach keeps both build and inference times bounded, in turn enabling the deployment of this pipeline in real-world scenarios.

Keywords: Localization · Camera Pose Estimation · Structure From Motion · Egocentric Vision

1 Introduction

Virtual Reality (VR) is becoming a game-changing technology in many scenarios – from gaming to medical applications – and is being increasingly applied to the exploration and preservation of cultural heritage sites [2,12]. Visual localization is a critical task to enable stable and reliable VR applications on these sites, where it can be used to enhance the visitor experience and receive contextualized information about the visited rooms and observed artworks [14]. Furthermore, visual localization is employed in other real-world applications – like in robot navigation, mixed reality, and self-driving vehicles – becoming a critical computer vision task. In recent years, a significant amount of research has focused on developing deep learning-based methods to directly regress 3D camera coordinates from raw images. However, such methods require extensive network training time and resources. Furthermore, this expensive training

G. L. Foresti et al. (Eds.): ICIAP 2023, LNCS 14233, pp. 512–524, 2023.
https://doi.org/10.1007/978-3-031-43148-7_43

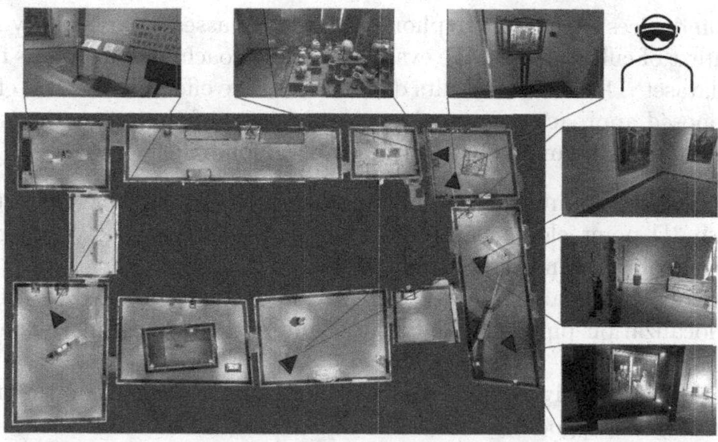

Fig. 1. The considered image-based localization problem consists in localizing a visitor of a museum from egocentric images collected through a wearable or mobile device. The figure shows some examples from the dataset proposed in [11], along with their positions in the map.

should be performed once for every scenario, limiting the applicability of these approaches to real-world cases.

Previous methods have proposed to use hierarchical localization (HLOC) approaches leveraging global feature matching and Structure from Motion (SfM) for fine-grained localization [15]. Despite their flexibility, these approaches have some known limitations in real environments, where image data can be noisy or blurred, and the presence of strong outliers invalidates their effectiveness.

In this paper, we propose an optimized pipeline for image-based localization based on the hierarchical localization idea. We test it in a museum environment, where the main downstream use case would be capturing user location through wearable devices or smartphones to optimize the visitor experience (see Fig. 1). Using a streamlined pipeline that discards blur images and duplicates, we propose a method that constructs a sparse 3D point cloud from raw reference images in under a couple of minutes, surpassing by a large margin long deep network training times and improving the HLOC framework by obtaining a 20x boost on build times with limited localization degradation. The 3D point model built using SfM allows us to perform both global image matching for coarse-grained localization and local matching for fine-grained localization using geometric camera pose estimation. One of the contributions proposed for mitigating the high variance of the geometric camera pose estimation is the fusion of fine- and coarse-grained localization pipelines. The choice of the localization method is simply driven by the estimated quality of the match between the query image and the ones registered in the 3D point cloud. The proposed approach efficiently queries the 3D point model and achieves high accuracy with less than one-second latency on a standalone desktop computer embedded with a mid-end graphic card, and it can be further engineered to run on portable localization devices such as smartphones or smart glasses – particularly relevant in the context of cultural sites. We evaluate the approach on four rooms from the Bellomo dataset [11], and the results demonstrate the effectiveness and efficiency of our proposed approach compared to current state-of-the-art methods.

To summarize, the main contributions are as follows:

- We propose a streamlined pipeline for efficiently and effectively constructing a sparse 3D point cloud using SfM from raw references, leveraging on the filtering of blurred and duplicated images.
- We employ both local and global localization outputs provided by the hierarchical localization pipeline to efficiently and effectively estimate the correct location.
- We perform extensive experimentation on four rooms from the Bellomo dataset, obtaining good accuracies with less than 1-second latencies.

2 Related Work

Localization methods based on classification rely on a discretization of the space in cells and train a classifier to infer the correct cell from an input RGB image. In this context, only the rough camera location is estimated, while its orientation is not predicted. Seminal works [21] considered the problem of inferring the room in which the user is located with classification approaches based on handcrafted features. More recent methods used Bag of Words representation [4,8], while others are based on CNNs [23]. Others [7,13] performed classification-based localization considering an *open-set* problem in which the camera may also acquire images of new locations not initially included at training time.

Among the approaches based on camera pose estimation, a line of works approximates the location of a test image assigning it the pose of the most similar image in the training set, as predicted by image retrieval methods [1,23].

Some methods treated camera pose estimation as a regression problem in which camera coordinates are predicted directly from monocular images. Most of these methods are based on a backbone CNN to extract features, later used to regress the camera pose [9,22]. Others predict relative camera pose [3,10]. (i.e., the pose of a test image relative to one or more training images).

Localization methods based on local feature matchings are the most accurate ones, as they directly link 2D local features extracted from the image to 3D scene coordinates. Matchings can be obtained with a descriptor matching algorithm [17] or regressing 3D coordinates from image patches [16,18].

Visual SLAM (Simultaneous Localization and Mapping) [6] is another widely studied set of methods for acquiring the location of a moving agent using camera sensors. However, SLAM makes assumptions quite different from our scenario and may present problems in our specific use case. In particular, i) SLAM assumes the environment is not known, while the model of our environment is always available and contains some labeled ground-truth positions; ii) SLAM has known issues when the camera is abruptly shaken – which often happens if the camera is from a smartphone or mounted on smartglasses iii) SLAM should rely on separate localization methods if the video stream is discontinuous, due to the so-called *kidnapped robot problem*. This problem may arise in our scenario, where the device may be activated only when the visitor needs it.

Particularly related to this paper are hierarchical localization (HLOC) works based on the combination of image retrieval approaches and geometric correspondences [15]. These approaches are based on a database of localized images for the first image retrieval step and the construction of a 3D model of the scene through Structure from Motion (SfM) to perform accurate camera pose estimation. Specifically, the work in [15] employs COLMAP [20] for performing SfM and constructing a 3D point cloud from a set of pictures taken in the environment. The method employs a monolithic approach, based on a shared CNN-based visual backbone, to produce both global descriptors through a NetVLAD head [1] and local descriptors using the efficient SuperPoint decoder [5].

3 Method

We rely on the hierarchical localization framework presented in [15], and we propose an optimized pipeline that obtains the best effectiveness and efficiency in cultural heritage sites like museums. In fact, although this approach obtains stable results, it adds some complexity that prevents its usability in real-time real-world scenarios where acquired data are noisy and redundant. Following, we describe in detail the improvements introduced in the pipeline to handle unclean data and enhance the framework for use in a specific downstream scenario.

3.1 Model Building

Usually, SfM is exceedingly expensive, and the construction time increases with the number of matching image pairs. To decrease the number of image matches, the hierarchical localization framework allows employing only the most similar images to a given one to check for local feature matches. This is done by performing k nearest neighbor search using the NetVLAD [1] global feature. In the experiments, we refer to the k used to build the model as k_{build}. Even if beneficial, this smart pair filtering procedure is often not sufficient for reaching competitive localization performance and build times. With datasets like the Bellomo dataset [11], where frames are sampled from a video acquired by a wearable device, the quality of the acquired images is often limited. Specifically, many frames from the video are blurred or near-duplicated. While near-duplication mostly causes problems in efficiency due to the increasing number of less informative images that have to be considered by SfM, frame blurring also has disadvantages in reconstruction accuracy, given that blurred images cause many false matches. Hence, we apply near-duplicate removal and blur image filtering to optimize the model creation process. We show in the experiments how these pre-processing steps help achieve higher performance and better efficiency.

Near-Duplicate Removal. Near duplicate removal relies on the similarity between low-level descriptors for finding almost identical keyframes. We can reuse the NetVLAD global descriptor used during the first image search stage for filtering out duplicated images at model construction time. This can be obtained by scanning the list of vectors starting from the last element $i = N-1$, and finding if there is at least of the previous elements $j < i$ such that $S(i,j) >= \delta_{\text{duplicate}}$, where $S(\cdot,\cdot)$ is the dot product in our case. If the above condition is met, element i is marked as duplicate. Notice that with this formulation, the first element $i = 0$ is never considered a duplicate, which makes sense in our scenario where images are sequentially obtained from a real-time acquisition device.

Blur Image Filtering. We rely on a simple approach based on the variance of the Laplacian of the image, which is computationally efficient and already suffices for our goal. In particular, to find blurred images, given the Laplacian of the pixel intensities I computed as $L(x,y) = \frac{\partial^2 I}{\partial x^2} + \frac{\partial^2 I}{\partial y^2}$, we can compute its variance across all the image pixels $\text{Var}[L]$ and check if this value is below a certain threshold δ_b. We then keep all the images such that $\text{Var}[L] > \delta_b$.

Model Geo-registration. SfM allows us to reconstruct a sparse point cloud of the environment, but it cannot infer the scale of the model until some of the points in the cloud are annotated with real-world coordinates. In order to estimate the location error in meters, we need to infer the correct scale of the point cloud. This can be achieved through the model geo-registration function of COLMAP, which takes the 3D coordinates of a subset of registered images as input and infers the model scale through a RANSAC estimator to be robust to possible outliers. Theoretically, only three images are sufficient for geo-registering a model. Nevertheless, usually, more images are used to diminish the effect of possible imprecise ground-truth annotations. Specifically, in our scenario, we used the 3D coordinates associated with the images in the training set to register the models of each room in the dataset.

3.2 Localization

Differently from the model building pipeline, the localization phase should happen in real-time. In this phase, we do not a-priori filter images based on their blurriness, as we could potentially ground every image in the 3D point cloud for deriving the user location. Drawing inspiration from the hierarchical localization method in [15], this method is based on a coarse-grained localization which uses an image-similarity-based approach, and a fine-grained localization, which instead relies on geometric pose estimation.

For coarse-grained localization, we employ k-nearest-neighbor search using global features to search the images registered in the 3D point cloud more similar to the query image. The images registered in the point cloud have been assigned 3D coordinates, so we can easily infer an approximate query location as follows:

$$\mathbf{x}_{\text{coarse}} = \frac{1}{k_{\text{infer}}} \sum_{i=1}^{k_{\text{infer}}} \mathbf{X}_i, \qquad \text{where} \quad \mathbf{X} = \{\mathbf{x}_i | i \in \text{search}(V_{\text{train}}, v_q, k_{\text{infer}})\}, \tag{1}$$

where search$(\cdot,\cdot,\cdot)$ finds the indexes of the k_{infer} most similar images to v_q in the training image set V_{train}, and $\mathbf{X}$ is the set of locations of the nearest neighbors registered images. Note that not all the training images are registered in the point cloud. Therefore, in some cases, it may happen that the resulting set $\mathbf{X}$ is empty. As it can be noticed, using $k_{\text{infer}} = 1$ we can simply localize the camera using the coordinates of the nearest neighbor only. If $k_{\text{infer}} > 1$, we are instead computing the centroid among the coordinates associated with the k-nearest-neighbors. We experimentally show that $k_{\text{infer}} = 1$ works the best in our scenario.

Fine-grained localization in the HLOC framework depends on the results from the coarse-grained method. Specifically, we employ k-nearest-neighbor search using global features to search among the training images the ones registered in the 3D point cloud more similar to the query image. Then, we can perform local feature matching between the query image and the local features already registered on the point cloud for the k images:

$$\mathbf{x}_{\text{fine}} = \text{pose\_estimation}(\mathcal{M}^{\text{3D}}, \mathbf{M}_q^{\text{2D}}, \mathcal{I}), \tag{2}$$

$$\text{where} \quad \begin{cases} \mathcal{M}^{\text{2D}}, \mathcal{M}^{\text{3D}} = \{(\mathbf{M}_i^{\text{2D}}, \mathbf{M}_i^{\text{3D}}) | i \in \text{search}(V_{\text{train}}, v_q, k_{\text{infer}})\} \\ \mathcal{I} = \text{match}(\mathcal{M}^{\text{2D}}, \mathbf{M}_q^{\text{2D}}) \end{cases} \tag{3}$$

$\mathbf{M}_i^{\text{2D}}$ and $\mathbf{M}_i^{\text{3D}}$ are the 2D and associated 3D coordinates of the found joints in the k_{infer} neighboring images, $\mathbf{M}_q^{\text{2D}}$ is the set of local features found in the query image, and $\mathcal{I}$ is the set of local feature inliers. The pose_estimation$(\cdot,\cdot)$ function is a COLMAP function[1] which performs geometric pose estimation using the matching 2D inliers to derive the actual pose, indicated as $\mathbf{x}_{\text{fine}}$. Note that $\mathcal{I}$ could be empty either if there are no retrieved images registered in the point cloud, or if there are no matching local features. In that case, the fine-grained position cannot be estimated, but we show in the experiments that this happens with an acceptable probability for a real-case scenario.

3.3 Mixing the Localization Outcomes

Although fine-grained localization has potentially higher accuracy, there may be some strong outliers due to failures in the geometric pose estimation. For this reason, we decided to prioritize the fine-grained over coarse-grained localization only if the number of inliers (indicated as $|\mathcal{I}|$) found from the local features matching phase is above a certain threshold τ. We argue that the number of matches is a good indicator of the quality of the fine-grained localization, and we prove it empirically in the experimental evaluation. Therefore, the final estimated position $\mathbf{x}$ is $\mathbf{x}_{\text{coarse}}$ if $|\mathcal{I}| < \tau$ and $\mathbf{x}_{\text{fine}}$ otherwise. The localization error is then computed using the standard Euclidean distance with the ground-truth position values $\mathbf{x}_{GT}$ provided within the dataset: $e = ||\mathbf{x} - \mathbf{x}_{GT}||_2$.

[1] https://github.com/colmap/pycolmap/blob/master/estimators/absolute_pose.cc.

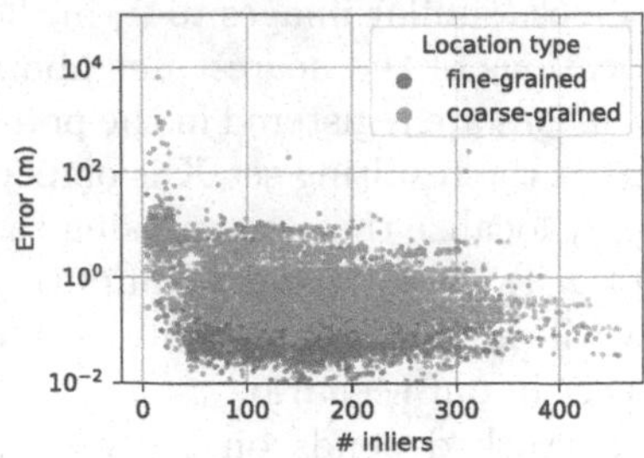

(a) Distribution of localization errors depending on the number of inliers for both coarse- and fine-grained estimated poses.

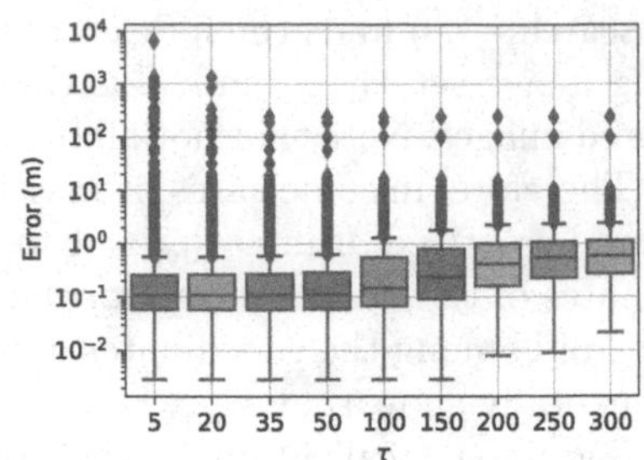

(b) Average localization error varying the threshold on the number of considered inliers.

Fig. 2. Analysis of the effectiveness of the mixing between fine- and coarse-grained pose estimations using the number of the inliers as the threshold.

4 Experiments

4.1 Dataset

The dataset used in this research has been introduced in [11]. It has been recorded in the Bellomo Palace Regional Gallery, a museum located in Syracuse, Italy. To capture the visitors' experiences, the authors recorded 10 videos using a GoPro Hero 4 wearable camera and Matterport 3D to create a 3D scan of the museum's environment. They selected four rooms within the museum to collect data, as they contained a variety of items such as statues, paintings, and display cases, which provide a representative sample of what a museum typically offers. The videos were extracted into image sequences. The obtained images are divided into three sets for training, testing, and validation. Specifically, all frames from the first to sixth video are used as the training set, frames from the seventh and eighth videos as the test set, and frames from the ninth and tenth videos as the validation set. We consider all position estimations further away than 1000m from the accessible area (far beyond the boundaries of a museum) as a failure in localization and discard them before computing the average localization errors.

We run our method on a mid-end desktop computer equipped with an RTX-2080Ti graphic card and an AMD Ryzen 7 1700 Eight-Core Processor.

4.2 Parameters Study

We run a preliminary analysis on the validation set to fix some of the system's hyper-parameters.

First, we focus on the model-building procedure for analyzing hyperparameters like k_{build}, δ_{blur}, $\delta_{duplicate}$. The results from the exploration of different build parameter configurations are reported in Table 1. We derive meaningful values for δ_{blur} and $\delta_{duplicate}$ from their distribution on the validation set. The

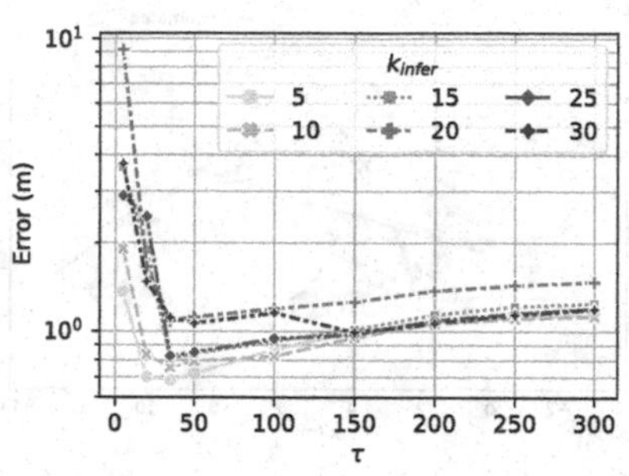

(a) Average localization error varying the threshold τ, for different values of k_{infer}.

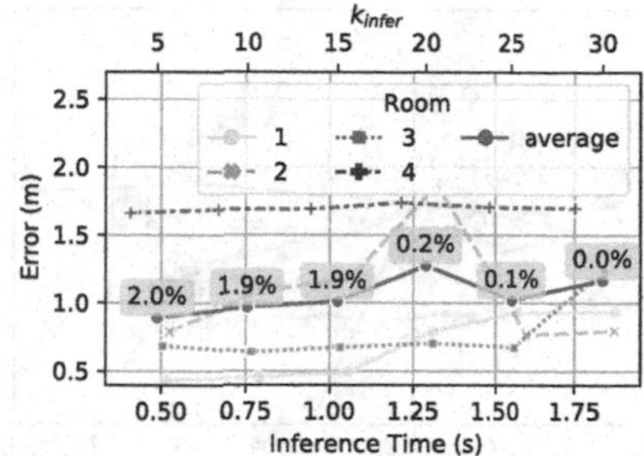

(b) Efficiency vs effectiveness, varying k_{infer}. In blue boxes, the avg. percentage of failure cases is reported.

Fig. 3. Effect of hyper-parameters k_{infer} and τ on effectiveness and efficiency, on the validation set.

Table 1. Mean location error and build times. *Survived images* are the images that remained after the blur and near-duplicate filtration, while *registered images* are the ones that were registered by COLMAP on the 3D point cloud.

δ_{blur}	$\delta_{\text{duplicate}}$	k_{build}	Error (m)	Time (s)	survived images (%)	registered images (%)
70	0.45	10	1.96	60.1	35.2	33.1
70	0.45	15	1.48	75.7	35.2	33.5
90	0.45	10	1.28	48.8	30.6	28.7
90	0.45	15	1.54	60.0	30.6	29.2
90	0.55	10	1.19	72.5	43.3	41.3
90	0.55	15	2.23	118.0	43.3	41.7
0	1.0	10	0.78	1524.1	100.0	98.8

localization error, measured in meters, is averaged through a selected range of values for the threshold τ and k_{infer} to give an overall estimate of the model's performance without a-priori setting any inference hyper-parameters. We leave the results obtained without any filtering ($\delta_{\text{blur}} = 0$, $\delta_{duplicate} = 1.0$) as the last row of the table. We can see how, using blur and near-duplicate filtering, we can obtain overall comparable error values with this original approach, in turn decreasing the build times with a speedup of more than 20x. Given its effectiveness-efficiency ratio, we consider the model built with $\delta_{\text{blur}} = 90$ and $\delta_{duplicate} = 0.55$ for further experiments on the inference parameters.

Next, we proceed by studying the hyper-parameters k_{infer} and τ. As previously hypothesized, while the advantage of fine-grained localization is the potential high accuracy, there are some strong outliers due to geometric estimation failures. This behavior is shown in Fig. 2a. If we apply the thresholding for deciding if either using fine-grained or coarse-grained localization outputs, we notice in Fig. 3b how we are able to diminish the number of outliers when the threshold

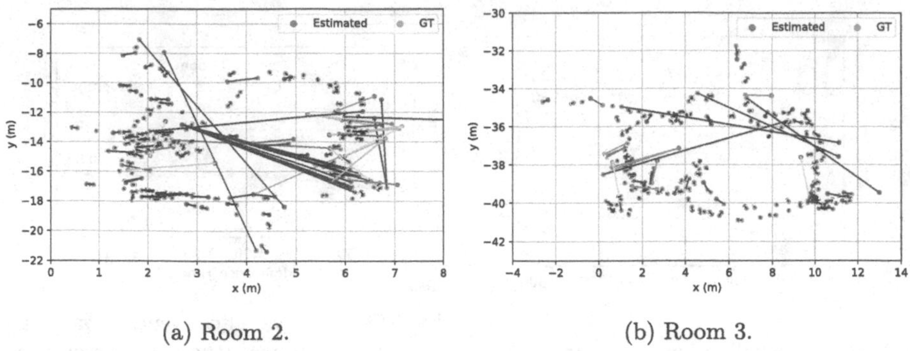

(a) Room 2.

(b) Room 3.

Fig. 4. Visualization of predicted with respect to ground-truth poses, projected on the XY plane. Lines represent corresponding ground truth and predicted poses. Pairs are colored with a gradient indicating the localization error (from blue – small error – to red – large error). (Color figure online)

τ is increased. This, in turn, validates our hypothesis that a higher number of inliers contributes to a better fine-grained localization. In Fig. 3a, we study how the localization error varies depending on the number of nearest neighbors. In particular, the lowest error is achieved for $k_{\text{infer}} \in [5, 10]$, for a relative small τ. This is reasonable since (i) too few or too many nearest neighbor registered images can provide noisy local matches that degrade the fine-grained geometric localization, and (ii) high thresholds τ inhibit the advantages introduced by the fine-grained localization. Analyzing the plots, we decide to fix $k_{\text{infer}} = 5$ and $\tau = 50$ in the rest of the experiments.

In Fig. 3b, we show how, varying k_{infer}, we also obtain different system latencies, as the process of geometric position estimation becomes more and more expensive with an increasing number of local feature matches. The choice of $k_{\text{infer}} = 10$ keeps the response time below 0.8 s, enabling a sufficient frame rate for localizing the user in real-time. The only drawback of keeping k_{infer} low is that we have, on average, 2% of query images that cannot be localized due to either (i) failure of coarse-grained localization – there are no images registered in the 3D point cloud among the first k_{infer} found – (ii) failure of fine-grained localization – there are no local features matches among the registered images found among the first k_{infer} ones – or (iii) the estimated location is beyond 1000m from the walkable area, which we consider a failure as well.

4.3 Results

We compare our method with the following state-of-the-art visual localization approaches: i) a SIFT-based image retrieval approach, called *Vote And Verify* [19], which tackles primarily image retrieval but enforces geometric verification constraints; ii) the PoseNet approach [9], which directly regress pose using a deep convolution network; iii) PAM-CAM [11], which also regresses the camera

Table 2. Final localization results on the Bellomo dataset.

(a) Results on the test sets, averaged among the four different rooms.

Method	Loc. error (m)
PoseNet (beta 100)	1.43
Vote-and-Verify	0.82
PAM-CAM	1.26
Our (only FG)	1.65
Our (only CG 1-nn)	1.16
Our (only CG 5-nn)	1.24
Our (FG + CG 1-nn)	**0.62**

(b) Localization error (mean and median, in meters) for each of the four rooms.

	R1	R2	R3	R4
mean	0.48	0.66	0.42	1.66
median	0.09	0.10	0.11	0.19

pose but using a more advanced deep network embodied with attention modules and trained by employing a self-supervision approach.

We report the results using the hyper-parameters set as explained in Sect. 4.2. We report four different variants: i) *only FG* is the model only employing fine-grained features matching, which downcasts the inference method to the one proposed in the HLOC framework [15]; ii) *only CG 1-nn* and iii) *only CG 5-nn* are the models employing only coarse localization – i.e., the position of the most-likely image registered in the 3D point cloud, using one nearest neighbor and the centroid among the five nearest neighbors respectively; iv) *FG + CG* is the final method employing both coarse-grained and fine-grained localization, using the thresholding method explained in Sect. 3.2.

Final results are reported in Table 2a. All the methods we use for comparison and reported in the table have been fine-tuned on the Bellomo dataset by the authors in [11]. The proposed method outperforms all the other ones on this challenging benchmark. Specifically, although either the FG or CG methods alone cannot improve over the state-of-the-art, we obtained the best results when employing both approaches jointly. These results prove that the non-regression-based approaches relying on geometric verification can obtain the best results by keeping the build (Table 1) and inference times (Fig. 3b) bounded for enabling real-time visitor localization. It is also interesting to note that 1-nn in the CG configuration obtains the best results over the 5-nn one, probably because the NetVLAD global features can retrieve with a high likelihood the most relevant registered images as the first result. In Fig. 2b, we also report localization errors for all four rooms. The highest contribution to the error comes from *Room 4*, probably due to scarce lighting and a big glass case in the middle, which creates false positive matching among local features. However, the median is far lower than the mean, suggesting that the relatively few outliers still have a strong impact which should be further mitigated in future works.

In Fig. 4, we show qualitative results of the estimated and ground-truth position pairs in two rooms of the Bellomo dataset. Lines indicate corresponding estimated and ground-truth positions. Some failure cases are particularly visible in Fig. 4a, where – due to failure in geometric estimation and retrieval of the correct 1-nn image – some query images are associated with the wrong registered camera. Apart from these edge cases, we can notice that the location is generally estimated with good accuracy.

5 Conclusions

This paper proposes an efficient pipeline based on the Hierarchical Localization (HLOC) framework for localizing egocentric video streams in interior cultural heritage sites, such as museums. The proposed method overcame some of the drawbacks of the original HLOC framework by filtering out uninformative visual inputs – near-duplicated or blurred images – and proposing a smart aggregation of localization information from both fine- and coarse-grained modules to mitigate the effect that strong outliers have in the geometric estimation. Thanks to extensive experimentation on the challenging Bellomo dataset, we were able to characterize the most influential factors affecting both fine- and coarse-grained localization outcomes, obtaining state-of-the-art results with respect to other approaches on the same dataset.

In future works, we plan to address 6D localization by including orientation estimation, and we plan to substitute the SfM construction pipeline by employing Matterport 3D scans to estimate the position using the depth information without relying on a 3D point cloud. This would further increase the efficiency and the overall usability of the proposed localization framework.

Acknowledgements. This research was supported by AI4Media – A European Excellence Centre for Media, Society, and Democracy (EC, H2020 No. 951911), by SUN – Social and hUman ceNtered XR (EC, Horizon Europe No. 101092612), and by VALUE – Visual Analysis For Location And Understanding Of Environments (PO FESR 2014/2020 program, action 1.1.5).

References

1. Arandjelovic, R., Gronat, P., Torii, A., Pajdla, T., Sivic, J.: NetVLAD: CNN architecture for weakly supervised place recognition. In: Proceedings of the IEEE Conference on Computer Vision and Pattern Recognition, pp. 5297–5307 (2016)
2. Arrighi, G., See, Z.S., Jones, D.: Victoria theatre virtual reality: a digital heritage case study and user experience design. Digit. Appl. Archaeol. Cult. Herit. **21**, e00176 (2021)
3. Balntas, V., Li, S., Prisacariu, V.: RelocNet: continuous metric learning relocalisation using neural nets. In: Ferrari, V., Hebert, M., Sminchisescu, C., Weiss, Y. (eds.) Computer Vision – ECCV 2018. LNCS, vol. 11218, pp. 782–799. Springer, Cham (2018). https://doi.org/10.1007/978-3-030-01264-9_46
4. Cao, S., Snavely, N.: Graph-based discriminative learning for location recognition. In: Proceedings of the IEEE Conference on Computer Vision and Pattern Recognition, pp. 700–707 (2013)
5. DeTone, D., Malisiewicz, T., Rabinovich, A.: Superpoint: self-supervised interest point detection and description. In: Proceedings of the IEEE Conference on Computer Vision and Pattern Recognition Workshops, pp. 224–236 (2018)

6. Durrant-Whyte, H., Bailey, T.: Simultaneous localization and mapping: part i. IEEE Robot. Autom. Mag. **13**(2), 99–110 (2006)
7. Furnari, A., Farinella, G.M., Battiato, S.: Recognizing personal locations from egocentric videos. IEEE Trans. Hum.-Mach. Syst. **47**(1), 6–18 (2016)
8. Ishihara, T., Vongkulbhisal, J., Kitani, K.M., Asakawa, C.: Beacon-guided structure from motion for smartphone-based navigation. In: 2017 IEEE Winter Conference on Applications of Computer Vision (WACV), pp. 769–777. IEEE (2017)
9. Kendall, A., Grimes, M., Cipolla, R.: PoseNet: a convolutional network for real-time 6-DoF camera relocalization. In: Proceedings of the IEEE International Conference on Computer Vision, pp. 2938–2946 (2015)
10. Laskar, Z., Melekhov, I., Kalia, S., Kannala, J.: Camera relocalization by computing pairwise relative poses using convolutional neural network. In: Proceedings of the IEEE International Conference on Computer Vision, pp. 929–938 (2017)
11. Mauro, D.D., Furnari, A., Signorello, G., Farinella, G.M.: Unsupervised domain adaptation for 6DoF indoor localization. In: International Conference on Computer Vision Theory and Applications - VISAPP (2021). https://iplab.dmi.unict.it/EGO-CH-LOC-UDA/
12. Milosz, M., Skulimowski, S., Kęsik, J., Montusiewicz, J.: Virtual and interactive museum of archaeological artefacts from Afrasiyab - an ancient city on the silk road. Digit. Appl. Archaeol. Cult. Herit. **18**, e00155 (2020)
13. Ragusa, F., Furnari, A., Battiato, S., Signorello, G., Farinella, G.M.: Egocentric visitors localization in cultural sites. J. Comput. Cult. Herit. (JOCCH) **12**(2), 11 (2019)
14. Ragusa, F., Furnari, A., Battiato, S., Signorello, G., Farinella, G.M.: Ego-CH: dataset and fundamental tasks for visitors behavioral understanding using egocentric vision. Pattern Recogn. Lett. **131**, 150–157 (2020)
15. Sarlin, P.E., Cadena, C., Siegwart, R., Dymczyk, M.: From coarse to fine: robust hierarchical localization at large scale. In: Proceedings of the IEEE/CVF Conference on Computer Vision and Pattern Recognition, pp. 12716–12725 (2019)
16. Sarlin, P.E., Debraine, F., Dymczyk, M., Siegwart, R., Cadena, C.: Leveraging deep visual descriptors for hierarchical efficient localization. In: Conference on Robot Learning, pp. 456–465. PMLR (2018)
17. Sarlin, P.E., DeTone, D., Malisiewicz, T., Rabinovich, A.: Superglue: learning feature matching with graph neural networks. In: Proceedings of the IEEE/CVF Conference on Computer Vision and Pattern Recognition, pp. 4938–4947 (2020)
18. Sarlin, P.E., et al.: Back to the feature: learning robust camera localization from pixels to pose. In: Proceedings of the IEEE/CVF Conference on Computer Vision and Pattern Recognition, pp. 3247–3257 (2021)
19. Schönberger, J.L., Price, T., Sattler, T., Frahm, J.-M., Pollefeys, M.: A vote-and-verify strategy for fast spatial verification in image retrieval. In: Lai, S.-H., Lepetit, V., Nishino, K., Sato, Y. (eds.) ACCV 2016. LNCS, vol. 10111, pp. 321–337. Springer, Cham (2017). https://doi.org/10.1007/978-3-319-54181-5_21
20. Schönberger, J.L., Zheng, E., Frahm, J.-M., Pollefeys, M.: Pixelwise view selection for unstructured multi-view stereo. In: Leibe, B., Matas, J., Sebe, N., Welling, M. (eds.) ECCV 2016. LNCS, vol. 9907, pp. 501–518. Springer, Cham (2016). https://doi.org/10.1007/978-3-319-46487-9_31
21. Starner, T., Schiele, B., Pentland, A.: Visual contextual awareness in wearable computing. In: Digest of Papers, Second International Symposium on Wearable Computers (Cat. No. 98EX215), pp. 50–57. IEEE (1998)

22. Walch, F., Hazirbas, C., Leal-Taixe, L., Sattler, T., Hilsenbeck, S., Cremers, D.: Image-based localization using LSTMs for structured feature correlation. In: Proceedings of the IEEE International Conference on Computer Vision, pp. 627–637 (2017)
23. Weyand, T., Kostrikov, I., Philbin, J.: PlaNet - photo geolocation with convolutional neural networks. In: Leibe, B., Matas, J., Sebe, N., Welling, M. (eds.) ECCV 2016. LNCS, vol. 9912, pp. 37–55. Springer, Cham (2016). https://doi.org/10.1007/978-3-319-46484-8_3

Annotating the Inferior Alveolar Canal: The Ultimate Tool

Luca Lumetti[1], Vittorio Pipoli[1,2], Federico Bolelli[1](✉), and Costantino Grana[1]

[1] University of Modena and Reggio Emilia, Modena, Italy
{luca.lumetti,vittorio.pipoli,federico.bolelli, costantino.grana}@unimore.it
[2] University of Pisa, Pisa, Italy
vittorio.pipoli@phd.unipi.it

Abstract. The Inferior Alveolar Nerve (IAN) is of main interest in the maxillofacial field, as an accurate localization of such nerve reduces the risks of injury during surgical procedures. Although recent literature has focused on developing novel deep learning techniques to produce accurate segmentation masks of the canal containing the IAN, there are still strong limitations due to the scarce amount of publicly available 3D maxillofacial datasets. In this paper, we present an improved version of a previously released tool, IACAT (Inferior Alveolar Canal Annotation Tool), today used by medical experts to produce 3D ground truth annotation. In addition, we release a new dataset, *ToothFairy*, which is part of the homonymous MICCAI2023 challenge hosted by the Grand-Challenge platform, as an extension of the previously released *Maxillo* dataset, which was the only publicly available. With ToothFairy, the number of annotations has been increased as well as the quality of existing data.

Keywords: IAC · CBCT · 3D Segmentation · Annotation Tool

1 Introduction

The placement of dental implants within the jawbone is a common surgical procedure that can raise different complications due to the presence of the Inferior Alveolar Nerve (IAN). Such nerve lies close to molars roots, necessitating meticulous preoperative planning to avoid causing any type of damage. Hence, an accurate segmentation of the bone cavity containing the IAN is crucial to avoid nerve injuries during surgery [8,13]. Such a cavity is identified as the Inferior Alveolar Canal (IAC).

The IAC segmentation, usually performed by radiologic technologists, is obtained from a 3D Cone Beam Computed Tomography (CBCT) scan by manually drawing a line on a 2D projection of the original volume. This type of annotation is referred to as *sparse* or 2D annotation (Fig. 1a) and provides medical experts with an *approximate localization* of the IAN's position and its distance

G. L. Foresti et al. (Eds.): ICIAP 2023, LNCS 14233, pp. 525–536, 2023.
https://doi.org/10.1007/978-3-031-43148-7_44

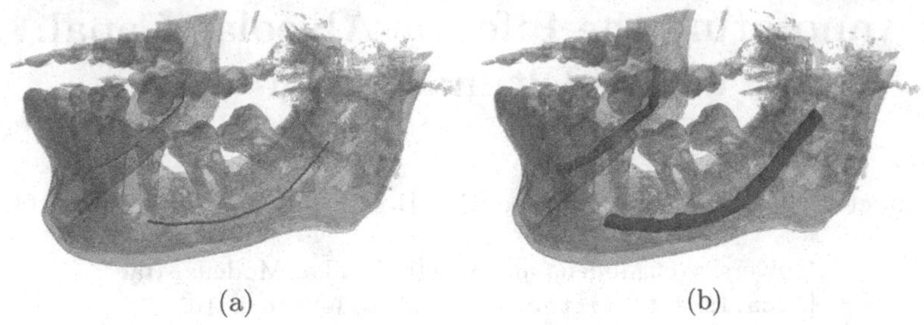

(a) (b)

Fig. 1. An example of sparse (a) and dense (b) ground truth annotations of the same volume. Both are obtained from (different) 2D views of the data and later re-projected to the 3D volume. More specifically, the sparse annotation is extracted from a single panoramic view (Fig. 2c), while the dense annotation is obtained from multiple cross sectional views (Fig. 3).

from the molars. Although 3D annotations (Fig. 1b) would provide exact knowledge about IAC shape and position, enabling meticulous surgical plan, they are often unavailable due to the burden of time and effort required to obtain them.

Therefore, the automatic segmentation of the IAC represents an active research field, as it holds the potential to revolutionize the surgical planning process, supporting the maxillofacial daily practice.

Although Convolutional Neural Networks (CNNs) have provided amazing results for both 2D and 3D segmentation, alongside several more computer vision and healthcare tasks [2–4, 17–19, 25–27], developing Deep Neural Networks (DNNs) for the automatic segmentation of the IAC is a challenging task due to the lack of publicly available 3D annotated data[1]. As the reader may know, collecting and labelling data is a time-consuming and resource-intensive process. Moreover, publicly releasing medical-related information raises privacy issues. To advance the research in IAC automatic segmentation and improve patient surgical outcomes, it is crucial to develop tools and applications that can support the creation of such high-quality datasets.

This paper addresses this issue by enhancing an existing software used by medical experts to produce IAC annotations. We demonstrate that the proposed additional features, later detailed in Sect. 3, allow us to both significantly reduce the annotation time and improve the overall quality of the resulting segmentation. Specifically, radiologists are now able to identify previously undetectable canal sections with an average increase in annotated voxels of 61.9%. Among others, one of the main features introduced in the software is the automatic segmentation of the canal from which technicians can start a simplified annotation process. More specifically, we integrate an improved version of the segmentation approach originally proposed in [7].

It is worth mentioning that the proposed tool, IACAT 2.0, led to the generation of a 3D-segmented-CBCTs dataset, *ToothFairy*, which is part of the homonymous MICCAI2023 challenge. Training state-of-the-art segmentation models

[1] At the moment of writing this paper, there is only one single dataset publicly available: https://ditto.ing.unimore.it/maxillo/.

with such annotations led to significant performance boost both on Dice and Intersection over Union (IoU) scores.

2 Related Work

Previous studies have proposed different architectures for the automatic segmentation of the inferior alveolar canal [7,9,14–16].

In [7] and [9], a three steps training procedure is proposed. During the first step, identified as *deep label expansion*, the network is fed with CBCT volumes paired with the corresponding sparse 2D labels and trained to generate 3D dense annotations. During the second step, the 3D synthetic labels are employed to perform a pre-training of the segmentation network that is finally fine-tuned with the 3D annotations performed by medical experts.

On the other hand, Kwak *et al.* [15], Jaskari *et al.* [14], and Lahoud *et al.* [16] proposed more straightforward approaches. More specifically, [15] simply compares different types of existing 2D and 3D architectures (*e.g.*, SegNet [1] and U-Net [28]) using a private dataset. Instead, a standard 3D-UNet [6] model with some slight improvements tailored for the specific task has been adopted by both [14] and [16].

A significant challenge in the field is represented by the lack of publicly available data and source code. Indeed, most of the aforementioned papers do not provide neither the annotations nor the information about how they have been obtained, which hinders the ability to reproduce their experiments and evaluate the effectiveness of the proposed models. The absence of open-source approaches for IAC segmentation is a major obstacle that must be addressed to facilitate progress in this area of research.

To the extent of annotating 3D data, previous works have relied on the use of proprietary software such as Photoshop[2] and Invivo[3] [23,30], which can be tedious, time-consuming, and not tailored for the specific task. Moreover, even when they propose a novel methodology to annotate such data, they do not release the source code of their implementation [14–16].

To address this issue, we present a new annotation tool specifically designed for the IAN canal. The tool provides the user with the capability of processing and visualize CBCT data exported in DICOM format and guides him toward the annotation of axial images, panoramic views, and cross-sectional images. The annotated data can be easily exported to be employed in different tasks, including the training of deep learning models.

3 The Ultimate Annotation Tool

With this work, we present IACAT 2.0, an improved version of the tool described in [20]. The proposed features are detailed in the following of this Section, highlighting the rationales behind the design choices and the benefit they introduce.

[2] https://www.photoshop.com.

[3] https://www.anatomage.com/invivo.

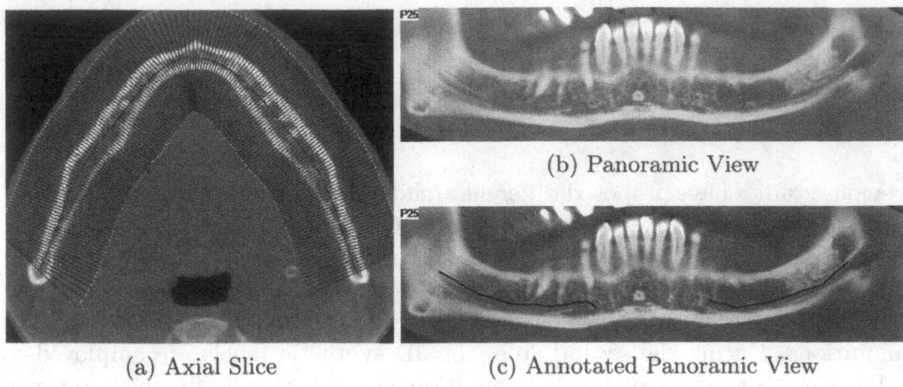

Fig. 2. 2D annotation of the IAC. (a) depicts an axial slice of the CBCT volume. The red curve, called panoramic base curve, identifies the jawbone. (b) is the panoramic view obtained from the CT-volume displaying voxels of the curved plane generated by the base curve and orthogonal to the axial view. (c) is the same view as (b) showing a manual annotation of the IAN performed by an expert technician. (Color figure online)

3.1 Preliminaries

To better introduce our proposals, we summarize the entire annotation flow:

1. After loading the input data, the arch approximation that better describes the canal course is identified inside one of the axial planes constituting the volume. The output is a one-pixel thick curve crossing the dental arch which is approximated with a polynomial (Fig. 2a in red). The curve is automatically generated by the tool and manually adjusted only when needed.
2. Sampling the polynomial, the tool thus generates a Catmull-Rom spline. For each point of the spline, a perpendicular line on the axial plane is computed (Fig. 2a in blue). These lines are identified as *Cross-Sectional Lines* (CSLs).
3. CSLs are used for *Multi Planar Reformations* (MPRs) to generate *Cross-Sectional Views* (CSVs). These views are 2D images obtained interpolating the values of the respective base lines (CSL) across the whole volume height. An example of CSV is provided in Fig. 3 (bottom-right), it corresponds to the plane identified in green on the panoramic view (top-right, same Figure).
4. For each CSV, a closed Catmull-Rom spline is finally drawn to annotate the position of the IAC (green closed lines of Fig. 3).
5. The splines are saved as the coordinates of their control points. The final smooth and precise ground-truth volume constituting the dataset is generated from this set of points by means of the α-shape algorithm [10].

Our contributions to the annotation pipeline can be summarized as follows. We integrated an automatic prediction of the IAC based on PosPadUNet3D [7], a state-of-the-art deep learning model for the task. Moreover, to improve the segmentation results, we introduced enhancements in the whole automatic annotation pipeline. This way, technicians are able to acquire annotations from model

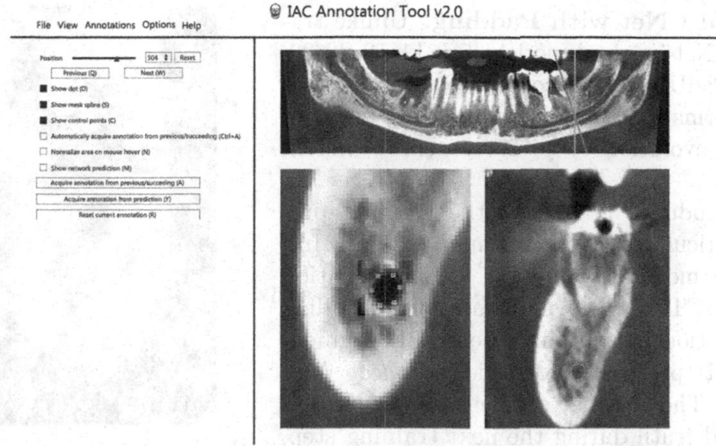

Fig. 3. A view of the IACAT 2.0. The panoramic view (top-right) roughly identifies the canal position (left branch in red, right branch in blue). On the same view, purple and green lines identify the position of a straight and tilted (to be orthogonal to the canal slope) CSVs, respectively. The tilted CSV corresponding to the green line, visualized in the lower part of the screen with two different zoom levels, is intended to produce the annotation by drawing a closed spline. The left-most part of the windows contains control buttons to perform different actions.

predictions, significantly reducing the annotation efforts to a mere adjustment. Additional mechanisms —*e.g.*, zoom in/out and local contrast-stretching— have also been introduced to improve eye-driven identification.

3.2 Automatic Segmentation of the IAC

Compared with the pipeline of PosPadUNet3D, the segmentation of the IAC has been improved through the implementation of ad-hoc pre- and post-processing techniques. Such additions are here introduced and detailed in the following Sections. The pre-processing technique, known as *Distance Transform*, mitigates the sparsity of the 2D annotation fed into the model during the *deep label propagation* step. On the other hand, the post-processing technique, called *Hann Windowing* [24], tackles the artifacts caused by patch-based learning. It achieves this by multiplying the frames with a window function. Together, these techniques have resulted in a more accurate and efficient automatic segmentation of the IAC.

Before digging into the details of the proposed improvements, it is worth noticing that the automatic segmentation of the IAN intervenes at point 4 of the annotation flow presented in Sect. 3.1. More specifically, technicians can visualize the prediction as depicted with red in Fig. 4 and chose whether to generate the closed spline automatically from it, or only use the annotation as a reference.

Positional UNet with Padding. Unlike traditional UNet-inspired models [5,6,11,12,22,28,29], PosPadUNet3D incorporates patch positional information in the bottleneck and employs padded convolution to preserve tensor dimensionality.

As introduced in Sect. 2, the training procedure articulates in three main stages. Initially, the model is fed with a concatenation of the CBCT patches and the corresponding 2D annotations to obtain a dense 3D segmentation, this process is identified as *deep label expansion*. The obtained segmentation is used as the ground truth during the next training step, enabling the use of weakly annotated volumes, such as scans with only 2D annotations. Finally, in the third step, the model is fine-tuned using scans that have 3D labels provided by technicians. When compared to existing state-of-the-art models, PosPadUNet3D demonstrate better performance in segmenting the IAC. Therefore it represents as an efficient and reliable method to implement the *acquire annotation from prediction* in IACAT 2.0.

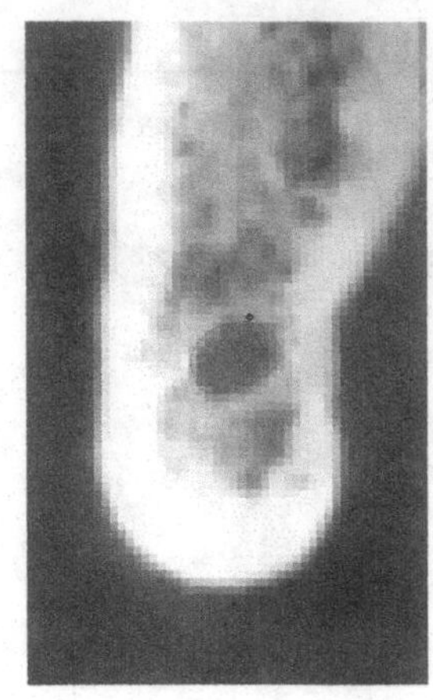

Fig. 4. CSV depicting the annotation performed by medical experts (in green), and the automatic prediction of the network (in red). The dark-red dot is the 2D sparse annotation. (Color figure online)

Distance Transform. Unfortunately, the sparse annotations employed for deep label expansion are scrimpy: more than 99.9% of the volume is annotated as background, thus it is not well suited for standard convolutional neural networks nor skip connections. Moreover, during patch-based training, the probability that a patch uniformly sampled from the original volume contains a relevant part of the annotation is low, and there will be mostly empty patches with no information at all, providing just useless computations in the pipeline.

To spread the information contained in the sparse annotation, we propose to add as pre-processing step the Euclidean distance transform. Such transformation takes a binary N-dimensional volume as input and, for each background element, efficiently computes its distance to the closest foreground label. This addition allows the network to gather critical information about the IAC position even in the early stage of the network, and also when dealing with patches that do not directly contain sparse labels. Such an approach enables faster convergence during training and improves final results. The distance transform is defined as:

$$d(x_i) = \min_{x_j} f(x_i, x_j)\ , \quad x_i \in \mathcal{M}_{\text{bg}},\ x_j \in \mathcal{M}_{\text{fg}} \tag{1}$$

where x_i refers to all the pixels of the background $\mathcal{M}_{\text{bg}}$ and x_j refers to all the pixels of the foreground $\mathcal{M}_{\text{fg}}$. The function f is the chosen distance function, which in our case corresponds to the standard Euclidean distance.

Hann Windows. Another tackled problem regards the artifacts produced during the patch-based training. When the threshold is removed from the output of the network, it is possible to better notice different inaccuracies produced by the network near the borders of each patch. Additionally, when all the patches are aggregated together, it is possible to notice that the distribution of errors lies exactly where the canal must be predicted close to the borders.

A similar, unrelated, problem also happens in audio encoding, where subframes of an audio track are encoded separately and merged afterwards. This procedure causes non-zero values to appear near sub-frame boundaries and is called *spectral leakage*.

As done for audio encoding, we propose to adopt the Hann window function to overcome such a limitation. The Hann window function is defined as follows:

$$W_{\text{Hann}}(i) = \frac{1}{2}(1 - \cos\frac{2\pi i}{I})$$

where i is an element in the interval I. The function is symmetric, with a maximum value of 1 in the middle of the window and a minimum value of 0 at the edges. Another interesting property is that the sum of two Hann windows shifted by $\frac{I}{2}$ (50%) is equal to a rectangular window of width I and height 1:

$$W_{\text{Hann}}(i) + W_{\text{Hann}}(i + \frac{I}{2}) = 1$$

Such a property is employed in audio encoding to remove the border artifacts by simply multiplying overlapped (by 50%) frames with the Hann function before summing them. This approach, which is defined in 1D for the audio, can be extended to be multi-dimensional and thus applied to 3D images:

$$W_{\text{Hann}}(i, j, k) = W_{\text{Hann}}(i)W_{\text{Hann}}(j)W_{\text{Hann}}(k)$$

where i, j, k identify a point in the space. To avoid numerical issues, the windows function applied in the image space must deal with border cases, ensuring that the sum is always one.

3.3 Localized Contrast Stretching and Zoom

The precise segmentation of the IAC in real-life scenarios can be challenging, even for experienced domain experts. This is mainly due to the low contrast in the area where the IAC is located. It is not rare to encounter CSV where, although present, the alveolar canal turns out to be indistinguishable or hard to be localized precisely (see Fig. 5b as an example). This phenomenon typically occurs because bone density is higher in specific locations, due to ageing or other patient-related conditions. However, the information underlying these regions can sometimes be revealed through the application of local contrast stretching. Therefore, we integrated into IACAT 2.0 the ability to increase the contrast range in a given area by applying the following formula:

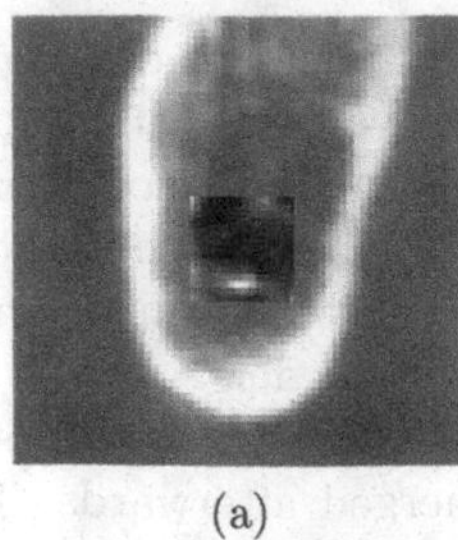

(a)

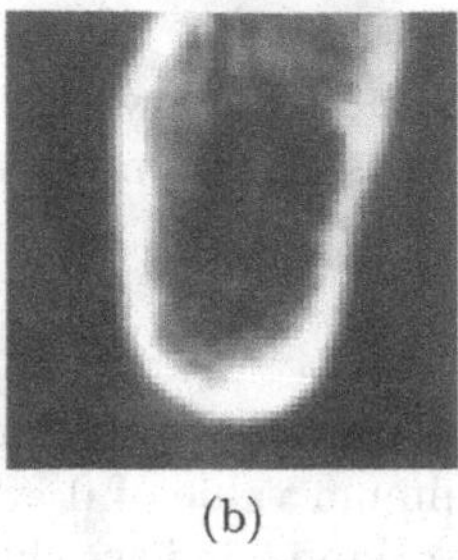

(b)

Fig. 5. CSVs with and without localized contrast stretching. (a) depicts a local contrast stretching function applied to the image (b).

$$v_{max} = \max_{i=x-w/2}^{x+w/2} \max_{j=y-h/2}^{y+h/2} I_{i,j} \tag{2}$$

$$v_{min} = \min_{i=x-w/2}^{x+w/2} \min_{j=y-h/2}^{y+h/2} I_{i,j} \tag{3}$$

where $I \in \mathcal{M}^{M \times N}$ is the matrix representing the original image, w and h are the width and height of the selected window, and (x, y) is the position of the window center. Given that, for each pixel $I_{i,j}$ of the window, the contrast stretching formula applied is:

$$I'_{i,j} = \frac{I_{i,j} - v_{min}}{v_{max} - v_{min}} \tag{4}$$

where $I'_{i,j}$ is the contrast-stretched pixel value at position (i, j). The window height and width are independently modifiable, to let the user fit the windows exclusively over the region of interest.

As the IAC might be small in some cases, a zoom functionality is also introduced to increase the precision of the carried-out annotation.

While the *contrast stretching* function allows the annotator to enhance the contrast of a specific region, the *Zoom* function facilitates the inspection of hard-to-annotate regions of the volumes.

The integration of the aforementioned functionalities provides the annotator with powerful tools to overcome the main challenges posed by difficult-to-detect IAC regions.

4 Evaluating IACAT 2.0

A team of 5 maxillofacial experts with more than 5 years of experience have been engaged to perform annotations using both the old and new version of the tool on 40 CBCTs of the public Maxillo dataset. The comparison underlines that annotations obtained with IACAT V1.0 suffered from several issues, including disconnected components and under-annotations that often occur near the terminal parts of the canal (see Fig. 6 as a reference). To produce a quantitative comparison, we introduce the following metric:

$$\text{Increase }\% = \frac{V_{i,\text{New}} - (V_{i,\text{Old}} \cap V_{i,\text{New}})}{V_{i,\text{Old}}} \cdot 100 \tag{5}$$

where $V_{i,\text{New}}$ and $V_{i,\text{Old}}$ are respectively the ground truth annotations of patient i in the corresponding dataset. This gives us a measure of how many more voxels

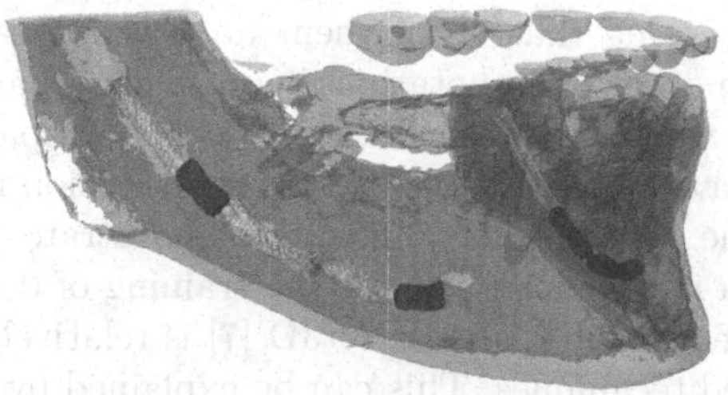

Fig. 6. Comparison of the annotation obtained using the two versions of the tool on volume P95 of the publicly available Maxillo dataset. In red the annotation obtained with the old tool, in green the voxels of the annotation that have been added thanks to the proposed improvements. (Color figure online)

have been annotated w.r.t. the annotations performed with the older version of the tool. On average, 61.9% more voxels per volume are selected as being part of the canal. In Fig. 6 an example of the aforementioned discrepancy is depicted.

Additionally, the time spent annotating each one of the 40 CBCTs has been recorded. The average time of 22 minutes per volume required when using IACAT V1.0 has been lowered to around 8 minutes per volume with IACAT 2.0, highlighting once again the benefits of the features introduced.

4.1 Inter-Agreements

To understand the room for improvement of any novel deep learning model, a human-baseline score must be defined. Such a score has been created by using two annotations of the same volume produced by different medical experts, and then computing the Dice and IoU scores among them. Indeed, we produced two annotations for 6 patients and obtained an average IoU of 0.70 and an average Dice of 0.81.

4.2 *ToothFairy* - A New Dataset

As previously stated, an additional contribution of this paper is the generation of a new dataset, ToothFairy, obtained by means of IACAT 2.0 using the data already released with the Maxillo dataset. ToothFairy counts 153 3D densely annotated CBCTs, *i.e.*, 62 more w.r.t. the original dataset[4]. Regarding the 91 volumes in common with the Maxillo

Table 1. Performance comparison of different models trained on Maxillo or ToothFairy dataset.

Model	Maxillo IoU	Maxillo Dice	ToothFairy IoU	ToothFairy Dice
AttentionUNet [22]	0.576	0.731	**0.612**	**0.759**
UNet++ [31]	0.542	0.703	**0.550**	**0.710**
UNet [28]	0.635	0.777	**0.643**	**0.783**
VNet [21]	0.524	0.688	**0.558**	**0.716**
PosPadUNet3D [7]	0.652	0.789	**0.663**	**0.797**

[4] Please note that the Maxillo dataset released a total of 343 CBCTs proving a 3D annotation only for 91 of them.

dataset, it is worth noting that 40 of them underwent re-segmentation using IACAT 2.0, while the other 51 annotations remained unchanged. With the aim of demonstrating the value of the new dataset, Table 1 compares the performance of different publicly-available state-of-the-art segmentation models trained with both the old and the new datasets. Results demonstrate that the use of the newly produced data is beneficial also for the training of deep neural networks. Notably, the improvement of PosPadUNet3D [7] is relatively modest compared to the other evaluated techniques. This can be explained by considering that its performance are close to the previously defined inter-agreement score (Sect. 4.1). Consequently, considering the human baseline which approximately corresponds to a Dice score of 0.81, the advancement achieved by PosPadUNet3D from 0.79 to 0.80 retains its significance.

5 Conclusion

In this work, we have introduced IACAT 2.0, an innovative tool for the annotation of the inferior alveolar canal in CBCT scans, which enhances the quality and expedites the annotation process. We have also presented ToothFairy, a new dataset of IAC 3D segmentation, which improves the quantity and quality of publicly available annotated CBCT scans. The proposed tool incorporates several novel functionalities, such as *acquire from prediction*, which uses the prediction of a state-of-the-art model to assist the annotation process, and *localized contrast stretching*, which enhances the contrast of dark regions to reveal hidden parts of the alveolar canal, simplifying the annotation process.

The carried out evaluation revealed that, by using IACAT 2.0, medical experts are able to identify previously undetectable canal regions. On average, the new annotations counts 61.9% more voxels and requires $\frac{1}{3}$ of time to be generated. Finally, to highlight the benefits introduced by our tool, we compared state-of-the-art segmentation models trained with and w/o ToothFairy dataset. Significant performance boost have been achieved by all the models on both Dice and IoU scores when using IACAT 2.0-generated data.

Acknowledgements. This project has received funding from the Department of Engineering "Enzo Ferrari" of the University of Modena through the FARD-2022 (Fondo di Ateneo per la Ricerca 2022).

References

1. Badrinarayanan, V., Kendall, A., Cipolla, R.: SegNet: a deep convolutional encoder-decoder architecture for image segmentation. IEEE Trans. Pattern Anal. Mach. Intell. **39**(12), 2481–2495 (2017)
2. Barraco, M., Stefanini, M., Cornia, M., Cascianelli, S., Baraldi, L., Cucchiara, R.: CaMEL: Mean Teacher Learning for Image Captioning. In: Proceedings of the International Conference on Pattern Recognition (2022)
3. Bolelli, F., Baraldi, L., Grana, C.: A hierarchical quasi-recurrent approach to video captioning. In: 2018 IEEE International Conference on Image Processing, Applications and Systems (IPAS), pp. 162–167. IEEE (Dec 2018)

4. Bontempo, G., Porrello, A., Bolelli, F., Calderara, S., Ficarra, E.: DAS-MIL: distilling Across Scales for MIL Classification of Histological WSIs. In: Medical Image Computing and Computer Assisted Intervention - MICCAI 2023 (2023)
5. Chen, J., et al.: TransUNet: Transformers Make Strong Encoders for Medical Image Segmentation. arXiv preprint arXiv:2102.04306 (2021)
6. Çiçek, Ö., Abdulkadir, A., Lienkamp, S.S., Brox, T., Ronneberger, O.: 3D U-Net: learning dense volumetric segmentation from sparse annotation. In: Ourselin, S., Joskowicz, L., Sabuncu, M.R., Unal, G., Wells, W. (eds.) MICCAI 2016. LNCS, vol. 9901, pp. 424–432. Springer, Cham (2016). https://doi.org/10.1007/978-3-319-46723-8_49
7. Cipriano, M., Allegretti, S., Bolelli, F., Pollastri, F., Grana, C.: Improving segmentation of the inferior alveolar nerve through deep label propagation. In: Proceedings of the IEEE/CVF Conference on Computer Vision and Pattern Recognition (CVPR), pp. 21137–21146. IEEE (2022)
8. Crowson, M.G., et al.: A contemporary review of machine learning in otolaryngology-head and neck surgery. The Laryngoscope **130**(1), 45–51 (2020)
9. Di Bartolomeo, M., et al.: Inferior alveolar canal automatic detection with deep learning CNNs on CBCTs: development of a novel model and release of open-source dataset and algorithm. Appl. Sci. **13**(5), 3271 (2023)
10. Edelsbrunner, H., Kirkpatrick, D., Seidel, R.: On the shape of a set of points in the plane. IEEE Trans. Inf. Theor. **29**(4), 551–559 (1983)
11. Guan, S., Khan, A.A., Sikdar, S., Chitnis, P.V.: Fully dense UNet for 2-D sparse photoacoustic tomography artifact removal. IEEE J. Biomed. Health Inf. **24**(2), 568–576 (2019)
12. Hatamizadeh, A., Nath, V., Tang, Y., Yang, D., Roth, H.R., Xu, D.: Swin UNETR: swin transformers for semantic segmentation of brain tumors in MRI images. In: Crimi, A., Bakas, S. (eds.) Brainlesion: Glioma, Multiple Sclerosis, Stroke and Traumatic Brain Injuries. BrainLes 2021. Lecture Notes in Computer Science, vol. 12962. Springer, Cham (2022). https://doi.org/10.1007/978-3-031-08999-2_22
13. Hwang, J.J., Jung, Y.H., Cho, B.H., Heo, M.S.: An overview of deep learning in the field of dentistry. Imaging Sci. Dent. **49**(1), 1–7 (2019)
14. Jaskari, J., et al.: Deep learning method for mandibular canal segmentation in dental cone beam computed tomography volumes. Sci. Rep. **10**(1), 5842 (2020)
15. Kwak, G.H., et al.: Automatic mandibular canal detection using a deep convolutional neural network. Sci. Rep. **10**(1), 5711 (2020)
16. Lahoud, P.: Development and validation of a novel artificial intelligence driven tool for accurate mandibular canal segmentation on CBCT. J. Dent. **116**, 103891 (2022)
17. Lovino, M., Ciaburri, M.S., Urgese, G., Di Cataldo, S., Ficarra, E.: DEEPrior: a deep learning tool for the prioritization of gene fusions. Bioinformatics **36**(10), 3248–3250 (2020)
18. Lovino, M., Urgese, G., Macii, E., Di Cataldo, S., Ficarra, E.: A deep learning approach to the screening of oncogenic gene fusions in humans. Int. J. Mol. Sci. **20**(7), 1645 (2019)
19. Marconato, E., Bontempo, G., Ficarra, E., Calderara, S., Passerini, A., Teso, S.: Neuro symbolic continual learning: knowledge, reasoning shortcuts and concept rehearsal. In: International Conference on Machine Learning (ICML) (2023)
20. Mercadante, C., Cipriano, M., Bolelli, F., Pollastri, F., Anesi, A., Grana, C.: A cone beam computed tomography annotation tool for automatic detection of the inferior alveolar nerve canal. In: 16th International Conference on Computer Vision Theory and Applications-VISAPP 2021. vol. 4, pp. 724–731. SciTePress (2021)

21. Milletari, F., Navab, N., Ahmadi, S.A.: V-Net: fully convolutional neural networks for volumetric medical image segmentation. In: 2016 Fourth International Conference on 3D Vision (3DV), pp. 565–571. IEEE (2016)
22. Oktay, O., et al.: Attention U-Net: learning where to look for the pancreas. In: Medical Imaging with Deep Learning (2022)
23. Park, J.S., Chung, M.S., Hwang, S.B., Lee, Y.S., Har, D.H.: Technical report on semiautomatic segmentation using the Adobe Photoshop. J. Dig. Imaging **18**, 333–343 (2005)
24. Pielawski, N., Wählby, C.: Introducing Hann windows for reducing edge-effects in patch-based image segmentation. PloS one **15**(3), e0229839 (2020)
25. Pollastri, F., et al.: A deep analysis on high resolution dermoscopic image classification. IET Comput. Vis. **15**(7), 514–526 (2021)
26. Porrello, A., et al.: Spotting insects from satellites: modeling the presence of Culicoides imicola through Deep CNNs. In: 2019 15th International Conference on Signal-Image Technology & Internet-Based Systems (SITIS), pp. 159–166. IEEE (2019)
27. Reyes-Herrera, P.H., Ficarra, E.: Computational Methods for CLIP-seq Data Processing. Bioinform. Biol. Insights **8**, BBI-S16803 (2014)
28. Ronneberger, O., Fischer, P., Brox, T.: U-Net: convolutional networks for biomedical image segmentation. In: Navab, N., Hornegger, J., Wells, W.M., Frangi, A.F. (eds.) MICCAI 2015. LNCS, vol. 9351, pp. 234–241. Springer, Cham (2015). https://doi.org/10.1007/978-3-319-24574-4_28
29. Sun, J., Darbehani, F., Zaidi, M., Wang, B.: SAUNet: shape attentive U-Net for interpretable medical image segmentation. In: Martel, A.L. (ed.) MICCAI 2020. LNCS, vol. 12264, pp. 797–806. Springer, Cham (2020). https://doi.org/10.1007/978-3-030-59719-1_77
30. Weissheimer, A., De Menezes, L.M., Sameshima, G.T., Enciso, R., Pham, J., Grauer, D.: Imaging software accuracy for 3-dimensional analysis of the upper airway. Am. J. Orthod. Dentofac. Orthop. **142**(6), 801–813 (2012)
31. Zhou, Z., Rahman Siddiquee, M.M., Tajbakhsh, N., Liang, J.: UNet++: a nested u-net architecture for medical image segmentation. In: Stoyanov, D. (ed.) DLMIA/ML-CDS -2018. LNCS, vol. 11045, pp. 3–11. Springer, Cham (2018). https://doi.org/10.1007/978-3-030-00889-5_1

A Learnable EVC Intra Predictor Using Masked Convolutions

Gabriele Spadaro[1(✉)], Roberto Iacoviello[2], Alessandra Mosca[3], Giuseppe Valenzise[4], and Attilio Fiandrotti[1]

[1] University of Turin, Turin, Italy
{gabriele.spadaro,attilio.fiandrotti}@unito.it
[2] Rai Radiotelevisione italiana, Rome, Italy
roberto.iacoviello@rai.it
[3] Sisvel Technology S.r.l., None, Italy
alessandra.mosca@sisveltech.com
[4] CNRS, Paris, France
giuseppe.valenzise@l2s.centralesupelec.fr

Abstract. The Enhanced Video Coding (EVC) workgroup of the Moving Picture, Audio and Data Coding by Artificial Intelligence (MPAI) organization aims at enhancing traditional video codecs by improving or replacing traditional encoding tools with AI-based counterparts. In this work, we explore enhancing MPEG Essential Video Coding (EVC) intra prediction with a learnable predictor: we recast the problem as a hole inpainting task that we tackle via masked convolutions. Our experiments in standard test conditions show BD-rate reductions in excess of 6% over the EVC baseline profile reference with some sequences in excess of 12%.

Keywords: EVC · intra prediction · learnable video coding

1 Introduction

Video content accounts for over 70% of Internet traffic volume [4], hence the interest in efficient video coding technologies. Recently, the trend has been leveraging recent advances in artificial intelligence and deep learning to improve the efficiency of video codecs and two distinct approaches have emerged. The first approach aims at integrating or replacing selected encoding tools of traditional codecs with learnable equivalents. The second approach aims at designing from scratch novel codecs with an end-to-end totally deep learning based architecture. The EVC project of the MPAI community[1] falls in the former category and aims at improving the efficiency of existing video codecs by at least 25% of BD-Rate. The MPEG-5 Essential Video Coding (EVC) [3,16] baseline profile has been chosen as reference as it relies on encoding tools that are at least 20 years mature, yet it shows compression efficiency comparable to H.265/HEVC [18].

[1] https://mpai.community/standards/mpai-evc/about-mpai-evc/.

G. L. Foresti et al. (Eds.): ICIAP 2023, LNCS 14233, pp. 537–549, 2023.
https://doi.org/10.1007/978-3-031-43148-7_45

The MPAI EVC project is currently studying a number of encoding tools based on deep learning, and this paper describes the ongoing activities on the intra prediction tool. Modern video codecs exploit the spatial correlation in pictures predicting each block to be encoded from a previously encoded area (*predictor*) of the same picture. The rationale behind intra prediction is that encoding the difference (the *residual*) between the block pixels and the one associated with its predictor is more efficient than encoding the block pixels themselves. Namely, the closer the predictor pixels are to the block ones, the fewer the bits to encode the residual and so the encoding rate. In MPEG-5 EVC, intra prediction consists in a set of 5 predefined linear functions where the mode yielding the best Rate-Distortion (RD) tradeoff is selected. However, not all contents (e.g., complex textures) can accurately be predicted by simple linear models, and in such cases the efficiency of intra prediction drops.

In this work we aim to improve intra prediction as specified by MPEG-5 EVC with a learnable predictor. We address the problem of predicting a block given its context as an image inpainting problem. Recently, deep convolutional generative neural networks have shown to outperform existing image inpainting methods thanks to their ability to learn highly non linear functions. Namely, masked convolutional neural networks have been recently proposed for image inpainting exploiting the a priori information on missing pixels that are weighted out from the context used to recover the missing image area. The method we propose relies on masked convolutions to generate the block predictor starting from the decoded context available at the receiver. In detail, we replaced the MPEG-5 EVC predictor mode 0 (i.e., the DC prediction) with a novel predictor that is computed by a masked convolutional autoencoder for each block to be encoded. Our encoding experiments in standard test conditions show Bjøntegaard Delta Rate (BD-Rate) reductions in excess of 6% over the MPEG-5 EVC Baseline Profile.

2 Background

This section first provides a primer to video coding, next reviews the state of the art in learnable intra-picture prediction.

2.1 Introduction to Video Coding

Existing video coding standards rely on a clever combination of hand-designed encoding tools, each bringing its own contribution to the overall codec performance as shown in Fig. 1. In state of the art video coding standards such as the H.265/HEVC or MPEG-5 EVC, the image is first recursively subdivided in blocks (*Coding Units - CUs*) of decreasing size, e.g. 64×64 down to 4×4 in the MPEG-5 EVC standard. Next, for each coding unit multiple encoding modes are evaluated by an algorithm aimed at finding the best RD tradeoff for a given Quantization Parameter (QP). Better RD tradeoffs can be achieved by predicting the coding unit from neighboring data within the same picture (intra-prediction)

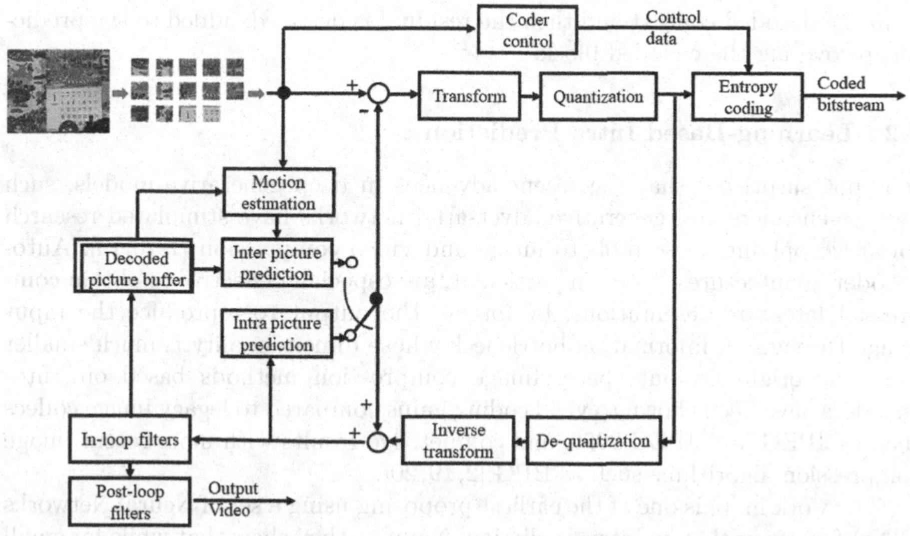

Fig. 1. Architecture of a traditional hybrid video codec with the main coding tools: this work deals with enhancing the intra tool with a learnable predictor.

or from previously encoded pictures if available (inter-prediction). Intra-frame prediction leverages the spatial correlation within the same picture generating a predictor for the CU to be encoded by extrapolating pixel values from a previously encoded neighborhood. The predicted block is then subtracted from the original block, producing a residual block that is transformed via discrete cosine transform, allowing low-pass filtering in the transformed domain by discarding and/or attenuating the coefficients in the high frequency values. The rationale behind intra prediction is that encoding the residual requires fewer bits than encoding the original block. The better the predictor, i.e. the closer to the block to be encoded, the lower the residual rate and the higher the coding efficiency. The MPEG-5 EVC Baseline profile includes 5 intra prediction modes: DC, horizontal, vertical and two diagonal modes for each CU. The encoder selects the intra mode that minimizes the residual rate, which may be then put into competition with other modes. Coefficient decimation and the subsequent quantization is the lossy part of the compression process that reduces the high frequency rate while keeping the resulting artifacts bearable to the human observer. The resulting signal is entropy encoded, via for example, arithmetic coded, which is a lossless form of compression. Within the encoder, a decoding part is implemented and the signal is reconstructed through a dequantization and inverse transformation step. By adding the predicted signal, the input data is reconstructed. Filters, such as a deblocking filter and a sample adaptive offset filter are used to improve the visual quality. The reconstructed picture is stored for future reference in a reference picture buffer to allow exploiting the similarities between two pictures. At the decoder side, the signaled predictor is generated from the decoded context and then the residual is decoded, added to the predictor, recovering the encoded block.

2.2 Learning-Based Intra Prediction

It is not surprising that the recent advances in deep generative models, such as auto-encoders and generative adversarial networks have stimulated research towards applying these tools to image and video compression [1,15,19]. Auto-encoder architectures [7,10], in particular, are especially effective to obtain compressed latent representations, by forcing the output to reproduce the input image through an information bottleneck whose dimensionality is much smaller than the original input space. Image compression methods based on auto-encoders have been shown to yield coding gains compared to legacy image codecs such as JPEG and JPEG 2000, and competitive results with more recent image compression algorithms such as BPG [2,19,20].

The work in [6] is one of the earliest proposing using a set of Neural Networks (NNs) for generating an intra prediction. Namely, they show that while for small sized blocks a fully connected NN gives best results, convolutional networks yield better predictors for large blocks with complex textures. Their experiments integrating a learnable predictor into H.265/HEVC show PSNR gains above 5% in some cases, depending on the content type and how the predictor is integrated into the codec. The authors attribute such gains to the improved ability to correctly predict complex textures.

The same authors propose an iterative approach to training a NN for intra prediction in [5]. First, a NN is trained on blocks and context extracted from a real partitioning of pictures as produced by the reference codec. Next, the NN is refined over the output of the same codec, yet this time the output includes the learnable intra predictor trained during the previous step. It is shown that this train-and-refine approach boosts further the performance of the learnable intra predictor with BD-rate reduction in excess of 4% BD-rate with H.265/HEVC and close to 2% for H.266/VVC.

The work in [9] tackles the same problem yet with a different approach that relies on recurrent NNs. Namely, they propose a recurrent architecture with three different spatial recurrent units that progressively generate predictor pixels by passing information exploiting the already encoded context. Beside MSE, they train their model keeping into account the Sum of Absolute Transformed Difference (SATD) as a proxy of the rate. They experimentally show that their approach yields bit rate reductions in excess of 2.5% when integrated into H.265/HEVC.

In [8], a NN that has multiple prediction modes and that co-adapts during training to minimize a loss function is proposed. The proposed loss function reflects the properties of the residual quantization of the typical hybrid video coding architecture by applying the ℓ_1-norm and a sigmoid-function to the prediction residual in the DCT domain. Furthermore, they reduce the complexity by pruning the resulting predictors in the frequency domain and by quantizing the network weights and utilizing fixed point arithmetic, thus allowing for a hardware-friendly implementation.

In [22], a slightly different approach is proposed, where a NN is used to refine the standard H.265/HEVC intra prediction modes rather than replacing them. Such approach builds upon a convolutional autoencoder that is trained to recover a missing area of an image by inpainting the masked pixels corresponding to the block to be predicted. The authors in [22] experimentally show that their approach reduces up to 25% the mean square error of the H.265/HEVC intra predictor without additional signalling in the bitstream.

So far, no one has yet evaluated a learnable predictor within the MPEG-5 EVC codec. To the best of our knowledge, this is the first work evaluating to which extent a learnable intra predictor can affect the efficiency of a royalty free codec.

3 Proposed Method

In this section, we first describe the architecture of the NN we use to generate an intra predictor from a decoded context and next we detail the training procedure. Indeed, generating an intra prediction given the previously decoded context is conceptually equivalent to inpainting an image region given the available neighbor pixels. Therefore, we recast intra picture prediction as an image inpainting problem, building upon the existing body of research on the topic. Recently, image inpaiting models adopted mechanisms of attention or trasformers to caputc long-range dependencies [11,21,23]. However, these models requires a large number of parameters and in some cases a minimum input size [11]. Since we have to work with small crops, however, we preferred to adopt a simpler convolutional network while keeping the number of parameters under control.

3.1 Network Architecture

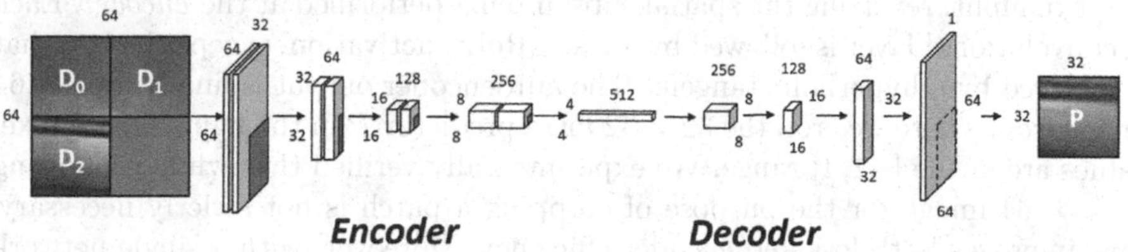

Fig. 2. Architecture and procedure for training the convolutional autoencoder used to generate a learnable intra predictor. In this example, a 32×32 prediction is generated from a 64×64 context.

Figure 2 shows the architecture of the convolutional autoencoder we propose to generate a predictor from a decoded context. For the sake of simplicity, we exemplify the case of a 32×32 predictor generated from a 64×64 context, however similar considerations hold for the other CU sizes supported by MPEG-5 EVC (16×16, 8×8, 4×4 CUs). The autoencoder receives as input a 64×64 patch representing the encoded context also available at the decoder (D_0, D_1, D_2) and outputs a 32×32 patch P corresponding to the intra predictor. The design is inspired by the context encoders for hole filling [14], yet with a number of significant differences and improvements tailored towards this task.

Concerning the *encoder*, it relies on masked convolutions where the convolution operator is constrained to valid input pixels (first layer) or features (following layers) [12]. In a nutshell, with masked convolutional layers learned filters operate only on pixels (or features) of the input image (or a feature map) that are not masked. For each convolutional layer, both a feature map and a binary mask are generated so that multiple masked layers can be stacked together. Second, we stack pairs of masked convolutional layers with 3×3 filters and leaky ReLUs where the filters of the first layer of the pair has 1 unit stride, whereas the second layer of the pair has stride of two and takes care of feature map downsampling replacing the pooling operator. We experimentally verified that this architecture reduces the number of learnable parameters as well as both the loss at training time and the intra-predictor rate at coding time. Third, rather than projecting the input image on a 1×1 latent space, we project it on a vector of feature maps sized 4×4 by dropping one convolutional layer. Again, we experimentally verified that this setup yields both lower losses at training time and better efficiency at encoding time. We attribute such improvements to a spatial-semantic depth tradeoff that is more appropriate for the purpose of our task. This result is in line with recent learnable video codecs such as [1,2], where the latent space encoded as bitstream is actually a serialization of a variable number of 4×4 feature maps. The *encoder* is thus composed by 4 blocks of masked convolutions where each block is composed of two stacked masked convolutional layers as detailed in Fig. 2. The output of the *encoder* is finally a collection of 512 feature maps sized 4×4 in the $[-1, 1]$ range.

Concerning the *decoder*, it is composed by a stack of 4 deconvolutional layers of size 4×4. Each deconvolutional layer doubles the resolution of feature maps in input, reversing the spatial subsampling performed at the *encoder*. Each deconvolutional layer is followed by a leaky ReLU activation, except the last that is followed by a hyperbolic tangent. The autoencoder output is finally a 64×64 image from where we crop the 32×32 intra predictor P in the figure where pixel values are in the $[-1, 1]$ range. We experimentally verified that while generating a 64×64 image for the purpose of cropping a patch is not strictly necessary, that improves both loss and encoder efficiency. Moreover, with a single network topology we can cope with CUs of different size (in our case, 16×16, 8×8 and 4×4) simply changing the crop operator geometry at the network output. Overall, the autoencoder counts about 6M parameters, where about 4M are for learning convolutional filters and 2M for learning the convolution masks.

3.2 Training

The autoencoder is trained by minimizing the error between the predictor P and the original patch O on a dataset of about 1000 images of different resolution and content type randomly sampled from the AROD dataset [17]. While these images are JPEG compressed, they are very high quality, and so they cannot be told from uncompressed images. We found that training the autoencoder on high quality images is of pivotal importance, even when the trained autoencoder receives in input a context encoded at high QPs. We also found out that training on larger datasets such Imagenet, Vimeo or BVI-DVC did not provide significant advances despite longer training times.

From each image, a 64×64 patch is cropped at a random position. The patch is then randomly flipped horizontally and vertically, followed by a 90° random rotation. Our experiments showed that this form of augmentation is key to prevent the network from overfitting on the training data. The bottom right 32×32 corner of the patch represents the original CU to recover, whereas the rest of the patch represents the (D_0, D_1, D_2) context. Prior to training, we prepare an appropriate binary mask that is provided in input to the first masked convolutional layer together with the context. The autoencoder is trained with SGD with a learning rate of 0.01 and over batches of 64 patches.

Ideally, the autoencoder shall be trained to minimize the linear combination of the rate and distortion terms corresponding to the operating point selected by the MPEG-5 EVC encoder [1,2]. However, for the sake of simplicity, we follow the approach used in other similar works such as [6] where the network is trained at minimizing the reconstruction loss only. In the original context encoder [14], the network is trained to minimize a linear combination of L2 loss (i.e., the mean square error) and an adversarial term. The adversarial term was shown to produce sharper and more visually pleasant results than a L2 loss alone. However, we found that the adversarial term yields artifacts that albeit visually pleasant do not help reducing the residual rate. Most important, we found out that minimizing the L1 loss (i.e., the absolute error) yields smaller residuals and thus lower rates. We hypothesize that the L2 term gives much more weight to a few training samples that yield a high loss value yet do not represent the average case for the MPEG-5 EVC encoder.

3.3 Integration into the MPEG-5 EVC Encoder

Once the autoencoder has been trained, it is interfaced with the MPEG-5 EVC encoder as follows. First, an external networked server process is started. The server loads the trained autoencoder into the GPU memory, sets up an UDP socket in listening mode and awaits for incoming messages. The EVC encoder is modified so that when an intra predictor has to be generated, the corresponding context D_1, D_2, D_3 is extracted from the currently encoded frame and is sent to the server above over an UDP socket. The server inputs such context to the trained autoencoder and returns the 32×32 output P, i.e. the learned predictor, to the encoder again via the UDP socket. The UDP socket scheme allows one to easily experiment with different neural network frameworks (PyTorch, TensorFlow, Keras, etc.) without modifying the encoder, thus simplifying the experiments. Finally, the MPEG-5 EVC encoder replaces the predictor with the autoencoder generated predictor and the encoding proceeds as usual, i.e., by putting the learned predictor in competition with other encoding modes.

Following the approach of [6], we consider two different approaches to integrate the trained autoencoder output within the MPEG-5 EVC intra prediction scheme.

The first approach consists in replacing the DC predictor (mode 0) with our learnable predictor for a total of 5 prediction modes. In [6] it is proposed to replace with the H.265/HEVC intra mode that is less likely to be selected due to the contextual intra mode signaling scheme H.265/HEVC employes. Conversely, in MPEG-5 EVC intra modes are simply signaled with variable length codes, so it is key that most probable modes are assigned shorted codes. Under the hypothesis that our learnable predictor is going to be picked by the RDO algorithm at least as frequently as the DC mode, we replaced the DC predictor with our learnable predictor.

The second approach consists in adding a sixth intra prediction mode for our learnable predictor aside the five MPEG-5 EVC intra modes. For the same reasons as above, we map our predictor to mode 0, whereas the DC predictor becomes mode 1, and so forth.

We point out that both the schemes above yield a completely decodable bitstream without the need for any side information under the reasonable assumption that the MPEG-5 EVC decoder has available the same autoencoder used by the encoder. Moreover, while the first approach is standard compliant as we do not change the bitstream, the latter only requires a simple modification at the decoder to parse the bitstream.

4 Experimental Results

4.1 Setup

We experiment encoding the first frame of the JVET CTC sequences at QP values in [22, 27, 32, 37, 42] as recommended by MPEG for their experiments with the NNVC reference software [13]. Our learnable intra predictor is applied to CUs of size 32×32, 16×16, 8×8 and 4×4; only CUs 64×64 do not enjoy our learnable predictor as our experiments showed no appreciable marginal gains.

As a preliminary experiment, we visually inspect the generated predictors for four different contexts in Fig. 3. The learnable predictor is able to inpaint the missing area of context with plausible predictions. With respect to the standard MPEG-5 EVC DC predictor, the learnable predictor yields better residual rates.

Table 1 shows the results of the encoding when our learnable predictor replaces the DC mode. The experiments report average BD-Rate improvements in excess of 6% and BD-PSNR improvements in excess of 0.5 dB for some

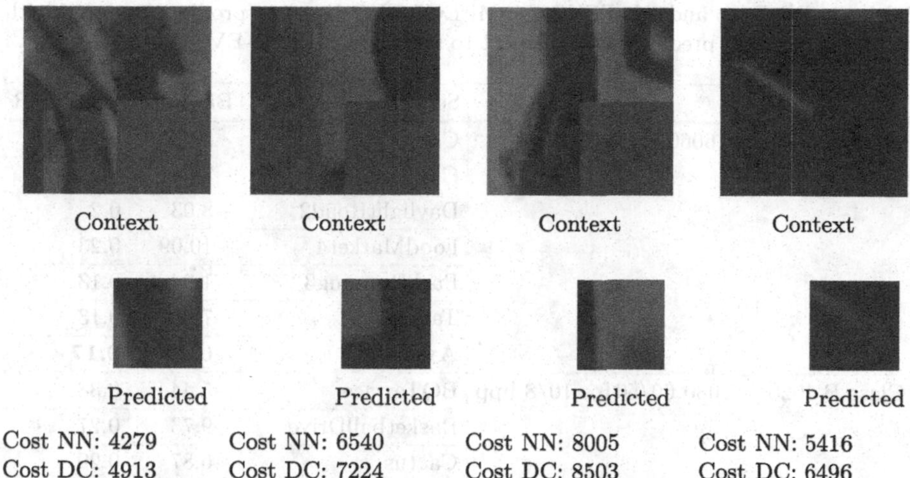

Fig. 3. Examples of 64 × 64 decoded context and 32 × 32 learnable predictor; the learnable predictor is capable of accounting also for complex texture patterns beyond simple linear interpolation. The reported "Cost" is the number of bits required to encode the residual.

sequences. The experiments show gains especially for sequences with spatial resolution above 720p: a plausible explanation may stem from the fact that most of the training images are above 600 pixels in height. We hypothesise that the addition of smaller images to the training set would boost the performance on video clips belonging to Classes C and D. The lowest performance is achieved for screen content (Class F), a non-unexpected result if we consider that text areas are more difficult to predict and our training set contains no computer screen images. A visual inspection of the decoded sequences shows no perceivable artefacts despite the learned intra predictor.

To gain a better understanding of these results, we performed a statistical analysis of the logs of the modified MPEG-5 EVC encoder for all the sequences in the table above. Table 2 shows the percentage of selection of each intra mode for Class A JVET sequences. With the reference MPEG-5 EVC encoder (left), the DC mode is selected about 51% of the times over the other 4 modes. When the DC predictor is replaced with our learnable predictor (center), this number increases to 62%, showing the advantage of a learnable predictor. That is, the intra mode indexed with code 0 is more likely to be signaled in the bitstream, and since it has the shortest code associated, the cost of intra signaling is reduced. Similarly, the residual rate for the learnable predictor was on average 9% lower than the equivalent rate of the EVC DC predictor. That is, replacing the DC predictor with our learnable predictor yields both lower residual and signaling rates if it is allocated mode 0, which explains the gains in Table 1.

Table 1. BD-Rate and BD-PSNR for MPEG5-EVC baseline profile integrated with our learnable intra predictor with respect to standard MPEG5-EVC baseline profile.

Class	Sequence	BD-Rate	BD PSNR
Class A3840 × 216060/50 fps10 bpp	Campfire	−2.96	0.11
	CatRobot	−7.6	0.23
	DaylightRoad2	−8.03	0.2
	FoodMarket4	−10.09	0.23
	ParkRunning3	−1.94	0.13
	Tango2	−7.96	0.13
	Average	**−6.43**	**0.17**
Class B 1920 × 1080 60/50 fps 10/8 bpp	BQTerrace	−5.44	0.38
	BasketballDrive	−9.73	0.27
	Cactus	−6.87	0.29
	MarketPlace	−5.69	0.21
	RitualDance	−11.85	0.67
	Average	**−7.92**	**0.36**
Class C 832 × 480 60/50/30 fps 8 bpp	BQMall	−5.39	0.36
	BasketballDrill	−7.52	0.41
	PartyScene	−2.99	0.26
	RaceHorsesC	−6.03	0.45
	Average	**−5.48**	**0.37**
Class D 416 × 240 60/50/30 fps 8 bpp	BQSquare	−2.06	0.19
	BasketballPass	−4.20	0.27
	BlowingBubbles	−4.06	0.28
	RaceHorsesD	−5.21	0.42
	Average	**−3.88**	**0.32**
Class E 1280 × 720 60 fps 8 bpp	FourPeople	−12.82	0.83
	Johnny	−12.50	0.58
	KristenAndSara	−11.11	0.64
	Average	**−12.14**	**0.68**
Class F Screen content 60 fps 8 bpp	ArenaOfValor	−5.14	0.33
	BasketballDrillText	−6.09	0.35
	SlideEditing	−1.17	0.18
	SlideShow	−1.44	0.18
	Average	**−3.46**	**0.24**
	Grand Average	**−6.55**	**0.36**

However, the analysis of the logs also revealed that the residual of the learnable predictor was lower than the residual of the DC predictor only in 53% of the cases. That is, in a significant number of cases the DC predictor is still a

Table 2. Percentage of intra modes selection for JVET Class A sequences. Left: 5 modes, reference. Center: 5 modes, proposed. Right: 6 modes, proposed.

Mode	%
-	-
0 DC	51.0
1 H	22.0
2 V	20.0
3 D1	4.7
4 D2	2.5

Mode	%
-	-
0 NN	62.0
1 H	19.0
2 V	14.0
3 D1	3.1
4 D2	1.6

Mode	%
0 NN	56.0
1 DC	25.0
2 H	9.7
3 V	6.1
4 D1	1.8
5 D2	1.4

better predictor than the learnable predictor and replacing this latter with the learnable predictor is suboptimal in terms of residual costs. For this reason, we added a sixth mode for our learnable predictor, encoded as mode 0, whereas the DC predictor was mapped to mode 1, and so on. In this scenario, the learnable predictor is put into competition with the 5 standard EVC intra prediction modes. We repeated the encodings and found that the overall BD-Rate and BD-PSNR improved only by 0.01 with respect to the numbers in Table 1. When the learnable predictor is put in competition with the other 5 modes (Table 2, right), it is selected only 56% of the times. We hypothesize that the 6-modes signaling rate leads to lower residual rates. However, the extra rate required for signaling the 6th mode counterbalances these gains, making this scheme less competitive than DC replacement in practice.

5 Conclusions and Future Works

We designed, trained and evaluated a learnable intra-picture predictor for a video codec compliant with the royalty free MPEG-5 EVC standard. Our experiments on standard test sequences show average BD-Rate gains in excess of 6% by replacing the standard DC predictor with our learnable predictor. When put into competition with the DC mode as an additional intra mode, our predictor still exhibits lower residual cost, however without appreciable gains in RD terms: we hypothesize that this is due to the increased signaling costs. Current endeavours of the MPAI EVC working group include enhancing the inloop filter with a learnable approach and resorting to a upsampling scheme outside the encoding loop.

References

1. Ballé, J., Laparra, V., Simoncelli, E.P.: End-to-end optimized image compression. In: International Conference on Learning Representations (ICLR), Toulon, France (2017)

2. Ballé, J., Minnen, D., Singh, S., Hwang, S.J., Johnston, N.: Variational image compression with a scale hyperprior. In: International Conference on Learning Representations (ICLR), Vancouver, CA (2018)
3. Choi, K., Chen, J., Rusanovskyy, D., Choi, K.P., Jang, E.S.: An overview of the MPEG-5 essential video coding standard [standards in a nutshell]. IEEE Signal Process. Mag. **37**(3), 160–167 (2020)
4. CISCO: Global 2021 forecast highlights (2021). https://www.cisco.com/c/dam/m/en_us/solutions/service-provider/vni-forecast-highlights/pdf/Global_2021_Forecast_Highlights.pdf
5. Dumas, T., Galpin, F., Bordes, P.: Iterative training of neural networks for intra prediction. IEEE Trans. Image Process. **30**, 697–711 (2020)
6. Dumas, T., Roumy, A., Guillemot, C.: Context-adaptive neural network-based prediction for image compression. IEEE Trans. Image Process. **29**, 679–693 (2019)
7. Goodfellow, I., Courville, A., Bengio, Y.: Deep Learning, vol. 1. MIT Press, Cambridge (2016)
8. Helle, P., et al.: Intra picture prediction for video coding with neural networks. In: 2019 Data Compression Conference (DCC), pp. 448–457. IEEE (2019)
9. Hu, Y., Yang, W., Li, M., Liu, J.: Progressive spatial recurrent neural network for intra prediction. IEEE Trans. Multimedia **21**(12), 3024–3037 (2019)
10. Kingma, D.P., Welling, M.: Auto-encoding variational Bayes. In: International Conference on Learning Representations (ICLR), Banff, CA (2014)
11. Li, W., Lin, Z., Zhou, K., Qi, L., Wang, Y., Jia, J.: Mat: Mask-aware transformer for large hole image inpainting. In: 2022 IEEE/CVF Conference on Computer Vision and Pattern Recognition (CVPR), pp. 10748–10758 (2022). https://doi.org/10.1109/CVPR52688.2022.01049
12. Liu, G., Reda, F.A., Shih, K.J., Wang, T.-C., Tao, A., Catanzaro, B.: Image inpainting for irregular holes using partial convolutions. In: Ferrari, V., Hebert, M., Sminchisescu, C., Weiss, Y. (eds.) ECCV 2018. LNCS, vol. 11215, pp. 89–105. Springer, Cham (2018). https://doi.org/10.1007/978-3-030-01252-6_6
13. MPEG: JVET common test conditions and evaluation procedures for neural network-based video coding technology. In: Output Document of JVET (2023)
14. Pathak, D., Krähenbühl, P., Donahue, J., Darrell, T., Efros, A.: Context encoders: feature learning by inpainting (2016)
15. Rippel, O., Bourdev, L.: Real-time adaptive image compression. arXiv preprint: arXiv:1705.05823 (2017)
16. Samuelsson, J., Choi, K., Chen, J., Rusanovskyy, D.: MPEG-5 EVC. In: SMPTE 2019, pp. 1–11. SMPTE (2019)
17. Schwarz, K., Wieschollek, P., Lensch, H.P.: Will people like your image? Learning the aesthetic space. In: 2018 IEEE Winter Conference on Applications of Computer Vision (WACV), pp. 2048–2057. IEEE (2018)
18. Sze, V., Budagavi, M., Sullivan, G.J.: High Efficiency Video Coding (HEVC). Integrated Circuit and Systems, Algorithms and Architectures, vol. 39, p. 40. Springer, Cham (2014)
19. Toderici, G., et al.: Full resolution image compression with recurrent neural networks. In: IEEE International Conference on Computer Vision and Pattern Recognition (CVPR), pp. 5435–5443, Honolulu, Hawaii, USA (2017)
20. Valenzise, G., Purica, A., Hulusic, V., Cagnazzo, M.: Quality assessment of deep-learning-based image compression. In: Multimedia Signal Processing, Vancouver, Canada (2018). https://www.hal.archives-ouvertes.fr/hal-01819588
21. Wan, Z., Zhang, J., Chen, D., Liao, J.: High-fidelity pluralistic image completion with transformers. CoRR abs/2103.14031 (2021). arxiv:2103.14031

22. Wang, L., Fiandrotti, A., Purica, A., Valenzise, G., Cagnazzo, M.: Enhancing HEVC spatial prediction by context-based learning. In: ICASSP 2019–2019 IEEE International Conference on Acoustics, Speech and Signal Processing (ICASSP), pp. 4035–4039. IEEE (2019)
23. Yu, Y., et al.: Diverse image inpainting with bidirectional and autoregressive transformers. CoRR abs/2104.12335 (2021). arxiv.org/abs/2104.12335

Enhancing PFI Prediction with GDS-MIL: A Graph-Based Dual Stream MIL Approach

Gianpaolo Bontempo[1,2], Nicola Bartolini[1], Marta Lovino[1], Federico Bolelli[1(✉)], Anni Virtanen[3], and Elisa Ficarra[1]

[1] University of Modena and Reggio Emilia, Modena, Italy
{gianpaolo.bontempo,nicola.bartolini,marta.lovino, federico.bolelli,elisa.ficarra}@unimore.it

[2] University of Pisa, Pisa, Italy
gianpaolo.bontempo@phd.unipi.it

[3] University of Helsinki, Helsinki, Finland
anni.virtanen@hus.fi

Abstract. Whole-Slide Images (WSI) are emerging as a promising resource for studying biological tissues, demonstrating a great potential in aiding cancer diagnosis and improving patient treatment. However, the manual pixel-level annotation of WSIs is extremely time-consuming and practically unfeasible in real-world scenarios. Multi-Instance Learning (MIL) have gained attention as a weakly supervised approach able to address lack of annotation tasks. MIL models aggregate patches (*e.g.*, cropping of a WSI) into bag-level representations (*e.g.*, WSI label), but neglect spatial information of the WSIs, crucial for histological analysis. In the High-Grade Serous Ovarian Cancer (HGSOC) context, spatial information is essential to predict a prognosis indicator (the Platinum-Free Interval, PFI) from WSIs. Such a prediction would bring highly valuable insights both for patient treatment and prognosis of chemotherapy resistance. Indeed, NeoAdjuvant ChemoTherapy (NACT) induces changes in tumor tissue morphology and composition, making the prediction of PFI from WSIs extremely challenging. In this paper, we propose GDS-MIL, a method that integrates a state-of-the-art MIL model with a Graph ATtention layer (GAT in short) to inject a local context into each instance before MIL aggregation. Our approach achieves a significant improvement in accuracy on the "Ome18" PFI dataset. In summary, this paper presents a novel solution for enhancing PFI prediction in HGSOC, with the potential of significantly improving treatment decisions and patient outcomes.

1 Introduction

High-Grade Serous Ovarian Cancer (HGSOC) is a form of ovarian cancer characterized by multiple treatment recurrences with variable response to platinum-based chemothereapy. The prediction of Platinum-Free Interval (PFI), defined as the time interval between the end of chemotherapy and disease recurrence [27], is determinant for treatment planning and is usually performed by analyzing

G. L. Foresti et al. (Eds.): ICIAP 2023, LNCS 14233, pp. 550–562, 2023.
https://doi.org/10.1007/978-3-031-43148-7_46

the histological tissue digitalized in Whole-Slide Images (WSIs). Unfortunately, NeoAdjuvant ChemoTherapy (NACT), recommended for HGSOC patients who are ineligible for Primary Debulking Surgery (PDS) [10,22], causes strong variable changes and heterogeneity in tumor morphology and composition, making the prediction of PFI from WSI extremely challenging.

Whole-slide imaging has emerged in recent years as a promising technology to enable the digitalization and the analysis of tissue sections [12]. The creation of multi-resolution gigapixel WSIs provides the opportunity of developing novel diagnostic tools for treatment and monitoring [6,25]. However, the manual pixel-level annotation of WSIs is a time-consuming and labor-intensive task. As an alternative, WSIs are often labelled with metadata (*e.g.,* genetic or other molecular features) characterizing the disease. In addition, because of their gigapixel size, WSIs are usually clipped into patches before being fed into a deep learning model. Given all these conditions, Convolutional Neural Networks (CNNs), which have provided amazing results for a multitude of tasks [13,16,21,31], cannot be directly applied to such data.

Consequently, Multi-Instance Learning (MIL) methods have gained considerable attention in WSI analysis [15,32], avoiding the need for pixel-level annotations. MIL is a weakly supervised learning approach used to assign a label to a set (or bag) composed of unlabelled instances. The label of the bag (*e.g.,* a WSI) is determined by the presence or absence of at least one positive instance (*e.g.,* patch containing tumour), so it is generally assumed that negative bags only contain negative instances (*e.g.,* patch not containing tumour), while positive bags contain at least one positive instance. When dealing with histological images, such assumption cannot be enough and Attention-Based MIL (AB-MIL) [2,9] should be employed to improve patch aggregations [32,35]. However, AB-MIL approaches do not exploit any spatial dependency between instances, which may be crucial in some application [7]. While some tasks can rely solely on morphology analysis (*e.g.,* tumor detection), others would benefit from a more comprehensive tissue analysis. An example of such a task is the aforementioned prediction of PFI on chemotherapy treated HGSOC tissue.

This paper proposes GDS-MIL, which integrates a state-of-the-art MIL model with Graph Neural Networks (GNNs) to contextualize patch local interactions better. Specifically, we use Graph ATtention networks (GATs) [33] to capture the spatial relationships between instances before MIL aggregation, introducing a local context into each instance. This approach has shown promising results, achieving a significant improvement on the "Ome18" PFI dataset. Our study provides a novel solution to improve the accuracy of PFI prediction in HGSOC, which could ultimately lead to better treatment decisions and improved patient outcomes [27].

2 Related Works

In this section, we briefly review recent developments in MIL models, as well as relevant studies that employ MIL for WSI analysis, and existing strategies for PFI prediction.

2.1 Multi-instance Learning for WSI Analysis

Consider a bag X^{bag} composed of a set of N feature vectors:

$$X^{bag} = \{x_1, x_2, ..., x_N\} \tag{1}$$

Each instance $x_i \in X^{bag}$, can be assigned to a class through a mapping process $f : X^{bag} \rightarrow \{0, 1\}$, where the negative and positive classes correspond to 0 and 1, respectively. While traditionally MIL approaches rely on simple aggregators like mean-pooling and max-pooling [8,24], recent studies have shown that there may be benefits in parameterizing the aggregation operator with neural networks [17,23]. The Attention-Based MIL (AB-MIL) [9] employs a side-branch network to calculate attention scores. Similarly, in [37], Zhang *et al.* apply an attention mechanism to support a double-tier feature distillation approach, where relevant features are distilled from pseudo-bags to the WSI using either "MaxMin" or Aggregated Feature Selection (AFS) [37]. Another approach, DS-MIL [15], applies non-local attention aggregation to measure the distance with the most relevant patch. In 2021, Lu *et al.* [18] propose an algorithm that applies a clustering loss to single or multiple branches (CLAM-SB and CLAM-MB), a variant of the classic AB-MIL. Shao *et al.* [28], instead, employ a transformer architecture named Trans-MIL.

2.2 PFI Prediction

A few algorithms for automatic PFI prediction have been proposed in the literature. Both Yu *et al.* [36] and Laury *et al.* [14] use pixel-level annotated WSI for their studies. Yu *et al.* propose a method based on a VGG [29], using portions of WSI for regression analysis finalized to PFI prediction, while Laury *et al.* develop a method based on multiple neural networks used in series, *i.e.*, the output of the first becomes the input of the following network, after human supervised rearrangements. The final aggregation is based on the ratio between digital biomarkers associated with a poor or good prognosis. Their approach employs WSI of treatment-naïve HGSOC. Only tumoral areas are analyzed for the PFI prediction, exploiting pixel-level annotations for the segmentation. Moreover, by focusing the method on treatment-naïve patients, the tumor tissue presents a higher homogeneity in its morphology and texture than tissues undergoing treatment.

Instead, our approach focuses on patients with HGSOC who underwent NACT therapy. Therefore, the WSIs analyzed in this paper are characterized by

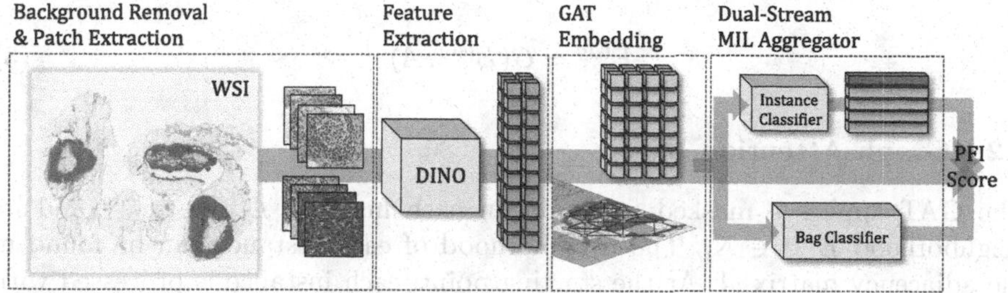

Fig. 1. DINO [4] features extractor is applied to patches tiled from the original WSI. The embeddings thus obtained are fed to a GAT module to capture patches' context and generate a more contextualized representation. A dual-stream MIL aggregation module is then employed to obtain the final prediction by averaging the scores of instance and bag classifiers.

unique morphological characteristics resulting from the treatment effects. Furthermore, to better understand the effects of the treatment, our method analyzes different tissues and compartments in the WSI (*e.g.*, tumor, stroma, inflammatory cells, etc.), and not only tumoral areas, increasing data heterogeneity.

Finally, our method does not require pixel-level annotations to predict the PFI score, relaying only on the global label. To achieve this goal, a graph attention layer has been incorporated into the model to analyze tissue as a complex system composed of multiple interconnected parts.

3 Model

In this study, we propose the use of a GAT to contextualize instances (WSI patches) through local interaction before MIL aggregation. Figure 1 summarizes the key elements of the proposed method, which are detailed in the following of this Section.

3.1 Graph Integration

Given the data as a set of instances $x_i^{ins} \in X^{bag}$ and a self-supervised feature extractor f, informative and discriminative embeddings are obtained as follows[1]:

$$E^{bag} = f(X^{bag}) = \{f(x_0^{ins}), ..., f(x_i^{ins})\} = \{E_0^{ins}, ..., E_i^{ins}\} \tag{2}$$

Each embedding contains important local information inside the patch (*e.g.*, representing the morphology). In order to also capture the micro and macro interaction between instances, we apply a GNN G [11,26,34], implemented with GATs. Given an adjacency matrix $\mathcal{A}$ considering the spatial coordinates of the instances (*e.g.*, each patch is connected to its at most 8 closest neighbors), a more contextualized instance representation is obtained as:

[1] x_i^{ins} represents a patch extracted from the X^{bag}, *i.e.*, the entire WSI.

$$\widehat{E}^{bag} = G(E^{bag}, A) \tag{3}$$

3.2 Graph Attention Layer

The GAT applies a masked attention on each instance $E_i^{ins} \in E^{bag}$ and its neighborhood $E_j^{ins} \in \mathcal{N}_i$. The neighborhood of each instance can be found in the adjacency matrix $\mathcal{A}$. At the starting point, each instance is processed with a shared weight matrix W$\in \mathbb{R}$ as $H^{ins} = W(E^{ins})$. The instance interaction is measured by an α_{ij} computed as:

$$\alpha_{ij} = \frac{\exp(\text{LeakyReLU}(a(H_i \parallel H_j))}{\sum_{k \in \mathcal{N}_i} \exp(\text{LeakyReLU}(a(H_i \parallel H_k))} \tag{4}$$

where $a \in \mathbb{R}^{2F}$ is a single-layer feedforward neural network and $\parallel$ is the concatenation operator. A multi-head attention produces a new instance representation as the average of the linear combinations of the neighborhood among each head $k \in K$:

$$\widehat{E}_i^{ins} = \sigma(\frac{1}{K} \sum_{k \in K} \sum_{j \in N_i} (\alpha_{ij}^k H_j^k)) \tag{5}$$

where σ is a softmax operation.

3.3 Bag-Level Representation

Taking inspiration from DS-MIL [15], the bag representation is built through a dual stream approach. In particular, starting from the graph output $\widehat{E}_i^{ins} \in \widehat{E}^{bag}$ a first patch classifier f_{patch} is used to identify the most critical patch instance as:

$$\widehat{E}_{crit}^{ins} = \underset{\widehat{E}_i^{ins}}{\text{argmax}}\, f(\widehat{E}_i^{ins}) \tag{6}$$

Given the most relevant instance, $\widehat{E}_{crit}^{ins}$, and a linear-layer neural networks, U, it is possible to build the attention scores of the current instance, $\widehat{E}_i^{ins}$, considering its similarity with $\widehat{E}_{crit}^{ins}$:

$$A_i = \text{softmax}(< U(\widehat{E}_i^{ins}), U(\widehat{E}_{crit}^{ins}) >) \tag{7}$$

After that, the bag label is obtained applying a classifier $\mathbf{W}_{\text{CLS}}$ over the bag embedding built as:

$$y_{\text{BAG}} = \mathbf{W}_{\text{CLS}} \sum_i^n \underbrace{A_i}_{\substack{\text{Attention scores w.r.t.} \\ \text{the critical patch.}}} * \underbrace{\text{V}(E_i^{ins})}_{\text{Patch-level value.}} \tag{8}$$

where V is another linear-layer neural networks, and n is $|\widehat{E}^{bag}|$.

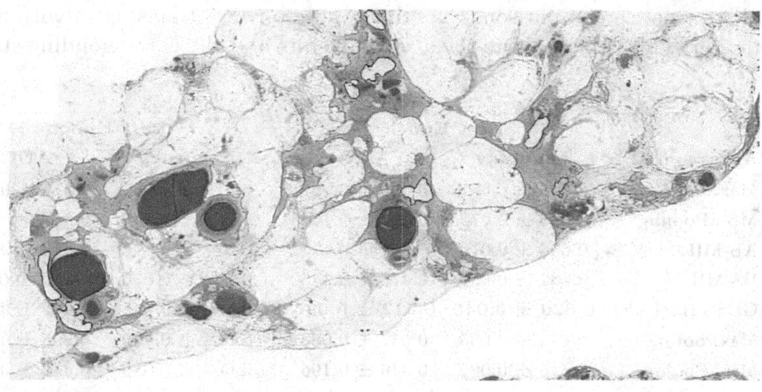

Fig. 2. Example of segmentation masks generated by the pre-processing algorithm. Green contours identify the considered tissue, blue ones are holes the algorithm will discard. The procedure allows for filtering out background, fat, and blood. (Color figure online)

4 Experimental Setup

4.1 Dataset

The dataset is composed by 176 omentum-tissue-WSIs [20] belonging to 77 different HGSOC patients who underwent NACT therapy. The staining procedure used for the WSIs was Hematoxylin and Eosin (HE) [19]. Images have been scanned by a Pannoramic SCAN 150 with a resolution of 0.22 μm/pixel at the 40× resolution. Each WSI is assigned a label based on the patients' PFI: those with a poor prognosis, *low-PFI* ($\leq$ 6 months), are 99 in total, while the other 77 scans have an *high-PFI* ($\geq$ 12 months). The dataset is split into 4-folds in order to perform cross-validation. For each split, a balance between low- and high-PFI was respected. We also ensured that WSIs from the same patient were not mixed between training and test sets.

4.2 Pre-processing

The state-of-the-art CLAM [18] framework has been employed to crop each WSI into multiple patches. This strategy involves selecting only relevant tissue by means of Otsu thresholding [38] and Connected Components Analysis [1].

Additionally, a red filter is used to remove blood[2]. An example of the resulting segmentation mask is shown in Fig. 2. The green contour delineates a portion of tissue that is preserved; the blue one indicates a removed area (holes). The preserved area is then cropped into non-overlapping 256×256 patches at different resolution scales. 20× and 5× resolutions were chosen to capture both micro and

[2] A fixed threshold is applied on the HSV (Hue Saturation Brightness) color space of the WSI thumbnail and later propagated to 5× and 20× resolutions.

Table 1. Performance comparison. Experiments were run 5 times, each with a 4-fold cross-validation. This table reports the average results and the corresponding standard deviation.

Scale	Approach	Best Epoch		Last Epoch	
		Accuracy	AUC	Accuracy	AUC
5×	MaxPooling	0.579 ± 0.067	0.432 ± 0.165	0.579 ± 0.055	0.419 ± 0.161
	MeanPooling	0.596 ± 0.072	0.427 ± 0.166	**0.594 ± 0.069**	0.413 ± 0.148
	AB-MIL	0.606 ± 0.076	0.467 ± 0.171	0.577 ± 0.076	0.413 ± 0.166
	DS-MIL	0.582 ± 0.090	0.478 ± 0.145	0.574 ± 0.075	**0.458 ± 0.136**
	GDS-MIL (our)	**0.620 ± 0.045**	**0.512 ± 0.096**	0.566 ± 0.045	0.402 ± 0.139
20×	MaxPooling	0.676 ± 0.055	0.637 ± 0.083	0.661 ± 0.072	0.598 ± 0.107
	MeanPooling	0.610 ± 0.089	0.446 ± 0.196	0.605 ± 0.086	0.443 ± 0.193
	AB-MIL	0.594 ± 0.095	0.510 ± 0.141	0.576 ± 0.090	0.438 ± 0.156
	DS-MIL	0.681 ± 0.033	0.650 ± 0.049	0.656 ± 0.028	0.572 ± 0.065
	GDS-MIL (our)	**0.704 ± 0.070**	**0.661 ± 0.099**	**0.663 ± 0.064**	**0.611 ± 0.092**

Table 2. Performance comparison on an Out of Distribution (OOD) testset.

Scale	Approach	Best Epoch		Last Epoch	
		Accuracy	AUC	Accuracy	AUC
5×	MaxPooling	0.552 ± 0.052	0.422 ± 0.102	**0.573 ± 0.021**	0.414 ± 0.103
	MeanPooling	0.556 ± 0.0001	0.385 ± 0.010	0.556 ± 0.0001	0.384 ± 0.009
	AB-MIL	0.563 ± 0.029	0.392 ± 0.065	0.517 ± 0.042	0.335 ± 0.062
	DS-MIL	0.486 ± 0.015	0.411 ± 0.015	0.500 ± 0.0001	0.405 ± 0.0124
	GDS-MIL (our)	**0.618 ± 0.036**	**0.490 ± 0.025**	0.566 ± 0.044	**0.429 ± 0.033**
20×	MaxPooling	0.646 ± 0.041	0.611 ± 0.048	0.625 ± 0.065	0.530 ± 0.067
	MeanPooling	0.580 ± 0.010	0.388 ± 0.010	0.580 ± 0.010	0.388 ± 0.010
	AB-MIL	0.510 ± 0.039	0.430 ± 0.008	0.500 ± 0.0001	0.386 ± 0.012
	DS-MIL	0.670 ± 0.023	0.632 ± 0.01	0.653 ± 0.015	0.540 ± 0.011
	GDS-MIL (our)	**0.764 ± 0.039**	**0.726 ± 0.042**	**0.712 ± 0.063**	**0.667 ± 0.082**

macro details in the dataset. On average, each WSI contains 5 960 patches at 20× resolution and 370 patches at 5× resolution.

DINO [4], a Vision Transformer (ViT) model [5], is then employed to produce high quality patch representations, while ensuring a fast processing with low computational resource requirements. This approach focuses on aligning exclusively the positive pairs by leveraging a teacher-student framework, which comprises two separate networks. We trained the model over the entire set of patches, separately for each resolution level.

4.3 Implementation Details

The optimization is performed using *Adam* with a learning rate of $2 * 10^{-4}$ and a weight decay of $5 * 10^{-3}$. The training is carried out for 200 epochs with the *CosineAnnealingLR* scheduler. We employ one single GAT layer with 3 heads used for multi-head attention. All the experiments are conducted using a unified codebase and under identical experimental conditions. Each bag is sub-sampled using a patch dropout probability of 0.5 to increase the number of bags and promote randomness during training. The Area Under the Curve (AUC) and the accuracy metrics are calculated as described in [30]. To ensure a fair comparison, all methods considered in our analysis are evaluated using the same metrics.

5 Results and Discussion

A comparison of the proposed solution with state-of-the-art MIL approaches is reported in Table 1 and Table 2. All the experiments have been performed on the previously described dataset and repeated 5 times to stress the robustness of the algorithms. Tables report the average performance and the associated standard deviation at 5× and 20× resolutions.

We compared the proposed model GDS-MIL with MaxPooling and MeanPooling to understand the effectiveness of patch-level classifiers, and AB-MIL [9] and DS-MIL [15] as state-of-the-art attention based MIL solutions. The performance of each approach is measured with average accuracy and average AUC, both at the best and last epoch. The best epoch is the one where the model obtains the best performance considering the average between accuracy and AUC on the test set, while the last epoch is the end of the training phase.

Experimental results demonstrate that GDS-MIL outperforms all the other approaches on both scales, achieving the highest accuracy and AUC scores at the best epochs. DS-MIL also performs well, achieving good scores on both scales, while MeanPooling and AB-MIL show moderate performance. Overall, the results suggest that integrating a graph-based solution improves our baseline (DS-MIL) by 3.5% on accuracy and 2% on AUC.

Even when considering only the last epoch, GDS-MIL outperforms the baselines improving DS-MIL by 1.3%.

In Table 2 we investigated a specific dataset split characterized by significant tissue heterogeneity. In this case, the contextualization introduced with the graph plays an even more relevant role: our model outperforms DS-MIL by 9.4% on accuracy and 9.3% on AUC.

A further analysis is reported in Table 3, stressing the relevance of graph (main) hyper-parameters such as layer type, number of sequential layers, and number of heads within the same graph layer.

5.1 Model Analysis

Experimental results demonstrate that the 20× scale resolution is the most effective when tackling the PFI prediction task on omentum WSIs taken from NACT patients. Specifically, a patch-level classifier such as MaxPooling can achieve surprisingly good performance at 20× resolution, with an accuracy of 0.676 and AUC of 0.637. This phenomenon implies the existence of morphology and patterns correlated to the PFI which can be exploited to solve the task. This conclusion is also supported by the effectiveness of DS-MIL which achieves an accuracy

Table 3. Performance comparison changing the type of graph layer (type), the number of layers ($\mathcal{L}$) and heads ($\mathcal{H}$) used by the graph neural network.

Type	$\mathcal{L}$	$\mathcal{H}$	AUC	Acc.	Type	$\mathcal{L}$	$\mathcal{H}$	AUC	Acc.	Type	$\mathcal{L}$	$\mathcal{H}$	AUC	Acc.
GCN	1	1	0.625	0.667	GAT	1	1	0.664	0.704	GATv2	1	1	0.657	0.657
GCN	1	2	0.623	0.648	GAT	1	2	0.667	0.732	GATv2	1	2	**0.734**	**0.732**
GCN	1	3	**0.634**	0.648	GAT	1	3	**0.726**	**0.764**	GATv2	1	3	0.667	0.704
GCN	2	1	0.602	**0.676**	GAT	2	1	0.634	0.722	GATv2	2	1	0.679	0.722
GCN	2	2	0.608	0.648	GAT	2	2	0.607	0.648	GATv2	2	2	0.628	0.694
GCN	2	3	0.591	0.657	GAT	2	3	0.619	0.694	GATv2	2	3	0.655	0.713
GCN	3	1	0.564	0.648	GAT	3	1	0.641	0.694	GATv2	3	1	0.641	0.685
GCN	3	2	0.595	0.648	GAT	3	2	0.639	0.704	GATv2	3	2	0.642	0.713
GCN	3	3	0.572	0.639	GAT	3	3	0.697	0.732	GATv2	3	3	0.660	0.713

of 0.681 and an AUC of 0.649. The attention mechanism used by DS-MIL allows to identify the most relevant WSI regions, guiding the PFI classification.

However, adding a graph attention layer can significantly improve the performance at both considered resolutions. This finding suggests that incorporating spatial context into each instance, including both neighborhood morphology and interaction, allows to change the meaning of critical patch. In GDS-MIL, the relevance score of each instance is not limited to the instance itself, but also influenced by the area where it is located, allowing for a more fine-grained criticality assessment. These results suggest that the proposed model is highly effective and can offer significant improvements over existing state-of-the-art approaches. The high standard deviation of all reported experiments is intrinsically connected to the small number of WSIs and to the high heterogeneity of the task.

5.2 Hyperparameter Analysis

To stress the contribution of different graph layers, Table 3 is reported. The results indicate that, in general, using layers of a Graph Convolutional Network [11] leads to worse performances compared to GAT [33] and GATv2 [3]. When relying on convolutional layers, the patch representation becomes similar to its neighborhood, resulting in a loss of important details. In contrast, leveraging an attention layer enables the patch to acquire context information, while preserving its own unique features. No significant difference can be observed between GAT and GATv2, with the latter performing slightly better than the former.

The experiments reported in Table 3 also reveal that a higher number of graph layers has a negative impact on the performance. This is mainly related to the smoothing operation performed by the graph on the patch representation. If the smoothing is too strong, it becomes challenging for the MIL module to distinguish what is actually important. Therefore, it is crucial to identify a trade-off between the number of layers and the overall performance.

Moreover, increasing the number of heads applied to the attention mechanism generally provide better performances. Indeed, using a multi-head approach enhances the ability to capture the most important information from the neighborhood and build a more contextualized representation of each instance.

In summary, our analysis highlights the importance of carefully selecting the graph hyper-parameters. Specifically, the adoption of attention layers usually provide better performance than convolutional graph layers. Limiting the number of graph layers, and considering an higher number of heads during the self-attention process can also improve the final results. This is the reason why we opted for a single GAT layer consisting of three heads.

6 Conclusions

This paper proposes GDS-MIL method which integrates a GAT into a MIL architecture for predicting the PFI of WSIs obtained from NACT patients. Our results demonstrate that introducing a spatial contextualization has beneficial effects on the MIL architecture. A future work will analyze what kind of biological patterns have major impact for the prediction in order to better explain the PFI task.

Acknowledgements. This project has received funding from DECIDER, the European Union's Horizon 2020 research and innovation programme under GA No. 965193, and from the Department of Engineering "Enzo Ferrari" of the University of Modena through the FARD-2022 (Fondo di Ateneo per la Ricerca 2022).

References

1. Allegretti, S., Bolelli, F., Cancilla, M., Pollastri, F., Canalini, L., Grana, C.: How does connected components labeling with decision trees perform on GPUs? In: Vento, M., Percannella, G. (eds.) CAIP 2019. LNCS, vol. 11678, pp. 39–51. Springer, Cham (2019). https://doi.org/10.1007/978-3-030-29888-3_4
2. Bontempo, G., Porrello, A., Bolelli, F., Calderara, S., Ficarra, E.: DAS-MIL: Distilling Across Scales for MIL classification of histological WSIs. In: Medical Image Computing and Computer Assisted Intervention - MICCAI 2023 (2023)
3. Brody, S., Alon, U., Yahav, E.: How attentive are graph attention networks? In: International Conference on Learning Representations (2022)
4. Caron, M., et al.: Emerging properties in self-supervised vision transformers. In: Proceedings of the IEEE/CVF International Conference on Computer Vision (ICCV), pp. 9650–9660 (2021)
5. Dosovitskiy, A., et al.: An image is worth 16x16 words: transformers for image recognition at scale. In: International Conference on Learning Representations (2021)

6. Evans, A.J., et al.: US food and drug administration approval of whole slide imaging for primary diagnosis: a key milestone is reached and new questions are raised. Arch. Pathol. Lab. Med. **142**(11), 1383–1387 (2018)
7. Fatemi, M., et al.: Inferring spatial transcriptomics markers from whole slide images to characterize metastasis-related spatial heterogeneity of colorectal tumors: a pilot study. J. Pathol. Inf. **14**, 100308 (2023)
8. Feng, J., Zhou, Z.H.: Deep MIML network. In: Proceedings of the Thirty-First AAAI Conference on Artificial Intelligence, pp. 1884–1890. AAAI Press (2017)
9. Ilse, M., Tomczak, J., Welling, M.: Attention-based deep multiple instance learning. In: Proceedings of the 35th International Conference on Machine Learning, pp. 2127–2136. PMLR (2018)
10. Kehoe, S., et al.: Primary chemotherapy versus primary surgery for newly diagnosed advanced ovarian cancer (CHORUS): an open-label, randomised, controlled, non-inferiority trial. Lanchet **386**(9990), 249–257 (2015)
11. Kipf, T.N., Welling, M.: Semi-supervised classification with graph convolutional networks. In: International Conference on Learning Representations. ICLR (2017)
12. Kumar, N., Gupta, R., Gupta, S.: Whole slide imaging (WSI) in pathology: current perspectives and future directions. J. Digit. Imaging **33**(4), 1034–1040 (2020)
13. Landi, F., Baraldi, L., Corsini, M., Cucchiara, R.: Embodied vision-and-language navigation with dynamic convolutional filters. In: Proceedings of the British Machine Vision Conference (2019)
14. Laury, A.R., Blom, S., Ropponen, T., Virtanen, A., Carpén, O.M.: Artificial intelligence-based image analysis can predict outcome in high-grade serous carcinoma via histology alone. Sci. Rep. **11**(1), 19165 (2021)
15. Li, B., Li, Y., Eliceiri, K.W.: Dual-stream multiple instance learning network for whole slide image classification with self-supervised contrastive learning. In: 2021 IEEE/CVF Conference on Computer Vision and Pattern Recognition (CVPR), pp. 14318–14328 (2021)
16. Lovino, M., Ciaburri, M.S., Urgese, G., Di Cataldo, S., Ficarra, E.: DEEPrior: a deep learning tool for the prioritization of gene fusions. Bioinformatics **36**(10), 3248–3250 (2020)
17. Lu, M.Y., Chen, R.J., Wang, J., Dillon, D., Mahmood, F.: Semi-supervised histology classification using deep multiple instance learning and contrastive predictive coding. arXiv preprint: arXiv:1910.10825 (2019)
18. Lu, M.Y., Williamson, D.F., Chen, T.Y., Chen, R.J., Barbieri, M., Mahmood, F.: Data-efficient and weakly supervised computational pathology on whole-slide images. Nat. Biomed. Eng. **5**(6), 555–570 (2021)
19. Martina, J.D., Simmons, C., Jukic, D.M.: High-definition hematoxylin and eosin staining in a transition to digital pathology. J. Pathol. Inf. **2**(1), 45 (2011)
20. Meza-Perez, S., Randall, T.D.: Immunological functions of the Omentum. Trends Immunol. **38**(7), 526–536 (2017)
21. Morelli, D., Fincato, M., Cornia, M., Landi, F., Cesari, F., Cucchiara, R.: Dress code: high-resolution multi-category virtual try-on. In: IEEE/CVF Conference on Computer Vision and Pattern Recognition Workshops (CVPRW) (2022)
22. Nikolaidi, A., Fountzilas, E., Fostira, F., Psyrri, A., Gogas, H., Papadimitriou, C.: Neoadjuvant treatment in ovarian cancer: new perspectives, new challenges. Front. Oncol., 3758 (2022)

23. Panariello, A., Porrello, A., Calderara, S., Cucchiara, R.: Consistency-based self-supervised learning for temporal anomaly localization. In: Karlinsky, L., Michaeli, T., Nishino, K. (eds.) ECCV 2022. Lecture Notes in Computer Science, vol. 13805, pp. 338–349. Springer, Cham (2023). https://doi.org/10.1007/978-3-031-25072-9_22
24. Pinheiro, P.O., Collobert, R.: From image-level to pixel-level labeling with convolutional networks. In: IEEE Conference on Computer Vision and Pattern Recognition (CVPR), pp. 1713–1721 (2015)
25. Ponzio, F., Urgese, G., Ficarra, E., Di Cataldo, S.: Dealing with lack of training data for convolutional neural networks: the case of digital pathology. Electronics **8**(3), 256 (2019)
26. Porrello, A., Abati, D., Calderara, S., Cucchiara, R.: Classifying signals on irregular domains via convolutional cluster pooling. In: The 22nd International Conference on Artificial Intelligence and Statistics, pp. 1388–1397 (2019)
27. Pujade-Lauraine, E., Combe, P.: Recurrent ovarian cancer. Ann. Oncol. **27**, i63–i65 (2016)
28. Shao, Z., et al.: TransMIL: transformer based correlated multiple instance learning for whole slide image classification. In: Advances in Neural Information Processing Systems (NeurIPS), vol. 34, pp. 2136–2147 (2021)
29. Simonyan, K., Zisserman, A.: Very deep convolutional networks for large-scale image recognition. In: 3rd International Conference on Learning Representations (ICLR 2015) (2015)
30. Sokolova, M., Japkowicz, N., Szpakowicz, S.: Beyond accuracy, F-Score and ROC: a family of discriminant measures for performance evaluation. In: Sattar, A., Kang, B. (eds.) AI 2006. LNCS (LNAI), vol. 4304, pp. 1015–1021. Springer, Heidelberg (2006). https://doi.org/10.1007/11941439_114
31. Tomei, M., Baraldi, L., Calderara, S., Bronzin, S., Cucchiara, R.: RMS-Net: regression and masking for soccer event spotting. In: 25th International Conference on Pattern Recognition (ICPR), pp. 7699–7706. IEEE (2021)
32. Tourniaire, P., Ilie, M., Hofman, P., Ayache, N., Delingette, H.: Attention-based multiple instance learning with mixed supervision on the camelyon16 dataset. In: Proceedings of the MICCAI Workshop on Computational Pathology, pp. 216–226. PMLR (2021)
33. Velickovic, P., et al.: Graph attention networks. Stat **1050**(20), 10–48550 (2017)
34. Wu, Z., Pan, S., Chen, F., Long, G., Zhang, C., Philip, S.Y.: A comprehensive survey on graph neural networks. IEEE Trans. Neural Netw. Learn. Syst. **32**(1), 4–24 (2020)
35. Yao, J., Zhu, X., Jonnagaddala, J., Hawkins, N., Huang, J.: Whole slide images based cancer survival prediction using attention guided deep multiple instance learning networks. Med. Image Anal. **65**, 101789 (2020)
36. Yu, K.H., et al.: Deciphering serous ovarian carcinoma histopathology and platinum response by convolutional neural networks. BMC Med. **18**(1), 1–14 (2020)
37. Zhang, H., et al.: DTFD-MIL: double-tier feature distillation multiple instance learning for histopathology whole slide image classification. In: IEEE/CVF Conference on Computer Vision and Pattern Recognition (CVPR), pp. 18802–18812 (2022)

38. Zhang, J., Hu, J.: Image segmentation based on 2D Otsu method with histogram analysis. In: International Conference on Computer Science and Software Engineering, vol. 6, pp. 105–108. IEEE (2008)

Author Index

G. L. Foresti et al. (Eds.): ICIAP 2023, LNCS 14233, pp. 563–567, 2023.
https://doi.org/10.1007/978-3-031-43148-7